Distributed Generation

Edited by
D. N. Gaonkar

Distributed Generation,
Edited by D. N. Gaonkar

Republished by InTech
Janeza Trdine 9, 51000 Rijeka, Croatia

Additional hard copies can be obtained from orders@intechopen.com

Distributed Generation,
Edited by D. N. Gaonkar
p. cm.
ISBN 978-953-307-046-9

For EU product safety concerns:
IN TECH d.o.o., Prolaz Marije Krucifikse Kozulić 3, 51000 Rijeka, Croatia,
info@intechopen.com or visit our website at intechopen.com.

Meet the editor

Dattatraya N. Gaonkar (dngaonkar@ieee.org) is with the National Institute of Technology Karnataka, Surathkal, India. P is a Student Member of the IEEE, and Gaonkar is a Senior Member of the IEEE. This article first appeared as "Voltage Estimation in Smart Distribution Networks with Multiple DG Systems" at the 2015 IEEE India Conference (INDICON-2015). This article was reviewed by the IAS Industrial Automation and Control Committee.

Preface

In the recent years the electrical power utilities are undergoing rapid restructuring process worldwide. Indeed, with deregulation, advancement in technologies and concern about the environmental impacts, competition is particularly fostered in the generation side thus allowing increased interconnection of generating units to the utility networks. These generating sources are called as distributed generators (DG) and defined as the plant which is directly connected to distribution network and is not centrally planned and dispatched. These are also called as embedded or dispersed generation units. The rating of the DG systems can vary between few kW to as high as 100 MW. Various new types of distributed generator systems, such as microturbines and fuel cells in addition to the more traditional solar and wind power are creating significant new opportunities for the integration of diverse DG systems to the utility. Inter connection of these generators will offer a number of benefits such as improved reliability, power quality, efficiency, alleviation of system constraints along with the environmental benefits. With these benefits and due to the growing momentum towards sustainable energy developments, it is expected that a large number of DG systems will be interconnected to the power system in the coming years.

Unlike centralized power plants, the DG units are directly connected to the distribution system; most often at the customer end. The existing distribution networks are designed and operated in radial configuration with unidirectional power flow from centralized generating station to customers. The increase in interconnection of DG to utility networks can lead to reverse power flow violating fundamental assumption in their design. This creates complexity in operation and control of existing distribution networks and offers many technical challenges for successful introduction of DG systems. Some of the technical issues are islanding of DG, voltage regulation, protection and stability of the network. Some of the solutions to these problems include designing of standard interface control for individual DG systems by taking care of their diverse characteristics, finding new ways to/or install and control these DG systems and finding new design for distribution system.

DG has much potential to improve distribution system performance. The use of DG strongly contributes to a clean, reliable and cost effective energy for future. However technical issues given above need to be resolved, to pave the way for a sustainable energy future based on a large share of DG. In this context investigation on the utility interconnection aspects of DG systems such as, development of interface model, their operation and control, planning and design becomes significant. Hence a lot of research effort is required.

This book deals with the several aspects of the DG systems such as, benefits, issues, technology interconnected operation, performance studies, planning and design. Several authors are contributed for this book aiming to benefit students, researcher's academicians, policy makers and professionals. We are indebted to all the people who either directly or indirectly contributed towards the publication of this book.

D. N. Gaonkar

Contents

1

Distributed Cogeneration: Modelling of Environmental Benefits and Impact

Pierluigi Mancarella° and Gianfranco Chicco*
°Imperial College London, UK
**Politecnico di Torino, Italy*

1. Introduction

Radical changes occurred in the energy scenario in recent years, with a clear trend towards shifting part of the energy production from large centralized plants to relatively small decentralized systems. The growing diffusion of distributed generation systems for Combined Heat and Power (CHP) production represents a significant part of these changes. In particular, CHP generation could bring substantial improvements in energy efficiency and energy saving, as well as economic benefits, with respect to the separate production (SP) of electricity in the centralized power system and of heat in local boilers (Horlock, 1997). The development of CHP systems is particularly relevant for relatively small-scale applications (e.g., below 10 MW_e) in urban areas, including potential coupling to heat networks for larger capacities as well as micro-cogeneration (Pehnt et al., 2006; Pehnt, 2008) for domestic applications. The adoption of CHP systems can even be more effective when it is possible to supply, in periods with little or no heat demand, absorption chillers to satisfy the cooling demand (for instance, for air conditioning), thus obtaining high-efficiency seasonal tri-generation systems (Meunier, 2002; Mancarella, 2006). Moreover, CHP systems can be conveniently used in a distributed multi-generation framework, to supply various types of chillers to better fit the overall characteristics of the demand of various energy vectors (Chicco & Mancarella, 2009a).
Higher energy efficiency can also correspond to lower environmental impact in terms of CO_2 emissions with respect to SP, mainly depending on the generation characteristics of the power system in the specific country (Mancarella & Chicco, 2008a), particularly where electricity generation prevailingly occurs from fossil fuels. On the other hand, distributed cogeneration could worsen the air quality on the local level, due to emissions of various hazardous pollutants such as NO_x, CO, SO_x, Particulate Matter (PM), Unburned Hydrocarbons (UHC), and further substances conveying the pollution into the human body. In particular, in urban areas the environmental pollution is more critical because of a host of reasons, among which:

a) high concentration of background pollutants, in particular due to road traffic pollution;
b) difficult dispersion in the atmosphere of the pollutants produced from small-scale generators located in urban sites, with respect to large power plants with high stacks;
c) relatively high number of receptors, due to the population density;

d) presence of relative weak receptors, such as children, elders and sick people; and,
e) detrimental effects on non-human receptors (monuments, green urban areas and ecosystems), that could also contribute to keeping the pollutants within the area.

On the above reasons, the local air quality regulation could often be quite stringent, especially in urban areas, with environmental assessments tending to be conservative and leaving reduced margins to the deployment of cogeneration in heavily polluted zones. These limitations call for a thorough appraisal at the cogeneration system planning stage. In addition, it is important to consider that the emissions of certain pollutants may worsen even significantly in the off-design operation at partial load of the cogeneration unit. Hence, the environmental assessment of cogeneration systems has to be carried out not only on the basis of the full-load performance, but in actual operating conditions. This aspect is even more relevant considering that in today's and emerging energy systems the cogeneration units are not generally used in on/off operation only, but can be controlled to achieve specific objectives of electrical or heating load tracking, or with more refined strategies in which the cogeneration unit is combined into trigeneration or more generally multi-generation systems (Mancarella & Chicco, 2009b) and microgrids (Hatziargyriou et al., 2007). Further environmental benefits could refer to micro-cogeneration solutions.

This chapter addresses the manifold sides of potential environmental benefits and impact related to modelling and analysis of distributed cogeneration solutions. Section 2 recalls the energy efficiency benefits of adopting cogeneration systems. Section 3 deals with the modelling of global and local emissions. Section 4 describes the characterization of the emissions from typical CHP technologies already widely applied today. Section 5 presents specific indicators for environmental impact assessment. Section 6 discusses the role of environmental impact in the formulation of optimization methods. Section 7 illustrates the identification and determination of the environmental external costs from distributed cogeneration. Section 8 addresses the potential deployment of cogeneration in energy-related markets. Section 9 draws the conclusions and indicates directions of future research.

2. Energy efficiency of distributed cogeneration

A suitable characterization of cogeneration equipment and systems is conducted by using a black-box modelling approach, in which the performance of CHP units is represented by relevant input-output efficiency models (Mancarella, 2006; Chicco & Mancarella, 2009b). In particular, a cogeneration prime mover is represented by using its electrical efficiency η_W, thermal efficiency η_Q, and their sum known as Energy Utilisation Factor (*EUF*) (Horlock, 1997). By denoting with *W* the electrical energy (kWh_e), *Q* the thermal energy (kWh_t), and *F* the fuel thermal input (kWh_t) in a specified time interval, the energy efficiency indicators are expressed as:

$$\eta_W = \frac{W}{F}; \ \eta_Q = \frac{Q}{F}; \ EUF = \eta_W + \eta_Q \qquad (1)$$

The terms *W*, *Q* and *F* can also be interpreted as average powers within the specified time interval. For instance, this interpretation is useful for the sake of comparison of the energy production (represented as average power within a time interval in which the power

variation is relatively low) with the rated power of the equipment, to check whether the operational limits are exceeded, provided that significant power variations during the specified time interval can be excluded.
The fuel thermal input F is generally based on the Lower Heating Value (*LHV*) of the fuel. The efficiencies indicated above depend on many factors, such as the equipment technology, the loading level, the outdoor conditions, the enthalpy level at which heat is produced, the characteristics of the heat recovery system, and so forth (Danny Harvey, 2006).
The outcomes of the CHP system energy efficiency assessment can be represented in a synthetic way through suitable indicators. A classical way to define such indicators is to compare the production of the same energy outputs (electricity and heat) from the cogeneration system and from conventional systems for separate production of electricity and heat used as reference (Horlock, 1997). The SP systems are typically the electricity distribution system (EDS) for electricity production (associated to a reference electrical efficiency η_e^{SP}), and conventional boilers for heat production (associated to a reference thermal efficiency η_t^{SP}).
Considering the fuel thermal input F^{SP} to the conventional separate production system, the resulting energy efficiency indicator is the Primary Energy Saving (*PES*), also known as Fuel Energy Savings Ratio (*FESR*), expressed as:

$$PES = \frac{F^{SP} - F}{F^{SP}} = 1 - \frac{F}{\frac{W}{\eta_e^{SP}} + \frac{Q}{\eta_t^{SP}}} = 1 - \frac{1}{\frac{\eta_W}{\eta_e^{SP}} + \frac{\eta_Q}{\eta_t^{SP}}} \tag{2}$$

Energy efficiency benefits of cogeneration appear for positive values of the *PES* indicator, and the break-even condition is found for $PES = 0$. The simple and meaningful structure of the *PES* indicator makes it particularly useful to quantify the energy efficiency of a cogeneration system for regulatory purposes (Cardona & Piacentino, 2005). Extensions of the *PES* indicator have been proposed to encompass trigeneration systems (Chicco & Mancarella, 2007a) and more general multi-generation systems (Chicco & Mancarella, 2008b). A further indicator characterising the operation of a cogeneration system is the heat-to-power cogeneration ratio, typically denoted with the letter lambda (Horlock, 1997), that according to (1) can be also seen as the ratio of the thermal to electrical efficiency:

$$\lambda = \frac{Q}{W} = \frac{\eta_Q}{\eta_W} \tag{3}$$

3. Modelling of global and local emissions

3.1 Emission factors

The emissions of a generic pollutant p from a combustion device can be characterised through suitable *emission factors*, referred to the useful energy produced by the generic energy vector X. The corresponding emission factor model is expressed in the form

$$m_p^X = \mu_p^X \cdot X \tag{4}$$

where the term m_p^X represents the mass [g] of the pollutant p emitted to produce the energy vector X [kWh], and μ_p^X is the *emission factor* or *specific emissions* [g/kWh] of the pollutant p referred to X.
The emission factor depends on the type of generator and varies in different operating conditions (e.g., at full load or partial load), with the equipment aging and with the state of maintenance of the generator. The emission performance of the generator can be characterised through dedicated measurements in actual operating conditions.
In the definition of the emission factors for cogeneration applications, the energy vector X can be chosen in different ways, thus originating different definitions of the emission factor. For instance, X can represent the input fuel energy F [kWh_t], or the electricity production W [kWh_e] or the heat production Q [kWh_t]. The corresponding formulations can be written by expressing the mass of pollutant m_p on the basis of the emission factor adopted in different ways:

$$m_p = \mu_p^F \cdot F = \mu_p^W \cdot W = \mu_p^Q \cdot Q \tag{5}$$

In particular, the emission factor μ_p^F depends primarily on the characteristics of the chemical reactions and on the type of fuel used (Cârdu & Baica, 2002). It is then possible to evaluate the emission factor referred to the energy produced through the expression

$$\mu_p^X = \frac{\mu_p^F}{\eta_X} \tag{6}$$

where η_X is the efficiency equivalent to the production of the useful energy X using the fuel F. For instance, for a cogeneration system the term η_X can be the electrical efficiency η_W for electricity production, or the thermal efficiency η_Q for heat production.

3.2 Emission balances

The emission factors can be used to formulate *global* or *local* emission balances (Mancarella & Chicco, 2009a).
The *global* emission balance does not take into account the location of the emission source with respect to the receptors. For a given time interval (e.g., one hour) in which the variation of the energy production from the cogenerator is low (in such a way to assume almost constant values of the variables involved in the analysis), it is possible to compare the mass of the pollutant emitted by the cogeneration system with the mass of pollutant emitted from separate production of the same electrical and thermal energy (Mancarella & Chicco, 2009a; Mancarella & Chicco, 2009b).
With reference to the electricity production, taking into account the cogeneration ratio (3) and considering the total mass of pollutant emitted in global separate production (GSP) as the sum of the mass of pollutant emitted in the production of electricity and heat, it is possible to elaborate the previous expressions to get (Mancarella & Chicco, 2009a):

$$m_p^{GSP} = m_p^{Q,SP} + m_p^{W,SP} = \mu_p^{Q,SP} \cdot \lambda \cdot W + \mu_p^{W,SP} \cdot W = \left(\mu_p^{Q,SP} \cdot \lambda + \mu_p^{W,SP}\right) W \quad (7)$$

From (7) it is possible to define the *global* equivalent emission factor referred to the electricity production

$$\mu_p^{W,GSP} = \mu_p^{Q,SP} \cdot \lambda + \mu_p^{W,SP} \quad (8)$$

The emission factor $\mu_p^{W,GSP}$ is directly comparable with the corresponding emission factor μ_p^{W} of the cogeneration system. This is a major upside of the emission balance model developed, which allows emission comparison on a common basis (for instance, *specific* emissions with respect to the same useful energy output) and thus unbiased environmental impact appraisal of different generation rationales such as cogeneration and separate production. In fact, by comparing the emissions per unit of useful kWh produced, the fact that cogeneration can produce the same amount of useful energy burning less fuel (thus producing less pollutants) is intrinsically made apparent in the model. Hence, the environmental impact benefits (in terms of reduced specific emissions) arising from enhanced cogeneration efficiency are explicitly acknowledged as well. Conversely, comparisons carried out by considering only the concentration of pollutant (in mg/m^3) contained in the gases exhausted off to the ambient would be unable to take into account the energy efficiency benefits of cogeneration, as further remarked in Section 5.4. These concepts form the basis to adopt energy-output related specific emissions to define the emission reduction indicators illustrated in Section 5.1.

The *local* emission balance takes into account that the large power plants for electricity production are typically located far from urban areas. In the analysis of the effects of emitting pollutants with propagation in a relatively limited area (for instance, NO_X, CO and volatile organic compounds), it is possible to adopt an approximated model, neglecting both the propagation of these substances outside the area and the possible introduction of the same substances emitted by sources located outside the area. For instance, for an urban area the local emission balance considers only the local separate production (LSP) from residential boilers belonging to the area. In this case, the mass of the pollutant *p* emitted is only the one originated by the thermal production and is expressed as (Mancarella & Chicco, 2009a)

$$m_p^{LSP} = m_p^{Q,SP} = \mu_p^{Q,SP} Q = \mu_p^{Q,SP} \lambda\, W \quad (9)$$

The local equivalent emission factor, referred to the electricity production, is then defined as

$$\mu_p^{W,LSP} = \frac{m_p^{LSP}}{W} = \mu_p^{Q,SP} \lambda \quad (10)$$

More detailed evaluations can be conducted with reference to specific models of the pollutant dispersion in the atmosphere (Arya, 1999). However, the local emission balance model is useful for a preliminary assessment of the local emission impact through the dedicated indicators illustrated in Section 5.2. Of course, the local emission approximation can be more or less relevant, also depending on the specific pollutant and dispersion conditions, and represents a conservative ("pessimistic") evaluation of the actual impact due to the distributed energy system. On the other hand, the global emission approximation may often represent an "optimistic" evaluation of the impact from distributed sources (apart from the evaluation of greenhouse gases (GHGs), whose effect is essentially global to every extent, neglecting possible contribution to micro-climates). However, the simultaneous appraisal of local and global emission balances provides meaningful insights on the upper and lower bounds of the real environmental pressure (Mancarella & Chicco, 2009a). In particular, the indications yielded by these models are independent of the specific site and can be used to compare different scenarios of development of the cogeneration systems. The information provided can be useful for regulatory purposes, with the aim of setting up local emission limits, assuming conventional values for the separate production efficiencies and emission factors, as discussed in Section 5.3.

4. Characterization of the emissions from cogeneration technologies

The analysis of the emissions of various pollutants from specific cogeneration technologies takes into account different types of fuel (mainly natural gas, or alternative fuels such as biomasses) and operation in partial-load conditions, with potential remarkable worsening of the emission of pollutants such as NO_x and CO at relatively low loading level. The illustrations included here refer to technologies that have reached a wide commercial stage, such as the ones based on microturbine (MT) or internal combustion engine (ICE) prime movers. The same concepts can be extended to other technologies, such as fuel cells.

Generally, different cogeneration units, also with the same type of technology, could exhibit emission characteristics quite different from each other, because of the specific design of the combustion device, the possible presence or abatement system, and so forth. It is then tough to draw general emission models. For general studies, it is typically preferred to consider average values of emissions taken from inventories prepared by the various environmental protection agencies and research groups worldwide (e.g., EPRI, 2009; US Environmental Protection Agency, 2009), or elaborations of data provided by manufacturers.

Operation at partial load can be determined by the implementation of specific control (or load tracking) strategies, such as electrical load-following or heat load-following ones, as well as economically driven strategies such as based on the evaluation of the "spark spread" between natural gas cost and electricity cost.

The emissions in real operation conditions depend on the characteristics of the combustion occurring in the cogeneration prime mover. Experimental results provided by the cogeneration unit manufacturers and obtained during specific researches and on-site measurements have shown that the emissions of some pollutants (for instance, NO_x and CO) can worsen significantly during partial load operation. Furthermore, at decreasing loading level the evolution of these emissions is not linear and in some cases could exhibit non-monotonic behaviour, especially for microturbine units (Canova et al., 2008; Mancarella & Chicco, 2009a). These aspects make the emission impact assessment of cogeneration systems more complicated. In addition, below a certain loading level (e.g., 50%) the performance of the cogeneration unit could become so worse to suggest switching the unit off. In this case, the domain of definition of the operating conditions of the cogeneration unit becomes non-connected, including the discrete switch-off condition and a continuous operation range between the technical limits of minimum and maximum loading. These aspects impact on the characterization of the cogeneration systems and call for adopting dedicated analysis techniques, for instance based on mixed integer and linear/nonlinear programming or heuristic methods, in some cases developed for energy efficiency analyses not including environmental aspects (Horii et al., 1987; Illerhaus & Verstege, 1999; Tsay et al., 2001; Gómez-Villalva & Ramos, 2003).

Concerning the types of analysis, one of the key distinctions occurs between time-domain simulations and methods in which the succession of the time instants in the cogeneration system operation is not exploited. Time-domain simulations are needed when it is important to consider the coupling-in-time of events, for instance to take into account the integral effect of the emissions within a specified period, or the operational limits dependent on the time domain, such as the maximum number of switch-on/switch-off operations of the cogeneration unit (Freschi & Repetto, 2008). If the coupling-in-time is not strictly relevant, general evaluations can be carried out by using integral models, such as for instance the *equivalent load* approach illustrated in Mancarella & Chicco, 2009a. This approach is based on the construction of a discrete (multi-level) version of the load duration curve representing the electricity demand, containing pre-defined levels of partial-load operation. Each loading level is represented by a pair hourly energy-duration (number of hours), from which the equivalent electrical load is calculated as the weighted average of the hourly energies, assuming the load level durations as weights. Each level of the duration curve is then associated to a value of specific emissions, used to determine the mass of pollutant and the equivalent emission factors for the cogeneration system and for separate production.

The equivalent load approach can be used for different time horizons and with different levels of detail of the load duration curve. This makes the approach suitable both for planning analyses over one year or more, as well as to represent the local emissions occurring in the short-term (e.g., minute by minute) during the application of load tracking strategies, in order to quantify the cumulative duration for which the emission limits have been exceeded. Furthermore, the structure of the equivalent load approach provides *smooth* trends of variation of the specific emissions when the equivalent load changes, even in the case of high non-linearity of the specific emissions from the cogeneration units. The equivalent load approach could be adopted by regulatory bodies to establish a conventional technique for taking into account partial-load operation of the cogeneration units.

5. Indicators for environmental impact assessment

Specific system-based indicators are aimed at promoting policy developments, as well as determining the break-even conditions for which CHP systems are equivalent to the conventional separate production in terms of global or local emissions. The use of synthetic indicators for assessing the benefits of exploiting cogeneration technologies with respect to separate production is common in energy efficiency studies, as recalled in Section 2. Comparison between distributed and centralized systems can be resorted to also in terms of emission analysis (Strachan & Farrell, 2006). Similar concepts can be extended and applied in the framework of the global and local emission balance approach (Section 3.2). In particular, the indicators listed in the following subsections are particularly useful to obtain general indications independent of the characteristic of the individual site, thus of possible interest for regulatory purposes.

5.1 Global emission indicators

Let us consider the case of CO_2 emission assessment as a relevant example of application of the global emission balance approach, given the global warming impact of CO_2 as GHG. The CO_2 emission reduction due to cogeneration is expressed in relative terms with respect to the mass of pollutant emitted in global separate production. The resulting CO_2 Emission Reduction (*CO2ER*) indicator applied to global CO_2 emissions (Mancarella & Chicco, 2009a) is a sub-case of the *PCO2ER* indicator introduced in Chicco & Mancarella, 2008a for multi-generation systems:

$$CO2ER = \frac{m_{CO_2}^{GSP} - m_{CO_2}}{m_{CO_2}^{GSP}} = 1 - \frac{\mu_{CO_2}^{F} \cdot F}{\mu_{CO_2}^{W,SP} \cdot W + \mu_{CO_2}^{Q,SP} \cdot Q} = 1 - \frac{\mu_{CO_2}^{F}}{\mu_{CO_2}^{W,SP} \cdot \eta_W + \mu_{CO_2}^{Q,SP} \cdot \eta_Q} \tag{11}$$

where the emission balance is carried out by considering the mass m_{CO_2} of CO_2 emitted from the combustion of the fuel F to cogenerate useful electricity and heat, and the mass $m_{CO_2}^{GSP}$ of CO_2 emitted by the separate production of the same useful outputs (electricity W and heat Q) from conventional technologies. Exploiting cogeneration is environmentally effective for positive values of *CO2ER*, while *CO2ER* = 0 indicates the break-even condition. As a further step, it is possible to introduce the *CO_2 emission equivalent efficiencies* (Chicco & Mancarella, 2008b)

$$\eta_{CO_2}^{W,SP} = \frac{\mu_{CO_2}^{F}}{\mu_{CO_2}^{W,SP}}, \text{ and } \eta_{CO_2}^{Q,SP} = \frac{\mu_{CO_2}^{F}}{\mu_{CO_2}^{Q,SP}} \tag{12}$$

thus expressing the *CO2ER* (11) as

$$CO2ER = 1 - \frac{1}{\dfrac{\eta_W}{\eta_{CO_2}^{W,SP}} + \dfrac{\eta_Q}{\eta_{CO_2}^{Q,SP}}} \tag{13}$$

and obtaining an expression with formal analogy with the *PES* indicator (2) used in energy efficiency studies.

Considering the *global* equivalent emission factor referred to the electricity production defined in (8), the *CO2ER* expression can be further written in terms of the electricity production

$$CO2ER = 1 - \frac{\mu_{CO_2}^{F} \cdot F}{\mu_{CO_2}^{W,GSP} \cdot W} = 1 - \frac{\mu_{CO_2}^{F}}{\mu_{CO_2}^{W,GSP} \cdot \eta_W} \tag{14}$$

Both equation (13) and the last expression in equation (14) show that the *CO2ER* indicator can be expressed in terms of cogeneration efficiencies and emission factors only. The emission factor $\mu_{CO_2}^F$, referred to the cogeneration thermal *input F*, can be considered at first approximation *independent* of the loading level, estimating it as a function of the fuel carbon content and of its *LHV* (Educogen, 2001). As an example, the value $\mu_{CO_2}^F \cong$ 200 g/kWh$_e$ can be assumed for natural gas referred to the *LHV*.
The electrical and thermal efficiencies of the cogeneration unit can be evaluated depending on the loading level, giving the possibility of applying the indicator for explicit assessments under actual operating conditions of the cogeneration unit.
Taking into account equation (6), a further relevant result can be obtained if the cogeneration system and the separate production use the *same fuel*. In this case, it is possible to write equation (13) as

$$CO2ER = 1 - \frac{1}{\frac{\eta_W}{\eta_e^{SP}} + \frac{\eta_Q}{\eta_t^{SP}}} \tag{15}$$

In this way, the *CO2ER* indicator becomes equal to the *PES* indicator (2), that is, the environmental benefits can be evaluated by using *only* energy efficiencies, providing emission reduction results numerically coincident with the ones obtained from the energy saving analysis, as widely discussed in Chicco & Mancarella, 2008a, and Chicco & Mancarella, 2008b.
The underlying hypothesis leading to (15) is that complete combustion occurs, which is an excellent approximation in most cases (Educogen, 2001) and leads to a conservative model of the CO_2 emissions. In fact, with incomplete combustion part of the hydrocarbons produce pollutants other than CO_2 (for instance, CO), so that the CO_2 produced is lower than the one estimated by using the emission factor model.
The generalisation of the indicators to assess the *global* emission reduction for a generic pollutant *p* is straightforward, yielding to the *class* of indicators

$$pER = 1 - \frac{\mu_p^F \cdot F}{\mu_p^{W,SP} \cdot W + \mu_p^{Q,SP} \cdot Q} = 1 - \frac{\mu_p^F}{\mu_p^{W,SP} \cdot \eta_W + \mu_p^{Q,SP} \cdot \eta_Q} = 1 - \frac{\mu_p^F}{\mu_p^{W,GSP} \cdot \eta_W} \tag{16}$$

thus obtaining for instance indicators named *NOXER* for the case of NO_X, *COER* for the case of CO, and so forth, with the same conceptual implications described above for the CO_2 case.
Concerning global warming impact, in cogeneration applications CO_2 is the main GHG of interest. However, in certain cases also methane emissions could be of concern, particularly because methane could represent up to 90% of the total UHC emitted in natural gas-fuelled units. Thus, a further formulation is presented to enable assessing the global emission reduction for a generic GHG or for a GHG set $\mathbf{G}$. This formulation is based on the fact that the effect of a generic GHG can be compared with the effect of CO_2 in terms of *Global Warming Potential* (*GWP*). Since by definition GWP_{CO_2} = 1 in the case of CO_2, the *GWP* for the other GHGs is expressed in relative terms with respect to CO_2 (see also Chicco &

Mancarella, 2008a, for details). The equivalent emission factor introduced for the generic GHG $p \in \mathbf{G}$ is defined as

$$\mu_{CO_2eq,p}^{X} = GWP_p \cdot \mu_p^X \tag{17}$$

where GWP_p represents the mass of CO_2 equivalent to the emission of a unity of mass of the GHG p, while μ_p^X is the emission factor defined in equation (6). The expression of the equivalent indicator *GHGER* containing the effect of a set of GHG referred to the cogeneration system is

$$GHGER = 1 - \frac{\sum_{p \in \mathbf{G}} \mu_{CO_2eq,p}^{F} F}{\sum_{p \in \mathbf{G}} \left(\mu_{CO_2eq,p}^{W,SP} W + \mu_{CO_2eq,p}^{Q,SP} Q \right)} \tag{18}$$

5.2 Local emission indicators

Since the local emission balance model of Section 3.2 neglects the amount of pollutants produced by the electrical system considered to be sufficiently "far" from the area of interest, the local emission reduction indicators are defined starting from the global emission indicators and deleting the amount referred to the separate production of electricity. The *class* of *generalised local* emission reduction indicators for a generic pollutant p is then expressed as

$$pLER = 1 - \frac{\mu_p^F \cdot F}{\mu_p^{Q,SP} \cdot Q} = 1 - \frac{\mu_p^F}{\mu_p^{Q,SP} \cdot \eta_Q} = 1 - \frac{\mu_p^F}{\mu_p^{W,LSP} \cdot \eta_W} \tag{19}$$

where the last expression is obtained by taking into account the definition of the local equivalent emission factor referred to the electricity production in (10). The corresponding indicators are named NO_XLER for the case of NO_X, *COLER* for the case of CO, and so forth.

5.3 Conventional separate production efficiencies and emission factors

The results obtained from the application of the emission reduction indicators depend on the choice of the efficiencies and emission factors referred to separate production of electricity and heat. The rationale for setting up the conventional values has to be addressed in a systematic way, highlighting the implications of different choices that could be adopted. Different settings of the values could lead to different numerical outcomes of the indicators introduced above. Since these indicators are expressed in relative values (per unit or per cent), the relevant shareholder typically pays attention to the resulting numerical outcome to get an idea of the potential emission reduction. For instance, a numerical outcome of 20% emission reduction has different meanings depending on the set of separate production efficiencies used to determine it.

Among the different ways to set up the conventional reference values, it is possible to mention:

1) The definition of the conventional values on the basis of the *average values* of the emission factors, that is, $\mu_p^{W,SP}$ for electricity production and $\mu_p^{Q,SP}$ for heat production. In this case, these values are assigned by considering on the electrical side the average emissions from the power plants used to produce electricity, and for the thermal side the average emissions from different boilers, also supplied with different fuels. The average values are calculated as weighted sums of the emissions from different units with respect to the unit sizes and types, and can in case refer to the marginal units operating in the bulk power system generation scheduling. This kind of definition allows obtaining indications on the real emission reduction that could occur in a given energy scenario, for instance in a given country (Meunier, 2002).
2) Considering cogeneration systems supplied by a given fuel (for instance, natural gas), the conventional values can be defined by taking into account the emission factors of technologies supplied by the same fuel (that is, in the case of CO_2, with the same carbon content and thus basically with the same CO_2 emissions per unit of burned fuel). This approach is aimed at assessing the emission saving potential of CHP systems *intrinsic* in the plant characteristics. It is then possible to adopt the model (6) with separate production efficiencies η_e^{SP} and η_t^{SP} for electricity and heat, respectively, to determine the equivalent emissions. In turn, the separate production efficiencies can be chosen according to different rationales, taking into account:
   a) *average technologies* for sizes *similar* to the one of the cogeneration system under analysis (ASST - Average Same-Size Technologies); or,
   b) the *best available technologies* for sizes *similar* to the one of the cogeneration system under analysis (BSST - Best Same-Size Technologies); or,
   c) the *best available technologies* (BAT) without size limits (with natural gas, corresponding to high-efficiency boilers and combined cycle power plants subtracting the electricity transmission and distribution losses).

The numerical values of the separate production efficiencies and emission factors are given by system-wide assessments, and need to be updated after some years in order to account for possible changes in the energy generation mix. An example of values referred to average CO_2 emissions in the Italian system (year 2003) yields $\mu_{CO_2}^{W,SP} \cong 525$ g/kWh$_e$ and $\mu_{CO_2}^{Q,SP} \cong 275$ g/kWh$_t$ (Mancarella & Chicco, 2009a). On the basis of equation (6), it can be noted that when increasing the SP efficiencies in the *CO2ER* indicator cogeneration loses competitiveness with respect to separate production. In fact, the adoption of the best technologies as references clearly penalizes the numerical outcome of the indicator, making cogeneration look less convenient. However, if the indicators are used for regulatory purposes, for instance setting thresholds above which it is possible to obtain incentives, the regulatory body can set the thresholds taking into account the conceptual meaning of the reference values. The above approaches are useful to boost the investments into high-efficiency cogeneration systems, with possible economic incentives within nation-wide energy and environmental policies (European Union, 2004; Cardona & Piacentino, 2005).

5.4 Emission limits and promotion of energy efficiency

The emissions measured on-site, or suitable emission reduction indicators, are typically compared with the limits to the various pollutants established by regulatory bodies. The rationale for setting up the emission limits can play a key role to promote or limit the diffusion of energy efficient technologies. For instance, the emission limits could be established considering the concentration of pollutant contained in the gas released to the ambient (e.g., expressed in mg/m^3). This approach could intuitively seem suitable to avoid exceeding the emission thresholds. However, it is not adequate to promote energy efficiency. In fact, a generator with high efficiency and a given concentration of pollutant in the exhaust gases would be penalized with respect to another generator less efficient but with slightly lower concentration of pollutant in the exhaust gases, regardless of the fact that the actual emissions per unit of output of the generator with higher efficiency could be lower than for the other unit.

A viable alternative consists of setting up emission limits on the basis of the specific emissions μ_p^X referred to the *useful energy output* (for instance, expressed in mg/kWh). In this case, it is possible to promote both reduction of the real environmental impact (referred to the useful outputs) and increase of energy efficiency.

Another limiting factor to the development of cogeneration solutions is the way in which the interactions among different causes of pollution are taken into account in the environmental regulation, especially at the local level. In the presence of a remarkably high level of pollution due for instance to road traffic, the strict application of the emission limits when planning the installation of new cogeneration systems would make it hard to promote the diffusion of new efficient technologies introducing (even relatively low) new local emissions, since the burden of exceeding the emission limits would be totally charged to the *marginal plants* to be installed. Promoting the diffusion of energy efficient technologies at the planning stage thus requires a comprehensive re-assessment of the causes of pollution and the identification of measures for limiting the impact of each of these causes.

5.5 Indicators for comparative emission assessment

Generally, the fuel adopted to supply the cogeneration system is different with respect to the fuel considered to represent the separate production, especially when taking into account the mix of fuels used to produce electricity in the power plants at regional or nation-wide level. Thus, focusing on CO_2 emissions, on average the same cogeneration technologies can be effective in terms of emission reduction in a system with prevailing production of electricity from fossil fuels (above all if with heavily polluting marginal plants), while they could provide no benefit in systems with prevailing production from hydroelectric or nuclear sources (Meunier, 2002; Chicco & Mancarella, 2008b). These aspects have been outlined in Mancarella & Chicco, 2008a, by introducing additional environmental impact indicators on the basis of which it is possible to provide a quantitative assessment of the effectiveness of adopting a certain type of cogeneration within a given regional or national context. In particular, the use of the indicators denoted as CO_2 Emission Equivalent Efficiency (*CO2EEE*) and CO_2 Emission Characteristic Ratio (*CO2ECR*) enables the determination of the break-even conditions for which CHP systems are equivalent to the conventional separate production in terms of GHG emissions. The *CO2ECR* indicator is defined as

$$CO2ECR = \frac{\mu_{CO_2}^{F}}{\left(\mu_{CO_2}^{W}\right)^{SP}} \tag{20}$$

The *CO2ER* expression (11) can be rewritten by explicitly showing the term *CO2ECR* as

$$CO2ER = 1 - \frac{CO2ECR \cdot F}{W + \frac{\mu_{CO_2}^{Q,SP}}{\mu_{CO_2}^{W,SP}} \cdot Q} = 1 - \frac{CO2ECR}{\eta_W + \frac{\mu_{CO_2}^{Q,SP}}{\mu_{CO_2}^{W,SP}} \cdot \eta_Q} \tag{21}$$

It is then possible to define the indicator *CO2EEE* as the value of *CO2ECR* obtained by applying the break-even condition *CO2ER* = 0 (Mancarella & Chicco, 2008a):

$$CO2EEE = \frac{W}{F} + \frac{\mu_{CO_2}^{Q,SP}}{\mu_{CO_2}^{W,SP}} \cdot \frac{Q}{F} = \eta_W + \frac{\mu_{CO_2}^{Q,SP}}{\mu_{CO_2}^{W,SP}} \cdot \eta_Q \tag{22}$$

In this way, the indicators *CO2ECR* and *CO2EEE* can be easily calculated on the basis of the emission factor of the fuel used, of the efficiencies of the cogenerator and of the emission factors in separate production. Adopting cogeneration is then convenient in terms of reducing the CO_2 emissions if the following inequality holds:

$$CO2EEE \geq CO2ECR \tag{23}$$

In analogy to equation (6), for separate production it is possible to define the expressions

$$\mu_{CO_2}^{W,SP} = \frac{\mu_{CO_2}^{F,SP}}{\eta_e^{SP}}; \qquad \mu_{CO_2}^{Q,SP} = \frac{\mu_{CO_2}^{F,SP}}{\eta_t^{SP}} \tag{24}$$

Assuming that the same fuel is used to supply the cogeneration unit, the external boiler and the power system generation mix, the *CO2EEE* indicator can be expressed in a way depending only on the cogeneration unit and separate production efficiencies:

$$CO2EEE = \eta_W + \frac{\eta_e^{SP}}{\eta_t^{SP}} \cdot \eta_Q \tag{25}$$

In the general case, in which the fuels are not the same, the general scheme of analysis can be applied with some practical adjustments. For this purpose, suitable correction factors can be defined. Considering the emission factor $\mu_{CO_2}^{F,SP,e}$ for the equivalent fuel supplying the electricity separate production system, and the emission factor $\mu_{CO_2}^{F,SP,t}$ for the fuel supplying the boiler for separate production of heat, the correction factors are defined as

$$\varepsilon_e = \frac{\mu_{CO_2}^{F}}{\mu_{CO_2}^{F,SP,e}}; \qquad \varepsilon_t = \frac{\mu_{CO_2}^{F}}{\mu_{CO_2}^{F,SP,t}} \tag{26}$$

Thus, the emission factors referred to the separate production of electricity and heat become, respectively,

$$\mu_{CO_2}^{W,SP} = \frac{\mu_{CO_2}^{F,SP,e}}{\eta_e^{SP}} = \frac{\mu_{CO_2}^{F}}{\varepsilon_e \eta_e^{SP}}; \qquad \mu_{CO_2}^{Q,SP} = \frac{\mu_{CO_2}^{F,SP,t}}{\eta_t^{SP}} = \frac{\mu_{CO_2}^{F}}{\varepsilon_t \eta_t^{SP}} \tag{27}$$

and the expression of the *CO2EEE* indicator becomes

$$CO2EEE = \eta_W + \frac{\varepsilon_e \eta_e^{SP}}{\varepsilon_t \eta_t^{SP}} \cdot \eta_Q \tag{28}$$

thus highlighting the use of the corrected electrical efficiency $\varepsilon_e \eta_e^{SP}$ for separate production of electricity, and of the corrected thermal efficiency $\varepsilon_t \eta_t^{SP}$ for separate production of heat. Comparisons at regional or nation-wide level by using the *CO2ECR* and *CO2EEE* indicators have been reported in Mancarella & Chicco, 2008a.

5.6 Emission maps

In order to represent effectively the outcomes from break-even analyses, specific *emission mapping models* can be introduced. For instance, it is possible to work out the break-even conditions (in terms of pollutant emissions) with reference to the unitary production of electricity or heat, in terms of emission factors as defined in equation (5), taking the electrical and thermal efficiencies of the CHP prime mover as variables. The relevant break-even emission values so obtained in these *emission break-even maps* can be then compared to the actual emissions from every specific cogeneration unit and for every given pollutant. In this way, it is straightforward to estimate the environmental impact of cogeneration on the basis of the relevant emission balance considered, so that the maps drawn allow for general (not only break-even) emission assessment, as illustrated below. For specific analyses it could be possible to simulate the actual dispatch of the relevant cogeneration units to assess *punctually* their environmental impact with respect to the marginal plants operating in the bulk power system, as done for instance in Hadley & Van Dyke, 2003, or for bigger systems in Voorspools & D'haeseleer, 2000, and Voorspools & D'haeseleer, 2003.
Apart from the specific emission balance considered, several different maps can be drawn depending upon the *reference* emission characteristics assigned to the separate generation. The conventional reference values can be set up as indicated in Section 5.3. The reference numerical values considered for separate production may change significantly the outcomes of the analysis, so that these values must be carefully selected according to the specific systems under study and to the specific goal to pursue, above all for policy purposes.
Alternative models could be developed to account for the marginal operation of cogeneration units in different hours in an equivalent fashion, as done for instance by Tsikalakis & Hatziargyriou, 2007. However, for general and synthetic assessments, such as

for general policy regulation development, simulation-based or equivalently detailed approaches seem less feasible, above all in the presence of large power systems.
Once drawn the relevant break-even emission maps, given the cogeneration efficiencies (1) and the corresponding emissions for *every* operating point of the cogeneration system under analysis, it is possible to evaluate the local and global emission balances (Section 3.2), and thus the environmental performance with respect to the conventional separate generation of the same amount of electricity and/or heat, according to the emission evaluation model used. More specifically, given a certain pollutant, to which corresponds a certain emission break-even map on the basis of the separate production emission references selected for the analysis, the environmental impact comparison between conventional generation and cogeneration can be carried out on the basis of the following steps:

- assign the electrical and thermal efficiency of the analysed cogeneration system under a determined operation condition;
- in correspondence of these values of η_W and η_Q, determine on the (local and/or global) emission break-even map the relevant (local and/or global) emissions from the conventional separate production technologies taken as references;
- compare the emission values found to the *actual* emissions from the considered cogeneration system under the same operational conditions;
- on the basis of the separate production emissions and the actual cogeneration emissions, evaluate the relevant emission balance.

A specific example of use of the emission maps is shown here, considering the case of NO_x with average reference emission factors $\mu_{NO_x}^{W,SP}$ = 0.5 g/kWh$_e$ for electricity generation and $\mu_{NO_x}^{Q,SP}$ = 0.5 g/kWh$_t$ for heat generation.

In the *local* emission balance analysis, Fig. 1 shows the emission break-even curves, in terms of specific emissions expressed in mg/kWh$_e$. The curves are drawn in function of the cogeneration electrical efficiency, for discrete values of the thermal efficiency used as the curve parameter.

The mapping in Fig. 1 can be exploited by following the procedure outlined above, with reference to specific cogeneration prime movers. For instance, let us consider a MT and an ICE with rated characteristics indicated in Table 1.

unit	rated power [kW$_e$]	η_W	η_Q	λ_y	NO_x specific emissions	
					[mg/kWh$_e$]	[mg/kWh$_t$]
MT	100	0.29	0.48	1.65	170	101
ICE	180	0.34	0.49	1.44	1500	1041

Table 1. Rated characteristics for MT and ICE units.

The location within the map of the *actual* specific emissions (in mg/kWh$_e$) of the MT and the ICE units is reported in Fig. 1. For each unit, the rated electrical and thermal efficiency pair of values corresponds, on the map, to the emission break-even condition, that is, the maximum specific emissions (referred in this case to the kWh$_e$) that the unit should feature in order to guarantee an environmental impact lower than the conventional heat generation reference in the local emission balance. In particular, while the emission map in Fig. 1 is drawn with reference to the specific NO_x emissions per kWh$_e$, the break-even conditions are worked out so as to compare the cogeneration heat production with the same thermal energy produced by local boilers. Indeed, an equivalent emission break-even map might be drawn as well with reference to specific emissions per kWh_t produced. As the last step in the evaluation, once pointed out in the map the relevant break-even unitary emissions, the local emission balance can be graphically worked out by pointing out the *actual* emissions from the machine under analysis.

Considering the MT, the values of electrical efficiency (0.29) and thermal efficiency (0.48) are entered in the emission map, providing a break-even emission factor value of about 330 mg/kWh_e. Comparing the break-even emission value to the actual full-load MT emissions of about 170 mg/kWh_e (Table 1), the MT brings NO_x emission reduction per kWh_e produced of about 330–170=160 mg/kWh_e. A similar calculation could be readily developed with respect to the kWh_t produced. In this case, the MT would emit 101 mg/kWh_t (Table 1), to be compared with 200 mg/kWh_t emitted by the reference boiler; this brings NO_x emission reduction equal to 97 mg/kWh_t, again corresponding to $97 \cdot 1.65=160$ mg/kWh_e.

Considering the ICE, entering the electrical efficiency (0.34) and the thermal efficiency (0.49) in the emission map, the break-even NO_x emissions value for the reference boiler is of about 280 mg emitted per kWh_e of electrical energy cogenerated. Considering actual specific emissions of 1500 mg/kWh_e (Table 1), the local emissions using the ICE unit *increase* with respect to the local emissions from conventional reference boilers by about 1220 mg/kWh_e.

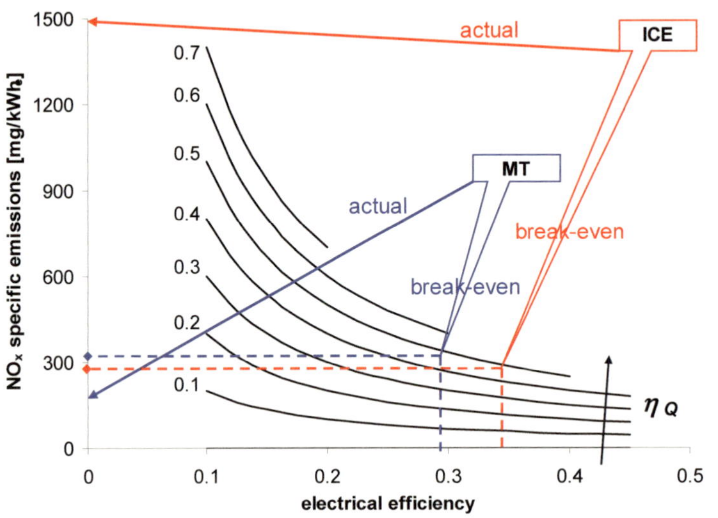

Fig. 1. Local NO_x emission balance assessment with MT and ICE (average separate production references).

The same considerations apply to the *global* emission map for NO_x, comparing the break-even conditions with the actual emission characteristics of the MT and ICE of Table 1. In this case, a similar representation (not shown here) can be adopted, in which the global emission balance yields a global emission *reduction* of about 650 mg/kWh_e for the MT, and a global emission *increase* of about 750 mg/kWh_e for the ICE.

Besides single cogeneration units, the emission mapping models can be used to evaluate more general solutions, such as particular scenarios of diffusion of cogeneration with different types of equipment. The general approaches introduced enable to undertake *scenario analyses* aimed at assessing the environmental impact from given types of equipment, number of units, and so forth. For instance, the scenario analyses can be formulated according to the lines indicated in Chicco & Mancarella, 2008c:

- different technologies are taken into account, with their energy and emission characteristics;
- a set of scenarios is defined, each of which is characterized by a given mix of technologies, that can be envisioned for the future, and by the level of penetration of cogeneration with respect to the (electrical and thermal) energy demand; each scenario also contains a specific set of reference values for separate production of electricity and heat;
- for each mix of technologies, the equivalent energy and emission performance is determined by weighting the contribution of each technology to the overall mix with a predefined model (e.g., linear);
- the *PES* and *CO2ER* indicators, as well as the local emission balance outcomes, can be used to evaluate the energy efficiency and the environmental impact for different pollutants.

In addition, a comprehensive environmental impact mapping of given cogeneration systems can be obtained by merging scenario analyses with off-design assessments.

6. Role of environmental impact in the formulation of optimisation methods

Minimization of the emissions from cogeneration plants has been included in the formulation of planning and operation problems in various literature studies. The simplest way to take into account emissions is to include the emission limits within the optimization problem constraints. More generally, emissions are taken into account in the definition of *multi-objective* optimisation problems of different types:

1. Problems in which an *equivalent objective function* is defined as the weighted sum of a set of objective functions; one of the objective functions (or more than one, for instance when local and global emissions are considered separately) refers to emission impact minimization.
2. Problems with *conflicting objectives* (typical problems in the case of considering different types of emissions) solved through the evaluation of compromise solutions with Pareto-front calculation techniques. The *Pareto front* contains all the non-dominated solution points of the multi-objective optimization problem. With reference to objective minimisations, a solution is *non-dominated* if no other solution exhibits lower values for *all* the individual objective functions. Non-dominated solutions in which none of the individual optima is achieved may be of interest because of providing compromise alternatives among the various objectives. For problems with non-convex Pareto front, the ε-constrained method (Yokoyama et al., 1988) optimizes the preferred objective by introducing the other objectives as constraints and leaving a margin of acceptable solutions bounded by a user-defined threshold ε. Conceptually, the components of the Pareto front could be obtained more extensively by varying the threshold ε. More recently, some literature methods have been proposed to find directly a number of compromise solutions belonging to the *best-known* Pareto front (Shukla & Deb, 2007).

3. Long-term or design problems solved through a *multi-criteria* approach (Giannantoni et al., 2005; Carpaneto et al., 2007). When the level of uncertainty becomes very large, the decision-maker can assume a wider discretion in defining a set of scenarios to be considered. These scenarios are then analysed by means of *multi-attribute* decision-making approaches, providing multiple results, among which the decision-maker can determine the preferred solution by evaluating the Pareto front through a suitable numerical technique, as described in Li, 2009, or exploiting risk-based tools (Carpaneto et al., 2008).

Some sample references including environmental aspects in cogeneration optimization are recalled here. Curti et al., 2000, introduce in the objective function a specific term representing the pollution cost rate determined for each pollutant, depending on the emission level, the specific damage cost referred to the pollutant, and a user-defined penalty factor. In the multi-objective approach with minimization of cost and multiple emissions presented in Tsay, 2003, the emissions of each pollutant (CO_2, SO_x, and NO_x) are modelled as a function of the fuel enthalpy dependent on the emission factor. In Aki et al., 2005, optimum energy pricing is obtained as a Pareto solution for a multi-objective model considering both CO_2 emissions and economic impact on consumers. The CO_2 emission limits are also used as a constraint in the optimization model to minimize the individual cost to the consumers. Pelet et al., 2005, introduce the CO_2 emissions among the multiple objectives of integrated energy systems, obtaining the best-known Pareto front through a dedicated heuristic. Boicea et al., 2009, determine the best-known Pareto front for a cluster of microturbines operating with electrical load-following control strategy.

Environmental objectives are also included in the optimization of tri-generation or multi-generation systems (for instance, Burer et al., 2003; Rong & Lahdelma, 2005; Li et al., 2006).

7. Identification and determination of the environmental external costs and role of Life Cycle Assessment

External costs can be defined as the costs determined by the activities of a subject that do not appear in the economic balance of that subject. Another definition refers to external costs as arising *"when the social or economic activities of one group of persons have an impact on another group and when that impact is not fully accounted, or compensated for, by the first group."* (Bickel & Friedrich, 2005).

For cogeneration plants, internal costs are typically referred to construction and operation of the plant. Cogeneration plant emissions impact on the environment and on the society because of the effects of the pollutants emitted on the human health or on other receptors (Gulli, 2006). Some of the costs to reduce the pollutant emissions, such as the ones for installing abatement systems, are included in the internal balance. The remaining costs referred to the impact on the environment and the society are external cost components. Environmental external costs are related to both local and global effects, and need to be compared to separate production externalities. External costs can generally be internalized, that is, included in the economic evaluation to complete the environmental analysis. When the net external cost balance is positive in favour of distributed systems, internalization can be carried out through fiscal incentives, discounts on purchasing of products, taxes, relaxation of the air quality constraints, and so forth. For instance, adopting classical economic indicators (Biezma & San Cristóbal, 2006), ICE technologies could result more convenient than microturbines on the basis of economic analysis, but MTs could exhibit emissions of CO and NO_X lower than the ones of the ICEs in a local emission balance. By internalizing the external costs due to global emissions, the margins of convenience of ICEs with respect to MTs could decrease substantially.

The analysis and assessment of external costs for energy system applications can be addressed as in the ExternE project (Bickel & Friedrich, 2005), based on a bottom-up approach in which the final impacts of the energy production are tracked back to their initial causes by determining the chain of events denoted as "impact pathways". The four stages of the ExternE model address the following aspects:

1. The description of the *technology* and characterisation of the related emissions. This requires the identification of the parameters referred to the emissions, like the flux of substances emitted in the environment and the concentration of pollutants inside these substances, as well as the stack height, and so forth. If the planning analysis refers to comparing different alternatives without specifying the technical details (Canova et al., 2008), it is possible to resort to the use of the emission factors within the emission balance approach illustrated in Section 3.
2. The analysis of the territorial *dispersion* of the emissions, with the objective of identifying the concentration of pollutants in the areas in which the receptors are located. The analysis is carried out by means of either *statistical* models based on the time series of environmental data measured in meteorology centres, or *deterministic* models, based on tracing the dispersion of the pollutants according to theoretical representation of the phenomena linked to their diffusion in the atmosphere (Arya, 1999).
3. The identification of the *receptors* and of the corresponding dose-response functions to assess the potential damages. The dose-response functions are defined in incremental terms, representing the increase of damage due to the increase of concentration of the pollutant. With reference to the human health, the dose-response functions can represent the increase in the number of subjects affected by a given pathology as a consequence of the exposure to the pollutants reaching the areas in which the subject is located (Bickel & Friedrich, 2005).
4. The determination of the *economic value* of the damages, taking into account permanent or temporary damages. Permanent damages refer to effects leading to premature death, generally evaluated by means of the Years of Lost Life (*YOLL*) to obtain the Value of Life Years Lost (*VOLY*) indicator (Krewitt et al., 1998; Bickel & Friedrich, 2005). Temporary damages refer to the quantification of illness, evaluated either by determining the cost incurred by the society to care the patients affected, denoted as Cost of Illness (*COI*), or by identifying the Willingness to Pay (*WTP*) of the individuals to avoid the occurrence or persistence of the causes, also taking into account further aspects of personal judgement like the opportunities lost because of illness (Dickie & Gerkin, 1989; Stieb et al., 2002).

A more comprehensive approach to the evaluation of the overall effects of introducing cogeneration systems in the energy context resorts to the concepts of Life Cycle Assessment (LCA), as discussed and applied in dedicated studies (Dincer, 1999; González et al., 2003; Chevalier & Meunier, 2005; Bickel & Friedrich, 2005).

For distributed cogeneration applications, in the overall LCA environmental balance the construction of MTs or ICEs (in terms of materials, manufacturing process, transport and installation) impacts to a minor extent with respect to the energy generated by the unit during its operational lifetime (Riva et al., 2006). This highlights the practical importance of studies addressing energy and environmental issues of cogeneration, especially when comparisons are made among technologies using the same fuel, in which further aspects concerning the construction and exploitation of the fuel transportation system have the same impact and thus can be removed from the analysis without changing the nature of the results. Conversely, LCA considerations could be needed when comparing technological alternatives adopting different fuels, especially when the fuel production and transportation infrastructure has different characteristics with respect to the one used for natural gas, as in the case, for instance, of biomasses or hydrogen (Chevalier & Meunier, 2005; Pehnt, 2001). More generally, the inclusion of external costs in the objective functions of regional energy planning studies could show more incisively the benefits of adopting cogeneration, as well as other energy efficient technologies, in the energy production system (Cormio et al., 2003).

8. Cogeneration deployment in energy-related markets

Cogeneration can already be exploited under specific tariff systems or within a competitive electricity market structure. In addition, it could be possible to trade energy-related commodities (Chicco & Mancarella, 2007b) relevant to cogeneration, such as:

- GHG *emission allowances*, introduced within an emission trading framework aimed at limiting the GHG emissions from energy consumers; currently, each entity participating in the emission trading mechanism is assigned a certain number of emission allowances, with the possibility of trading the positive or negative allowance spread on the relevant market (Boonekamp, 2004); the unitary price ρ_{CO_2} of the allowances is expressed in m.u./tonCO_2eq, where m.u. means monetary units; the allocation of CO_2 emissions for energy systems with multiple products and multiple inputs is illustrated and discussed in Rosen, 2008.
- *Energy efficiency (white) certificates*, corresponding to acknowledged primary energy saving obtained from actions aimed at reducing the electricity and/or gas consumption; the unitary price ρ_y of the white certificates is expressed in m.u./toe.

Currently, a limited number of relatively large actors can participate in such markets, also depending on the country-specific applications. However, it can be envisaged that in the future participation will be enlarged to smaller producers. Profitability of potentially deploying CHP technologies within such market frameworks (provided that the policy structure allows it) should be evaluated through energy-environmental economic models. If the focus is specifically set on the cost of electricity production, an application example is illustrated in Mancarella & Chicco, 2009a, whose approach defines an average production cost of electricity (in m.u./kWh$_e$) based on the average fuel cost for electricity production. For this purpose, the fuel component related to the electricity produced is discounted by an equivalent amount relevant to the cogenerated heat in an *incremental* fashion, according to the classical incremental heat rate model (Horlock, 1997). The average production cost of electricity defined by this approach can be compared to the actual electricity prices, providing preliminary hints to assess profitability of exploiting the cogeneration system. Furthermore, multi-scenario analyses are run to calculate the sensitivities of electricity production cost to emission allowance prices, white certificate prices, gas prices, and conventional separate production references (Chicco & Mancarella, 2007b). The outcomes from such an exercise enlighten how the competitiveness of distributed cogeneration could increase substantially if adequate pricing or market framework were set up to acknowledge the positive environmental externalities brought about by the enhanced performance intrinsic in the combined production.

9. Conclusions

This chapter has illustrated a number of aspects referred to the environmental impact of cogeneration systems, mainly focused on recent literature references and on the authors' work. In particular, specific models for evaluation of global and local pollutants have been discussed, introducing relevant emission reduction indicators to quantify the potential benefits of distributed cogeneration relative to classical separate production of heat in boilers and electricity in centralised power plants. Openings aimed to internalise environmental externalities within an LCA framework or potential energy-related markets, have also been illustrated. Starting from the concepts presented, there are several extensions for present and future research at both theoretical and application levels. On the technology side, the diffusion of new solutions of different type (for instance, fuel cells) and/or supplied by different fuels (e.g., biomasses) can change the scenario of convenience and profitability of adopting cogeneration in evolving energy systems. Other overall benefits could come from the interactions of various types of cogeneration prime movers with district heating, storage, and more generally multi-generation solutions for simultaneous production of different energy vectors. Broader availability and interaction of technologies can be accompanied by their more flexible exploitation under off-design conditions, introducing new uncertainty in the energy system analysis. Large uncertainty also appears at the planning and design stages because of the need of making hypotheses and of constructing very different scenarios of evolution of the electricity and gas prices in time horizons spanning over at least one decade. Further elements of uncertainty are introduced by the presence of energy-related markets for trading white certificates or emission allowances, and by a continuously changing policy framework, whose evolution depends on political decisions and on arbitrariness at the regulatory level. From the point of view of the decision-maker, helpful responses could come from the development of tools exploiting the concept of risk and formulating suitable strategies to hedge risks.

10. References

Aki, H.; Oyama, T. & Tsuji, K. (2005). Analysis of energy pricing in urban energy service systems considering a multiobjective problem of environmental and economic impact. *IEEE Transactions on Power Systems*, Vol. 18, No. 4, (November 2005) 1275-1282, ISSN 0885-8950

Arya, S.P. (1999). *Air pollution meteorology and dispersion*, Oxford University Press, ISBN 978-0195073980, New York

Bickel, P. & Friedrich, R. (ed.) (2005). *ExternE: Externalities of Energy, Methodology 2005 update.* European Communities, 2005, http://www.externe.info/brussels/methup05.pdf

Biezma, M.V. & San Cristóbal, J.R. (2006). Investment criteria for the selection of cogeneration plants — a state of the art review. *Applied Thermal Engineering*, Vol. 26, No. 5-6, (April 2006) 583-588, ISSN 1359-4311

Boicea, A.-V.; Chicco, G. & Mancarella, P. (2009). Optimal Operation of a Microturbine Cluster with Partial-Load Efficiency and Emission Characterization, *Proceedings of IEEE Power Tech 2009*, paper 109, ISBN 978-1-4244-2235-7, Bucharest, Romania, 28 June - 2 July 2009, The IEEE, Piscataway, NJ

Boonekamp, P.G.M. (2004). Energy and emission monitoring for policy use. Trend analysis with reconstructed energy balances. *Energy Policy*, Vol. 32, No. 8, (June 2004) 969-988, ISSN 0301-4215

Burer, M.; Tanaka, K.; Favrat, D. & Yamada, K. (2003). Multi-criteria optimization of a district cogeneration plant integrating a solid oxide fuel cell-gas turbine combined cycle, heat pumps and chillers. *Energy*, Vol. 28, No. 6 (May 2003) 497 - 518, ISSN 0360-6442

Canova, A. ; Chicco, G. ; Genon, G. & Mancarella, P. (2008). Emission Characterization and Evaluation of Natural Gas-Fueled Cogeneration Microturbines and Internal Combustion Engines. *Energy Conversion and Management*, Vol. 49, No.10, (October 2008) 2900-2909, ISSN 0196-8904

Cardona, E. & Piacentino, A. (2005). Cogeneration: a regulatory framework toward growth. *Energy Policy*, Vol. 33, No. 16, (November 2005) 2100-2111, ISSN 0301-4215

Cârdu, M. & Baica, M. (2002). Regarding the greenhouse gas emissions of thermopower plants. *Energy Conversion and Management*, Vol. 43, No. 16, (November 2002) 2135-2144, ISSN 0196-8904

Carpaneto, E.; Chicco, G.; Mancarella, P. & Russo, A. (2007). A decision theory approach to cogeneration planning in the presence of uncertainties, *Proceedings IEEE Power Tech 2007*, paper 592, ISBN 978-1-4244-2189-3, Lausanne, Switzerland, 1-5 July 2007, The IEEE, Piscataway, NJ

Carpaneto, E.; Chicco, G.; Mancarella, P. & Russo, A. (2008). A risk-based model for cogeneration resource planning, *Proceedings 10th International Conference on Probabilistic Methods Applied to Power Systems (PMAPS 2008)*, paper 083, ISBN 978-1-9343-2521-6, Rincón, Puerto Rico, May 25-29, 2008

Chevalier, C. & Meunier, F. (2005). Environmental assessment of biogas co- or tri-generation units by life cycle analysis methodology. *Applied Thermal Engineering*, Vol. 25, No. 17-18, (December 2005) 3025-3041, ISSN 1359-4311

Chicco, G. & Mancarella, P. (2007a). Trigeneration Primary Energy Saving Evaluation for Energy Planning and Policy Development. *Energy Policy*, Vol. 35, No. 12, (December 2007) 6132-6144, ISSN 0301-4215

Chicco, G. & Mancarella P. (2007b). Exploiting Small-Scale Cogeneration in Energy-Related Markets, *Proceedings IEEE PowerTech 2007*, paper 319, ISBN 978-1-4244-2189-3, Lausanne, Switzerland, 1-5 July 2007, The IEEE, Piscataway, NJ

Chicco, G. & Mancarella, P. (2008a). Assessment of the Greenhouse Gas Emissions from Cogeneration and Trigeneration Systems. Part I: Models and Indicators. *Energy*, Vol. 33, No. 3 (March 2008) 410-417, ISSN 0360-0442

Chicco, G. & Mancarella, P. (2008b). A Unified Model for Energy and Environmental Performance Assessment of Natural Gas-Fueled Poly-Generation Systems. *Energy Conversion and Management*, Vol. 49, No. 8 (August 2008) 2069-2077, ISSN 0196-8904

Chicco, G. & Mancarella, P. (2008c). Environmental Sustainability of Distributed Cogeneration Systems, *Proceedings IEEE Melecon 2008*, paper t5-sf0030, ISBN 978-1-4244-1633-2, Ajaccio, France, May 5-7, 2008, The IEEE, Piscataway, NJ

Chicco, G. & Mancarella, P. (2009a). Distributed Multi-Generation: a Comprehensive View. *Renewable and Sustainable Energy Reviews*, Vol. 13, No. 3 (April 2009) 535-551, ISSN 1364-0321

Chicco, G. & Mancarella, P. (2009b). Matrix Modelling of Small-Scale Trigeneration Systems and Application to Operational Optimization. *Energy*, Vol. 34, No. 3 (March 2009) 261-273, ISSN 0360-6442

Cormio, C.; Dicorato, M.; Minoia, A. & Trovato, M. (2003). A regional energy planning methodology including renewable energy sources and environmental constraint. *Renewable and Sustainable Energy Reviews*, Vol. 7, No. 2,(April 2003) 99-130, ISSN 1364-0321

Curti, V.; von Spakovsky, M.R. & Favrat, D. (2000). An environomic approach for the modeling and optimization of a district heating network based on centralized and decentralized heat pumps, cogeneration and/or gas furnace. Part I: Methodology. *International Journal of Thermal Sciences*, Vol. 39, No. 7, (July 2000) 721–730, ISSN 1290-0729

Danny Harvey, L.D. (2006). *A handbook on low energy buildings and district energy systems*. Quicksilver Drive, ISBN 9781844072439, Sterling, USA

Dickie, M. & Gerkin, S. (1989). Benefits of reduced mordidity from air pollution control – A survey, Chapter 6 in *Valuation methods and policy making in environmental economics*, Folmer H. & van Ierland E. (editors) , 105-122, Elsevier Science, ISBN 0-444-87382-1, Amsterdam, The Netherlands

Dincer, I. (1999). Environmental impacts of energy. *Energy Policy*, Vol. 27, No. 14 (December 1999) 845-854, ISSN 0301-4215

Educogen (2001). The European Educational Tool on Cogeneration, December 2001, web site (available October 2009): www.cogen.org/projects/educogen.htm

EPRI (2009). Distributed Energy Resources Emissions Survey and Technology Characterization, www.epriweb.com/public/000000000001011256.pdf

European Union (2004). Directive 2004/8/EC of the European Parliament and of the Council of 11 February 2004 on the promotion of cogeneration based on a useful heat demand in the internal energy market and amending Directive 92/42/EEC. *Official Journal of the European Union*, 21.2.2004; L 52: 50–60, http://europa.eu.int/eur-lex/pri/en/oj/dat/2004/l_052/l_05220040221en00500060.pdf

Freschi, F. & Repetto, M. (2008). Artificial Immune System in the management of complex small scale cogeneration systems. In *Handbook of research on Artificial Immune Systems and Natural Computing: Applying Complex Adaptive Technologies*, Hongwei Mo (Ed.), 141-158, Medical Information Science Reference, ISBN 978-1-60566-310-4

Giannantoni, C.; Lazzaretto, A.; Macor, A.; Mirandola, A.; Stoppato, A.; Tonon, S. & Ulgiati, S. (2005). Multicriteria approach for the improvement of energy systems design. *Energy*, Vol. 30, No. 10, (July 2005) 1989–2016, ISSN 0360-6442

Gómez-Villalva, E. & Ramos, A. (2003). Optimal Energy Management of an Industrial Consumer in Liberalized Markets. *IEEE Transactions on Power Systems*, Vol. 18, No. 2, (May 2003) 716-723, ISSN 0885-8950

González, A.; Sala, J.M.; Flores, I. & López, L.M. (2003). Application of thermoeconomics to the allocation of environmental loads in the life cycle assessment of cogeneration plants. *Energy*, Vol. 28, No. 6, (May 2003) 557-574, ISSN 0360-6442

Gulli, F. (2006). Social choice, uncertainty about external costs and trade-off between intergenerational environmental impacts: the emblematic case of gas-based energy supply decentralization. *Ecological Economics*, Vol. 57, No. 2, (May 2006) 282-305, ISSN 0921-8009

Hadley, S.W. & Van Dyke, J.W. (2003). Emissions benefits of distributed generation in the Texas market. Report prepared for the Gas Technology Institute. Oak Ridge National Laboratory, Oak Ridge, Tennessee, USA, http://www.ornl.gov/sci/btc/apps/Restructuring/Texas_DG_AnalysisFinal.pdf.

Hatziargyriou, N.; Asano, H.; Iravani, R. & Marnay, C. (2007). Microgrids. An overview of ongoing research, development, and demonstration projects. *IEEE Power and Energy Magazine*, Vol. 5, No. 4, (July/August 2007) 78-94, ISSN 1540-7977

Horii, S.; Ito, K.; Pak, P.S. & Suzuki, Y. (1987). Optimal planning of gas turbine co-generation plants based on mixed-integer linear programming. *International Journal of Energy Research*, Vol. 11, No. 4, (October/December 1987) 507–518, ISSN 0363-907X

Horlock, J.H. (1997). *Cogeneration-Combined Heat and Power (CHP)*. Krieger, ISBN 0-89464-928-0, Malabar, FL

Illerhaus, S.W. & Verstege, J.F. (1999). Optimal operation of industrial CHP-based power systems in liberalized energy markets. *Proceedings IEEE Power Tech.1999*, ISBN 0-7803-5836-8, paper BPT99–352-13, (29 August-2 September 1999), The IEEE, Piscataway, NJ

Kreider J.F. (ed.) (2001). *Handbook of Heating, Ventilation and Air Conditioning*, CRC Press, ISBN 978-0849395840, Boca Raton, FL

Krewitt, W.; Hurley, F.; Trukenmüller, A. & Friedrich, R. (1998). Health Risks of Energy Systems. *Risk Analysis*, Vol. 18, No. 4, (August 1998) 377-383, ISSN 1539-6924

Li, H.; Nalim, R. & Haldi, P.-A. (2006). Thermal-economic optimization of a distributed multi-generation energy system - a case study of Beijing. *Applied Thermal Engineering*, Vol. 26, No. 7, (May 2006) 709–719, ISSN 1359-4311

Li, X. (2009). Study of multi-objective optimization and multi-attribute decision-making for economic and environmental power dispatch. *Electric Power Systems Research*, Vol. 79, No. 5, (May 2009) 780-795, ISSN 0378-7796

Mancarella, P. (2006). *From cogeneration to trigeneration: energy planning and evaluation in a competitive market framework*. Doctoral thesis, Politecnico di Torino, Torino, Italy, April 2006.

Mancarella, P. & Chicco, G. (2008a). Assessment of the Greenhouse Gas Emissions from Cogeneration and Trigeneration Systems. Part II: Analysis Techniques and Application Cases. *Energy*, Vol. 33, No. 3 (March 2008) 418-430, ISSN 0360-0442

Mancarella, P. & Chicco, G. (2008b). CO_2 Emission Reduction from Multigeneration Systems: Evaluation of Combined Cooling Heat and Power Scenarios. *Energetica*, Vol. 56, No. 8 (August 2008) 326-333, ISSN 1453-2360

Mancarella, P. & Chicco, G. (2009a). Global and Local Emission Impact Assessment of Distributed Cogeneration Systems with Partial-load Models. *Applied Energy*, Vol. 86, No. 10 (October 2009) 2096-2106, ISSN 0306-2619

Mancarella, P. & Chicco, G. (2009b). *Distributed Multi-Generation Systems. Energy Models and Analyses*, Nova Science Publishers, ISBN 978-1-60456-688-8, New York

Meunier, F. (2002). Co- and tri-generation contribution to climate change control. *Applied Thermal Engineering*, Vol. 22, No. 6, (April 2002) 703-718, ISSN 1359-4311

Pehnt, M. (2001). Life-cycle assessment of fuel cell stacks. *International Journal of Hydrogen Energy*, Vol. 26 , No. 1, (January 2001) 91-101, ISSN 0360-3199

Pehnt, M. (2008). Environmental impacts of distributed energy systems - The case of micro cogeneration. *Environmental Science & Policy*, Vol. 11, No. 1, (February 2008) 25-37, ISSN 1462-9011

Pehnt, M.; Cames, M.; Fischer, C.; Praetorius, B.; Schneider, L.; Schmacher, K. & Voß, J.-P. (2006). *Micro Cogeneration. Towards decentralized energy systems*. Springer-Verlag, ISBN 978-3540255826, Berlin Heidelberg, Germany

Pelet, X.; Favrat, D. & Leyland, G. (2005). Multiobjective optimisation of integrated energy systems for remote communities considering economics and CO_2 emissions. *International Journal of Thermal Sciences*, Vol. 44, No. 12, (December 2005) 1180–1189, ISSN 1290-0729

Rosen, M.A. (2008). Allocating carbon dioxide emissions from cogeneration systems: descriptions of selected output-based methods. *Journal of Cleaner Production*, Vol. 16, No. 2, (January 2008) 171-177, ISSN 0959-6526

Riva, A.; D'Angelosante, S. & Trebeschi, A. (2006). Natural gas and the environmental results of life cycle assessment. *Energy*, Vol. 31, No. 1, (January 2006) 138-148, ISSN 0360-6442

Rong, A. & Lahdelma, R. (2005). An efficient linear programming model and optimization algorithm for trigeneration. *Applied Energy*, Vol. 82, No. 1, (September 2005) 40-63, ISSN 0306-2619

Rong, A. & Lahdelma, R. (2007). CO2 emissions trading planning in combined heat and power production via multi-period stochastic optimization. *European Journal of Operational Research*, Vol. 176, No. 3, (February 2007) 1874-1895, ISSN 0377-2217

Shukla, P.K. & Deb, K. (2007). On finding multiple Pareto-optimal solutions using classical and evolutionary generating methods. *European Journal of Operational Research*, Vol. 181, No. 3, (September 2007) 1630-1652, ISSN 0377-2217

Stieb, D.M.; De Civita, P.; Johnson, F.R.; Manary, M.P.; Anis, A.H.; Beveridge, R.C. & Judek, S. (2002). Economic evaluation of the benefits of reducing acute cardiorespiratory morbidity associated with air pollution. *Environmental Health: A Global Access Science Source*, Vol. 1, No. 1, (18 December 2002)

Strachan, N. & Farrell, A. (2006). Emission from distributed vs. centralized generation: The importance of system performance. *Energy Policy*, Vol. 34, No. 17, (November 2006) 2677-2689, ISSN 0301-4215

Tsay, M.-T.; Lin, W.-M. & Lee, J.-L. (2001). Application of evolutionary programming for economic dispatch of cogeneration systems under emission constraints. *Electrical Power and Energy Systems*, Vol. 23, No. 8, (November 2001) 805-812, ISSN 0142-0615

Tsay, M.-T. (2003). Applying the multi-objective approach for operation strategy of cogeneration systems under environmental constraints. *Electrical Power and Energy Systems*, Vol. 25, No. 3, (March 2003) 219-226, ISSN 0142-0615

Tsikalakis, A.G. & Hatziargyriou, N.D. (2007). Environmental benefits of distributed generation with and without emissions trading, *Energy Policy*, Vol. 35, No. 6, (June 2007) 3395-3409, ISSN 0301-4215

US Environmental Protection Agency (2009). Catalogue of CHP Technologies, www.epa.gov

Voorspools, K.R. & D'haeseleer, W.D. (2000). The influence of the instantaneous fuel mix for electricity generation on the corresponding emissions. *Energy,* Vol. 25, No. 11, (November 2000) 1119-1138, ISSN 0360-6442

Voorspools, K.R. & D'haeseleer, W.D. (2003). The impact of the implementation of cogeneration in a given energetic context. *IEEE Transactions on Energy Conversion,* Vol. 18, No. 1, (March 2003) 135-141, ISSN 0885-8969

Yokoyama, R.; Bae, S.H.; Morita, T. & Sasaki, H. (1988). Multiobjective generation dispatch based on probability security criteria. *IEEE Transactions on Power Systems,* Vol. 3, No. 1, (February 1988) 317–324, ISSN 0885-8950

11. Acronyms

ASST	Average Same-Size Technologies
BAT	Best Available Technologies
BSST	Best Same-Size Technologies
CHP	Combined Heat and Power
CO2ER	CO_2 Emission Reduction
COI	Cost of Illness
DG	Distributed Generation
FESR	Fuel Energy Savings Ratio
GHG	GreenHouse Gases
GSP	Global Separate Production
ICE	Internal Combustion Engine
LCA	Life Cycle Assessment
LHV	Lower Heating Value
LSP	Local Separate Production
MT	Microturbine
m.u.	monetary units
NG	Natural Gas
PCO2ER	Poly-generation CO_2 Emission Reduction
PES	Primary Energy Saving
PM	Particulate Matter
SP	Separate Production
toe	tonne of oil equivalent
UHC	Unburned Hydro-Carbons
VOLY	Value of Life Years Lost
WTP	Willingness to Pay
YOLL	Years of Lost Life

2

Distributed generation and the regulation of distribution networks[1]

Jeroen de Joode[1,2], Adriaan van der Welle[1] and Jaap Jansen[1]
[1]*Energy research Centre of the Netherlands (ECN)*
Petten
[2]*Faculty of Technology, Policy and Management, Delft University of Technology*
Delft
The Netherlands

1. Introduction

Electricity systems around the world are experiencing an increase in the amount of distributed electricity generation for a variety of reasons. In general the two main driving forces for a further penetration of distributed electricity generation in energy systems are market liberalisation and environmental concerns (Pepermans et al., 2005; Peças Lopes et al., 2007). Distributed electricity generation allows electricity system actors more flexibility in dealing with changing market conditions and can have favourable environmental characteristics due to its renewable character (e.g. small-scale wind turbines and photo-voltaics) or the efficient combination of electricity and heat production (e.g. small-scale combined heat and power (CHP) units). With an increasing attention for renewable electricity generation and energy efficiency improvements in energy policy the future will continue to see an increase in the share of distributed electricity generation in electricity systems.

In projections on future electricity system developments the amount of distributed electricity generation is generally not explicitly presented. As an approximation we can use projections on the amount of specific generation technologies of which at least part is integrated in the electricity system at the distributed level. Figure 1 shows projections for generation capacity based on renewable electricity technologies, as well as CHP, for the European Union (EU) (EC, 2008). Although the actual amount of distributed generation in

---

[1] Parts of this contribution are based on De Joode et al. (2009). We thank Martin Scheepers for cooperation in the research that led to this contribution. Furthermore we acknowledge the support of the IEE programme of the European Commission (contract no. EIE/04/015/S07.38553) and the Next Generation Infrastructures Foundation (www.nginfra.nl). The sole responsibility for the content of this contribution lies with the authors. It does not represent the opinion of the Community. The European Commission is not responsible for any use that may be made of the information contained therein.

the EU in the period until 2030 is uncertain, it is highly likely that the increase in renewable-based and CHP-based electricity generation will also be reflected in the developments on the distributed level.

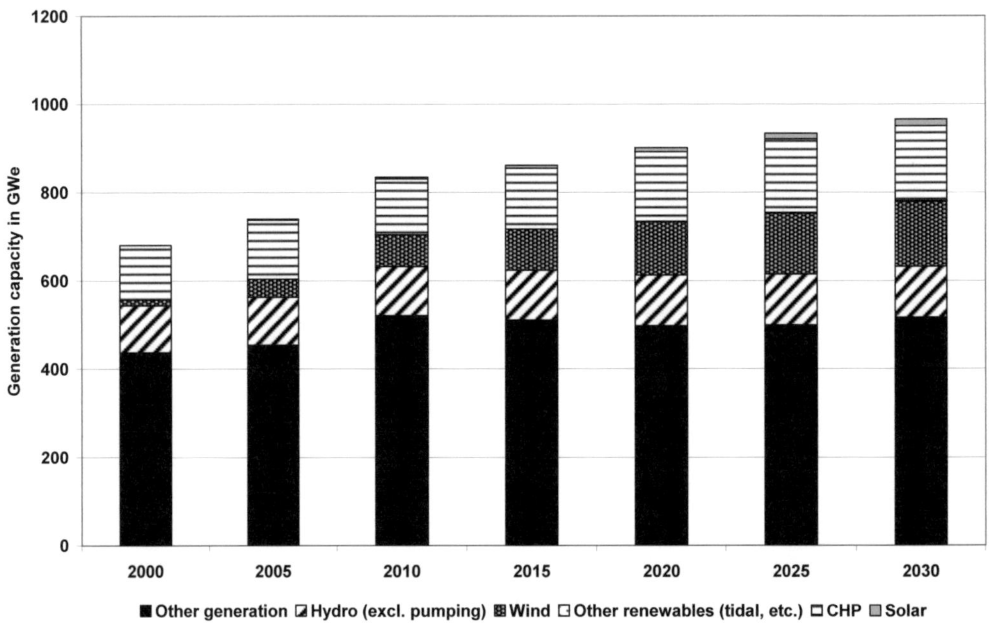

Fig. 1. Projections for future electricity generation capacity in the EU-27 (EC, 2008)

The trend towards a larger share of distributed electricity generation in electricity systems creates both opportunities and challenges (Peças Lopes et al., 2007) and impacts different electricity system actors (e.g. consumers, producers, network operators) in different ways.[2] A successful transformation of the current electricity system designed around large-scale centralised electricity generation towards a future electricity system with a large role of small-scale distributed electricity generation requires a efficient integration of distributed generation from two perspectives: market integration and network integration. Market integration issues surrounding distributed generation relate to the access of producers of distributed electricity generation to electricity markets and institutional and contractual arrangements of the producers with other electricity market actors. Network integration of distributed generation is about operating the electricity system, including network flows from the connection of new distributed generation units, in an efficient manner.

Challenges in successfully integrating distributed generation in electricity systems are technical, commercial or regulatory in nature (Peças Lopes et al., 2007; Strbac, 2002). In this contribution we focus on the regulatory challenges. Most electricity systems throughout the world have been liberalised. European electricity markets have been in a process of liberalisation since 1998. However, due to natural monopoly characteristics the network

[2] Barth et al. (2008) discuss the allocation of costs due to the integration of an increasing amount of renewable electricity generation over electricity market actors.

related part of the system has remained regulated (Joskow, 2005). This includes the lower voltage distribution networks where distributed generation is connected. We explain the concept of natural monopolies in more detail in Section 2.4. The Distribution System Operator (DSO) has the task to facilitate both the connection of distributed generation to the network, and the feed-in of electricity from distributed generation in the network. More in general the DSO bears responsibility for planning, operating and maintaining the distribution network. The DSO is not responsibly for the supply of electricity to end-consumers. This task rests with the electricity retail companies. A large(r) cost burden for the DSO could very well act as a barrier in the continuing integration of distributed generation in electricity systems if distributed generation is not explicitly included in the design and application of network regulation (Woodman and Baker, 2008). It is therefore important to have a good understanding of how distribution network regulation affects the further integration of distributed generation. Just as the electricity system itself, distribution network regulation was designed in an era where large-scale centralised electricity generation was the standard and distribution networks were primarily used for the distribution of electricity flows from higher voltage networks to end-consumers.

An important aspect of analysing the regulatory challenge in integrating more and more distributed generation is the way in which distribution networks are managed. There is a distinction between passive network management and active network management. The current network management approach is based on an equal treatment of both consumers and distributed generators that want to be connected to the distribution network. New distributed generation connections influence electricity flows in the distribution network. The network needs to be dimensioned on peak generating capacity of the distributed generation unit resulting, in some cases, in incremental reinforcement in the network by the DSO. This is also called the passive network management approach, or 'fit and forget' approach: when distributed generation enters the distribution network, upgrade investments based on peak distributed generation output are undertaken and for the remainder any electricity production from the unit is taken as given. An alternative approach is provided with an active network management philosophy (McDonald, 2008). This approach is based on the concept of intelligent networks where technological innovations on power equipment and ICT are combined to allow for a more efficient use of distribution network capacity, for instance by network reconfiguration. In addition, it is characterised by active involvement of both consumers and distributed generators: load and generation characteristics are taken into account in network operation and planning. When confronted with new connections the active DSO explicitly recognizes the network contribution of electricity consuming and producing entities in its network planning and includes this in investment decision-making. The hypothesis is that integrating distributed generation in distribution networks can be done more cost effective in networks operating under active network management compared to networks operating under passive network management. Therefore, adapting distribution network regulation to efficiently accommodate more and more distributed generation might include the facilitation of a transformation in network management approach as well as a potential compensation for financial impacts.

The main hypothesis in this contribution is that current regulation of distribution networks needs to be revised in order for a continuing integration of distributed generation in electricity systems in an efficient manner. We support this hypothesis by providing a systematic overview of the financial impact of distributed generation on the DSO using available literature and by providing a model-based analysis.

In the remainder we proceed as follows. Section 2 presents definitions. Section 3 discusses the impact of an increasing amount of distributed generation in the distribution network on the financial position of the DSO. In Section 4 we deal with methodological issues. We more precisely define the object of study and scope of this contribution and present the methodology applied. Section 5 presents the results of our research on the economic impact of distributed generation on the owner and operator of the distribution network and the impact of changes in distribution network regulation on this economic impact. Section 6 discusses the results and Section 7 provides some recommendations for both policy and regulation and future research.

2. Definitions

Our focus in this contribution is on the relationship between the integration of distributed generation and distribution network regulation. We first of all need to provide a more precise definition of distributed generation, distribution networks, regulation, and passive and active network management.

2.1 Distributed generation

Note that distributed generation may also be referred to as 'dispersed generation', 'embedded generation', or 'decentralised generation'. In combination with demand response and energy storage it is sometimes referred to as 'distributed energy resources'.

	Combined Heat and Power (CHP)	Renewable Energy Sources (RES)
Large scale generation	• Large district heating [a] • Large industrial CHP [a]	• Large hydro [b] • Offshore wind • Onshore wind (partly) • Co-firing biomass in coal power plants • Geothermal energy
Distributed Generation (DG)	• Medium district heating • Medium industrial CHP • Commercial CHP • Micro CHP	• Medium and small hydro • On-shore wind (partly) • Tidal energy • Biomass and waste incineration/gasification • Solar energy (PV)

Table 1. Characterisation of electricity generation units (Scheepers, 2004)

[a] typically > 50 MW_e

[b] typically > 10 MW_e

We follow Ackermann et al. (2001) in defining distributed generation as: "Distributed generation is an electric power source connected directly to the distribution network or on the customer site of the meter". It is difficult to provide a universal definition in a quantitative sense because this is country specific and relates to characteristics of the centralised electricity system. Co-generation (or CHP) and electricity generated from renewable energy sources (RES) are often considered as distributed generation. However, as is shown in Table 1, only a part of CHP and RES can be considered as distributed generation.

2.2 Distribution networks

Ackermann et al. (2001) discuss the concept of distribution networks. They note that distribution networks have been designed with a different purpose than transmission networks: they were generally not designed to connect power generation devices. Another general characteristic of distribution networks as in contrast to transmission networks is the often radial or loop design. Transmission networks generally have a meshed design. What constitutes a distribution network from an institutional / regulatory perspective is less clear: the term distribution network is not fixed to specific voltage levels. A specific voltage level network that is considered to be part of the distribution network in the one jurisdiction can be considered to be part of the transmission network in another.

2.3 Distribution network regulation

The distribution of electricity is considered to be a natural monopoly due to economic characteristics. The distribution of electricity is highly asset-specific involving a large share of capital expenditures relative to operational expenditures, and concerns long lifetime of investment. In other words: general textbook economics does not apply here (Joskow, 2005). Prevailing distribution network regulation regimes across EU countries prevent DSOs from acting as a monopolist. Among others, tariff conditions for distribution network services are under regulatory scrutiny via revenue of price cap regulation. For a discussion on the types of network regulation in electricity distribution we refer to Joskow (2006) and Rudnick and Donoso (2000). When talking about regulation in the context of electricity distribution networks we distinguish two basic elements: the regulation of access conditions and the regulation of distribution network connection conditions. Both types of conditions relate to the distribution network service provision to both consumers and producers.

Firstly, access conditions in general encompass third party access to distribution networks. E.g. all parties wishing to contract distribution network services should be allowed to do so on a non-discriminatory basis. Basically, it is the regulatory authority that approves the level of charges. Charges for the use of the distribution network are generally referred to use of system (UoS) charges. An important aspect of distribution network regulation is the differential allocation of overall system costs to producers and consumers. For example, in some jurisdictions electricity producers are not charged use of system charge. In fact, in about half of the EU countries distributed generation does not pay UoS charges (Cossent et al, 2009). Furthermore there are differences in the design of UoS charges across electricity systems, for example in the degree in which system costs are covered by fixed UoS charges (e.g. capacity dependent charge) or variable UoS charges (e.g. charges dependent on feed-in or consumption).

Secondly, there are different approaches with respect to connection conditions. In principle, a DSO is always obliged to connect consumers or producers to its distribution network, but different approaches exist in the allocation of costs incurred by the new connection. Three different approaches exist: shallow connection charging, shallowish connection charging, and deep connection charging. Under a regime of shallow connection charging the consumer or producer connected to the network is only charged a limited part the costs related to connection, for example only the on-site costs of connection. Under a deep connection regime the consumer or producer is charged not only the full cost of connecting to the existing distribution network, but if applicable also the costs of network upgrades elsewhere in the distribution network that are required to accommodate the future electricity flow resulting from the new connection. A shallowish connection regime is somewhere in-between the former two categories: the consumer or producer is charged only part of the deep network cost. Connection costs that are not covered by charges to the consumer or producer are socialised, meaning that costs are recovered via general UoS charges. Regulation may differentiate in the connection charging regime to be applied to particular consumers or producers, for example based on yearly off take or the electrical capacity of the connection. In our model-based analysis in Section 4 we assume a regulatory regime based on shallow connection charging since the majority of EU member states apply shallow or 'shallowish' connection charging on the distribution network level (Cossent et al, 2009).

3. Background

The increasing penetration of distributed generation impacts the financial position of the DSO in a number of ways. Figure 2 provides an overview of the economic position of the DSO by listing its revenues and expenditure items.

Revenues for the DSO are generally regulated via either revenue or price cap regimes and consist of revenues from connection charges and revenues from UoS charges. The expenditures of the DSO are divided into two categories: (a) capital expenditures, and (b) operational expenditures. Capital expenditures include investments in distribution network assets such as transformers, switchboards, and cables, and the consequential depreciation costs and remuneration of debt. Operational expenditures encompass costs due to use of the transmission network, distribution losses, costs of ancillary services, and operational and maintenance costs of assets.

Connecting distributed generation units to the existing distribution network implies an increase in the direct (investment) costs of connecting. Under a shallow connection charge regime this can imply an increase in capital expenditures of the DSO. Under a deep or shallowish connection charging regime there is no impact on the DSO because costs are passed-through to the owner of the distributed generation unit. The increased number of distributed generation units in the distribution network can also give rise to additional network investment elsewhere in the network, especially in a situation where the DSO is operating under a passive network management approach where the network is designed based on deterministic rules for minimum and maximum generation and consumption. When an active network management approach is adopted the investment cost impact of distributed generation on distribution networks might change compared to the passive network management approach. When distributed generation is located near final consumers, the net electricity demand to be accommodated via the transmission and distribution networks decreases. This implies a reduced need for existing network capacity. In this respect we refer to the 'deferred investment value' of distributed generation. It can be defined as the value of postponing the need to reinforce the system in case of load growth or reducing the investment required in case of equipment replacement (Cao et al., 2006; Méndez et al., 2006a; Jil & Goos, 2006). In short, the impact of distributed generation on DSO investment expenditures may be positive or negative dependent on network characteristics, type of network management operation and dynamics in the distribution network (e.g. electricity demand growth and need for asset replacement).

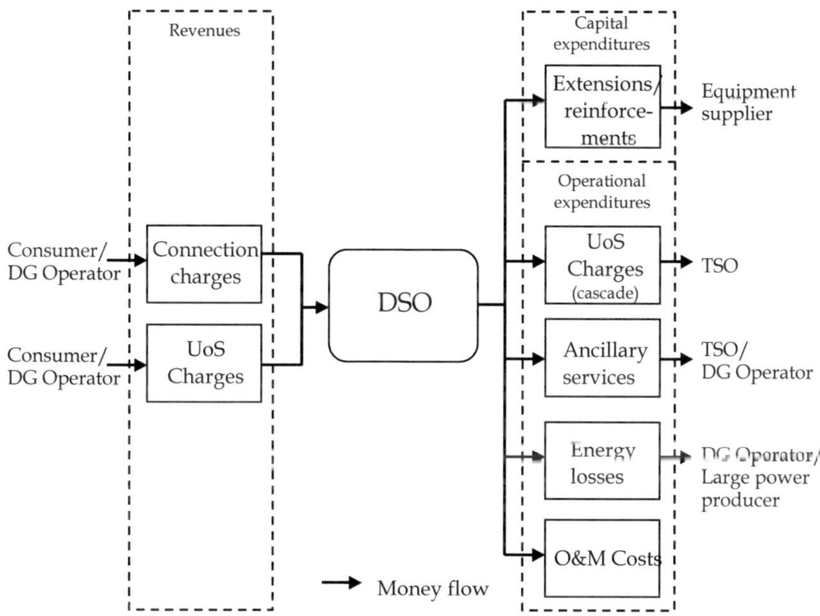

Fig. 2. Overview of DSO revenues and expenditures

The DSO expenditures on UoS charges for the transmission system are a matter of accounting. Final electricity consumers and in some countries also (distributed) generators connected to the distribution network pay for the use of the electricity transmission system to the DSO. The DSO passes-through these revenues to the transmission system operator. An increase in the amount of distributed generation may indirectly affect the level of the charge of the use of the transmission system but will not give rise to an improved or worsened financial position of the DSO.

For the procurement of ancillary services the DSO pays charges to the Transmission System Operator (TSO). Procured ancillary services can include the provision of frequency response, reactive power and regulating and standing reserve (Peças Lopes et al., 2007). Increasing penetration of distributed generation in distribution networks can negatively impact the power quality within the network and thus create a larger need for particular ancillary services.

Dependent on the total amount of distributed generation, the size of distributed generation units and the dispersion of the units within a distribution network an increase in distributed generation can either decrease or increase the level of distribution losses. Méndez et al. (2006b) find that in general distributed generation decreases distribution losses when the overall level of distributed generation is low and when distribution networks are characterised by large distances between connections, as is for example the case for distribution networks in rural areas. For higher levels of distributed generation the impact on distribution losses can actually turn negative. The same tends to hold for distribution networks where the distances between connections are shorter, as is for example the case for distribution networks in urban areas.

The level of maintenance costs can be related to the amount of distribution network assets of the DSO. Analogue to the impact of distributed generation on the need for additional distribution network investment the level of maintenance costs may de- or increase with the amount of distributed generation in the distribution network. It is more difficult to assess the impact of an increasing penetration of distributed generation on operational cost. Furthermore, the impact can differ between networks operated under passive and active network management. In a more actively managed distribution network where there is a larger role for intelligent metering and ICT assets the operational costs might increase.

4. Methodology

In order to evaluate the impact of an increasing penetration of distributed electricity generators on the DSO we deploy a combination of two different models; a load flow-based generic distribution network model and a financial model representing the financial position of the DSO. Figure 3 shows the basic input-output relation between the generic distribution network model and the DSO financial model.

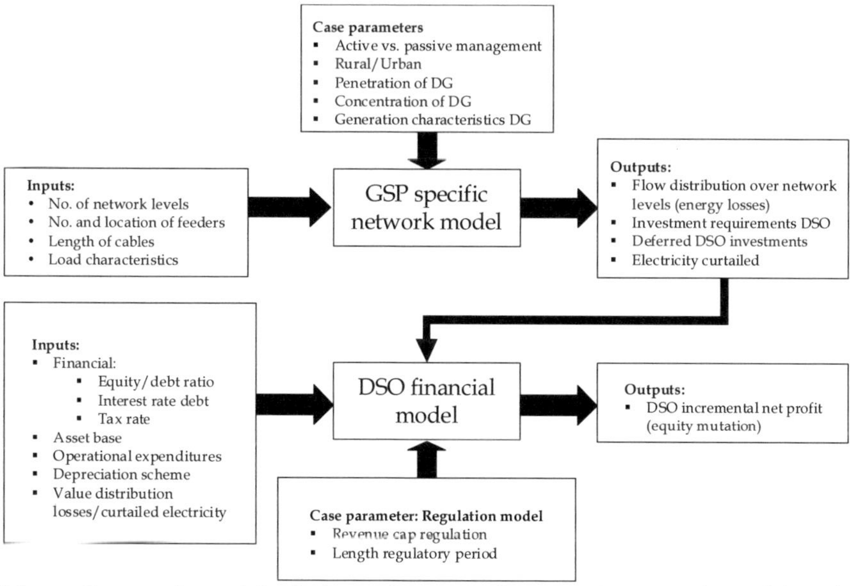

Fig. 3. Schematic overview of the linkages between the generic network model and the DSO financial model

4.1 Distribution network model (GSP)

The generic distribution network model is based on the topology and design of an 'average' UK distribution network that is connected to a 'grid supply point' (GSP). The distribution network consists of various cascading low voltage level networks with substations and transformers between each of these networks. The voltage networks are modelled as radial networks from the GSP to the end-users. A graphical representation of a generic distribution network, with various loads and distributed generators connected, is shown in Figure 4. In our analysis, the distribution network is dimensioned at a total load of 1155 GWh per year (being equal to the load of one grid supply point). Here we do not discuss the model further in depth. For an extensive model and data input description we refer to Pudjianto et al. (2006).

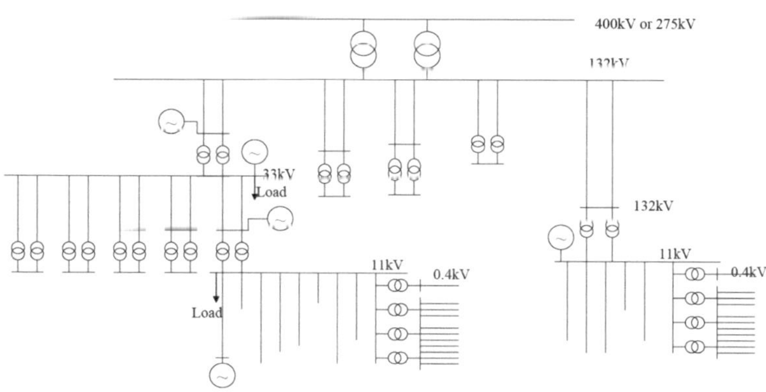

Fig. 4. Representation of the module-based distribution network design in the generic distribution network model (Pudjianto et al., 2006)

4.2 DSO financial model

The DSO revenue and expenditure model contains the revenue and expenditure items of the DSO and calculates net profit over a longer period of time. The model has the following basic characteristics. First, the model is incremental in the sense that it does not explicate the costs and revenues related to 'business as usual' but only the incremental costs and revenues caused by the increase of distributed generation in the distribution network. This enables a strict assessment of the impact of DG on the DSO. Second, the model is dynamic: it captures a number of regulatory periods with each covering a number of years. This approach is in line with current practice where the DSO, being a regulated actor, is confronted with a network performance assessment by the regulatory authority once per regulatory period. Third, the specific regime of distribution network regulation assumed to apply for the studied DSO is based on incentive based regulation and more in particular, based on a revenue-cap system. Our choice for this specific regime can be justified by an EU wide observed tendency to move from traditional cost-plus regimes to incentive based regimes (Skytte and Ropenus, 2007). Below we go into more detail on this type of distribution network regulation since this is vital for the remainder of this contribution.

Incentive based regulation trough a revenue cap basically implies the following: the DSO is only allowed to receive a maximum total allowed revenue (TAR) in return for the provision of distribution network services in a year, with the TAR in one year being equal to the TAR in the previous period corrected for (i) a requirement on improved efficiency performance, (ii) change in overall price level (inflation), and (iii) optional compensation schemes for adverse developments in demand . Since the DSO model uses and presents real prices the revenue cap scheme included in the DSO model does not contain a correction for inflation. The foregoing results in the following formula:

$$TAR_t = TAR_{t-1}\left(1 - X\right) \pm AF \qquad (1)$$

where:

TAR_t = Total allowed revenue in year t

X = Required yearly improvement in efficiency performance

AF = Adjustment factor

The total allowed revenue in the starting year is dependent on the capital expenditures consisting of the total regulated asset base (RAB) and the weighted average cost of capital (WACC), as well as the operational expenditures. The RAB represents the value of the DSOs asset base.

$$TAR_0 = f\left(RAB, WACC, OPEX\right) \qquad (2)$$

In traditional incentive-based regulation DSOs are remunerated according to their outputs (the services they deliver). These outputs are usually compared to each other with benchmarking. Often these benchmarking exercises do not take into account the structural differences between regions with respect to distributed generation penetration, concentration and the distance between distributed generation and demand. In other countries not all new investments are fully included in the regulatory asset based: i.e. no automatic pass-through of investment costs to end-consumers is allowed for. The only exception is the UK, which disposes of a special rule regarding the remuneration of distributed generation related investment. Consequently, in the remaining analysis we assume that DSOs do not fully recover the costs of distributed generation related incremental network reinforcement costs.

4.3 Case parameters

In order to explore the impact of the increasing penetration of distributed generation on the financial position of the DSO we identify different case parameters that simulate different conditions under which DSOs are facing increasing penetration of distributed generation. We use the following case parameters: different levels of distributed generation in the network, different concentration levels for distributed generation units, different levels of less controllable type of distribution generation, different network topologies and different types of network management. Below, we briefly comment on each case parameter.

Firstly, the amount of distributed generation in the network indicates the amount of peak distributed generation capacity in the network. This amount ranges from 50 to 100 to 200 MW. Given the total level of demand of 1155 GWh in the distribution network, and assumed load factors for intermittent (30%) and non-intermittent (60%) distributed electricity generation, this results in a penetration rate in the range of 10 to 90% of total load. Secondly, the concentration of distributed generation capacity indicates whether the spatial concentration of distributed generation within the distribution network is either high or low, where concentration refers to the physical location of the distributed generation unit. Thirdly, the share of intermittent generation in total distributed generation capacity is considered an important case parameter. Distributed electricity producing units vary in electricity generation profile; whereas some distributed generation units can produce at relatively constant rate, others are volatile in electricity generation. The former category of units is referred to as non-intermittent distributed generation; the latter is referred to as intermittent distributed generation. A related distinction is between controllable and non-controllable distributed generation units in the sense that distributed generation units producing electricity intermittently can in principle be controllable. For example, small-scale CHP units at industrial sites are in principal controllable (and hence can respond to changes in electricity prices) but since their electricity production follows industrial heat demand, the resulting electricity production profile is referred to as being intermittent. Within the generic distribution network model, all distributed generation connected to the distribution network is either fully non-intermittent or fully intermittent and there is no distinction between controllable and non-controllable, which is a simplification of reality where an indefinite portfolio of distributed generation units with different electricity generating profiles exist. Hence, in our analysis we explore the two extremes of 0% intermittency of total distributed generation capacity and 100% intermittency of total distributed generation capacity. Fourthly, we consider the type of network to be of influence when assessing the impact of distributed generation penetration on the distribution network. A rural network typically consists of longer lines and less concentrated loads than urban networks. Jenkins et al. (2000) mention two different technical barriers for increasing levels of distributed generation. In rural areas voltage management and thermal rating issues can hinder further deployment of distributed generation, whether urban areas experience system fault level issues. This implies different requirements for DSO investments in both networks. Fifthly and lastly the type of network management approach, passive or active bears impact on the way in which distributed generation is integrated in the distribution system. Passive network management can be described as a 'fit and forget' network management approach where network requirements caused by new distributed generation connections are met before distributed generation installation by 'simple' capacity expansion. Active network management on the other hand uses 'intelligence' in the network to cope with the operation of new distributed generation in real-time.

We have analyzed the impact of distributed generation penetration on the financial position of the DSO for a limited number of cases. Table 2 gives an overview of the studied cases and the taken values for the earlier described case parameters. The different cases can be interpreted as different type of distribution networks / distribution system operators.

Case parameter	Values	Unit
1. Level of distributed generation	[50, 100, 200]	[MW]
2. Concentration of distributed generation	[low, high]	[discrete]
3. Share of intermittent generation	[0, 100]	[%]
4. Type of network	[rural, urban]	[discrete]
5. Type of network management	[passive, active]	[discrete]

Table 2. Overview of analysed cases and their parameter values

5. Analysis and results

5.1 Impact of distributed generation on the DSO

The impact of increasing the amount of distributed generation in a distribution network on the (regulated) financial position of the DSO is indicated as a percentage change in the financial position (net profit) of the DSO as compared to a situation where no distributed generation enters the distribution network. Table 3 presents the results of our analysis.
Recall that costs and revenues resulting from 'normal' business, i.e. a distribution network where no distributed generation is present in the network, are not analyzed: the financial model is incremental in nature. For example, for the first case we find that an integration of 100 MW of concentrated and intermittent distributed generation capacity in a rural distribution network operated under a passive network management philosophy, results in a 7.8% increase in net profit for the DSO compared to the no distributed generation penetration case.[3] The table reports two results: one including and one excluding the potential replacement value of integrated distributed generation. The replacement value is created when distributed generation penetration at one point in time, prevents or defers the (replacement) investment in network capacity upgrades at a later point in time. For example, continuous load growth due to increasing electricity demand would in due time

[3] We do not necessarily need to have full knowledge of all costs and revenues in the reference case as long as we do include all relevant cost and revenue items (i.e. items that change as a result of the increasing penetration of distributed generation).

require new investments in the substation that connects the distribution with the transmission network, unless increasing distributed generation levels within the network could cover the additional load growth. Hence, due to the uncertain development of load growth in the distribution network and the technical state of electrical equipment we choose to present a result including 0% and 100% of the potential replacement value. In reality, the actual value will be somewhere in-between.

Results show that, if the potential value of deferred investments is not taken into account, DSOs operating under a passive network management regime generally do not profit from the presence of distributed generation in their distribution network. Although low distributed generation penetration levels do benefit the DSO somewhat, higher penetration levels result in a negative overall impact. The concentration of distributed generation within the network is a particular influential factor: the more concentrated the presence of distributed generation in the distribution network, the more negative the impact. The driver for the generally positive results for low penetration levels and the generally negative results for high penetration levels are distribution losses.

DSOs operating under an active network management philosophy are generally confronted with comparable results as the passive network management case. Penetration of distributed generation in the network is favourable for the DSO for low penetration levels, but becomes unfavourable the higher the penetration rate and the more concentrated the distributed generation in the network. However, it should be noted that the negative results are relatively small for the majority of the cases analyzed: the net impact of distributed generation penetration is mostly within the range of 8% of the 'business as usual' profit DSOs make.

Parameter					Impact on DSO net profit	
Level DG (MW)	Network type	Concentration DG	Type of DG	Network management type	Excluding potential deferred investment	Including potential deferred investment
100	R	H	I	P	7.8%	29.6%
100	R	H	I	A	9.2%	31.0%
100	R	H	NI	P	-7.0%	14.8%
100	R	H	NI	A	-6.7%	15.1%
50	R	L	I	P	2.1%	12.9%
50	R	L	I	A	2.1%	12.9%
50	R	L	NI	P	5.2%	15.9%
50	R	L	NI	A	5.2%	15.9%
200	R	H	I	P	-16.2%	1.0%
200	R	H	I	A	-21.6%	-4.4%
200	R	H	NI	P	-44.7%	-27.5%
200	R	H	NI	A	-57.3%	-40.0%
100	R	L	I	P	-4.3%	17.6%
100	R	L	I	A	-4.5%	17.4%
100	R	L	NI	P	0.3%	22.2%
100	R	L	NI	A	0.3%	22.2%
100	U	H	I	P	-1.2%	20.6%
100	U	H	I	A	4.6%	26.4%
100	U	H	NI	P	-10.5%	11.3%
100	U	H	NI	A	-3.8%	18.0%
50	U	L	I	P	-8.4%	2.3%
50	U	L	I	A	-1.6%	9.1%
50	U	L	NI	P	2.6%	13.4%
50	U	L	NI	A	0.4%	11.2%
200	U	H	I	P	-26.4%	-9.2%
200	U	H	I	A	-32.9%	-15.7%
200	U	H	NI	P	-41.1%	-23.9%
200	U	H	NI	A	-51.9%	-34.6%
100	U	L	I	P	-10.6%	11.3%
100	U	L	I	A	-2.3%	19.6%
100	U	L	NI	P	1.2%	23.2%
100	U	L	NI	A	0.1%	22.1%

Table 3. Incremental impact of the integration of distributed generation on the financial position of a DSO (DG = distributed generation, U = urban network, R = rural network, H = high penetration, L = low penetration, I = intermittent, NI = non-intermittent, P = passive network management, A = active network management)

The added value of distributed generation with respect to the investment deferral for connections to the higher voltage network levels can be substantial. Considering the maximum replacement values of distributed generation, it can be expected that the overall impact of distributed generation penetration on the DSO business can be neutral or positive in the majority of cases. However, the realization of this positive value for the DSOs is dependent on a larger number of non-distributed generation related factors and is beyond the scope of this investigation (e.g. load growth dynamics and the status of interconnection equipment). However, we hypothesize that it is unlikely that the full replacement value is going to be realised in all cases. This is something to be tested in further research. Observing the differential impact on the DSO under passive and active network management we conclude that there is an implicit incentive for the DSO to adopt an active network management approach in a number of cases, in particular the case where distributed generation penetration is low or medium.

The observation that an increased penetration of distributed generation is negative in cases where there is no actual replacement value for the DSO does not automatically justify adaptation of distribution network regulation. When assessing the necessity of such adaptation one also needs to take into account potential positive impacts on the DSO brought about by increased distributed generation penetration. These could for example emerge in the field of ancillary services provision and demand response mechanisms (Strbac 2008, Jansen et al. 2007). The DSO can for example act as an intermediary in the provision of these network related types of services due to the operational network information acquired through implementation of active network management. These services would imply a new source of revenues for the DSO that might compensate for negative impact with regard to network operations and investment due to increased distributed generation penetration. Additional research on the significance of this type of business opportunities for DSOs is required. For now we assume that these opportunities do not (fully) compensate the negative impact of the type shown above and therefore proceed to propose possible improvements for current network regulation that can mitigate the negative impact.

5.2 Impact of adaptations to distribution network regulation

The negative impact of distributed generation integration on the DSO's financial position as reported above may hamper the deployment of distributed generation resulting in a 'conflict' with the national and European policy objectives for CHP and RES-E. To solve this problem the additional costs of distributed generation integration should be socialized among all customers connected to the network, i.e. electricity consumers and generators alike. The network costs for connecting and integrating distributed generation is then treated in the same way as network costs related to electricity consumption. This reflects the role of the distribution network: providing access to the electricity market for consumers and (distributed) generators under similar conditions. The extra network costs induced by distributed generation connections can be allocated to consumers and distributed generation operators through the use of system charges. These tariffs (connection charges and use of system charges) are calculated from the TAR by taking into account the number of connections, size of connections, amounts of kWh and kW_{peak}, et cetera.

With the revenue cap formula (Equation 1) as a starting point, five alternative ways to compensate for the negative impact of distributed generation penetration have been tested. The DSO financial model was applied to test the effectiveness of four of these alternatives. A fifth alternative is to consider distributed generation as a cost driver in the DSO benchmarking exercise. However, the model is not capable of analysing this option. The four alternatives for adaptation of distribution network regulation are:

1. Automatically allowing for investments due to distributed generation in the regulated asset base;
2. Compensation related to an indicator for the presence and impact of distributed generation in the distribution network;
3. Including a direct revenue driver in the regulatory formula based on distributed generation;
4. A combination of alternatives (1) and (3)

Below we briefly expand on these alternatives.

An automatic allowance of investment expenditures related to the integration of distributed generation compensates for the possible negative impact of the penetration of distributed generation with respect to capital expenditures but not with respect to operational expenditures. Theoretically, the pass-through of distributed generation related capital expenditures can lie between 0 and 100%. A 100% pass-through in fact implies a cost-plus arrangement for distributed generation related expenditures. Such a high pass-through factor might not be desirable in a large number of cases since especially for low levels of distributed generation the DSO can in fact benefit from the presence of distributed generation. Moreover, a pass-through factor of less than 100% will provide an incentive to limit these investments. We propose a partial pass-through of expenditures related to the integration of distributed generation with the pass-through factor increasing with the level of distributed generation in the distribution network. In our analysis we apply a 30% pass-through is for a low distributed generation penetration rate, 70% for a medium and 90% for a high distributed generation penetration rate. This type of compensation measure may be described by the following formula:[4]

$$TAR_t = TAR_{t-1}(1 - X) + y\% \cdot I_t^{DG} \tag{3}$$

Where

y = Share of eligible distributed generation related investments in distribution network assets

I_t^{DG} = Total eligible distributed generation related investments in distribution network assets in year t

A second alternative is correcting the increase in DSO efficiency requirements over time (the X-factor) with an individual DSO based factor representing the presence and impact of distributed generation in its network. Such a factor could be based on an index capturing the

[4] Assuming that demand growth is zero so that the adjustment factor AF (see equation 1) is equal to zero.

impact of distributed generation on distribution networks. In such an index the different context specific parameters, such as amount, concentration, type of network and the like, determining the actual impact of distributed generation could be included. Formula-wise this would encompass the following:

$$TAR_t = TAR_{t-1}\left(1 - X + K_{Ind}\right) \tag{4}$$

For our exploratory analysis we have set the values for K_{Ind} at 0.75% for a distributed generation penetration level of 11%, 1.5% for a penetration level of 23%, 5% for a penetration level of 46%, and 10% for a penetration level of 91%.

A third possible alternative is the straightforward inclusion of a revenue driver based on distributed generation in the revenue cap formula, where the revenue driver can be based on the amount of distributed generation capacity in the distribution network, or energy delivered by distributed generation to the distribution network. This results in the following formula:

$$TAR_t = TAR_{t-1}\left(1 - X\right) + F_1 \cdot kW^{DG} + F_2 \cdot MWh^{DG} \tag{5}$$

In our exploratory example the additional revenue driver is based on both the total amount of distributed generation capacity in the network (F_1=2.5 €/kW for a low, F_1=2 €/kW for a medium and F_1=1 €/kW for high a distributed generation penetration) and the electricity generated by distributed generation units in the network (F_2=0 €/MWh for a low, F_2=2.5 €/MWh for a medium and F_2=3.5 €/MWh for a high distributed generation penetration).

Finally, as a fourth alternative we test a combination of alternatives 1 and 3. The DSO is allowed to pass-through a certain percentage of additional capital expenditures due to increased penetration of distributed generation in the network and is allowed additional revenue based on an 'energy' based revenue driver:

$$TAR_t = TAR_{t-1}\left(1 - X\right) + y\% \cdot I_t^{DG} + F \cdot MWh^{DG} \tag{6}$$

In our analysis the rate for total eligible distributed generation related investments (I_t^{DG}) is 50% and the direct revenue driver (F) has the value of 2 €/MWh.

Table 4 shows the incremental financial impact for the DSO of increasing the amount of distributed generation in the network in the reference case, where regulation is not adapted, and the potential deferred investment value is excluded, and the four alternative cases.

Parameter					Reference	Regulatory adaptations			
Level DG (MW)	Net-work type	Concentration DG	Type of DG	Management type	Impact on DSO	Allowance in RAB	Quality indicator	Direct revenue driver	RAB and direct revenue driver
100	R	H	I	P	7.8%	10.6%	12.7%	9.9%	17.2%
100	R	H	I	A	9.2%	10.6%	14.1%	11.3%	16.3%
100	R	H	NI	P	-7.0%	1.0%	11.5%	9.3%	9.1%
100	R	H	NI	A	-6.7%	-1.8%	11.7%	9.5%	7.3%
50	R	L	I	P	2.1%	2.1%	4.6%	3.2%	4.5%
50	R	L	I	A	2.1%	2.1%	4.6%	3.2%	4.5%
50	R	L	NI	P	5.2%	5.2%	10.1%	6.2%	9.8%
50	R	L	NI	A	5.2%	5.2%	10.1%	6.2%	9.8%
200	R	H	I	P	-16.2%	2.8%	4.7%	5.4%	8.9%
200	R	H	I	A	-21.6%	-6.6%	-0.2%	1.6%	1.7%
200	R	H	NI	P	-44.7%	-18.3%	-1.8%	3.0%	-3.8%
200	R	H	NI	A	-57.3%	-36.4%	-13.6%	-8.2%	-19.0%
100	R	L	I	P	-4.3%	-2.4%	1.9%	-1.3%	4.1%
100	R	L	I	A	-4.5%	-3.2%	1.7%	-1.5%	3.1%
100	R	L	NI	P	0.3%	3.7%	16.7%	14.5%	12.0%
100	R	L	NI	A	0.3%	2.5%	16.6%	14.5%	11.2%
100	U	H	I	P	-1.2%	3.8%	4.1%	1.3%	11.7%
100	U	H	I	A	4.6%	5.6%	9.5%	6.7%	10.8%
100	U	H	NI	P	-10.5%	2.7%	9.0%	6.8%	9.8%
100	U	H	NI	A	-3.8%	-1.0%	13.7%	11.5%	8.2%
50	U	L	I	P	-8.4%	-2.3%	-5.0%	-6.9%	3.9%
50	U	L	I	A	-1.6%	0.6%	1.3%	-0.1%	4.3%
50	U	L	NI	P	2.6%	2.6%	5.3%	1.5%	7.3%
50	U	L	NI	A	0.4%	2.3%	5.3%	1.5%	8.2%
200	U	H	I	P	-26.4%	7.2%	-4.7%	-2.5%	9.7%
200	U	H	I	A	-32.9%	4.2%	-10.9%	-9.0%	6.2%
200	U	H	NI	P	-41.1%	3.4%	1.6%	5.5%	8.6%
200	U	H	NI	A	-51.9%	-2.8%	-8.4%	-2.8%	1.6%
100	U	L	I	P	-10.6%	-4.4%	-4.0%	-7.6%	4.8%
100	U	L	I	A	-2.3%	-1.7%	3.3%	0.5%	3.8%
100	U	L	NI	P	1.2%	1.2%	17.6%	15.4%	10.6%
100	U	L	NI	A	0.1%	1.2%	16.5%	14.3%	10.2%

Table 4. Incremental impact of the integration of distributed generation on the financial position of a DSO for in the case of no regulatory adaptation (reference case) and in the case of four alternative adaptations to regulation (DG = distributed generation, U = urban

network, R = rural network, H = high penetration, L = low penetration, I = intermittent, NI = non-intermittent, P = passive network management, A = active network management)

Our results indicate that there is no 'one size fits all' solution for neutralizing the negative impact of distributed generation penetration on the financial position of a DSO. Since the negative impact of distributed generation through either operational expenditures (distribution losses) or capital expenditures (network upgrades) is dominant, a specific regulatory arrangement with compensatory elements based on either 'energy produced by distributed generation' or 'distributed generation capacity connected' can not fully compensate the one DSO without unnecessarily 'overcompensating' other DSOs. The most successful regulatory improvement seems to be the combination of a special allowance and a direct revenue driver. When applying this regulatory alternative DSOs will be able to recover their costs. It should be noted that a minor 'overcompensation' of DSOs for the negative impact they experience from distributed generation penetration in the network might effectively work as an incentive to optimally facilitate the integration of additional distributed generation within the distribution network.

6. Discussion

Our analysis has shown that increasing presence of distributed generation in distribution networks is not always favourable where specific network and distributed generation characteristics play a crucial role. We analyzed the economic impact of an increasing penetration of distributed generation in distribution networks on the financial position of the DSO taking into account a number of network and distributed generation characteristics. In addition, we included two different network management philosophies (passive and active network management) in the analysis. We find that DSOs operating under a passive network management regime generally do not profit from the presence of distributed generation in their distribution network, except at low distributed generation penetration levels: network reinforcement costs increase with the level of distributed generation penetration. DSOs adopting active network management are generally faced with similar general results as DSOs adopting passive network management. Penetration of distributed generation in the network is favourable for the DSO for low penetration levels, but becomes unfavourable the higher the penetration rate and the more concentrated the presence of distributed generation in the distribution network.

The added value of distributed generation with respect to the deferral of investments at the grid supply point (the connection with the higher voltage level network) can be substantial. However, the realisation of this positive value for DSOs is dependent on a larger number of factors that are beyond the scope of our research. These relate to the load growth dynamics and the status of interconnection equipment (economic lifetime and depreciation). Considering the maximum replacement values of distributed generation, it might be expected that the overall impact of distributed generation penetration on the DSO business, can be neutral or positive in the majority of cases. Observing a differential impact on the DSO under passive and active network management we conclude that there is an implicit incentive for the DSO to adopt an active network management approach in a number of cases, in particular the case where distributed generation penetration is low or medium. For cases with a high penetration of distributed generation definitive conclusions cannot be drawn: more research is needed on the costs for the DSO when combining passive and active network management elements.

Regarding the implications of the analysis for the need for alternative regulatory arrangements we note that there are several cases where the DSO is negatively impacted by the increasing penetration of distributed generation, regardless of the network management philosophy. This implies that an alternative regulatory arrangement compensating the DSO for the negative impact might be warranted. Moreover, we have seen that in such cases the incentive for DSOs to move from a passive network management approach to an active network management approach is not always present.

From the analysis on regulatory improvements we found that an alternative regulatory arrangement based on a combination of a special regulated asset base allowance and a direct revenue driver seems most successful in compensating the negative distributed generation impact on DSOs. When applying this regulatory improvement DSOs will be able to recover their costs and at the same time are stimulated to integrate existent and new types of distributed generation without at the same time largely 'overcompensating' some DSOs that experience little or even no negative impact.

The most sophisticated revenue driver seems to be a driver that accounts for all connected kW of distributed generation capacity as well as for all kWh of distributed generation electricity fed into the grid. Regarding the possible 'overcompensation' caused by regulatory improvements we noted that a minor 'overcompensation' in practice might work as an incentive to fully facilitate distributed generation integration within their distribution network.

Finally, it should be noted that the analysis in this paper is based on shallow network charging and constant UoS charging principles. Theoretically, the operators of distributed generation units could be given more efficient signals regarding siting and timing of production of distributed electricity generation. This principle could be implemented through a move to (i) shallowish or deep connection cost charging or (ii) time and location differentiated UoS charges[5]. This could substantially reduce the costs of network integration for the DSO, both in operational and capital costs. However, successful implementation of time and location dependent network charges is not to be expected in the short to medium-run for technological, political and regulatory reasons.

7. Recommendations

Our analysis allows us to draw some policy recommendations. Firstly, current regulation of DSOs should recognise the differential impact of an increasing level of distributed generation on the financial position of the DSO and therefore regulators should investigate alternative regulation that can sufficiently take into account the drivers behind this impact. Secondly, as a specification of the first recommendation, we suggest to implement a

[5] The latter is basically a nodal pricing system (Schweppe et al., 1988).

regulatory formula where the impact of distributed generation on both operating expenditures (with the main driver being electricity generated by distributed generation units, impacting distribution losses) and capital expenditures (with the main drivers the amount of distributed generation units connected to the distribution network) are included. Thirdly, in determining the specific values in the alternative regulatory formula we would recommend to aim for some 'overcompensation' of any negative impact of distributed generation penetration on DSOs in order to provide them with an explicit incentive to facilitate and accommodate new distributed generation connections in their distribution networks. Fourthly, when considering the above recommendations on the neutralization and possible incentivisation of DSOs it should be kept in mind that the desirability of regulation aimed at these aspects is intertwined with the developments in the field of ancillary services provision by DSOs. This was discussed in Jansen et al. (2007). On the one hand, DSOs might need to be compensated for the negative impact following the penetration of distributed generation in distribution networks. On the other hand, if DSOs benefit from the provision of ancillary services enabled by the increasing presence of distributed generation, the DSO should compensate the distributed generation operators for this. Netting both impacts implies that valuable economic signals are lost. Changes in regulation should therefore be targeted at either one, or both, described impacts. In the long-run, implementation of location and time dependent network charges could counter the negative impacts for DSOs described in this paper. However, considering the fact that DSOs in some countries are already experiencing the negative impact of large shares of distributed generation in the network, solutions to this particular problem need to be found in adaptation of current regulation in the short-run.

In addition to the recommendations for policy and regulation we can also draw some recommendations for future research. Firstly, further research is needed into the issue of deferred investment value. Our analysis showed that this is a very important variable in any financial analysis on the integration of distributed generation in electricity systems. Secondly, the focus of our analysis was on the integration of distributed generation and its impact on the (financial position of the) DSO. The underlying goal of the research was the efficient integration of renewable and distributed generation in current electricity systems. Since this process of integration affects all actors in the electricity system a focus on just the distribution network and its operator is not sufficient. In the end what should be assessed is the overall socio-economic optimal integration of renewable and distributed generation. In other words: how can we optimally facilitate the transition to an electricity system with a larger share of renewable and distributed electricity generation at the least cost for society as a whole? This requires an integral impact analysis of an increase in the penetration of distributed generation on society as a whole, including the impact on networks (distribution and transmission), markets (wholesale, balancing, ancillary services) and the environment (environmental externalities.

8. References

Ackermann, T.; Andersson, G. & Söder, L. (2001). Distributed Generation: A Definition. *Electric Power Systems Research*, Vol. 57, No. 3, April 2001, 195-204.

Barth, R.; Weber, C.; & Swider, D.J. (2008). Distribution of Costs Induced by the Integration of RES-E Power. Energy Policy, vol. 36, No. 8, August 2008, 3107-3115.

Cao, D.M.; Pudjianto, D.; Grenard, S.; Strbac, G.; Martikainen, A.; Kärkkäinen, S. & Farin, J. (2007). Costs and Benefits of DG Connections to Grid System. DG Grid report D8.

Cossent, R.; Gomez, T.& Frias, P. (2009). Towards a Future with Large Penetration of Distributed Generation: Is the Current Regulation of Electricity Distribution Ready? Regulatory Recommendations under a European Perspective, *Energy Policy*, Vol. 37, No. 3, March 2009, pp. 1145–1155.

de Joode, J.; Jansen, J.C.; van der Welle, A.J. & Scheepers, M.J.J. (2009). Increasing Penetration of Renewable and Distributed Electricity Generation and the Need for Different Network Regulation. *Energy Policy*, Vol. 37, No. 8, August 2009, 2907-2915.

European Commission (EC) (2008). *European Energy and Transport Trends to 2030-update 2007.* European Commission, Directorate-General for Energy and Transport, ISBN 978-92-79-07620-6, Office for Official Publications of the European Communities, Luxembourg, April 2008.

Gil, H.A. & Joos, G. (2006). On the Quantification of the Network Capacity Deferral Value of Distributed Generation, *IEEE Transactions on Power Systems*, Vol. 21, No. 4, November 2006, Pages 1592-1599.

Jansen, J.C.; Van der Welle, A.J.; & De Joode, J. (2007). The evolving Role of the DSO in Efficiently Accommodating Distributed Generation. ECN report ECN-E--07-063, September 2007, Petten.

Joskow, P.L. (2005). Regulation of Natural Monopolies. Massachusetts Institute of Technology, Centre for Energy and Environmental Policy Research. Working Paper 0508.

Joskow, P.L. (2006). Incentive Regulation in Theory and Practice: Electricity Distribution and Transmission Networks. Cambridge Working Papers in Economics CWPE0607, Electricity Policy Research Group Working Paper EPRG 0511.

McDonald, J. (2008). Adaptive Intelligent Power Systems: Active Distribution Networks. *Energy Policy*, Vol. 36, No. 12, December 2008, 4346-4351.

Méndez, V.H.; Rivier, J.; de la Fuente, J.I.; Gomez, T.; Arceluz, J.; Marin, J. & Madurga, A. (2006a). Impact of Distributed Generation on Distribution Investment Deferral, *Electrical Power and Energy Systems*, Vol. 28, No. 4, May 2006, 244-252.

Méndez, V.H.; Rivier, J. , Gómez, T. (2006b). Assessment of Energy Distribution Losses for Increasing Penetration of Distributed Generation. *IEEE Transactions on Power Systems*, Vol. 21, No. 2, May 2006, 533-540.

Peças Lopes, J.A.; Hatziargyriou, N.; Mutale, J.; Djapic, P. & Jenkins, N. (2007). Integrating Distributed Generation into Electric Power Systems: A Review of Drivers, Challenges and Opportunities. *Electric Power Systems Research*, Vol. 77, No. 9, July 2007, 1189-1203.

Pepermans, G.; Driesen, J.; Haeseldonckx, D.; Belmans, R. & D'haeseleer, W. (2005). Distributed Generation: Definition, Benefits and Issues. *Energy Policy*, Vol. 33, No. 6, April 2005, 787-798.

Pudjianto, D.; Cao, D.M..; Grenard, S. & Strbac, G. (2006). Method for Monetarisation of Cost and Benefits of DG Options. University of Manchester and Imperial College London, January 2006. DG-Grid project D7.

Rudnick, H. & Donoso, J.A. (2000). Integration of Price Cap and Yardstick Competition Schemes in Electrical Distribution Regulation. *IEEE Transactions on Power Systems*, Vol. 15, No. 4, November 2000, 1428–1433.

Scheepers, M.J.J. (2004). Policy and Regulatory Road Maps for the Integration of Distributed Generation and the Development of Sustainable Electricity Networks. Final report of the Sustelnet project, ECN-C--04-034, August 2004, Petten.

Schweppe, F.C.; Caramanis, M.C.; Tabors, R.D. & Bohn, R.E. (1988). *Spot Pricing of Electricity*, Kluwer Academic Publishers.

Skytte, K. & Ropenus, S. (2007). Regulatory Review and Barriers for the Electricity Supply System for Distributed Generation in the EU-15. *International Journal of Distributed Energy Resources*, Vol. 3, No. 3, July-September 2007, 243-257.

Strbac, G. (2002). Impact of Dispersed Generation on Distribution Systems: A European Perspective. *Power Engineering Society Winter Meeting*, Vol. 1, August 2002, 118-120.

Woodman, B. & Baker, P. (2008). Regulatory Frameworks for Decentralised Energy. *Energy Policy*, Vol. 36, No. 12, December 2008, 4527-4531.

Steady-State Assessment of the DG Impact on Voltage Control and Loss Allocation

Enrico Carpaneto and Gianfranco Chicco
Politecnico di Torino
Italy

1. Introduction

The introduction of distributed generation (DG) affects the operational characteristics of the distribution systems. The impact depends on the level of penetration of DG, as well as on the possibility of operating significant portions of the distribution system as micro-grids, or to allow temporary operation of intentional islands.
In the electrical sector, the major changes occurred in the last two decades led to modify the structure of the electricity business from a vertically integrated utility (VIU) system, in which the four major functions of the electrical chain (that is, generation, transmission, distribution and retail) were operated by the same company, to an unbundled system in which these functions are separated and are performed by different companies.
One of the main effects of unbundling has been the decoupling of the function of generator with respect to the one of the distribution network operator (DNO). A plurality of local generators connected to the today's distribution systems are owned and managed independently of the DNO. The objectives themselves at the local generation site management may differ from the ones of the DNO. More specifically:

- at the local generation site, the main goals are to provide energy to the local users, to sell excess electricity to the grid when the electricity price makes it convenient, and to reduce the internal losses;
- the DNO's main objectives are to run the distribution system in a reliable, safe and secure way, without exceeding the operational limits on voltages and currents, maintaining reduced losses and satisfactorily high power quality levels;
- emerging options include the role of energy service companies or organised consortia in managing local generators located at different points in the networks in an integrated way, mainly for economic purposes, or the comprehensive management of portions of the networks with generators and loads as micro-grids.

This chapter illustrates and discusses some specific aspects concerning voltage control, reactive power support and loss allocation. Other operational issues such as short-circuit capability and protection, DG dynamics, possible DG contribution to ancillary services, reliability and power quality, interactions with heat/cooling equipment, storage units and other components, and the economics of DG operation, are not addressed here.

2. Voltage control and reactive power support

2.1. General aspects

The increased presence of DG calls for revisiting the current distribution system operation practices (Borbely & Kreider, 2001; Pepermans et al., 2005). The time and spatial variation of generation and loads and the contribution of different types of voltage-controllable local generators to distribution system voltage control need to be addressed under a comprehensive approach. This section recalls the characteristics and modelling of the voltage controllers, including the standard voltage controller and the modified combined voltage/reactive power controller for synchronous machines, the grid interconnection through induction generators, and different types of static converter-interfaced DG. Furthermore, this section illustrates and discusses the general formulation of the voltage control as an optimisation problem, by using an objective function based on the voltage deviations with respect to given voltage references, taking into account methods and variants proposed in the literature. Specific aspects include the conceptual challenges of voltage control with DG and the discussion on the peculiarities of voltage control in urban and rural areas.

In traditional distribution systems, without DG, the voltage variations at the network nodes are mainly due to the evolution of the voltage at the supply side or to load variations. For a MV distribution system, voltage control is typically centralized at the HV/MV supply substation level, with voltage controller with load compensation, which drives the under-load tap changer (ULTC) of the substation transformer, and with possible power factor correction capacitors connected at the MV busbars. Under this type of control, if all the network loads are passive and of resistive-inductive type, the active and reactive power flows in the network branches are typically unidirectional from the supply node to the loads, and the voltage profile is decreasing in each path starting from the supply node to reach a terminal node. The presence of a significant amount of local power factor correction capacitors at some network nodes could reverse the direction of the reactive power flows in some network branches, but the voltage profile in the distribution system normally remains decreasing from the supply node the terminal nodes, with the exception of very particular cases with branches supplying highly capacitive loads. For decreasing voltage profiles, voltage control can be set up at centralized level on the basis of the study of reference cases for the distribution network with maximum and minimum loading levels.

The inclusion of DG in a radial distribution system could change the situation (Hadjsaid et al., 1999). A low/moderate amount of local generation reduces the net amount of local load, without changing the direction of the power flow, and this reflects in improving the voltage profile. Yet, an increasing amount of local generation may change the direction of the power flow in the line connecting the local generator, and in other distribution system branches. In the branches in which the power flow has been reverted, the voltage rises rather than dropping, leading in general to a voltage profile in the network non-monotonically decreasing. Voltage rise has then to be considered as a specific issue (Section 2.5).

For voltage control purposes, a relevant factor is the X/R ratio between the series parameters of the system branches. Local voltage control is more effective when the X/R ratio is high, such as in aerial lines. For instance, considering aerial lines with series reactance X_A and resistance R_A, and cable lines with corresponding X_C and R_C parameters, indicatively $X_A \sim 4\, X_C$ and $R_A << R_C$.

Furthermore, local voltage control is more effective in rural than in urban distribution systems. In fact, in *rural* distribution systems the network is weaker than in urban areas, with lower short-circuit capacity, the lines are mainly aerial and relatively long, and potentially large customers could be located far from the HV/MV substation. This leads to relatively large variations in the voltage profile and high sensitivity to power quality aspects. On these considerations, exploiting local voltage control could be useful to alleviate the effects on the voltage variations, but voltage controllability is limited by the reactive power capabilities of the local generators (Section 2.3.2). Conversely, *urban* distribution systems are typically robust, with relatively short lines (mainly cables) and voltage control prevailing from the HV/MV substations. This makes local voltage control generally not efficient when the size of the local generator is much smaller than the short circuit power of the supply grid.

The specific objectives of the local DG units connected to the grid are of different types, (Vovos et al., 2007) including:

- *Participation in voltage control*: this is the basic requirement in transmission system operation, where primary and/or secondary voltage control are of concern. Similar possibilities for coordinated control are now emerging also in distribution systems with multiple distributed generators, in the light of possible operation as virtual power plants (Pudjianto et al., 2007) or as parts of micro-grids (Nikkhajoei & Lasseter, 2009).
- *Fixed power factor operation*: maintaining the power factor (seen at the network side) at a given value corresponds to consider the local generator as a negative load, with no participation in voltage control. The power factor value depends on possible needs for reactive support provision (in the point of view of the DNO), while power factor close to unity is typically preferred by the operator managing the local unit, because of the corresponding reduction of the internal losses.

However, since the local generation units are generally owned and managed independently of each other, there is little or no coordination of the local generators with the centralized distribution system controls. The objectives of managing the local units (maximization of efficiency and profitability of the local system) could be to some extent conflicting with those of the distributor (system losses minimization or voltage support optimization).

In the presence of DG, generation and load patterns exhibit variability in time and space, leading to various operating conditions, whose range of variation cannot be simply synthesised on the basis of reference cases with maximum and minimum loading levels. Moreover, the output from some DG sources (such as wind and photovoltaic systems) depends on random parameters, making it necessary to extend the tools used for evaluating the voltage profiles to the use of probabilistic power flow calculation techniques.

2.2. The distribution system seen from the local generator terminals

2.2.1. External characteristics at the network connection node

The main concepts are illustrated with reference to a single local generator with transformer connected to the MV distribution network. Fig. 1 shows the network structure and the local system model.

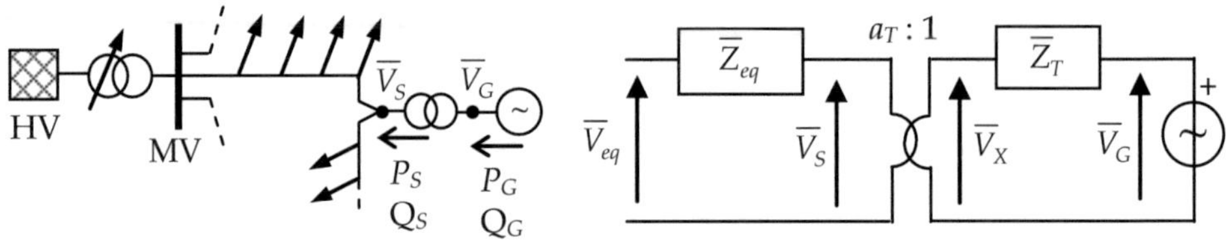

a) network structure b) local system model

Fig. 1. Model of the local generator and its connecting transformer.

For a radial network, the equivalent impedance $\overline{Z}_{eq} = R_{eq} + jX_{eq}$ indicated in Fig. 1 can be approximated in different ways. For relatively fast variations, namely, faster than the response of the automatic voltage regulator (AVR) acting on the tap changer of the HV/MV substation transformer, the equivalent impedance includes the HV/MV transformer impedance $\overline{Z}_H = R_H + jX_H$, and the sum $\overline{Z}_{line} = R_{line} + jX_{line}$ of the line impedances between the MV terminals of the HV/MV transformed and the MV side of the local generation site, thus obtaining $\overline{Z}_{eq} = \overline{Z}_H + \overline{Z}_{line}$. For relatively slow variations, the AVR acting on the tap changer moves the voltage controlled point to the MV busbars of the HV/MV substation transformer, and the equivalent impedance includes only the line impedance $\overline{Z}_{line}$.

In order to get a significant voltage variation, the reactance X_{eq} has to be relatively high, that is, the grid should be relatively weak to get the grid voltage affected by the voltage control of the local generator. Conversely, low X_{eq} values require high reactive generation capability by the DG unit in order to provide adequate voltage control, otherwise the generator operates at its reactive power limits (see Section 2.3.2) and under these conditions loses the possibility of playing a role in voltage control.

The external system seen from the local generator terminals can be represented by its *external characteristic* on a plane with axis given by the reactive power generation Q_G and the voltage at the generator terminals V_G, for a generator producing a specified value of active power P_G and current $\overline{I}_G$, taking into account the impedance $\overline{Z}_T = R_T + jX_T$ of the transformer connecting the local generator to the grid. The network side is represented by the voltage $\overline{V}_S$ and by the current $\vec{I}_S$ injected into the grid. The transformation ratio is $a_T = I_G / I_S$. The basic equations to be considered are the active and reactive power balances

$$P_S = P_G - R_T I_G^2 \tag{1}$$

$$Q_S = Q_G - X_T I_G^2 \tag{2}$$

and the expressions related to the definition of the apparent power

$$V_G^2 I_G^2 = P_G^2 + Q_G^2 \tag{3}$$

$$V_S^2 I_S^2 = P_S^2 + Q_S^2 \tag{4}$$

By elaborating these expressions, the external characteristic on the $V_G(Q_G)$ plane is represented as a family of curves depending on the voltage magnitude V_S at the grid side. The general formulation of the external characteristic is

$$Q_G = \frac{X_T V_G^2}{Z_T^2} - \sqrt{\frac{V_G^2 V_S^2}{a_T^2 Z_T^2} - P_G^2 + 2\frac{R_T V_G^2}{Z_T^2} P_G - \frac{R_T^2 V_G^4}{Z_T^4}} \qquad (5)$$

For a generation systems with given parameters, the shape of the curves depends on the active power generation P_G. Typically, the curves exhibit a nearly-linear shape. Fig. 2 shows the family of curves obtained for a 500 kVA transformer by assuming 1 MVA as base power, with R_T = 0.024 p.u., X_T = 0.1176 p.u., a_T = 1.0 and P_G = 0.4 p.u., for a set of values of the curve parameter V_S. For instance, considering null reactive power generation the voltage magnitude V_G to be imposed at the generator terminals to get a given voltage magnitude V_S is higher than V_S, because of the needed compensation of the voltage drop occurring on the transformer series impedance in the generation of the active power P_G.

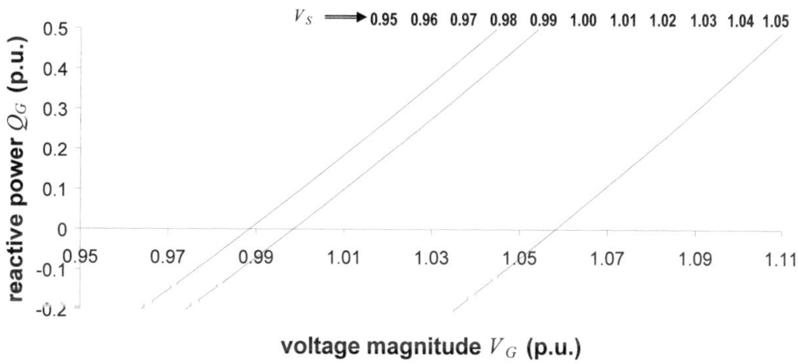

Fig. 2. Family of curves representing the external characteristic seen from the local generator terminals.

Equation (5) can also be used to update the reactive power Q_G (and the voltage V_G, if the reactive power limits are violated) during the iterative process of the backward-forward sweep algorithm (Carpaneto et al., 2008b).

2.3. Connection of synchronous generators to the grid

2.3.1. Transformer-based network connection

Local synchronous generators are used in applications like hydro power units. The traditional solution used in the VIU system to connect the local generators to the grid adopts a single transformer with fixed transformation ratio a_T. The value of the transformation ratio a_T depends on the transformer short-circuit impedance X_T and on the nominal power factor of the generator, in order to make it possible the injection in the grid of the nominal reactive power at nominal voltage. From the circuit in Fig. 1, the link among voltages and reactive power generation is expressed in an approximated form as

$$V_G = \frac{V_S}{a_T} + a_T X_T \frac{Q_G}{V_S} \tag{6}$$

The expression (6) clearly shows the strong coupling between the reactive power Q_G and the generator voltage magnitude V_G. Decoupling between Q_G and V_G can be enhanced by making the transformation ratio a_T variable, by means of a tap changer-under load (TCUL). The TCUL provides better operational flexibility at the expense of increased investment and operational costs.

2.3.2. Voltage controllers and reactive power limits

Different types of *voltage controllers* can be used to act on the local generator:

1. A *standard* voltage controller operating in *voltage-support* mode. The voltage magnitude at the generator terminals can be maintained at the predefined value until the excitation limits of the synchronous machine are reached. The excitation limits depend on the excitation current, but at first approximation it is possible to consider constant reactive power generation limits, with minimum value $Q_G^{\min}$ and maximum value $Q_G^{\max}$. The reactive power limit $Q_G^{\min}$ is usually time-independent (stability limit), while the limit $Q_G^{\max}$ is time-inverse (rotor thermal limit). When the local generator operates at its reactive power limits, the generator voltage is imposed by the network, taking into account the transformation ratio and impedance of the local transformer. The local generator at its terminals is modelled as a classical PV generator with reactive power limits. Fig. 3 shows the PV controller characteristic for a 500 kVA generator with reactive power generation limits $Q_G^{\min}$ = -0.125 p.u., $Q_G^{\max}$ = 0.3 p.u., and a voltage control range extended from 0.98 p.u. to 1.02 p.u., with reference value V_R = 1 p.u..

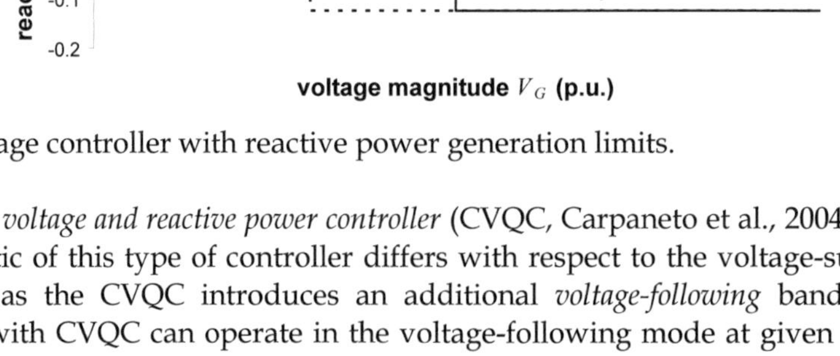

Fig. 3. PV voltage controller with reactive power generation limits.

2. A *combined voltage and reactive power controller* (CVQC, Carpaneto et al., 2004). The static characteristic of this type of controller differs with respect to the voltage-support type controller, as the CVQC introduces an additional *voltage-following* band. The local generator with CVQC can operate in the voltage-following mode at given (e.g., unity) power factor when the voltage at its terminals falls within the range from $V_G^{\min}$ to $V_G^{\max}$. In this voltage range, operation of the local generator and of its transformer can

be enhanced by keeping the control at null reactive power, with the effect of reducing the power losses in the local generator-transformer unit. In the other parts of the characteristic, the excitation control holds the generator voltage at the corresponding voltage value, until the reactive power limits are reached. Exploiting the CVQC guarantees then a good compromise between the needs of fixed reactive power/voltage control and the system generator requirements for limiting the internal losses. Fig. 4 shows an example of the CVQC characteristic for a 500 kVA generator with reactive power generation limits $Q_G^{\min}$ = -0.125 p.u., $Q_G^{\max}$ = 0.3 p.u., and two voltage control ranges extended from 0.96 p.u. to 1.0 p.u. (regulated at V_R^{low} = 0.99 p.u.) and from 1.0 p.u. to 1.04 p.u. (regulated at V_R^{high} = 1.02 p.u.), respectively.

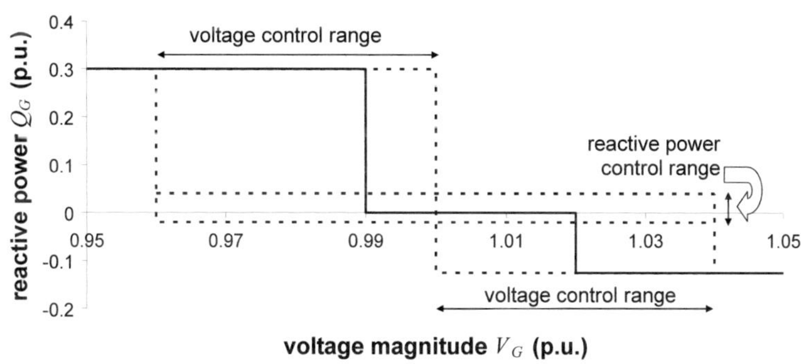

Fig. 4. CVQC voltage controller with limits and external characteristics.

In the CVQC case, the *transition* from the voltage following to the voltage-support control mode is practically driven by the network-side voltage, by properly choosing the regulated values V_R^{low} and V_R^{high} and of the tap position of the local transformer. When the local generator operates at *unity power factor*, the generator output current and the voltage drop on the transformer connecting the local generator to the network are relatively low, at the benefit of the local system. The proposed regulator automatically guarantees its *participation* to the system voltage control only when the network voltage tends to move out of the prescribed voltage range. For a predefined the range of network-side voltages (from V_S^{low} to V_S^{high}) at which operation is required in the internal voltage-following band, the voltage limits are approximated by calculating $V_R = V_R^{\text{high}}$ with $V_S = V_S^{\text{low}}$ (and $V_R = V_R^{\text{low}}$ with $V_S = V_S^{\text{high}}$) from the equation

$$V_R - \frac{V_S}{a_T} - R_T \frac{P_G}{V_G} - X_T \frac{Q_G}{V_G} = 0 \tag{7}$$

where a_T corresponds to one of the available tap positions of the off-line tap changer of the local transformer.

2.3.3. Operating points

The operating point at a voltage-controllable node is given by the intersection of the $Q_G(V_G)$ characteristic of the external system (like in Fig. 2) and of the local voltage controller, including the reactive power limits (Fig. 3 or Fig. 4). The analysis of the intersection points (as shown in Fig. 5 in the case with CVQC voltage controller) provides interesting hints on the voltage controllability of the distribution system. In particular:

1. It is possible to control the voltage V_S at the network side by means of the local generator only for a limited range of values of the voltage V_S. For local generators of relatively small size with respect to the HV/MV substation, the voltage control range is thus very limited.
2. Conceptually, the voltage control band of the local generator should be chosen in such a way to make voltage control effective, taking into account that the voltage V_S changes during the day. Thus, generators with the same characteristics but located in different network nodes could need different settings of the voltage controller.
3. In the presence of multiple voltage-controllable local generators, the settings should be defined by some coordinated control. However, if the local generators are owned and managed by different entities, there is no such coordination. Coordinated control can be attempted within a micro-grid (Nikkhajoei & Lasseter, 2009).

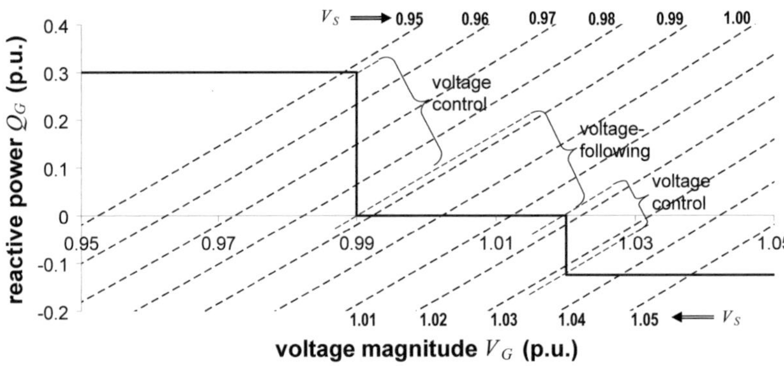

Fig. 5. Operating points obtained from the intersection of the characteristics of external system and CVQC voltage controller for a given DG active power generation.

A detailed analysis of the voltage controller operation with variable daily load patterns is presented in Carpaneto et al., 2004. The periods of time in which the controllers operate in voltage-support mode, voltage-following mode or at the maximum/minimum reactive power generation limits are identified for different setting of the controllers and transformation ratio, with the aim of endeavouring to find the most promising control settings. An approximated $V_R(a_T)$ representation is found as a straight line of the type

$$V_R = k_a\, a_T + V_0 \tag{8}$$

in which the parameters k_a and V_0 depend on the type of voltage controller and on the location of the generator in the distribution system. In particular, from the results found in Carpaneto et al., 2004, it emerges that the best conditions generally occur for k_a = -1. Hence, for a given generation unit it is possible to choose the tap setting a_T among the available ones, thus obtaining from Equation (8) the value of V_R for which the time of voltage-controlled operation is maximum. For the CVQC, a similar representation can be considered by using V_R as the central value of the two voltage controlled bands and taking into account the width of the voltage control band as additional variable.

2.3.4. Capability curves

Considering the nominal apparent power S_N and the nominal power factor $\cos(\varphi_N)$ of the synchronous machine, the maximum reactive power can be approximated as $Q_G^{\max} \approx S_N \sin(\varphi_N)$.

For a local generation unit composed of the generator and the interconnection transformer, the significant notion to represent the boundaries of the active and reactive power that can be injected into the grid interconnection point is the *capability chart* (Losi et al., 1998), that is, the capability curve referred to the grid connection point, including the effects of all components in the local system generation unit and their specific settings.

2.4. Other generators and network interfaces

Besides synchronous machines, the basic types of connection of local generation units to the distribution network are:

- *Induction generators* (without converters), used in very small hydro power plants and in some types of wind systems.
- *Power electronic inverter-based* grid interface, with different types of inverters. Blaabjerg et al., 2006 present the basic control structures of the inverters, the control strategies adopted in case of faults in the grid and the methods for DG synchronization with the grid. In many cases, the inverters are controlled to inject power into the grid at unity conventional power factor (calculated with the waveform components at fundamental frequency). The inverter control can also be adapted for to provide enhanced power conditioning, with various objectives such as injecting/absorbing reactive power, providing harmonic filtering, three-phase system balancing, mitigating the effects of voltage dips or short interruptions, eliminating zero-sequence components. Voltage-source inverter (VSI), that can be voltage-controlled or current-controlled (Ko et al., 2006). Voltage control can be aimed at reproducing reactive power control characteristics similar to a synchronous generator. In this case, additional functions of the inverter control include local voltage stabilization, reactive power support/power factor correction, improvement of local voltage quality at low-order harmonics, uninterruptible power supply (UPS), and active power support by using a bi-directional VSI to control the active power flow between the DC bus and the grid. However, current is not controlled, and since a VSI inverter has a short-time overload capacity much lower than a synchronous machine, the inverter rating has to be augmented (with increased cost) to provide the extra current. With current control, it is possible to obtain currents with nearly sinusoidal waveform, also with benefits to protection system operation, but with no contribution to improve the local voltage quality. Finally, Z-source inverters (ZSI), are a different category, whose applications to DG are described in Gajanayake et al., 2009.
- *Doubly-Fed Induction Generators* (DFIG), widely used in wind systems, as a particular case in which the rotor-side converters are added to the wound-rotor induction machine (Baroudi et al., 2007). One of the advantages of the DFIG structure is that the size of the

rotor-side converter is relatively low with respect to the rated power of the induction machine. However, this leads to a relatively small impact of the converter on voltage and reactive power control.

2.5. Basics of the voltage control with DG

In the traditional systems without DG, it was generally possible to set up a suitable value of supply voltage V_{MV} at the MV busbars in order to obtain all voltage magnitudes at the distribution system network nodes within the acceptable range around the rated voltage of the system. Possible cases with excessive voltage drops could be solved through appropriate location of power factor correction capacitors at some nodes of the distribution system.
The presence of DG units makes it necessary to reformulate the voltage control problem. In particular, the general idea of finding a value of the supply voltage V_{MV} at the MV busbars could lead to the impossibility of obtaining all voltage magnitudes at the distribution system network nodes within the acceptable range. In order to illustrate this point with a simple qualitative example, let us consider the two-feeder "fork" system (Fig. 6), composed of a passive feeder, represented by an equivalent load at the end of the feeder, and another feeder with uncontrollable DG (negative PQ load) at its end, With respect to the voltage profiles, the passive feeder experiences a voltage drop, while the DG feeder experiences a voltage rise. The amounts of the voltage variations depend on the load and local generation power, respectively, and change in time. Practically, it may happen that the voltage profiles exceed both the voltage limit on the passive feeder and the upper voltage limits in the DG feeder. In this case, no solution can be found by only varying the voltage V_{MV} from centralised voltage control. This explains the need for adding further voltage controllers in the distribution systems. Local voltage controllers embedded in the DG interface with the grid can fit this need.

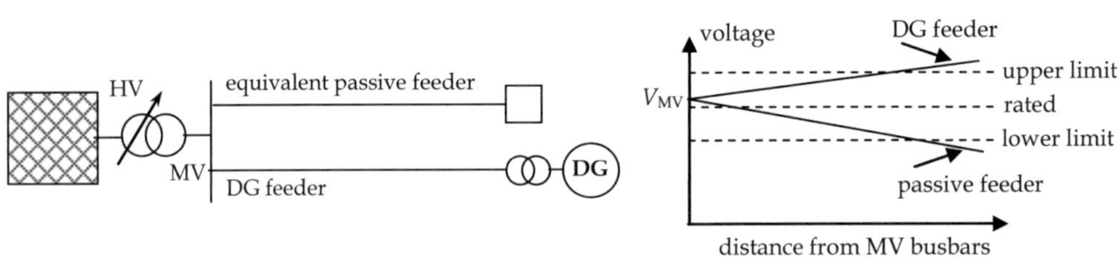

Fig. 6. The two-feeder "fork" system and the voltage profiles along the feeders.

However, the interaction among different voltage-controlled DG units operated independently of each other could create confusion and be ineffective. Suitable solutions can be found looking for coordinated voltage control strategies involving both the DNO and the operators of the DG units, based on agreed objectives and reliable communications (Baran & El-Markabi, 2007). In order to avoid the complexity of coordination, alternative solutions have been proposed to provide local solutions by making the DG unit injecting in the network, in addition to active power, independently-controlled reactive power (Bollen & Sannino, 2005; Carvalho et al., 2008) to compensate for the voltage difference between the voltages at the generator terminal and at the network terminal. For this purpose, imposing $V_G = V_S$ in Equation (5) yields the reactive power Q_G that the generator should provide for a given active power production P_G to compensate for the local voltage variation. These solutions require on the one hand sufficiently high reactive power capability in the DG unit, and on the other hand the control is time-dependent (with time-variable tap changing at the local transformer), in order to follow the voltage variations at the network side. In particular, the reactive power capability of the local generator decreases when the active power production is relatively high.

2.6. General formulation of the voltage control problem

The voltage control problem in the distribution systems with DG can be formulated with the objective of maintaining the voltage close to a reference value at any node of the system. This objective can be used either individually, to set up the voltage controls for a given system structure, or as part of a multi-objective optimization problem together with other objectives such as losses, operational costs, reliability indicators, energy efficiency indicators, environmental impact indicators, duration of the voltage-controlled operation through local DG groups, and others.

Voltage control optimisation can be considered as an operational planning problem, in which all the system components are already located in the network and the focus is set on their operational strategy. The period of analysis can be of the order of one day, or one week. Within this period, the time evolution of typical generation and load profiles at each node is assumed to be known. The input data include the reference values of the control systems to be considered. The constraints are given by the power flow equations, the minimum and maximum voltage limits, the thermal limits of the branches and of the MV/LV transformers, the reactive power limits of the local generators, where applicable, and other operational limits of the equipment.

Besides the modelling of the different types of voltage controllers, the key aspect to be considered in voltage control optimisation is the dependence on time of the electrical variables and controls. In this respect, the possible solutions also depend on the economic aspects linked to the acquisition of the control systems. With reference to relatively large public distribution networks, the controllers can be adjusted on-load (such as the centralised control at the HV/MV substation, driving the ULTC), or off-load (such as the tap positions of the MV/LV transformers). For cost reasons, off-load tap changers can be considered also at the local transformers connecting the DG units to the grid. According to these concepts, the following classification can be used for the voltage controls (Carpaneto et al., 2004):

- *time-independent*: tap positions of the transformers located at each MV/LV node and of the transformers connected to the local generation groups, voltage set points of the voltage-controllable local generators, and width of the internal voltage-following band for each CVQC;
- *time-dependent*: voltage reference of the centralised voltage control (the "control law").

A general mathematical formulation of the voltage control objective function, based on combining various literature results, is presented here for a voltage control optimization refers to the MV distribution system. A radial system is considered, with each branch numbered according to its ending node. Let us denote as a_k the transformation ratio deriving from the transformer tap positions of the MV/LV transformers. The explicit model of all the MV/LV transformers at the load or local generation nodes is provided, in order to set the target voltage magnitude $\hat{V}_{k,LV}$ at each node belonging to the set $\boldsymbol{K}_{LV} = \{k = 1,\ldots K_{LV}\}$ of the distribution system. The set $\boldsymbol{K}_{LV}$ also includes voltage-uncontrollable generation nodes. The upper voltage limits at the generation nodes cannot be exceeded, as the voltage control settings of the voltage-controllable generation units are constrained to avoid exceeding these limits.

The time domain is represented by a set of discrete points $j = 1,.., J$. The variable of interest is the voltage magnitude V_{kj} at each MV node corresponding to a LV node $k \in \boldsymbol{K}_{LV}$, for each point in time $j = 1,.., J$. In order to be compared to the target voltage $\hat{V}_{k,LV}$, the MV node voltage is first subject to the transformation ratio a_k, then the voltage drop in the transformer windings is modelled, in terms of the loading condition, as a fraction of the short-circuit voltage Δv_k given by the ratio between the current I_{kj} circulating in the transformer and the corresponding thermal limit $I_k^{\max}$. Other terms appearing in the objective function are the consumer energy E_{kj} introduced as a weighting factor to give more importance to the nodes $k \in \boldsymbol{K}_{LV}$ and to the time intervals $j = 1, ..., J$ in which the energy consumption is higher, and the consumer damage constant C_k (Strezoski et al., 2001). The final form of the objective function is:

$$f = \sum_{j=1}^{J} \sum_{k \in \boldsymbol{K}_{LV}} \left(C_k E_{kj} \left(\frac{V_{kj}}{a_k} - \frac{I_{kj}}{I_k^{\max}} \Delta v_k - \hat{V}_{k,LV} \right)^2 \right) \tag{9}$$

The formulation can be easily adapted to take into account the MV consumers ($\Delta v_k = 0$). The objective function (9) has to be minimized, and the related constraints have to be met at any time interval.

A further variant consists of defining a "customer voltage quality" index (Choi & Kim, 2000) in which the maximum deviation with respect to the voltage reference is over-penalised any time it exceeds the limits of the acceptable voltage range. On the structural point of view, in some cases step voltage regulators could be installed into the distribution system feeders (Roytelman & Ganesan, 2000), further increasing the number of control variables.

In practical systems, the number of time-independent and time-dependent control variables to be determined is significantly high, so that performing exhaustive search on the discrete variables would be prohibitive, and the shape of the objective function is unknown. Hence, heuristic techniques are typically used for the solution of the optimization problem (Carpaneto et al., 2004; Senjyu et al., 2008).

3. Evaluation and allocation of distribution system losses

3.1. Generalities on distribution system losses

In a VIU structure, the cost of losses was included into the overall electricity production costs, and there was no need for determining it specifically. In the restructured electricity business, the specific costs associated to any individual aspect of the business need to be identified. In particular, the system losses are an additional component with respect to the actual energy consumption (for loads) or energy generation (for local generators) indicated in the economic transactions between the distribution system operator and the consumer/producer located at a specific node in the network. Evaluation and allocation of the system losses to suppliers and consumers are key issues to be addressed, in order to set up appropriate economic penalties or rewards for suppliers and consumers.

In general terms, for a distribution system the total energy losses are determined as the difference between the measured energy output from the HV/MV substation and the measured energy input to the load points (distributed generation can be treated as a negative load). The quantification of the total losses is affected by uncertainty, because the measuring instruments are characterised by their intrinsic accuracy, the synchronization of the remote meters is not guaranteed, and the measured data could be affected by errors in the communication system.

The total losses include technical and non-technical losses (NTL) (Taleski & Rajicic, 1996). The technical losses occur in the circuits in their normal operation, have a non-linear dependence on currents and powers, and cannot be easily assessed, especially in situations with scarcity of data on the electrical network parameters and on the system operation.

The concept of NTL encompasses various components typically referred to frauds (e.g., from meter tampering, meter by-passing or illegal connections), billing and measurement errors (e.g., human errors during meter readings, meter or data communication failures, and metering equipment deterioration or ageing). Other causes, such as imperfect electrical contacts and current tappings due to local isolation failures, can be attributed to NTL as well. For billing purposes, the NTL are typically allocated to the rate classes in proportion to their energy consumption.

This section deals with the *technical losses*, whose value is determined by running a power flow under the hypothesis that the parameters of the distribution system, the local generations and the loads are known without uncertainty. Specific aspects concerning loss evaluation and loss allocation in balanced and unbalanced distribution systems with distributed generation are addressed. Loss evaluation is dealt with in the general case of unbalanced systems, highlighting specific loss partitioning aspects. Loss allocation refers to assign to each supplier and consumer in the distribution system a portion of the system losses, to be taken into account in the payments in addition to the components of the electricity tariff (such as a fixed component, a component related to the contract power, a component related to the energy consumption, and further components for exceeding specific thresholds set on maximum power or reactive energy).

3.2. Loss evaluation in three-phase systems

Loss analysis in general three-phase systems is useful to point out the basic aspects of loss allocation, and is relevant to distribution systems with single-phase lines and loads.

The network branch modelling in unbalanced multi-wire distribution systems is typically done by using the Carson's equations to calculate the self and mutual impedances for an arbitrary number of conductors, and by applying the Kron reduction to determine the 3x3 reduced impedance matrix $\mathbf{Z}_{abc}$ of each branch, referred to the phases a, b and c (Kersting, 2001). This reduced branch representation is particularly useful to carry out three-phase power flow calculations without introducing a detailed model of the return path (composed of the neutral conductor and the ground).

In Kersting, 2001, it is indicated to calculate the real power losses of a line segment as the difference (by phase) of the input power in the line segment minus the output power of the line segment. This classical technique is used for computing the total losses in a branch represented by its reduced 3x3 matrix, as the difference between the input (node m) and output (node h) power of the branch (Fig. 7), corresponding to write, considering $\mathbf{v}_m = [\bar{V}_{m,a}, \bar{V}_{m,b}, \bar{V}_{m,c}]^{\mathrm{T}}$, $\mathbf{v}_h = [\bar{V}_{h,a}, \bar{V}_{h,b}, \bar{V}_{h,c}]^{\mathrm{T}}$ and $\mathbf{i} = [\bar{I}_a, \bar{I}_b, \bar{I}_c]^{\mathrm{T}}$ (the superscript T indicates transposition):

$$\Delta P_{tot} = \Re e\{(\mathbf{v}_m - \mathbf{v}_h)^{\mathrm{T}} \mathbf{i}^*\} = \Re e\{\mathbf{i}^{\mathrm{T}} \mathbf{Z}_{abc} \mathbf{i}^*\} \tag{11}$$

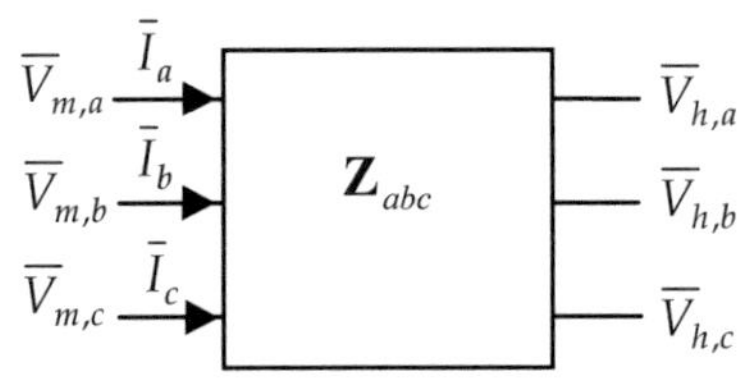

Fig. 7. Representation of a branch reduced to the three-phase branch impedance matrix.

An alternative way for computing the total losses can be defined by taking into account the real part of the branch impedance matrix, $\mathbf{R}_{abc} = \Re e\{\mathbf{Z}_{abc}\}$. In fact, as demonstrated in Appendix B of Carpaneto et al., 2008a, the following equivalence holds:

$$\Re e\{\mathbf{i}^{\mathrm{T}} \mathbf{Z}_{abc} \mathbf{i}^*\} = \mathbf{i}^{\mathrm{T}} \Re e\{\mathbf{Z}_{abc}\} \mathbf{i}^* \tag{12}$$

so that the expression of the total losses becomes

$$\Delta P_{tot} = \mathbf{i}^{\mathrm{T}} \mathbf{R}_{abc} \mathbf{i}^* \tag{13}$$

However, as remarked in Carpaneto et al., 2008a, the formulations (12) and (13) are not generally equivalent for the purpose of partitioning the total losses among the three phases. Differences in loss partitioning occur in branches with non-zero current in the return path. In these cases, only the Resistive Component-based Loss Partitioning (RCLP) method, defined in Carpaneto et al., 2008a, by using the matrix $\mathbf{R}_{abc}$, provides the correct decomposition of the currents in the neutral conductor and in the ground into various components to be associated with the phase currents. The RCLP method provides a meaningful representation of the Joule losses in each physical conductor (phases and neutral) and in the ground.

Considering the vector $\Delta\mathbf{p} = [\Delta P_a \; \Delta P_b \; \Delta P_c]^{\mathrm{T}}$ containing the losses associated to the phase currents, the RCLP method provides the partition of the total losses as

$$\Delta\mathbf{p} = \Re e\{\mathbf{i} \otimes (\mathbf{R}_{abc} \mathbf{i}^*)\} \tag{14}$$

where $\otimes$ denotes the component-by-component vector product. In this way, for each component the associated losses are proportional to the projection of the phasor representing that current component onto the phasor representing the specified current (Carpaneto et al., 2008a).

Application of (14) with the matrix $\mathbf{Z}_{abc}$ instead of $\mathbf{R}_{abc}$ with non-zero current in the return path would result in the loss partitioning paradox identified and explained in Carpaneto et al., 2008a. Occurrence of this paradox leads to partition the total losses in uneven way, for instance with heavily loaded phases associated to losses even higher than the total losses, or with negative losses that can be associated to lightly as well as to heavily loaded phases. The correct partitioning obtained by using the RCLP method still admits negative losses to occur as a result of the decomposition of the return path currents into the components associated with the phase currents, but is able to fully explain the individual terms of such decomposition.

3.3. Loss allocation concepts and principles

The main difficulty of setting up loss allocation techniques in distribution networks depends on the fact that the branch losses are expressed as non-linear (nearly quadratic) functions of the current or power generations and loads. Furthermore, cross-terms appear, due to the interaction between power injections in different nodes.

The loss allocation concepts have become much more important because of the growing presence of distributed generation and resources in the distribution systems. In fact, the presence of a relatively significant amount of distributed generation may reverse the power flows in some branches of the distribution systems. Thus, a local generator operating at a specific location in the distribution network and with a given output may provide benefits to the network depending on the system structure and on the location and amount of every generator and load. The need for taking into account the full power flow solution also indicates that it is not correct to use the substitution method that considers the difference of the total losses in the presence or absence of a single unit (local generator or load) for determining the effects of that unit on loss allocation (for further details, see Section 8.4 of Jenkins et al., 2000). In addition, the diffusion of distributed generation has made it inappropriate to use methods based on uniform or demand-squared loss allocation formulated by taking into account only the demand side.

The loss allocation results should reflect the contribution of each supplier/consumer to the system losses, taking into account the active power and reactive power sides, as well as incorporating the effects of voltage controls. Variation during time of generation, load patterns, system structure and control settings has to be appropriately taken into account.

The possible benefits of loss allocation are determined on the basis of the concept of *marginal losses*. Conceptually, for a local generator (or load) the marginal losses, determined for a certain shapshot in time, are defined by checking whether a small (theoretically infinitesimal) increase of the amount of active or reactive generation (or load) increases or reduces the system losses P_L. In general, the distribution system losses depend on the net power (i.e., generation minus load) connected to a node. At a generic node k, considering the net active power P_k and the net reactive power Q_k, the effects of marginal losses can be expressed by introducing the marginal loss coefficients $\rho_{Pk} = \partial P_L / \partial P_k$ and $\rho_{Qk} = \partial P_L / \partial Q_k$, acting as sensitivity factors. Considering for instance the net active power and using the superscript 0 to denote the reference conditions (i.e., the present power flow solution), it is possible to write, for small deviations:

$$P_L - P_L^{(0)} = \rho_{Pk}\left(P_k - P_k^{(0)}\right) \tag{15}$$

On these bases, loss reduction benefits for the system occur when $\rho_{Pk} > 0$ and the net power is decreased, as well when $\rho_{Pk} < 0$ and the net power is increased.

These concepts can be taken into account by the regulation in order to set up a system of incentives and penalties. As such, the owner of a local unit could incur an incentive if the unit exploitation determines marginal loss reduction, or a penalty if the unit exploitation causes a marginal loss increase. If each node of the system contains either a load or a generator, on the basis of the above concepts, penalties would occur for load nodes with negative marginal loss coefficients or generation nodes with positive marginal loss coefficients, while incentives would occur for load nodes with positive marginal loss coefficients or generators with negative marginal loss coefficients (Mutale et al, 2000). This conceptual structure, in which loss allocation reflects the contribution of each supplier/consumer to active and reactive losses, provides correct signals to the electricity business, as it stimulates the introduction of new local generation only in the locations and with the amounts for which a benefit on network efficiency could occur.

In addition, an effective loss allocation method should be formulated according to easily understandable principles, based on the real data of the networks, and should recover the total amount of losses. The loss allocation results should be economically efficient, avoiding discrimination and cross-subsidization among users (Jenkins et al., 2000).

After calculating the loss allocation terms for all network components (generators, loads, compensation devices, and shunt parameters of the network), these terms have to be attributed to specific entities in order to clear the loss allocation economic issues. For this purpose, the losses allocated to each load or compensation device (such as power factor correction capacitor) are attributed to the specific owner, whereas the losses allocated to the shunt parameters of the branch model are attributed to the DNO.

3.4. Network-related aspects impacting on the formulation of loss allocation techniques

In principle, the loss allocation problem has a different formulation for transmission systems or for radial distribution systems.

In transmission systems, a (large) generator is connected to the slack node of the power flow equations, then part of the losses are allocated also to the slack node. Furthermore, a transmission network with K nodes is represented by using the full KxK bus impedance matrix $\mathbf{Z}_{bus}$, or the full KxK bus admittance matrix $\mathbf{Y}_{bus}$.

In distribution systems, the supply node is uniquely determined by the connection to a higher voltage network through the power substation and corresponds to the slack node of the system. The supply side, managed by the DNO, typically supplies the largest part of the power, but it does not correspond to a physical generator. As such, the loss allocation methods have to be carefully designed to take into account the specific characteristics of the slack node. Furthermore, in distribution systems local loads or generators are owned and managed by different entities, and operated within specific regulatory frameworks.

In order to allocate the distribution system losses, the DNO is considered as the subject undertaking bilateral contracts for loss allocation with the distribution system entities (Carpaneto et al., 2006a). The distribution system losses are allocated to the local generators and loads connected to the distribution system nodes, *excluding* the slack node. For this reason, a distribution network with K nodes is represented by using the reduced (K-1)x(K-1) bus impedance matrix $\mathbf{Z}_{bus}$, or the reduced (K-1)x(K-1) bus admittance matrix $\mathbf{Y}_{bus}$.

As a consequence of the conceptual differences indicated above, it is possible to apply the loss allocation methods formulated for transmission systems to the distribution systems, provided that the losses allocated to the slack node are redistributed among the other nodes (Carpaneto et al., 2006a). A simple way to avoid allocating losses to the slack node is to connect the slack node to the reference node with a null impedance, that is, to impose the slack voltage magnitude to zero in the network used for determining the loss allocation; in this case, it is possible to use the loss allocation methods defined for transmission systems for allocating losses in distribution systems, as shown in the examples presented in Carpaneto et al., 2006b.

A significant case in the formulation of loss allocation methods is the one in which the distribution network has negligible shunt parameters in the branch model, such as for most aerial distribution networks and for low voltage cables. In this case, if no generation nor load is modelled as impedance or admittance component, the network has no connection to the system reference node (*floating* network). This causes the impossibility of constructing the full *KxK* bus impedance matrix and to use the loss allocation methods based on this matrix. However, the technique of connecting the slack node to the reference node for distribution systems in this case enables obtaining meaningful loss allocation also for a floating network. Strictly speaking, for a floating network correct results would be obtained also with non-null impedance for the slack node connection to the reference node (Carpaneto et al., 2006b).

3.5. Loss allocation techniques

The most appropriate loss allocation techniques start from the power flow results and exploit the concept of marginal losses to formulate suitable indicators to express the positive or negative contribution of generators and loads to reduce the system losses. These techniques can generally be partitioned into *derivative-based* and *circuit based*. The correctness in the formulation of the computational techniques depends on avoiding the occurrence of the loss allocation paradox identified in Carpaneto et al., 2006a.

3.5.1. Derivative-based methods

A general expression for the derivative-based methods can be built by approximating the total losses L in function of the net node power vector $\mathbf{p}$ in quadratic form (Carpaneto et al., 2008b)

$$L = L_0 + \mathbf{b}^{\mathrm{T}}\mathbf{p} + \frac{1}{2}\mathbf{p}^{\mathrm{T}}\mathbf{A}\,\mathbf{p} \qquad (16)$$

where $L_0 = L\big|_{\mathbf{p}=0}$ represents the no-load losses, the column vector $\mathbf{b} = \left.\frac{\partial L}{\partial \mathbf{p}}\right|_{\mathbf{p}=0}$, and $\mathbf{A}$ is a symmetric matrix. Higher-order terms are neglected. The terms L_0 and $\mathbf{b}$ are null in the absence of shunt circuit components (e.g., shunt line or transformer parameters) and of circulating currents depending on different voltage settings at different PV nodes (as the ones introduced by local generators operating in the voltage control range).

By indicating the derivative of the total losses with respect to the vector $\mathbf{p}$ as $L_{\mathbf{p}} = \frac{\partial L}{\partial \mathbf{p}}$, from analytical elaborations it is possible to obtain

$$2L = \left(L_{\mathbf{p}}^{\mathrm{T}} + \mathbf{b}^{\mathrm{T}}\right)\mathbf{p} \tag{17}$$

from which the vector $\boldsymbol{\psi}$ containing the loss allocation coefficients is defined in such a way to represent the total losses as $L = \boldsymbol{\psi}^{\mathrm{T}}\mathbf{p}$:

$$\boldsymbol{\psi} = \frac{L_{\mathbf{p}} + \mathbf{b}}{2} \tag{18}$$

The expression (18) indicates how the loss allocation vector depends on the derivative of the total losses with respect to the load vector. In particular, the derivative $L_{\mathbf{p}}$ alone is unable to provide a loss allocation vector, and reconciliation to the total losses is needed by dividing by 2 even in the case in which $\mathbf{b} = \mathbf{0}$.
From another point of view, the exact variation of the losses defined in the quadratic form (16) with respect to load power variations can be expressed by considering two generic net power vectors $\mathbf{p}_1$ and $\mathbf{p}_2$, leading to the total losses $L^{(1)}$ and $L^{(2)}$, respectively, by using the average value of the derivatives calculated in the two configurations (Carpaneto et al., 2008b):

$$L^{(2)} - L^{(1)} = \left(\frac{L_{\mathbf{p}}\big|_{\mathbf{p}_1} + L_{\mathbf{p}}\big|_{\mathbf{p}_2}}{2}\right)^{\mathrm{T}}\left(\mathbf{p}_2 - \mathbf{p}_1\right) = \left[\mathbf{b} + \frac{1}{2}\mathbf{A}\left(\mathbf{p}_1 + \mathbf{p}_2\right)\right]^{\mathrm{T}}\left(\mathbf{p}_2 - \mathbf{p}_1\right) \tag{19}$$

The expression (19) is independent of L_0. If $\mathbf{p}_1 = \mathbf{0}$ and $\mathbf{p}_2 = \mathbf{p}$, the equation providing the total losses $L = L^{(2)}$ becomes

$$L = \left[\mathbf{b} + \frac{1}{2}\mathbf{A}\mathbf{p}\right]^{\mathrm{T}}\mathbf{p} \tag{20}$$

If $\mathbf{b} = \mathbf{0}$, the loss coefficient vector $\boldsymbol{\psi} = \frac{1}{2}\mathbf{A}\,\mathbf{p}$ can be directly used to represent the total losses as $L = \boldsymbol{\psi}^{\mathrm{T}}\mathbf{p}$, with no need of reconciliation. If $\mathbf{b} \neq \mathbf{0}$, the product $\boldsymbol{\psi}^{\mathrm{T}}\mathbf{p}$ gives an approximation of the total losses.
The above illustration of the properties of the total losses is useful to discuss the formulation of some derivative-based methods proposed in the literature. In general, the matrix $\mathbf{A}$ and the vector $\mathbf{b}$ are not known. The methods are then elaborated by using the power flow state variables $\mathbf{x}$ (voltage magnitudes at the PQ nodes and voltage phase angles at all nodes, slack node excluded), and the Jacobian matrix $\mathbf{J}_{\mathbf{x}}$ containing the derivatives of the power flow equations with respect to the vector $\mathbf{x}$. Two methods based on expressing the total losses in quadratic form have been presented in Mutale et al., 2000:

1) the Marginal Loss Coefficients (MLC) method, in which an auxiliary vector $\boldsymbol{\sigma}$ is calculated by solving the linear system

$$\mathbf{J}_{\mathbf{x}}^{\mathrm{T}}\boldsymbol{\sigma} = \frac{\partial L}{\partial \mathbf{x}} \tag{21}$$

and reconciliation is needed to get $L = \boldsymbol{\psi}^{\mathrm{T}}\mathbf{p}$, since the product $\boldsymbol{\sigma}^{\mathrm{T}}\mathbf{p}$ in real systems approximately represents half of the total losses, on the basis of the same concepts discussed in (18), obtaining the loss allocation vector

$$\boldsymbol{\psi} = \boldsymbol{\sigma}\frac{L}{\boldsymbol{\sigma}^{\mathrm{T}}\mathbf{p}} \approx \frac{\boldsymbol{\sigma}}{2} \tag{22}$$

2) the Direct Loss Coefficients (DLC) method, using the Taylor series expansion of the total loss equation around the no-load conditions, in which an auxiliary vector $\boldsymbol{\gamma}$ is calculated by solving the linear system

$$\bar{\mathbf{J}}_{\mathbf{x}}^{\mathrm{T}}\boldsymbol{\gamma} = \frac{1}{2}\mathbf{H}\,\Delta\mathbf{x} \tag{23}$$

in which the rationale of using the average Jacobian matrix $\bar{\mathbf{J}}_{\mathbf{x}}$ calculated from the Jacobian matrices in two configurations (the current operating point and no-load) is based on the same concepts discussed in (19), and the Hessian matrix $\mathbf{H}$ of the loss equation is calculated at the current operating point. If the conditions corresponding to $\mathbf{b} = 0$ are satisfied, there is no need for reconciliation ($\boldsymbol{\psi} = \boldsymbol{\gamma}$), otherwise the product $\boldsymbol{\gamma}^{\mathrm{T}}\mathbf{p}$ gives an approximation of the total losses.

In distribution systems with voltage-controllable distributed generation, the definition of the MLC and DLC methods is affected by the fact that the voltage magnitude of a PV node is not a state variable in the power flow equations, thus the loss allocation coefficients are undefined for PV nodes. However, as remarked in Section 2.3.3 for synchronous generators, the local generators could operate in voltage control mode only for a portion of the total time interval of operation, being constrained to the reactive power limit in other time periods. These aspects may cause a discontinuity in the time evolution of the MLC and DLC coefficients. However, a voltage-controllable local generation unit is typically connected to the grid through a local transformer, that can be considered as integral part of the local system. As such, it is possible to adopt a two-step technique of analysis (Carpaneto et al., 2008b). In the first step, the power flow is solved by taking into account the detailed characteristics of the local generator (including its voltage control system) and of the local transformer. In the second step, each generation unit (generator and transformer) is replaced by the net power injected into the distribution network calculated from the power flow, thus constructing a *reduced network* in which no PV node appears. The losses in the local transformer are part of the local system and are correctly excluded from the loss allocation. Any possible reactive power limit enforced or other specific modelling details are implicitly embedded in the net power representation. The loss allocation is then calculated for the reduced network by using derivative-based or other methods. For the derivative-based methods, the Jacobian matrix to be used for loss allocation purposes has to be recalculated also when the power flow solution has already used a method requiring the construction of a Jacobian matrix, since the number of nodes in the reduced network is different with respect to the one of the original network. One critical aspect is the calculation of the no-load configuration to be used in equation (23) when one or more local generator operate as a PV nodes. In this case, being the loss allocation calculated on the reduced network, the effect of the voltage setting at PV nodes cannot be taken into account.

3.5.2. Circuit-based methods

The derivative-based methods illustrated in the previous subsection require the calculation of the Jacobian matrix (and in one case of the Hessian matrix) of the power flow equations. Indeed, the power flow for radial distribution networks is typically solved by methods like the backward/forward sweep, that exploit the network structure and circuit equations and do not require the construction of the Jacobian nor Hessian matrices. This fact leads to the definition of circuit-based loss allocation methods, in which no information on the derivatives is needed. One aspect to be verified for a circuit based-method is the possibility of reproducing the sensitivity information needed for representing the marginal losses, in order to ensure that the method effectively provides the correct signals, as discussed in Section 3.3. In the methods indicated in this section, this sensitivity information is implicitly provided by verifying that, for variable loads and in comparable cases (concerning the presence of voltage controls), the allocated losses computed with the circuit-based methods exhibit the same behaviour as those obtained by using derivative-based methods.

The formulation of the circuit-based methods could be affected by a conceptual paradox (Carpaneto et al., 2006a), based on concepts similar to those discussed in Section 3.2 for phase loss partitioning. In particular, the total active power losses on a branch can be calculated in two alternative ways, for a given branch current $\overline{I}$:

a) considering the branch impedance $\overline{Z}$ (the asterisk denotes conjugation):

$$L = \Re e\left(\overline{I}^{*}\,\overline{Z}\,\overline{I}\right) \tag{24}$$

b) considering the real part R of the branch impedance:

$$L = \overline{I}^{*} R\,\overline{I} \tag{25}$$

However, using the above formulations leads to different loss partitioning results between the active and reactive power flows in the branch. In particular, if the characteristic angle of the branch impedance $\overline{Z}$ is higher than the characteristic angle of the load, the loss allocation carried out by using (25) fails to provide meaningful results. This fact can be easily highlighted by considering the presence in the same node of two loads with the same active power but different reactive powers, resulting in allocation of more losses to the load

with lower reactive power (Carpaneto et al. 2006a). This paradox never occurs by using (25) for loss allocation purposes.
The rationale for interpreting the direct use of the sole resistive components in (25) rather than the real part of the terms in (25) containing the complex impedance can be shown by considering a branch with series impedance $\overline{Z} = R + \mathrm{j}X$, supplying a load with complex power $\overline{S} = P + \mathrm{j}Q$, with voltage magnitude V at the load terminal. The total losses are

$$L = \left(\frac{RP + XQ}{V^2} \right) P + \left(\frac{RQ - XP}{V^2} \right) Q \qquad (26)$$

The interdependence between active and reactive flows can be eliminated by simplifying the terms depending on the product of the active and reactive power, obtaining:

$$L = \left(\frac{RP}{V^2} \right) P + \left(\frac{RQ}{V^2} \right) Q \qquad (27)$$

The multipliers of P and Q are the loss allocation coefficients applied to the active and reactive load, respectively. The use of the coefficients determined from (26) leads to the loss allocation paradox mentioned above. The coefficients defined in (27) with reference to the sole resistive parameter of the branch allow obtaining meaningful and paradox-free loss allocation.
In order to show the formulations of some efficient circuit-based loss allocation methods proposed in the literature to be used in distribution networks with DG, let us consider a distribution system with K+1 nodes and B branches. The slack node is denoted as node 0. For any node k = 0,..., K, the power flow data and results include the complex node voltage $\overline{V}_k$, the net input complex power $\overline{S}_k = P_k + \mathrm{j}Q_k$ and the net input current $\overline{I}_k$.
On the basis of these definitions, four circuit-based methods are presented below.

1. *Z-bus loss allocation* (Conejo et al., 2000). The application of this method in its classical form requires the construction, for a non-floating system, of the matrix $\mathbf{Z}_{bus}$ with dimensions (K+1)x(K+1), slack node included. The elements of the $\mathbf{Z}_{bus}$ matrix of the distribution system are indicated as $\overline{z}_{km} = r_{km} + \mathrm{j}x_{km}$. The expression of the losses L_k allocated to each node k = 0,..., K is:

$$L_k = \Re e \left(\overline{I}_k^* \sum_{m=0}^{K} r_{km} \overline{I}_m \right) \qquad (28)$$

In order to get meaningful results for distribution systems in which the slack node does not participate to the loss allocation, the losses allocated to the slack node in (28) are redistributed among the other nodes. For this purpose, the technique of connecting the slack node to the reference node is used, as discussed in Section 3.4. This connection has a similar effect to partitioning the slack node losses among the other nodes and extends the application of the method to floating systems (Carpaneto et al., 2006b).

2. *Loss allocation through a modified* $\mathbf{Y}_{bus}$ *matrix* (Daniel et al. 2005). The original method described in the paper is formulated to be applied to transmission systems, in which loss allocation is carried out separately for generation nodes (sources) or load nodes (sinks), by decomposing the system currents. In the application to distribution systems, a modified matrix $\mathbf{Y}_{bus}$ with dimensions $(K+1)$x$(K+1)$, included the slack indicated at node $k = 0$ for the sake of representation, is constructed by including the equivalent admittances at the generation nodes, and the loads are treated as current injections. For the load node $k = 1,\ldots, K$, the losses L_k are allocated on the basis of the load current $\bar{I}_k$ by taking into account the resistance $R^{(b)}$ of a branch $b = 1,\ldots,B$ of the network, the column vector $\mathbf{i}$ containing all the load node currents and the column vector $\mathbf{c}^{(b)}$ containing the b-th column of the incidence matrix containing the relations among node current injections to branch currents:

$$L_k = \Re e\left(\bar{I}_k \sum_{b=1}^{B} \mathbf{i}^{*\mathrm{T}} \left(\mathbf{c}^{(b)^{*\mathrm{T}}} R^{(b)} \mathbf{c}^{(b)} \right) \right) \tag{29}$$

This method is applicable to any type of distribution system, considering the distributed generation as a negative load, also in floating cases, being it possible to construct the bus admittance matrix in any case.

3. *Branch Current Decomposition Loss Allocation* (BCDLA, Carpaneto et al., 2006a). This method is defined for a radial distribution system. Let us consider the set $\boldsymbol{K}^{(b)}$ of the downward nodes supplied from branch b, the set $\boldsymbol{B}_k$ of the branches belonging to the unique path from node k to the slack node, and the net input current $\bar{I}_k$ at node k (including the contribution of the loads and of the shunt elements connected to node k). The current $\bar{I}^{(b)}$ passing in the series element of the π–model of branch $b = 1,\ldots, B$ with branch resistance $R^{(b)}$, is

$$\bar{I}^{(b)} = \sum_{k \in \boldsymbol{K}^{(b)}} \bar{I}_k \tag{30}$$

The losses allocated to node k are calculated as

$$L_k = \Re e\left(\bar{I}_k^{*} \sum_{b \in \boldsymbol{B}_k} R^{(b)} \bar{I}^{(b)} \right) \tag{31}$$

The BCDLA method has been extended to perform loss allocation in three-phase unbalanced systems (Carpaneto et al., 2008a), exploiting the effective loss partitioning determined by the application of the RCLP method discussed in Section 3.2. The three-phase net input current at node k is represented by the column vector $\mathbf{i}_k = \left[\bar{I}_{k,a}, \bar{I}_{k,b}, \bar{I}_{k,c}\right]^{\mathrm{T}}$, and the losses allocated to the three phases of the load connected to node k are denoted as $\boldsymbol{\lambda}_k = \left[L_{k,a}\, L_{k,b}\, L_{k,c}\right]^{\mathrm{T}}$. The current flowing in the series elements of branch $b = 1,\ldots, B$ is

$$\mathbf{i}^{(b)} = \sum_{k \in \boldsymbol{K}^{(b)}} \mathbf{i}_k \tag{32}$$

and the losses allocated to the three phases of node k are

$$\boldsymbol{\lambda}_k = \Re e\left[\mathbf{i}_k \otimes \sum_{b \in \boldsymbol{B}_k} \left(\mathbf{R}_{abc}^{(b)} \left(\mathbf{i}^{(b)}\right)^*\right)\right] \tag{33}$$

The BCDLA method is particularly effective for various reasons. First, it is paradox-free and uses directly the power flow results, with no approximation required. In particular, it can be conveniently used in association to efficient implementations of the backward/forward sweep algorithm for solving the power flow in which there is no need of storing the bus impedance matrix or bus admittance matrix coefficients. Furthermore, it is applicable to any kind of radial system, either with non-negligible shunt parameters or floating. Then, it defines the loss allocation factors for all active and reactive power components at any node. In particular, all shunt components (such as shunt branch parameters and power factor correction devices) are treated in a consistent way, making it possible to identify their specific contribution depending on their location in the distribution system. Finally, the sum of the allocated losses equals the total losses, with no need of reconciliation.

4. *"Succinct" method for loss allocation* (Fang & Ngan, 2002). The original formulation of the "succinct" method assumes that the losses associated to the shunt admittance branches can be allocated in average terms to all users, and focuses on the allocation of the losses due to the series branch impedance, thus being affected by the paradox discussed above. A revisited paradox-free version of the method has been formulated in (Carpaneto et al. 2006a), expressing the losses allocated to node $k = 1,\ldots,K$ as

$$L_k = \Re e\left\{\bar{I}_k^* \sum_{b=1}^{B}\left[\left(r_{ik} - r_{qk}\right)\bar{I}^{(b)}\right]\right\} \tag{34}$$

where i and q are the sending and ending node of branch b, respectively, whereas r_{ik} and r_{qk} are resistances taken from the real part of the bus impedance matrix.
In the application of the method to distribution systems, also in this case the slack node is connected to the reference node, also enabling the application of the method for floating systems, as discussed in Section 3.4.

3.5.3. Other methods

A multi-stage loss allocation scheme has been formulated by Costa & Matos, 2004. Within this scheme, the first stage consists of loss allocation to the consumers or to their providers,

the second stage to loss allocation to the distributed generators, and the third stage to allocation of the remaining loss variations referred to voltage profile variations.
Further methods rely upon the concept of tracing the electricity flow, attributing the flows in the network branches to the nodal injections. Application of these method could be complicated in meshed networks, but in radial distribution systems each branch flow is given by the sum of the shunt contributions in the downward portion of the network. In this case, the electricity flow tracing is uniquely defined, and loss allocation procedures such as the one presented in Bialek and Kattuman, 2004, can be directly used.

4. Conclusions

This chapter has recalled the basic aspects and some specific details of the steady-state assessment of topics like voltage control, reactive power support and loss allocation in distribution networks with distributed generation. Voltage control is one of the issues that needs to be addressed in a dedicated way in the presence of distributed generation. One of the open fields of research in this area is the promotion of a coordinated voltage control (Viawan & Karlsson, 2007; Nikkhajoei & Lasseter, 2009; Madureira & Peças Lopes, 2009) in the distribution network, or in specific portions of the network that could be managed as micro-grids. Coordinated voltage control could provide benefits for the interaction of multiple local generators scattered in the distribution network. Within a micro-grid, the impedance between the local generators is relatively small. If multiple local generator are connected to the micro-grid, trying to perform voltage control, inaccurate setting of the voltage set points of the local generators can cause the presence of circulating currents in the network. In order to prevent this effect, voltage versus reactive power droop control can be exploited (Nikkhajoei & Lasseter, 2009). Additional possibilities may come from the adoption of secondary voltage control, fast with respect to the TCUL and slow with respect to the local controllers. The control strategies have generally to be evaluated in conditions of time-dependent variation of generation and load. Variability in time of the local controls requires additional investments, justifiable when sufficient benefits can be guaranteed. Indeed, the effects of applying any type of voltage control have to be checked against their impact on other objectives such as loss or operating costs reduction.
Concerning loss allocation, the circuit-based techniques presented in this chapter have been shown to be effective for radial systems. The presence of micro-grids and more generally the perspective of increased diffusion of distributed generation may suggest the adoption of non-radial network structures. Further analyses have to be carried out to extend the circuit-based loss allocation methods to operation of non-radial structures.

5. References

Baran, M.E. & El-Markabi, I.M. (2007). A multiagent-based dispatching scheme for distributed generators for voltage support on distribution feeders, IEEE Trans. on Power Systems, Vol. 22, No. 1, (February 2007) 52-59, ISSN 0885-8950

Baroudi, J.A.; Dinavahi, V. & Knight, A.M. (2007). A review of power converter topologies for wind generators. Renewable Energy, Vol. 32, No. 14, (November 2007) 2369-2385, ISSN 0960-1481

Bialek, J.W. & Kattuman, P.A. (2004). Proportional sharing assumption in tracing methodology, IEE Proc.-Gener. Transm. Distrib., Vol. 151, No. 4, (July 2004) 526–532, ISSN 1350-2360

Blaabjerg, F.; Teodorescu, R.; Liserre, M. & Timbus, A.V. (2006). Overview of Control and Grid Synchronization for Distributed Power Generation Systems. IEEE Trans. on Industrial Electronics, Vol. 53, No. 5, (October 2006) 1398-1408, ISSN 0278-0046

Bollen, M.H.J., & Sannino, A. (2005). Voltage control with inverter-based distributed generation. IEEE Trans. on Power Delivery, Vol. 20, No. 1, (January 2005) 519-520, ISSN 0885-8977

Borbely, A.M. & Kreider, J.F. (ed.) (2001). Distributed generation - The power paradigm of the new millennium. CRC Press, ISBN 9780849300745, Boca Raton, FL

Carpaneto, E.; Chicco, G.; De Donno, M. & Napoli, R. (2004). Voltage controllability of distribution systems with local generation sources, Proc. Bulk Power System Dynamics and Control - VI, pp. 261-273, ISBN 88-87380-47-3, Cortina d'Ampezzo, Italy, August 22-27, 2004

Carpaneto, E.; Chicco, G. & Sumaili Akilimali, J. (2006a). Branch current decomposition method for loss allocation in radial distribution systems with distributed generation, IEEE Trans. on Power Systems, Vol. 21, No. 3, (August 2006) 1170–1179, ISSN 0885-8950

Carpaneto, E.; Chicco, G. & Sumaili Akilimali, J. (2006b), Application of the circuit-based loss allocation techniques to radial distribution systems, Proc. VI World Energy System Conference, pp. 127-132, ISBN 978-88-87380-51-4, Torino, Italy, July 10-12, 2006

Carpaneto, E.; Chicco, G. & Sumaili Akilimali, J. (2008a). Loss partitioning and circuit-based loss allocation in three-phase radial distribution systems, IEEE Trans. on Power Systems, Vol. 23, No. 3, (August 2008) 1039-1049, ISSN 0885-8950

Carpaneto, E.; Chicco, G. & Sumaili Akilimali, J. (2008b). Characterization of the loss allocation techniques for radial systems with distributed generation, Electric Power Systems Research, Vol. 78, No. 8, (August 2008) 1396-1406, ISSN 0378-7796

Carvalho, P.M.S.; Correia, P.F. & Ferreira, L.A.F. (2008). Distributed Reactive Power Generation Control for Voltage Rise Mitigation in Distribution Networks. IEEE Trans. on Power Systems, Vol. 23, No. 2, (May 2008) 766-772, ISSN 0885-8950

Choi, J.-H. & Kim, J.-C. (2000). Network reconfiguration at the power distribution system with dispersed generations for loss reduction, Proc. IEEE PES Winter Meeting 2000, ISBN 0-7803-5935-6, Vol. 4, pp. 2363-2367, Singapore, 24 27 January 2000

Conejo, A.J.; Galiana, F.D. & Kockar, I. (2000). Z-bus loss allocation, IEEE Trans. on Power Systems, Vol. 16, No. 1, (May 2000) 105-109, ISSN 0885-8950

Costa, P.M. & Matos, M.A. (2004). Loss allocation in distribution networks with embedded generation, IEEE Trans. on Power Systems, Vol. 19, No. 1, (February 2004) 384–389, ISSN 0885-8950

Daniel, J.S.; Salgado, R.S. & Irving, M.R. (2005). Transmission loss allocation through a modified Ybus, IEE Proc.-Gener. Transm. Distrib., Vol. 152, No. 2, (March 2005) 208-214, ISSN 1350-2360

Fang, W.L. & Ngan, H.W. (2002). Succinct method for allocation of network losses, IEE Proc.-Gener. Transm. Distrib., Vol. 149, No. 2, (March 2002) 171–174, ISSN 1350-2360

Gajanayake, C.J.; Vilathgamuwa, D.M.; Loh, P.C.; Teodorescu, R. & Blaabjerg, F. (2009). Z-source-inverter-based flexible distributed generation system solution for grid power quality improvement. IEEE Trans. on Energy Conversion, Vol. 24, No. 3, (September 2009) 695-704, ISSN 0885-8969

Hadjsaid, N.; Canard, J.-F. & Dumas, F. (1999). Dispersed generation impact on distribution networks. IEEE Computer Applications in Power, Vol. 12, No. 2, (April 1999) 22-28, ISSN 0895-0156

Jenkins, N.; Allan, R.; Crossley, P.; Kirschen, D. & Strbac, G. (2000). Embedded generation, IEE Power and Energy Series 31, The IEE, ISBN 0-85296-7748, London, UK

Kersting, W.H. (2001). Distribution systems modeling and analysis. CRC Press, ISBN 0-8493-0812-7, Boca Raton, FL

Ko, S.-H.; Lee, S.R.; Dehbonei, H. & Nayar, C.V. (2006). Application of voltage- and current-controlled voltage source inverters for distributed generation systems. IEEE Trans. on Energy Conversion, Vol. 21, No. 3, (September 2006) 782-792, ISSN 0885-8969

Losi A.; Mangoni V,; Russo M. (1998). Optimal exploitation of generator-transformer units. IEEE Trans. on Energy Conversion, Vol. 13, No. 1, (March 1998) 90-95, ISSN 0885-8969

Madureira, A.G. & Peças Lopes, J.A. (2009). Coordinated voltage support in distribution networks with distributed generation and microgrids. IET Renewable Power Generation, Vol. 3, No. 4, (December 2009) 439-454, ISSN 1752-1416

Mutale, J.; Strbac, G.; Curcic, S. & Jenkins, N. (2000). Allocation of losses in distribution systems with embedded generation. IEE Proc.-Gener. Transm. Distrib., Vol. 147, No. 1, (January 2000) 7-14, ISSN 1350-2360

Nikkhajoei, H. & Lasseter, R.H. (2009). Distributed generation interface to the CERTS microgrid. IEEE Trans. on Power Delivery, Vol. 24, No. 3, (July 2009) 1598-1608, ISSN 0885-8977

Pepermans, G.; Driesen, J.; Haeseldonckx, D.; Belmans, R. & D'haeseleer, W. (2005). Distributed generation: definition, benefits and issues. Energy Policy, Vol. 33, No. 6, (April 2005) 787-798, ISSN 0301-4215

Pudjianto, D.; Ramsay, C. & Strbac, G. (2007). Virtual power plant and system integration of distributed energy resources. IET Renewable Power Generation, Vol. 1, No. 1, (March 2007) 10-16, ISSN 1752-1416

Roytelman, I. & Ganesan, V. (2000). Coordinated local and centralized control in distribution management systems, IEEE Trans. on Power Delivery, Vol. 15, No. 2, (April 2000) 718-724, ISSN 0885-8977

Senjyu, T.; Miyazato, Y.; Yona, A.; Urasaki, N. & Funabashi, T. (2008). Optimal Distribution Voltage Control and Coordination With Distributed Generation. IEEE Trans. on Power Delivery, Vol. 23, No. 2, (April 2008) 1236-1242, ISSN 0885-8977

Strezoski, V.C.; Katic, N.A. & Janjic, D.S. (2001). Voltage control integrated in distribution management system, Electric Power Systems Research, Vol. 60, No. 2, (December 2001) 85-97, ISSN 0378-7796

Taleski, R. & Rajicic, D. (1996). Energy summation method for energy loss computation in radial distribution networks. IEEE Trans. on Power Systems, Vol. 11, No. 2, (May 1996) 1104–1111, ISSN 0885-8950

Viawan, F.A. & Karlsson, D. (2007). Combined local and remote voltage and reactive power control in the presence of induction machine distributed generation. IEEE Trans. on Power Systems, Vol. 22, No. 4, November 2007, 2003-2012, ISSN 0885-8950

Vovos, P.N.; Kiprakis, A.E.; Wallace, A.R. & Harrison, G.P. (2007). Centralized and Distributed Voltage Control: Impact on Distributed Generation Penetration. IEEE Trans. on Power Systems, Vol. 22, No. 1, (February 2007) 476-483, ISSN 0885-8950

4

Voltage Variation Analysis of Normally Closed-Loop Distribution Feeders Interconnected with Distributed Generation

Tsai-Hsiang Chen[1], Wen-Chih Yang[2], Yi-Da Cai[1] and Nien-Che Yang[1]
[1] *Dept. of Electrical Engineering, National Taiwan University of Science and Technology*
[2] *Dept. of Electrical Engineering, Technology and Science Institute of Northern Taiwan*
Taiwan, Republic of China

1. Introduction

The Kyoto Protocol went into effect on February 16, 2005. The need to reduce greenhouse gases has led to growing worldwide interest in renewable energy generation, especially wind power. Due to the desire for more renewable energy, many small power sources have been hooked up to distribution systems. The penetration of distributed generation (DG) is fast increasing in distribution networks throughout the world, especially in Europe. It is predicted that DG will account for more than 25% of new generation being installed by 2010. The major part of the increasing DGs should be covered by wind power. Wind energy is a type of clean energy, produces no air pollution, and therefore has rapidly become the most competitive energy resource among the renewable energy resources. Wind Force 12 points out that 12% of the world's electricity needs will be from wind power by 2020. As outlined in GWEC's Global Wind 2008 Report, global wind energy capacity could reach more than 1,000 GW by the end of 2020. Wind power could produce about 2,600 TWh of electricity per year, which is to supply 10-12% of global electricity demand by 2020. And, this would save as much as 1,500 million tons of CO_2 every year.

Distribution feeders have various arrangements, examples of which are radial, loop, mesh, and spot network (Chen et al., 2004), (Huang & Chen, 2002) and (Lakervi & Holmes, 1995). In Taiwan, most primary distribution feeders are in radial arrangement due to the simple construction and low installation cost involved for this arrangement. However, the reliability of radial distribution feeders is too low to meet the critical customers, such as hospitals, skyscrapers, and factories having sensitive loads. In order to solve this problem, the utility here set to upgrade the arrangements of primary distribution feeders supplying critical customers from radial into normally closed-loop arrangement. Fig. 1 shows the schematic diagram of a distribution feeder in a normally closed-loop arrangement. Two radial distribution feeders supplied by the same main transformer are connected to each other at their ends by a tie breaker and a tie line. The tie breaker is normally closed. In normal operating conditions, the main transformer supplies its loads via two paths. Hence, a normally closed-loop distribution feeder does have a higher reliability than a radial one.

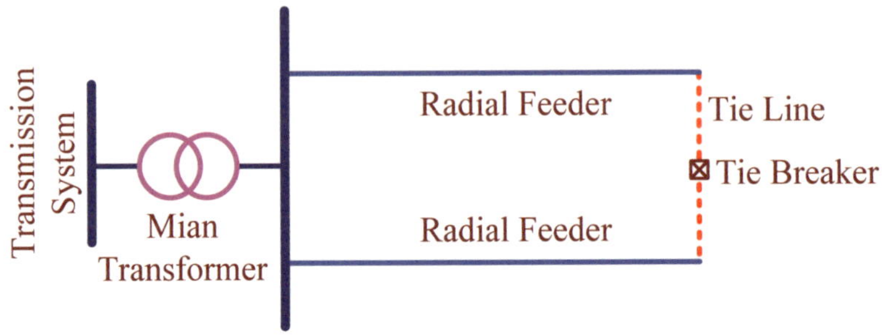

Fig. 1. Schematic diagram of a distribution feeder in a normally closed-loop arrangement

Furthermore, distribution generation has become a trend in Taiwan in recent years. More and more wind turbines, solar cells, small hydro generators, and biomass power plants are installed to generate green electric power and reduce greenhouse gas. In Taiwan, distribution generation sources (DGSs) are allowed to interconnect with the utility's distribution feeders so long as they obey the interconnection rules. When a DGS is interconnected with a normally closed-loop distribution feeder and starts generating electric power, the feeder voltages vary (Kojovic, 2002). Actually, the feeder voltages may vary largely. Once the feeder voltage deviates largely from the nominal voltage, this becomes harmful to the customers of the feeder. Hence, it is important for the utility's power engineers to understand the voltage variation of a normally closed-loop distribution feeder interconnected with DGSs. Hence, the impact of DG interconnection on the power system is a hot subject in recent years. In this chapter, the steady-state voltage variations of a normally closed-loop distribution feeder interconnected with a DGS are analyzed.

In this chapter, the voltage variation of a normally closed-loop distribution feeder interconnected with a distributed generation source is analyzed. First, the relevant background information of normally closed-loop distribution systems and DGS are introduced. Second, the interconnection rules for the steady-state voltage deviation caused by DG grid-connection in the US, Germany, and Denmark are introduced. Third, a model of a sample system involving a normally closed-loop distribution feeder and a DGS is constructed and the maximum allowable capacity of the DGS is evaluated. The relevant rules for DG interconnection in Taiwan are also introduced. Moreover, two formulas are presented, one for evaluating the voltage variation at the point of common coupling of a DGS connected to a distribution system, and another one for calculating the maximum allowable capacity of DGSs under a voltage variation limit. Forth, the definitions and purposes of each simulation scenario and sub-scenario are described. There are many factors affecting distribution feeder voltage variation. In this chapter, the major factors, such as interconnection location and the operating power factor of a DGS, plus the loading, power factor, and operating condition of a distribution feeder are taken into account. Finally, the steady-state voltage variation of a normally closed-loop distribution feeder interconnected with a DGS is analyzed. And, some valuable simulation results are summarized.

2. Interconnection rules for distributed generations

Table 1 outlines the requirements for the steady-state voltage deviation caused by DG grid-connection in the US, Germany, and Denmark. IEEE Std. 1547 states that the DR (distributed resource) unit shall parallel with the area electrical power system (Area EPS) without causing a voltage fluctuation at the point of common coupling (PCC) greater than ±5% of the prevailing voltage level of the Area EPS at the PCC, and meet the flicker requirements. The cumulative influence of the existing DR units parallel with the same Area EPS must be taken into account in the evaluated value of the voltage variation. The relevant codes of Germany and Denmark were established by considering single wind turbines and whole wind farms separately to bound the voltage variation at the PCC. Although the system characteristics, voltage levels and considerations are different from country to country, the requirements of maximum permissible steady-state voltage deviation caused by DG grid-connection are commonly bounded within 1 to 5% (IEEE 1547, 2003), (VDEW), (VDN, 2004), (Eltra: a, 2004) and (Eltra: b, 2004).

	Area, Regulation and Scope		Voltage Deviation
US	IEEE Std. 1547		± 5 %
Germany	VDEW	Medium voltage network	< 2 %
	VDN	Individual generating unit (wind turbine)	≤ 0.5 %
	VDN	Entire plant (wind farm)	≤ 2 %
	VDN	System faults	≤ 5 %
Denmark	DEFU	10 ~ 20kV grid	≤ 1 %
	Eltra Transmission grid	General constraint (wind farm)	< 3 %
		Until a freaquency of 10 per hour (wind farm)	< 2.5 %
		Until a freauency of 100 per hour (wind farm)	< 1.5 %
	Eltra Distribution grid	10 ~ 20kV grid (wind turbine)	≤ 1 %
		50 ~ 60kV grid (wind turbine)	≤ 3 %

Table 1. Overview of common requirements for voltage deviation

3. Description of the sample system

3.1 Structure of the sample system

Fig. 2 shows the one-line diagram of the sample system constructed by the present work.. The sample feeder is in a normally closed-loop arrangement. It consists of two radial feeders, A and B. The radial feeders A and B have lengths of 10 and 6 km, respectively, and the tie line is 1 km. Hence, the sample feeder has a length of 17 km in total. The impedance of the sample feeder is 0.0901+j 0.1325 Ω. Buses are set on the sample feeder at intervals of 1 km, and a lumped load is connected to each bus. Every bus has the same loading. Moreover, the rating of the main transformer is 60 MVA. The nominal voltages on the primary and secondary sides are 161 kV and 22.8 kV, respectively. Short-circuit capacity on the primary side is 7500 MVA.

There is a DGS interconnected with the sample feeder. The interconnection location and operating power factor of the sample DGS are changed according to the needs of the current study. The maximum allowable capacities of the sample DGS under different operating conditions are calculated and presented in Section 4.

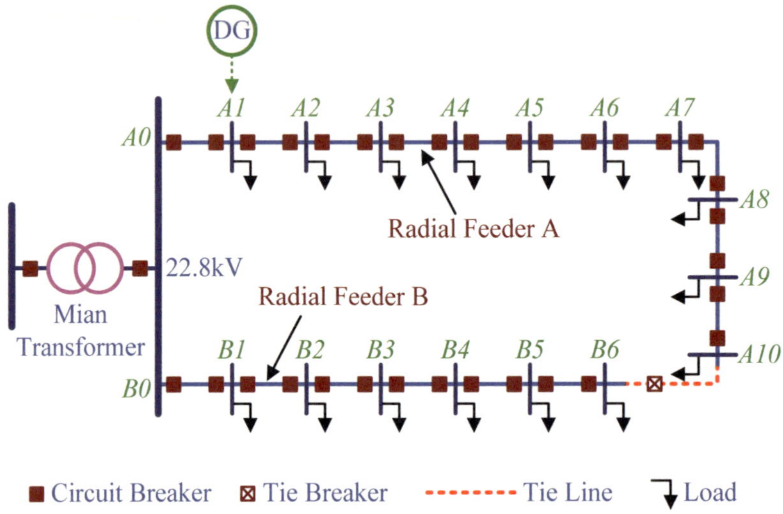

Fig. 2. One-line diagram of the sample system

3.2 Operating conditions of the sample feeder

A normally closed-loop distribution feeder can be operated under two conditions to supply its customers. The two conditions are normal and abnormal operating conditions. The normal condition is that when no fault occurred on the sample feeder. Hence, the sample feeder is operated normally and is kept in closed-loop arrangement. An abnormal condition occurs when a fault occurred on the sample feeder. Under this condition, the sample feeder cannot be kept in a closed-loop arrangement. It might become one or two radial feeders after the fault is cleared by the circuit breakers neighbouring it. In this chapter, the fault is assumed to have occurred on the segment of the sample feeder between buses A0 and A1. Under this condition, the sample feeder became a long radial feeder with the length of 16 km.

4. Calculation of the maximum allowable capacity of the sample DGS

4.1 Rules for DG interconnection

The interconnection of DGSs can surely affect the operation of a distribution system (Persaud et al., 2000), (Yang & Chen, 2005) and (Lopes, 2002). Hence, the maximum allowable capacity of a DGS should be limited in order to assure that the distribution system interconnected by it is safe. In Taiwan, the rules for DG interconnection in distribution systems have been made. Some of the rules related to this work are outlined in the following:

1. The voltage at the point of common coupling (PCC) of a DGS should not deviate more than ±2.5% while the DGS interconnects to the PCC.
2. The voltage profile along a distribution feeder should be kept within ±5% of the nominal voltage.
3. The maximum current generated by a DGS that flows into a distribution feeder should not exceed 300A.

4. The operating power factor of a DGS should be kept within 0.85 lagging and 0.95 leading.

4.2 Maximum allowable capacity of the sample DGS

Equation (1) is the formula for evaluating the voltage variation at the PCC of a DGS interconnected with a power system (Papathanassiou & Hatziargyriou, 2002). Rearranging this formula can obtain the formula for calculating the rated apparent power of a DGS under a voltage variation limit as shown in (2).

$$VR\% = \frac{S_G}{S_S} cos(\varphi_S + \varphi_G) \times 100 \tag{1}$$

where

VR is the voltage variation at PCC.
S_G is the apparent power of a DGS.
S_S is the short-circuit capacity at the PCC.
φ_S is the phase angle of the system impedance at PCC.
φ_G is the phase angle of the DGS's output power.

$$S_G = \frac{VR \times S_S}{cos(\varphi_S + \varphi_G)} \tag{2}$$

Based on (2) and the rules for DG interconnection as described above, the maximum allowable capacities of the sample DGS while the sample feeder operated under normal and abnormal operation conditions are calculated and are shown in Tables 2 and 3, respectively. These two tables represent that the maximum allowable capacity of a DGS will change with its interconnection location and operating power factor. Additionally, the maximum allowable capacity of a DGS under normal operating conditions is larger than the one under abnormal operating condition.

Interconnection Location	Operating Power Factor		
	0.85 lagging	1.0	0.95 leading
A1	11.91	12.00	12.00
A5	8.20	12.00	12.00
A10	7.60	12.00	12.00

Table 2. Maximum allowable capacities of the sample DGS under normal operating conditions of the sample feeder.

Interconnection Location	Operating Power Factor		
	0.85 lagging	1.0	0.95 leading
A1	3.23	5.13	10.00
A5	4.00	6.68	12.00
A10	4.55	7.85	12.00

Table 3. Maximum allowable capacities of the sample DGS under abnormal operating conditions of the sample feeder.

5. Definitions of the simulation scenarios

There are many factors affecting the voltages of a distribution feeder. The interconnection location and operating power factor of a DGS, and the loading and operating condition of a distribution feeder are the major factors. They are all taken into account in this chapter. In order to present the voltage variation of a normally closed-loop distribution feeder interconnected with a DGS completely, five simulation scenarios, in which each simulation scenario included four sub-scenarios, are selected and simulated. The definitions and purposes of each scenario and sub-scenario are described in the following.

5.1 The Simulation scenarios

The first scenario is used as a basic scenario. In this scenario, the sample feeder is operated with a normal load, and the loads along the sample feeder are uniformly distributed. The second scenario is used to present the effect of loading change on feeder voltage variation. In this scenario, the sample feeder is operated with a heavy load. The third scenario is used to present the effect of load power factor change on feeder voltage variation. In this scenario, the average load power factor of the sample feeder is changed to 0.8 lagging. The fourth and fifth scenarios are used to present the voltage variation of a normally closed-loop distribution feeder operated under abnormal conditions and with normal and heavy loadings, respectively.

Table 4 shows the operating conditions and load parameters of the sample feeder defined in each scenario. Observing this table in detail, the differences among the six scenarios can be easily determined.

Simulation Scenario Number	Parameters of the Sample Feeder		
	Total Load	Load Power Factor	Operating Condition
1	8MVA	0.95lagging	normal
2	12MVA	0.95lagging	normal
3	8MVA	0.80lagging	normal
4	8MVA	0.95lagging	abnormal
5	12MVA	0.95lagging	abnormal

Table 4. Simulation scenarios selected by this chapter

5.2 The Sub-scenarios

The four sub-scenarios are used to present the effect of the change in interconnection location and maximum allowable capacity of a DGS on distribution feeder voltages. In this chapter, the first sub-scenario is used to present the voltage variations along the sample feeder without DGS. Meanwhile, the second to fourth sub-scenarios are used to present the voltage variations along the sample feeder while the sample DGS was interconnected to the front, middle, and end, that is, bus A1, A5, and A10, of the sample feeder, respectively. For convenience, the four sub-scenarios are named "w/o DG", "A1," "A5," and "A10," respectively.

6. Analysis of feeder voltage variation

In this chapter, the power system simulation software CYME is used to simulate the sample system operated under various operating conditions as defined in the simulation scenarios. The simulation results are presented and discussed in the following.

6.1 Effect of the interconnection of a DGS

Figs. 3 to 5 show the voltage profiles along the sample feeder in scenario 1 while the sample DGS is operated with 1.0, 0.85 lagging, and 0.95 leading power factor, respectively. The three figures pointed out three phenomena. They are stated as follows:

1. The interconnection of the sample DGS made the voltages along the sample feeder go up. The maximum voltage variation appeared at the PCC of the sample DGS.
2. The voltage variation at PCC will become larger, while the electrical distance between the PCC of the sample DGS and the main transformer is becomes farther.
3. The voltage variations along the sample feeder are all within the limits of ±2.5% wherever the sample DGS is interconnected to.

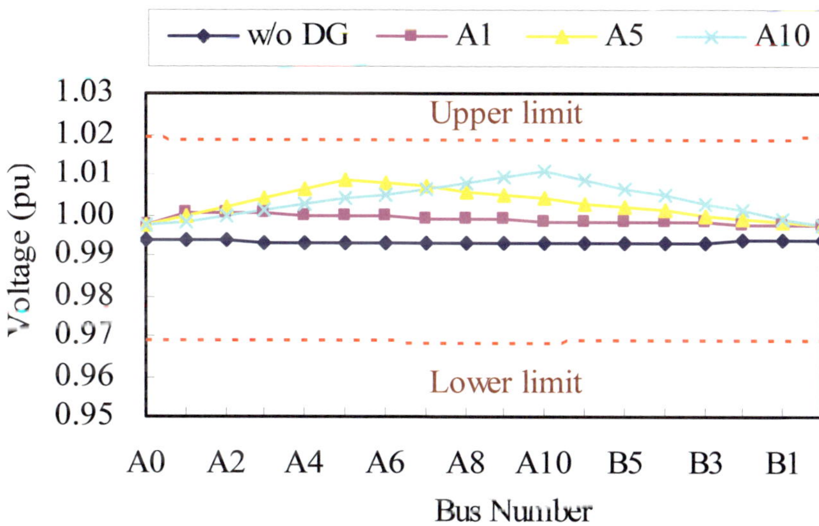

Fig. 3. Voltage profiles along the sample feeder in scenario 1 while the sample DGS is operated with 1.0 power factor

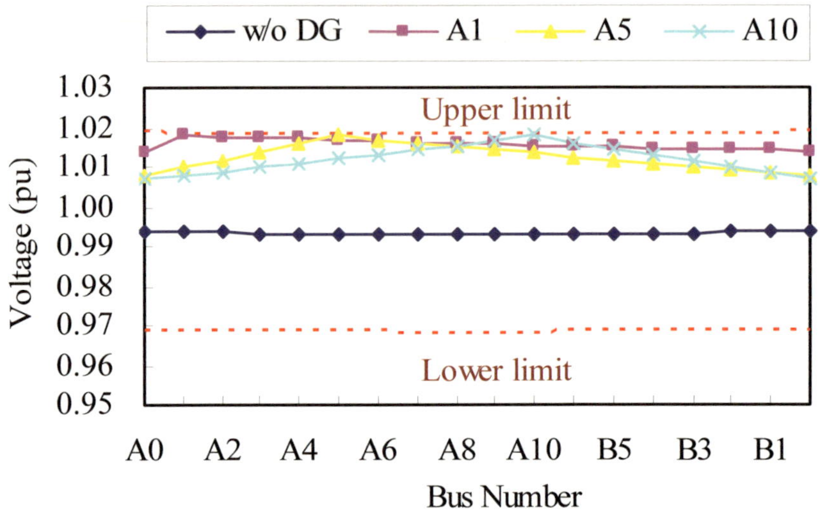

Fig. 4. Voltage profiles along the sample feeder in scenario 1 while the sample DGS is operated with 0.85 lagging power factor

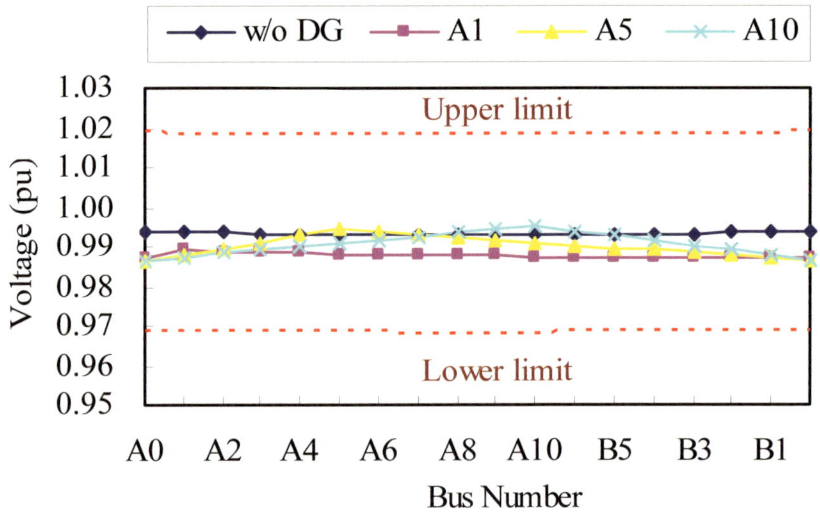

Fig. 5. Voltage profiles along the sample feeder in scenario 1 while the sample DGS is operated with 0.95 leading power factor

6.2 Effect of the change in feeder loading

Figs. 6 to 8 show the voltage profiles along the sample feeder in scenario 2, while the sample DGS is operated with 1.0, 0.85 lagging, and 0.95 leading power factor, respectively. The three figures pointed out two phenomena. They are stated as follows.

1. If the feeder loading is increased, the voltage variations along the sample feeder will be increased.

2. The voltage variation at PCC of the sample DGS will exceed the upper limit while the sample DGS is operated with 0.85 lagging power factor. The results demonstrate that the evaluation of the maximum allowable capacity of a DGS as described in Section 4 is not very precise because (2) does not involve the uncertainty of feeder loading.

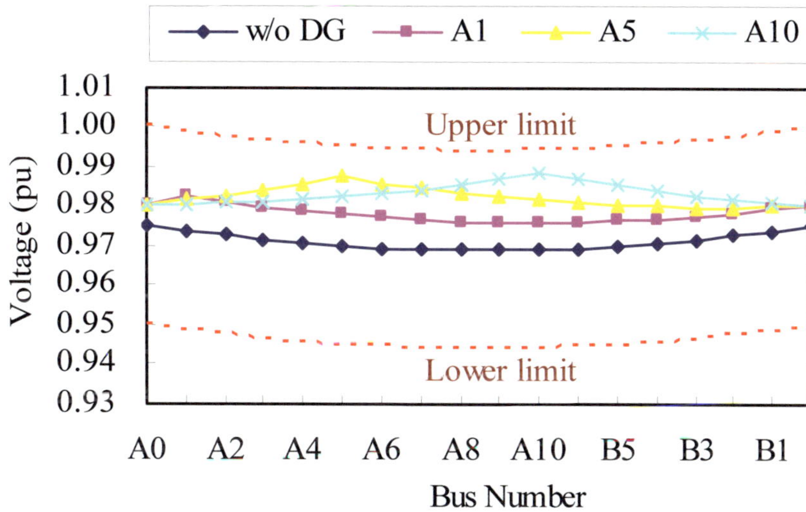

Fig. 6. Voltage profiles along the sample feeder in scenario 2 while the sample DGS is operated with 1.0 power factor

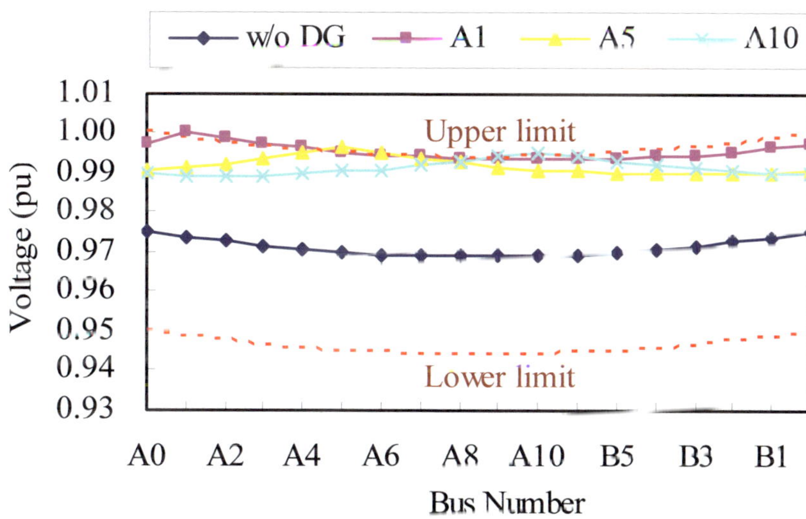

Fig. 7. Voltage profiles along the sample feeder in scenario 2 while the sample DGS is operated with 0.85 lagging power factor

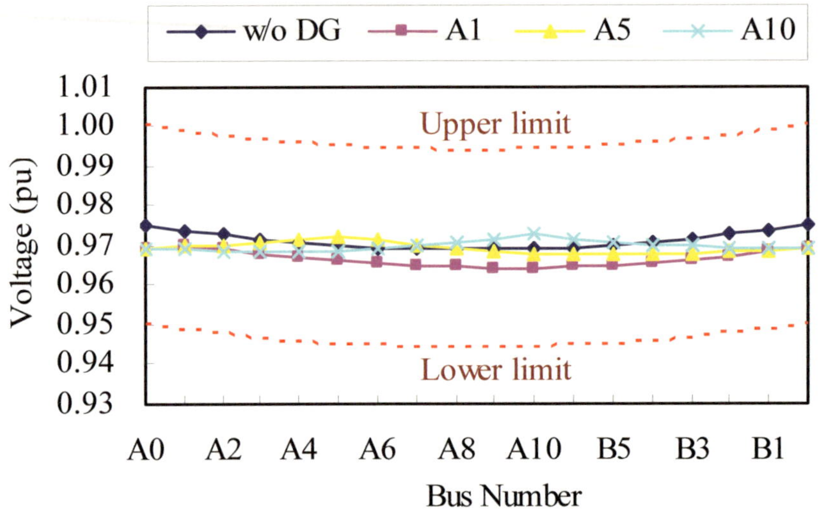

Fig. 8. Voltage profiles along the sample feeder in scenario 2 while the sample DGS is operated with 0.95 leading power factor

6.3 Effect of the change in load power factor

Figs. 9 to 11 show the voltage profiles along the sample feeder in scenario 3 while the sample DGS is operated with 1.0, 0.85 lagging, and 0.95 leading power factor, respectively. Comparing the three figures with Figures 3 to 5, we can find that the load power factor do not have much effect on the voltage variations along the sample feeder. This means that the voltage variation of a normally closed-loop distribution feeder does not exceed the voltage variation limits so long as the load power factors of feeder are kept in reasonable values.

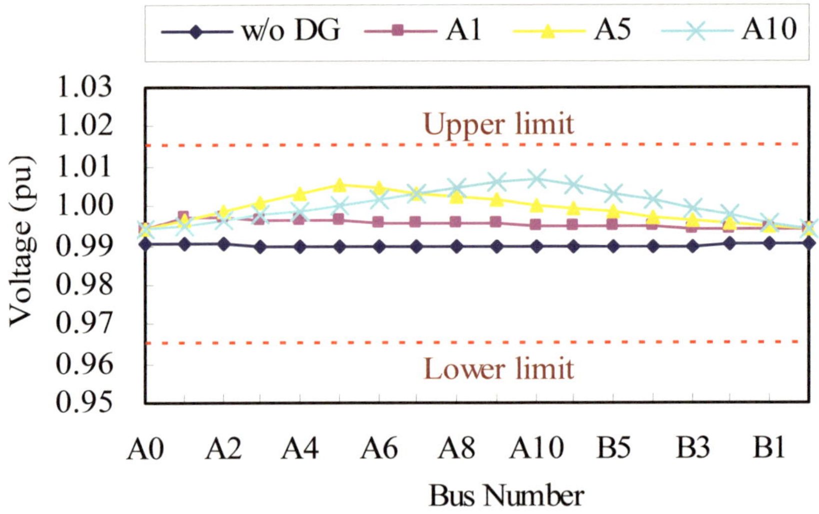

Fig. 9. Voltage profiles along the sample feeder in scenario 3 while the sample DGS is operated with 1.0 power factor

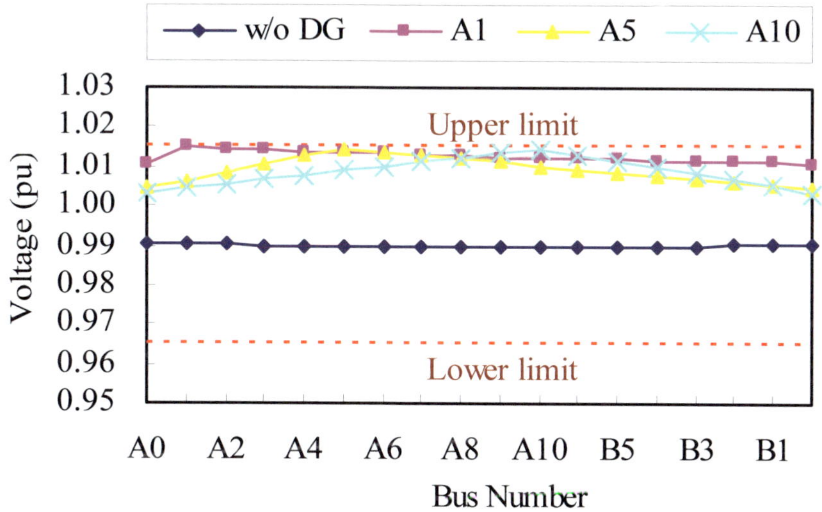

Fig. 10. Voltage profiles along the sample feeder in scenario 3 while the sample DGS is operated with 0.85 lagging power factor

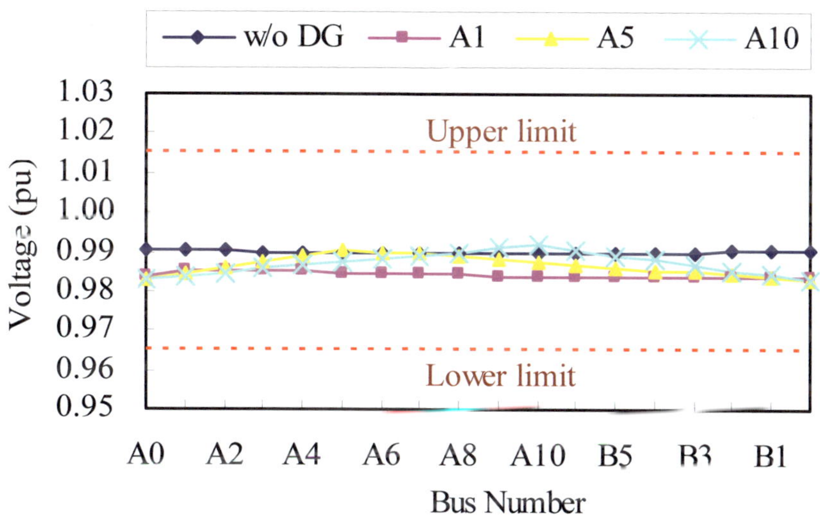

Fig. 11. Voltage profiles along the sample feeder in scenario 3 while the sample DGS is operated with 0.95 leading power factor

6.4 Effect of the change in feeder operating condition

Figs. 12 to 14 show the voltage profiles along the sample feeder in scenario 4 while the sample DGS is operated with 1.0, 0.85 lagging, and 0.95 leading power factor, respectively. In this scenario, the sample feeder is under abnormal operating conditions and becomes a long radial feeder. Under this circumstance, its voltage profiles may change largely. Fig. 13 shows that the voltages along the sample feeder have exceeded the voltage tolerance limits of ±5% while the sample DGS is interconnected to bus A1 and operated with 0.85 lagging power factor. The reason is that the sample DGS is interconnected to the end of the sample feeder and produced a lot of reactive power into the sample feeder.

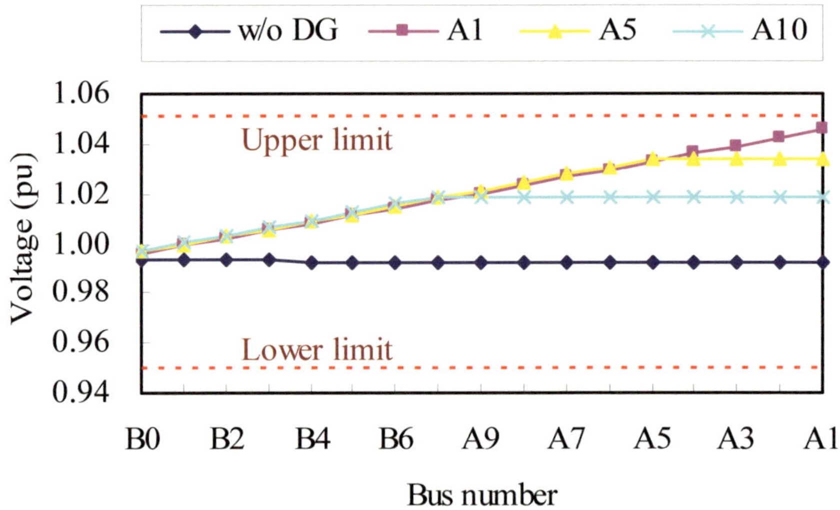

Fig. 12. Voltage profiles along the sample feeder in scenario 4 while the sample DGS is operated with 1.0 power factor

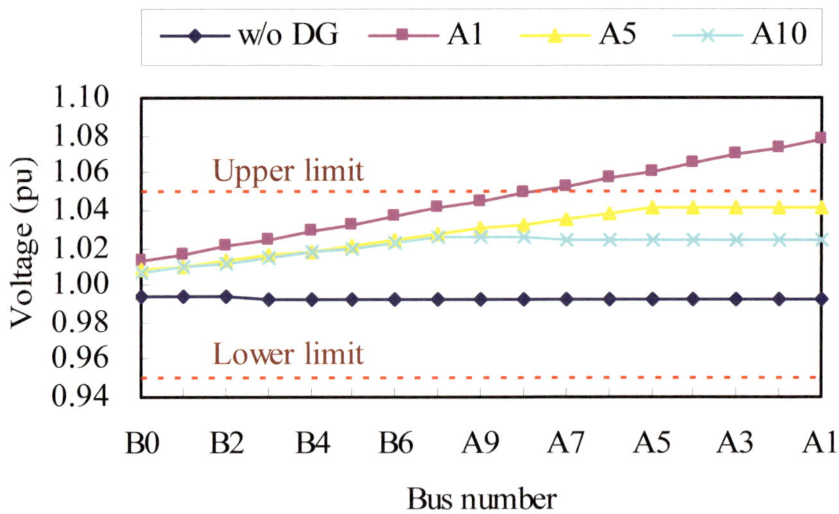

Fig. 13. Voltage profiles along the sample feeder in scenario 4 while the sample DGS is operated with 0.85 lagging power factor

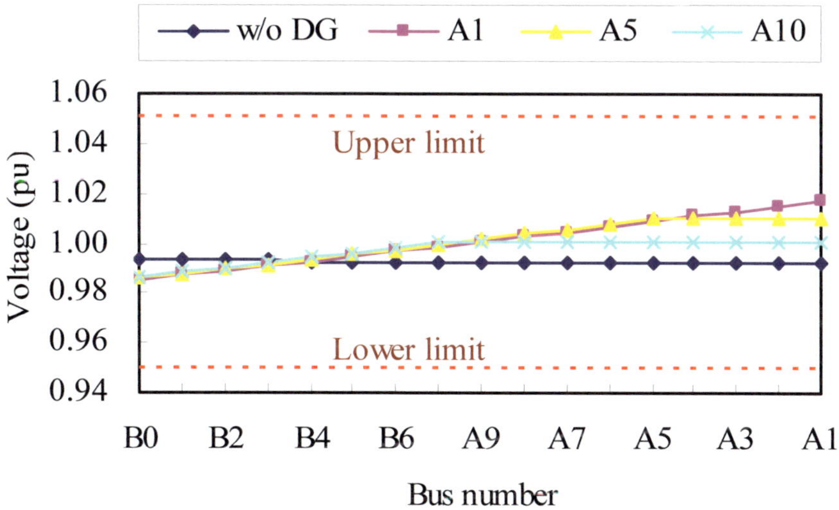

Fig. 14. Voltage profiles along the sample feeder in scenario 4 while the sample DGS is operated with 0.95 leading power factor

Figs. 15 to 17 show the voltage profiles along the sample feeder in scenario 5 while the sample DGS is operated with 1.0, 0.85 lagging, and 0.95 leading power factor, respectively. In this scenario, the sample feeder is still operated under abnormal conditions. However, the voltages along the sample feeder do not exceed the voltage tolerance limits of ±5%. The reason is that the sample feeder is operated with heavy loading. The reactive power produced by the sample DGS is all absorbed by the sample feeder loads.

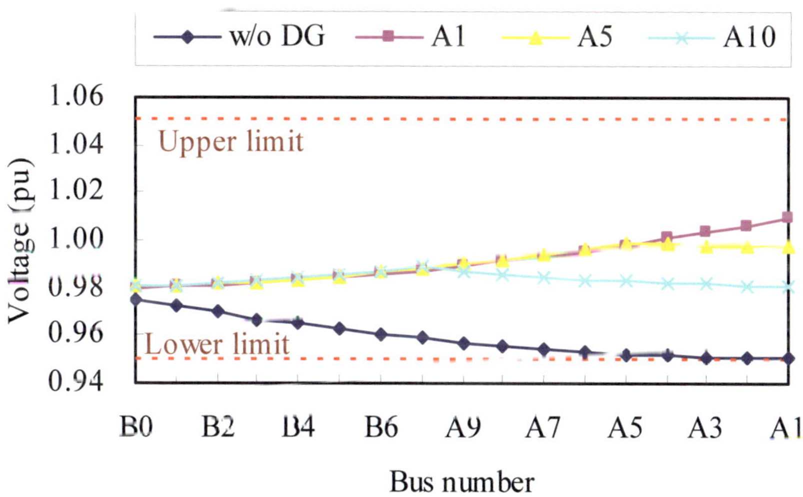

Fig. 15. Voltage profiles along the sample feeder in scenario 5 while the sample DGS is operated with 1.0 power factor

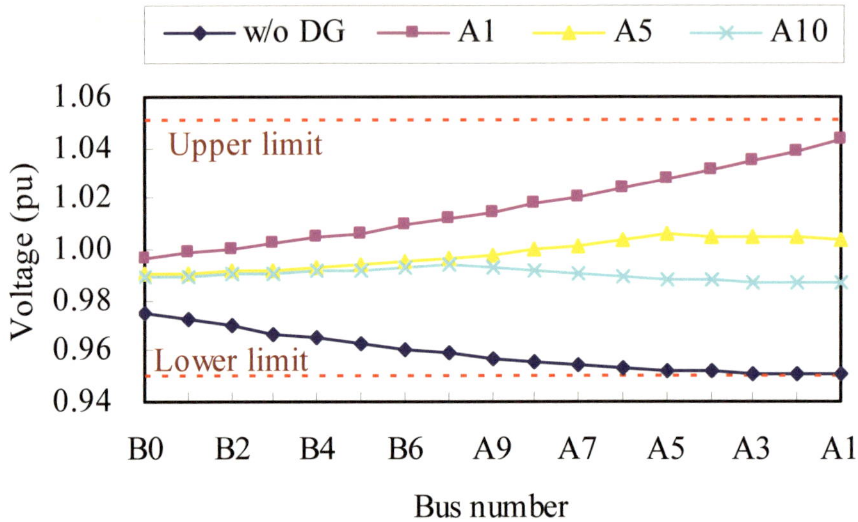

Fig. 16. Voltage profiles along the sample feeder in scenario 5 while the sample DGS is operated with 0.85 lagging power factor

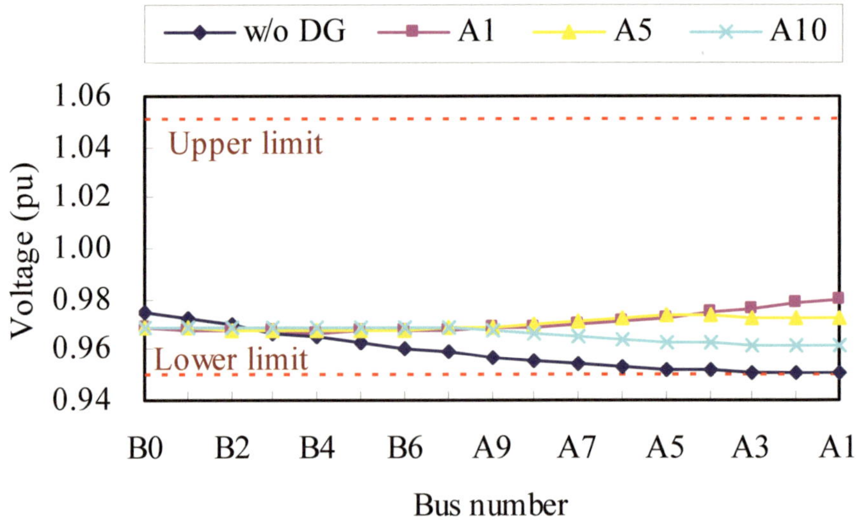

Fig. 17. Voltage profiles along the sample feeder in scenario 5 while the sample DGS is operated with 0.95 leading power factor

7. Conclusions

The steady-state voltage variation of a normally closed-loop distribution feeder interconnected with a DGS has been analyzed in this chapter. The simulation results of the sample system are also presented in this chapter. Some important results are summarized in the following:

1. The interconnection location of a DGS is an important factor affecting the voltage variation of a normally closed-loop distribution feeder. The voltage variation at the PCC increases with the electric distance of a DGS from the main transformer to the PCC.
2. When a DGS is operated with lagging power factor, reactive power is produced. A DGS with low lagging power factor can enlarge the voltage variation of the distribution feeder interconnected by it.
3. The loading of a normally closed-loop distribution feeder is an important factor affecting its voltage variation as well. When a normally closed-loop distribution feeder is operated with heavy loading, the voltage variation along the feeder may exceed voltage variation limits, especially when the DGS is operated with lagging power factor.
4. When a normally closed-loop distribution feeder is operated in a normal condition and the interconnection location of a DGS is nearer to the main transformer, the voltage variation along the feeder is smaller. In contrary, when a normally closed-loop distribution feeder is operated in an abnormal condition and the interconnection location of a DGS is nearer to the end of the feeder, the voltage variation along the feeder is larger.

Upgrading distribution feeders from radial into a normally closed-loop arrangement is an important measure to heighten power supply reliability. However, the interconnections of DGSs make the voltage variations of normally closed-loop distribution feeders to become complex. The results of this chapter are particularly useful to power engineers so they can have a better understanding of the effects of DGSs on the voltages of normally closed-loop distribution feeders and so that they can be helped also in operating their distribution systems safely.

8. Acknowledgements

The authors would like to thank the National Science Council of the Republic of China, Taiwan, for financially supporting this research under Contract No. NSC 95-2221-E-149 - 018-MY2.

9. References

Chen, T.H.; Huang, W.T.; Gu, J.C.; Pu, G.C.; Hsu, Y.F. & Guo, T.Y. (2004). Feasibility study of upgrading primary feeders from radial and open-loop to normally closed-loop arrangement, *IEEE Transactions on Power Systems*, Vol. 19, No. 3, (August 2004), (1308-1316), ISSN 0885-8950.

Eltra: a (2004). Wind turbines connected to grids with voltages below 100 kV, May 2004.

Eltra: b (2004). Wind turbines connected to grids with voltages above 100 kV, November 2004.

Huang, W.T. & Chen, T.H. (2002). Assessment of upgrading existing primary feeders from radial to normally closed-loop arrangement, *Proceedings of IEEE Transmission and Distribution Conference and Exhibition 2002: Asia Pacific*, 2123-2128, ISBN 0-7803-7525-4, Tokyo, Japan, October 2002, IEEE, New York, USA.

IEEE 1547 (2003). IEEE Standard for Interconnecting Distributed Resources with Electric Power System, 2003.

Kojovic, L. (2002). Impact of DG on voltage regulation, *Proceedings of IEEE Power Engineering Society Summer Meeting*, 97-102, ISBN 0-7803-7518-1, Chicago, USA, July 2002, IEEE, New York, USA.

Lakervi, E. & Holmes, E.J. (1995). *Electricity distribution network design*, Institution of Electrical Engineers, ISBN 0863413099, London, UK.

Lopes, J.A.P. (2002). Integration of dispersed generation on distribution networks - impact studies, *Proceedings of IEEE Power Engineering Society Winter Meeting*, 323-328, ISBN 0-7803-7322-7, Columbus, USA, January 2002, IEEE, New York, USA.

Papathanassiou, S.A., & Hatziargyriou N.D. (2002). Technical requirements for the connection of dispersed generation to the grid, *Proceedings of IEEE Power Engineering Society Summer Meeting*, 749-754, ISBN 0-7803-7173-9, Vancouver, Canada, July 2002, IEEE, New York, USA.

Persaud, S.; Fox, B. & Flynn, D. (2000). Impact of remotely connected wind turbines on steady state operation of radial distribution networks, *IEE Proceedings-Generation, Transmission and Distribution*, Vol. 147, No. 3, (May 2000), (157-163), ISSN 1350-2360.

VDEW, Guidelines for the parallel operation of own energy generation systems with the middle voltage grid of the utility company.

VDN (2004). REA generating plants connected to the high and extra-high voltage network, August 2004.

Yang, W.C. & Chen, T.H. (2005). Analysis of interconnection operation of a radial feeder with a cogeneration plant, *Proceedings of IEEE Power Engineering Society General Meeting*, 1815-1820, ISBN 0-7803-9156-X, San Francisco, CA, June 2005, IEEE, New York, USA.

5

Effect of DG on distribution grid protection

Edward Coster
Stedin, Eindhoven University of Technology
The Netherlands

Johanna Myrzik and Wil Kling
Eindhoven University of Technology
The Netherlands

1. Introduction

Up till recently past electric power systems were characterized by centralized production units, a high voltage transmission grid for the bulk energy transmission and medium and low voltage distribution grids to bring the energy to the consumer. Traditionally no generation sources were connected to the distribution grid, however, this has changed significantly the past decade. Nowadays various types of small generation sources, better known as distributed generation (DG), are connected to the distribution grid. Due to CO_2 reduction goals many of the small units integrated in the distribution grid are renewable energy sources, such as wind turbines, small scale hydro plants and photovoltaic panels but also high efficient non-renewable energy sources, such as small Combined Heat and Power (CHP) plants are implemented.
Connection of DG not only alters the load flow in the distribution grid but can also alter the fault current during a grid disturbance. Most distribution grid protective systems detect an abnormal grid situation by discerning a fault current from the normal load current. Because DG changes the grid contribution to the fault current, the operation of the protective system can be affected. This is reported in (Deuse et al., 2007; Doyle, 2002; Kauhamieni & Kumpulainen, 2004), however, in these papers the protection problems are discussed in general terms. In this chapter a detailed analysis of possible protection problems is given. It starts with an analytical description of fault currents in distribution grids including DG. With the aid of the analytical equations the effect of DG on the fault current is studied and key parameters are identified. This chapter also provides an equation to calculate the location where the DG-unit has the most effect on the grid contribution to the fault current. During the design stage of the protective system for a distribution feeder including DG this equation can be applied to determine if protection problems are to be expected. The application of the derived equations are demonstrated on a generic test feeder.
An overview of all possible protection problems is presented and a classification of the protection problems is given. Furthermore these protection problems are linked to the theoretical background which is discussed in the beginning of the chapter. In this part of the chapter solutions for the possible protection problems are presented as well as new developments in protective systems which enables a further integration of DG in distribution grids.

The chapter ends with a case study on a benchmark network which demonstrates the fault detection problem. Dynamic simulations show how the fault detection problems arises and what remedies can be taken to prevent these.

2. Fault currents in faulted distribution feeders including DG

The connection of DG to distribution feeders changes the fault currents in faulted feeders. The rate of change of the fault currents strongly depends on the ability of the DG to contribute to the fault current. DG based on an asynchronous generator does not provide a sustainable fault current during a grid-disturbance. The same holds mostly for inverter-connected DG such as micro-turbines, fuel-cells and PV-systems, from which the fault current contribution can be neglected (Jenkins et al., 2000; Morren & de Haan, 2008). However, in (Baran & El-Markabi, 2004) it is demonstrated that in weak systems, during a high resistive fault, inverter-connected DG although, change the grid contribution to the fault current. This is also reported in (Kumpulainen et al., 2005). The reference (Baran & El-Markabi, 2005) proposes for weak systems an extension of the conventional fault analysis method to include the effect of inverter-connected DG. A generator type that contributes a sustainable fault current is the synchronous generator (Jenkins et al., 2000). These type of generators can be found in small combined heat and power plants. In this section the effect of synchronous generators on the grid contribution to the fault current is considered.

2.1 Theoretical background

To analyse the effect of DG on the fault current in a feeder, a generic feeder is taken as a reference as shown in figure 1. At distance d a DG-unit is connected and at the end of the feeder a three-phase fault is present. For the analysis it is convenient to use a distance parameter to indicate the location of the DG which is relative to the total feeder length. This parameter is defined as:

$$l = \frac{d}{d_{tot}} \tag{1}$$

In equation (1) d is the distance to the DG-unit and d_{tot} is the total feeder length.

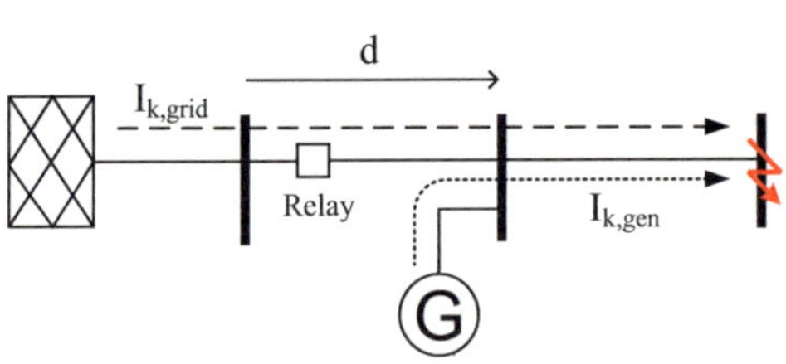

Fig. 1. Short-circuit current contribution of both grid and DG-unit

An electric equivalent of the feeder shown in figure 1 is given in figure 2. In this figure Z_L is the total line-impedance, Z_g the generator-impedance and Z_s the source-impedance. The voltages of the grid and generator are denoted as U_s and U_g.

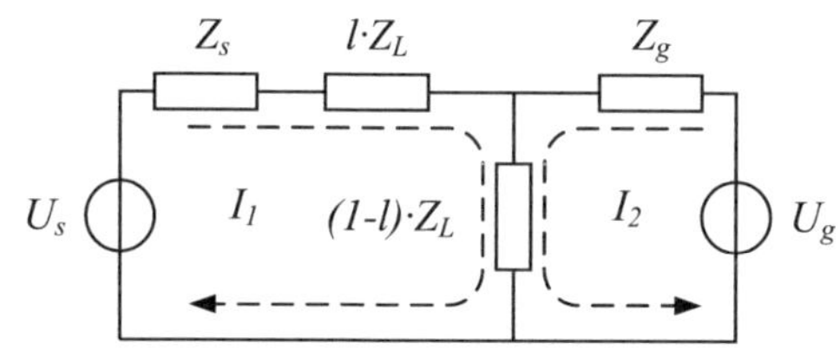

Fig. 2. Network equivalent of figure 1

Defining the mesh currents I_1 and I_2 and applying the Kirchhoff Voltage Law (KVL) for U_s and U_g can be found:

$$\begin{bmatrix} U_s \\ U_g \end{bmatrix} = \begin{bmatrix} Z_s + Z_L & (1-l)\cdot Z_L \\ (1-l)\cdot Z_L & Z_g + (1-l)\cdot Z_L \end{bmatrix} \cdot \begin{bmatrix} I_1 \\ I_2 \end{bmatrix} \tag{2}$$

In figure 2 and equation (2) I_1 is the grid contribution $I_{k,grid}$, and I_2 is the DG-contribution, $I_{k,gen}$, to the total fault current. An analytical expression for I_1 and I_2 can be found by solving equation (2). Because of the strong relation with the IEC60909 fault-analysis method, in this chapter Thevenin's Theorem is applied on the network of figure 2 to find a analytical expression for $I_{k,grid}$ and $I_{k,gen}$. In figure 3 the Thevenin equivalent of the network of figure 2 is shown.

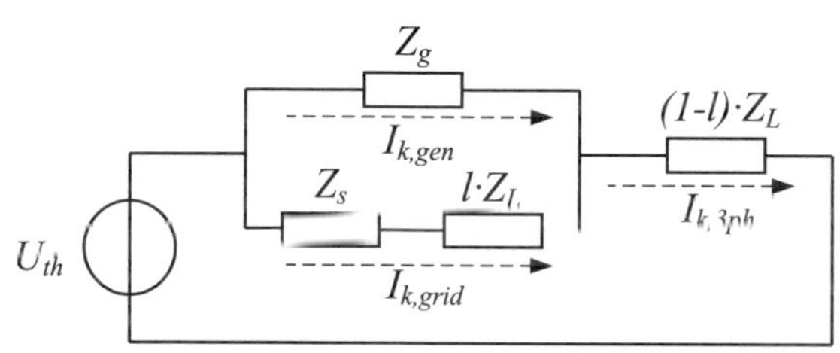

Fig. 3. Thevenin equivalent of figure 1

For this figure the Thevenin impedance is:

$$Z_{th} = \frac{(Z_s + l\cdot Z_L)\cdot Z_g}{Z_s + l\cdot Z_L + Z_g} + (1-l)\cdot Z_L \tag{3}$$

In equation (3) $Z_s = jX_s$ is the grid impedance, $Z_g = jX_g$ is the generator impedance and $Z_L = R_L + jX_L$ is the total line or cable impedance. l is the relative generator location as defined in equation (1). The total three-phase short circuit current can be calculated by:

$$I_{k,3ph} = \frac{U_{th}}{\sqrt{3}\cdot Z_{th}} \tag{4}$$

Combining equation (3) and equation (4) yields:

$$I_{k,3ph} = \frac{U_{th}\cdot(Z_g + l\cdot Z_L + Z_s)}{\sqrt{3}\left[(Z_L\cdot Z_g + Z_s\cdot Z_g + Z_s\cdot Z_L) + l\cdot Z_L(Z_L - Z_s) - l^2 {Z_L}^2\right]} \tag{5}$$

For the grid contribution holds:

$$I_{k,grid} = \frac{Z_g}{(Z_g + l \cdot Z_L + Z_s)} \cdot I_{k,3ph} \quad (6)$$

Substituting equation (5) in equation (6) gives for the grid contribution:

$$I_{k,grid} = \frac{U_{th} \cdot Z_g}{\sqrt{3}\left[(Z_L \cdot Z_g + Z_s \cdot Z_g + Z_s \cdot Z_L) + l \cdot Z_L(Z_L - Z_s) - l^2 {Z_L}^2\right]} \quad (7)$$

The total short-circuit current, $I_{k,3ph}$, is determined by equation (5) which is a non-lineair equation, so $I_{k,grid}$ is non-lineair as well. In case of a weak grid, Z_s can be as large as Z_g and due to the contribution of the generator, the grid contribution to the short-circuit current decreases.

2.2 Simulation of a 3-bus test network

In the previous section equation (7) describes the grid contribution to the fault current in a distribution feeder including a synchronous generator. This equation shows that the grid contribution will be determined by the total feeder impedance, the local short-circuit power at the substation, the generator size and location. To determine the impact of the synchronous generator on the short-circuit current a 3-bus test network is defined and modeled in simulation software. The test grid consists of an external grid, three MV-nodes which are connected by two connections. At busbar 2 a synchronous generator is connected. The test network is depicted in figure 4 and is used to illustrate the theoretical background.

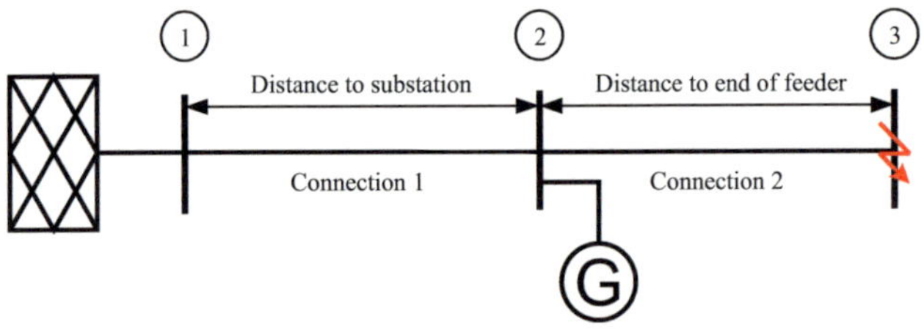

Fig. 4. 3-bus network

To demonstrate the effect of the generator size and its location in the test network these two parameters are modified. For that repetitive calculations have to be performed. The calculations are executed for a regular cable and an overhead line type from what the parameters are shown in table 1.

Type	R [Ω/km]	X [Ω/km]	$I_{nom}[A]$
XLPE 630 mm^2 Al	0.063	0.109	575
DINGO 19/.132	0.218	0.311	525

Table 1. Cable and overhead line parameters of the 3-bus test network

The length of both connections is adjusted by 10% of the total feeder length. To keep an equal total feeder length, connection 1 is increasing and connection 2 is decreasing with the same

step size. In this way node 2 is shifting from node 1 to node 3 and the effect of the location of the generator can be observed. After the modification of the length of the connections a three-phase fault calculation at busbar 3 is performed. This fault calculation is based on the well known IEC 60909 method and per step size the grid contribution is stored. For each step the generator size is increased with steps of 2 MW with a maximum of 10 MW.
The result is shown in figure 5 and it can be seen clearly that the generator has an effect on the grid contribution to the fault current. As expected, large size generators influence the grid contribution more than small size generators.

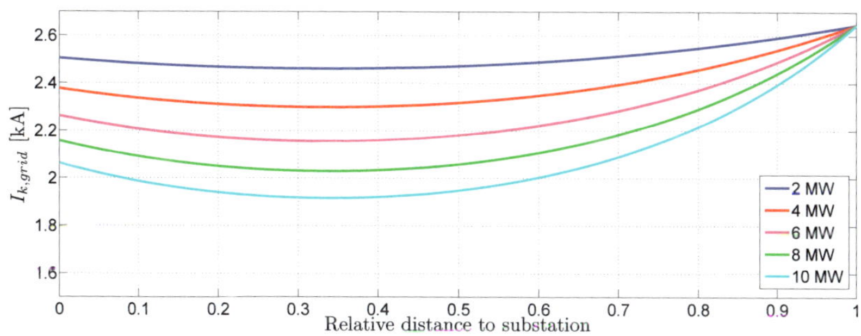

(a) Results for the cable case

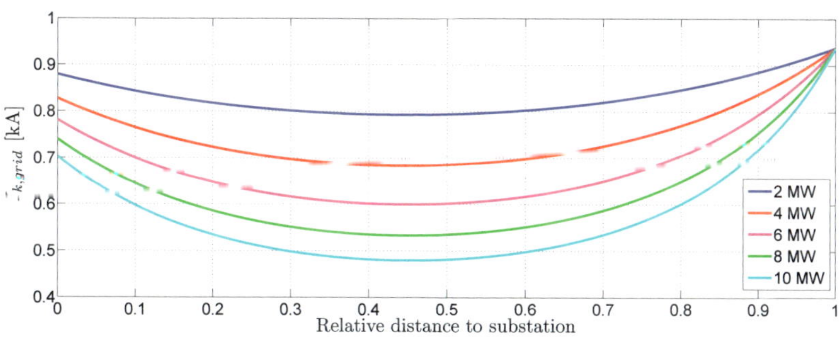

(b) Results for the overhead line case

Fig. 5. Simulation results for the 3-bus test network

Due to the difference in design the impedance of a cable system differs from the impedance of overhead lines. In table 1 it can be seen that the impedance of the overhead line is larger than the impedance of the cable. Hence the results in figure 5 show that the grid contribution to the fault current is affected most when the generator is connected to an overhead line feeder.

2.3 Maximum generator impact

The maximum DG impact on the grid contribution to the short-circuit current occurs when the grid contribution is the minimum. Hence the minimum of equation (7) has to be determined. This is done by taken the derivative of equation (7). This leads to:

$$\frac{dI_{k,grid}}{dl} = \tag{8}$$

$$\frac{jX_g\left(R_L^2 - X_L(X_L - X_s) - 2l(R_L^2 - X_L^2) + j(R_L(2X_L - X_s))\right)}{-X_g\left(X_L + X_sR_L\right) - l^2(R_L^2 - X_L^2) + j\left(X_g(R_L - X_s(X_g - X_L)) + lR_L(2X_L - X_s) + 2l^2R_LX_L\right)^2}$$

The minimum of $I_{k,grid}$ can be found with:

$$\frac{dI_{k,grid}}{dl} = 0 \tag{9}$$

Which yields for l:

$$l = \frac{1}{2} \cdot \frac{R_L^2 - X_L(X_L - X_s) + jR_L(2X_L - X_s)}{(R_L^2 - X_L^2) - 2jX_LR_L} \tag{10}$$

With this equation the location of maximum generator impact can be calculated which can be helpful at the planning stage.

2.4 Definitions used in distribution grid protection

The goal of a protective system is to recognize certain system abnormalities which, if undetected, can lead to damage of equipment or extended loss of service (Anderson, 1999). The protective system takes corrective actions for instance isolating a faulted component of the system and restoring the rest of the grid to normal operating conditions. Two important aspects of protective systems are:

1. Reliability
2. Security

Reliability is the probability that the system will function correctly when required to act. Security is the ability of a system to refrain from unnecessary operations. The optimal protection settings are a trade-off between reliability and security. Improving the reliability of a protection scheme by applying more sensitive settings can lead to a reduction of the security of the protection scheme. These definitions will be applied throughout the chapter and it will be discussed how reliability and security is influenced when DG is integrated in the distribution grid.

3. Protection problems

Connection of small generators to distribution grids is not new at all. But in the recent past the number of small generators has been increased rapidly and the effect on distribution grid operation has become noticeable. Concerns have arisen if the distribution system including distributed generation is still protected properly. In (Mäki et al., 2004) it is stated that protection issues might become one of the biggest technical barriers for wide-scale integration of distributed generation in the Nordic distribution grids. Extensive research is done to address possible protection problems in distribution grids including distributed generation. For instance, in (Deuse et al., 2007; Doyle, 2002; Driesen & Belmans, 2006; Driesen et al., 2007; Hadjsaid et al., 1999; Kauhamieni & Kumpulainen, 2004) it is discussed that the following protection problems might appear:

- Blinding of protection
- False tripping
- Lost of fuse-recloser coordination
- Unsynchronized reclosing
- Prohibition of automatic reclosing

These problems strongly depend on the applied protective system and consequently on the type of distribution grid. Blinding of protection and false tripping are protection problems which can occur in distribution grids built of cables as well as overhead lines while fuse-recloser coordination problems and recloser problems only appear in distribution grids which (partly) consist of overhead lines. In general the mentioned protection problems can be divided into two categories:

1. Fault detection problems
2. Selectivity problems

In the next subsections all mentioned protection problems will be categorized in these two categories and will be discussed in detail. In the description the protection problems are also linked to the theoretical background.

3.1 Blinding of protection

As discussed and demonstrated in section 2 the grid contribution to the total fault current will be reduced because of the contribution of distributed generation. Due to this reduction it is possible that the short-circuit stays undetected because the grid contribution to the short-circuit current never reaches the pickup current of the feeder relay. Overcurrent relays as well as directional relays and reclosers rely their operation on detecting an abnormal current. Hence, all protective systems based on these protection devices can suffer malfunctioning because of the reduced grid contribution. This mechanism is called blinding of protection and belongs to the first category of protection problems.
In (Chilvers et al., 2004; 2005) distance protection is applied to increase the amount of distributed generation connected to the distribution grid. Distance protection is a zone protection and the protected feeder is divided into a number of zones. The first zone covers approximately 85% of the line length while zone 2 and 3 are used for the rest of the line length and as a backup protection for subsequent distance protections. Faults in zone 2 and 3 are cleared with a time delay in order to obtain selectivity with the subsequent distance protections. This protection type acts more or less independent of the size of the fault current. However, due to the reduced grid contribution the impedance calculated to the fault location will increase and causes protection underreach. Faults normally cleared in zone 1 might then be cleared in zone 2 with subsequently a longer fault clearing time. The seriousness of this problem depends on local short-circuit power, X/R ratio of the distribution feeder and size of the generator which are the key parameters mentioned in section 2.
It can be concluded that distributed generation with a relevant contribution to the fault current directly affects the sensitivity of a protective system and therefore the reliability of the protective system.

3.2 False tripping

False tripping, also known as sympathetic tripping, is possible when a generator which is installed on a feeder, contributes to the fault in an adjacent feeder connected to the same substation. The generator contribution to the fault current can exceed the pick-up level of the overcurrent protection which can lead to a trip of the healthy feeder before the actual fault is cleared. This mechanism can be categorized to the category of selectivity problems. In figure 6 the principle of false tripping is shown schematically.

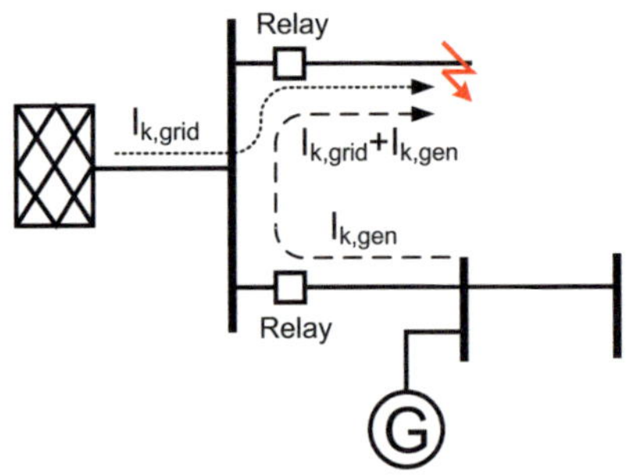

Fig. 6. Principle of false tripping

The generator has a major contribution to the fault current when the generator and/or the fault are located near the substation. Especially in weak grids with long feeder length which are protected by definite overcurrent relays false tripping can occur. In this case the settings of the protection relays have to ensure that faults at the end of the feeder are also detected which lead to a relatively small pick-up current. Here DG affects the security of the protective system.

In (Kauhamieni & Kumpulainen, 2004) it is discussed that in some cases false tripping can be prevented by finding another suitable relay setting. Practically it means that the fault clearing time has to be increased rather than the pick-up current. Increase of the pick-up current results in a less sensitive feeder protection and probably not all faults will be cleared anymore. Hence, the security of the protective system increases, but the reliability of the protective system decreases. Changing the fault clearing time lead to the disconnection of the faulted feeder first and prevent the healthy feeder from false tripping. When selectivity cannot be reached by changing the protection settings the application of directional overcurrent protection can solve the problem (Kumpulainen & Kauhaniemi, 2004). However, directional protection is slower, more expensive and usually not the standard solution of grid operators.

3.3 Recloser problems

Protection of overhead distribution feeders with automatic reclosers is a very efficient way to protect against temporary disturbances and minimize the number of supply interruptions. Because of the coordination between the reclosers and the lateral fuses permanent faults are cleared in a selective way. Connection of DG to these type of feeders causes several protection problems at the same time. First of all the fault current detection by the recloser is affected by the generator contribution and can lead to a detection problem. Secondly the coordination between reclosers or fuse and recloser can be lost which directly causes selectivity problems. This is explained in more detail with the feeders shown in figure 7.

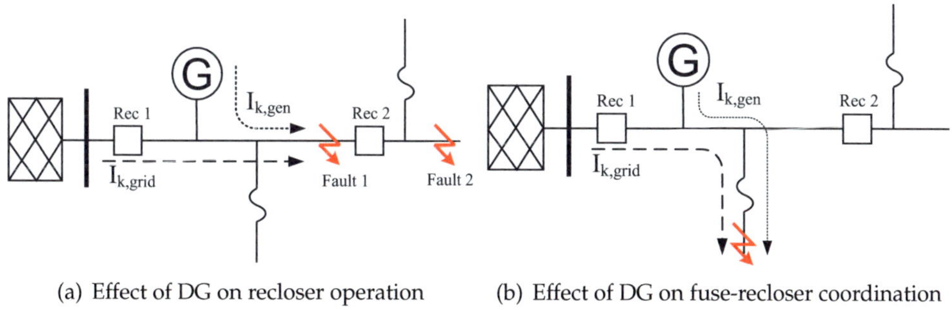

Fig. 7. Radial distribution feeder including protection devices

In figure 7(a) for the fault location *Fault 1* holds:

$$I_{k,tot} = I_{k,grid} + I_{k,gen} \tag{11}$$

For the indicated fault location *Fault 1* the short-circuit current sensed by *Rec 1* is $I_{k,grid}$. As demonstrated in section 2 the grid contribution reduces and lead to a delayed fault detection or in the worst case to no detection at all. This is an example of a fault detection problem. For fault location *Fault 2* the short-circuit current seen by *Rec 2* is $I_{k,tot}$ which is larger than the current sensed by *Rec 1*. Most reclosers are equipped with a dependent time-current characteristic and the coordination between *Rec 1* and *Rec 2* still holds. Because of the connection of the generator to the feeder the total short-circuit current is increasing and for end-of-line faults the maximum interrupting rating of *Rec 2* has to be checked.
In (Anderson, 1999) the coordination between a fuse and a recloser is explained in detail. As shown in figure 8 the fuse and recloser are coordinated such that there is selective fault clearing for the fault currents $I_{k,min} < I_{fault} < I_{k,max}$.

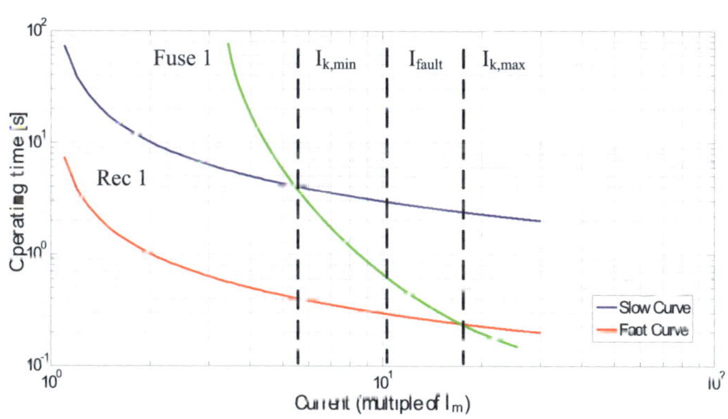

Fig. 8. Coordination between a lateral fuse and a recloser

For the situation in figure 7(b) the coordination between the fuse and recloser is lost when $I_{k,tot} > I_{k,max}$. In that case the curve of the fuse is under the curve of the recloser and the

fuse clears the fault before the recloser operates. Hence, temporary faults will be cleared permanently and lead to unnecessary interruptions.
Besides detection problems and lost of coordination DG also causes unsynchronized reclosing. During the recloser's dead time a part of the feeder is disconnected from the main system to allow the arc to deionize. The connected generator tends to keep the disconnected feeder part energized and maintains the arc at the fault-location. Hence the temporary fault becomes permanent. Moreover, due to unbalance between load and generation the generators will drift away from synchronism with respect to the main grid which results in a unsynchronized reclosing action. This can seriously damage the generator and causes high currents and voltages in neighboring grids (Kauhamieni & Kumpulainen, 2004).

4. Solutions and alternative protective systems

Installing small generation in the distribution grid has become popular since the mid eighties and the protection problems caused by DG has been studied accordingly (Dugan et al., 1984; Rizy et al., 1985). In the literature for the problems mentioned in the previous section a wide pallet of solutions is offered. These solutions vary between a simple change in relay settings to a complete new adaptive protective system. In this section an overview of possible solutions is given.

4.1 Prevention of detection and selectivity problems

Fault detection problems do have a relation with the amount of generation connected to the distribution grid and the local short-circuit power. To prevent fault detection problems a first attempt is to modify the relay settings of the relays and reclosers (Baran & El-Markabi, 2004; Hadjsaid et al., 1999; Kumpulainen et al., 2005; Mäki et al., 2004). The generator contribution leads to a reduction of the grid contribution to the fault current hence the pick-up current of the relays has to be reduced. However, fault detection problems might be solved by reducing the pick-up current, the sensitivity and security of the protective system is decreased and might lead to false tripping in case of a fault in an adjacent feeder. In (Mäki et al., 2004) an example of a weak network is given where blinding of protection occurs due to the connection of a small wind farm. By reducing the pick-up current blinding of protection is solved but at the same time it introduces for faults in a certain area false tripping. A proposed solution is to install protection devices with an additional time delay to give the feeder including the wind farm a longer fault clearing time. These type of solutions also discussed in (Deuse et al., 2007). Another example of changed protection settings is discussed in (Baran & El-Markabi, 2004). Here an adaptive overcurrent relay is proposed which decreases the pick-up current as the output of the local generation increases. This is also studied in (Vermeyen, 2008) where it is stated that a continuous adaptation of the pick-up current as function of the generator output results in less superfluous disconnection of the feeder.

4.2 Mitigation of recloser problems

Fault detection problems also occur in distribution feeders with overhead lines including DG, which are protected by reclosers. As a result fuse-recloser coordination can be lost. In (Brahma & Girgis, 2002) modern microprocessor-based reclosers are applied to restore the coordination between the fuse and recloser. In the microprocessor several trip curves can be programmed and the microprocessor keeps track which curve is in use. As explained in figure 8 the recloser is equipped with a fast and a slow curve. In the microprocessor the fast curve should be programmed in such a way that this curve is selective with the lateral fuses, especially in presence of DG. To prevent unsynchronized reclosing, DG has to be disconnected as soon as possible which brings the grid back in the situation without DG. Hence, the fast curve has to be active only during the first reclosing action. In the second reclosing cycle, the slow curve is active which is selective with the lateral fuses and the fault can be cleared in a selective way. A different approach to solve the fuse-recloser problem is to limit the infeed of the DG. Therefore for laterals where DG is connected onto, the protective scheme is modified by replacing the lateral fuse for a high speed recloser (Funmilayo & Butter-Purry, 2009). The reclosers are coordinated in such a way that the lateral recloser operates before the recloser in the main feeder. Furthermore at the coupling point the DG-unit is equipped with an overcurrent relay. The modifications are shown in figure 9.

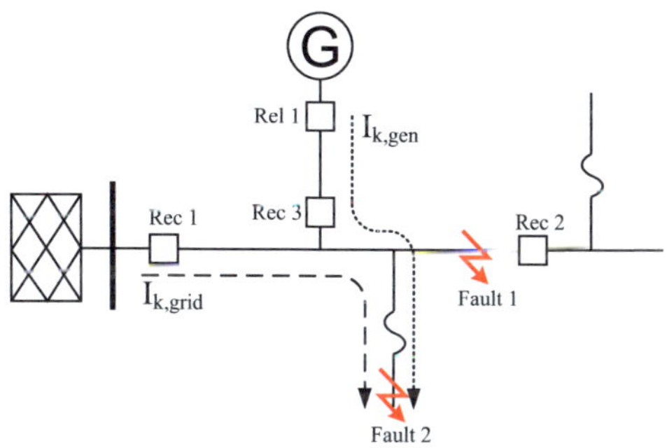

Fig. 9. Modified protective scheme for a radial feeder protected by a recloser

For fault location *Fault 1* the recloser *Rec 1* operates on the fast curve and clears the fault. The lateral recloser *Rec 3* also operate at his fast curve and limits the infeed of the DG-unit. Both reclosers are high speed reclosers and the main feeder as well as the lateral are reconnected after a short time period. When it concerns a temporary fault the feeder can stay in operation. For a permanent fault, after the reclosing action of both reclosers, the overcurrent relay *Rel 1* of the DG-unit disconnects the DG-unit. This occurs before the delayed operation of recloser *Rec 1*. Since the DG-unit is removed from the system, the fault clearing can proceed as it normally does for distribution grids without DG. For fault location *Fault 2* a lateral recloser *Rec 3* and the main feeder recloser *Rec 1* clear the fault. In case of a permanent fault the reclosing action of both reclosers will again lead to a fault current which will blow the fuse. Now the fault is removed and the feeder can stay in operation. When the fuse fails to clear the fault the overcurrent relay *Rel 1* will disconnect the DG-units and the recloser at the main feeder will lock-out after one or two delayed reclosing attempts.
The idea of the previous solution is to limit the DG infeed and restore the radial nature of the distribution grid. The most effective way solving all protection problems is a fast disconnection of all involved DG during a disturbance. In some connection standards, e.g. (IEEE1547, July, 2003), this is even obliged. As discussed in (Tailor & Osman, 2008) the disconnection of the DG-units has to take place before the fuses or reclosers can operate. For this purpose a regular circuit breaker is relatively slow and it is proposed to replace the mechanical circuit breaker by a semiconductor switch. The semiconductor switch contains two anti-parallel connected Gate Turn Off (GTO) thyristors and a current sensing unit with a microprocessor. The current sensing unit is set with a preset value and monitors continuously the DG phase currents. If the threshold is exceeded it indicates that a fault has occurred and the current sensing

unit is sending blocking signals to the GTOs. Within a few milliseconds the DG-units are removed from the distribution grid and the radial nature is restored before a fuse or recloser has operated.

4.3 Strategies for solving unsynchronized reclosing and islanding

The most challenging protection problem is unsynchronized reclosing and islanding. Unsynchronized reclosing only occurs in distribution grids protected by reclosers while islanding can occur in grids with a conventional overcurrent protection as well. Unsynchronized reclosing is already discussed in section 3.3. The islanding problem has a strong relation with unsynchronized reclosing. During the reclosers dead-time, a DG-unit can be still connected to the isolated part of the feeder. The DG-unit tends to feed the local connected load and the isolated part can be considered in island operation. In case of a large unbalance between load and generation the speed of the generator will in- or decrease and the voltage and frequency will exceed the allowable tolerances mentioned in the standards. Because of the violation of these tolerances the DG-unit will be disconnected by its own voltage or frequency protection. This action should take place before the reclosing action to prevent unsynchronized reclosing. The most effective solution to prevent islanding and subsequently unsynchronized reclosing is the disconnection of the DG-unit before the reclosing action takes place. The challenge in here is to detect the formed island fast enough. Island detection methods can be divided into three categories (Abarrategui et al., 2007; Mahat et al., 2008):

1. Passive methods
2. Active methods
3. Traditional methods or remote techniques

The passive methods make use that when an island is formed some important parameters, such as voltage, current, frequency and harmonic distortion, changes. Monitoring the change of these parameters can lead to a detection of an island. The difficulty of these methods is defining suitable threshold values to differentiate islanding from other disturbances. An example of a popular passive islanding protection is the rate of change of frequency (ROCOF). For systems with a load or generation surplus the ROCOF protection works well. However, in a perfect match of generation and load the rate of change of frequency is small and island detection will be quite cumbersome.
Active islanding detection methods intentional create a small disturbance in the system which results in a significant change in system parameters in an islanded situation. In case the feeder is connected to the main grid the effect of these small disturbances are hardly noticeable. An example of a intentional disturbance is an introduction of a voltage fluctuation applied through a small change of the AVR of the DG-unit. For an islanded feeder the effect of the AVR is much larger than for feeders which are connected to the main system. The active methods are able to detect an island even when the load matches still the generation however, setting up an intentional disturbance needs some time and therefore these methods are slower than the passive methods.
Traditional methods or remote techniques are based on communication between the utility and the DG-unit. In the substation the position of the circuit breakers which can cause the island are monitored and when one or more of these circuit breakers opens a transfer trip signal is sent to the DG-unit. For the monitoring system a Supervisory Control and Data Acquisition

(SCADA) system can be used. For the transmission of the trip signal a dedicated communication channel has to be present which is often expensive to implement and hence uneconomical. The investment in communication channels can be prevented by using power-line carrier (PLC) communication (Benato et al., 2003; Ropp et al., 2000). This islanding detection system uses a ripple control signal which is superimposed on the medium voltage. The signal is detected via a sensor which is located at the DG-site. Opening the circuit breaker not only interrupts the load current but also the ripple control signal. The loss of the signal is sensed by the sensor and subsequently the DG-unit is disconnected. In (Kumpulainen et al., 2005) the PLC transfer trip method is considered reliable and selective. However, further studies as well as field tests are needed to verify the feasibility of the method. Efficient and reliable islanding protection methods are necessary to remove barriers which nowadays limits the integration of DG in distribution grids.

4.4 Developments in protective systems

As discussed in the previous sections integrating DG in distribution grids can lead to serious protection problems. Now the tendency is, in case of a grid disturbance, to disconnect the DG as soon as possible in order to restore the original nature of the distribution grid. Restoring the original nature of distribution grids results in a unidirectional fault current and the traditional protective system has proven its capability to clear the fault in a selective way. For remote faults disconnection of DG is, however, not always necessary and a waste of useable energy. Recent developments in protective systems are focused on adaptive protection schemes which can distinguish grid disturbances in distribution grids including DG. The papers written on these developments are numerous and some interesting and promising results are reported in (Brahma & Girgis, 2004; Perera & Rajapakse, 2006; Perera et al., October 2008). Traditional protective systems make use of locally measured quantities and react if one of these quantities is violating a certain threshold while new adaptive protective systems rely on information obtained by specific measurement systems. The protective strategy divides the distribution grid of a certain area into zones rather than protecting a single component or feeder. An example of this strategy is given in (Brahma & Girgis, 2004) where the distribution grid is split into zones which are able to run in island operation. The protective system is based on a centralized computer wherein the grid topology is programmed. Via communication channels all actual breaker positions are known. The computer executes off-line load flow and short-circuit current calculations and stores the results in a database. Topology changes due to switching actions will update the tables in the database. The central computer uses synchronized current vector measurements at the main source, distributed generators and breakers. In case of a fault these measurements are compared with the values in the database to identify the faulted section or zone. A trip signal is sent to the breakers which interconnect the various zones and the faulted zone is isolated. The remaining zones return to normal operation and in the faulted zone the fault is cleared. A drawback of this system is its heavy dependence on a centralized processing system and the communication links between the zones.

In (Perera & Rajapakse, 2006; Perera et al., October 2008) an agent-based protective system is discussed which overcomes the drawback of the previous system. This protective system also splits up the distribution grid into zones and exchanges data for these zones via communication links. The agents are located at strategic locations and make use of local current measurements. Via a wavelet transformation the current signal is processed and the fault direction is determined. All agents are equipped with a fault locating algorithm and with the aid of the data exchange between the agents and the fault locating algorithm the faulted zone is cleared. It is demonstrated that this protective system also works for high impedance faults and for distribution grids including DG.

Adding communication links to existing distribution grids is costly and hardly justified in comparison with the benefits of improving the availability of DG. However, the need for more data for accurate grid operation, smart metering and introduction of micro grids and virtual power plants, also asks for communication links. The installation of communication channels can become attractive when these links are shared by these processes. In this way new developments become possible which are hardly feasible when each individual development is justified by its own.

5. Case study on a benchmark network

In this section the fault detection and selectivity problems are evaluated with the aid of a generic benchmark network. First, the fault detection problem is studied with static calculations while the effect of the generator dynamics on the performance of the protective system is examined with dynamic simulations. In the case study it will be demonstrated how the protective system can be modified to mitigate and prevent fault detection problems.

5.1 Test system topology

In (Strunz, 2009) benchmark systems for network integration of distributed generation including all grid parameters and a reference load flow are provided. A distinction is made between benchmark networks which are common in North America and benchmark networks which are typical for Europe. In this chapter the European medium voltage benchmark network is used for which the single line diagram is depicted in figure 10. All network data is given in (Strunz, 2009). The topology of the benchmark network consists of the feeder systems Feeder 1 and Feeder 2 which are indicated in the dashed boxes. Both feeders are operated at 10 kV and are fed via separate transformers from the 110 kV transmission system. The configuration of the network can be modified by means of the switches S1, S2 and S3. Via these coupling switches radial, ring and meshed operation of the benchmark network is possible.

5.2 Fault detection problems

The theory of fault currents in faulted distribution feeders including DG is discussed in section 2. In this case study the effect of DG on the fault currents of Feeder 2 of the benchmark network is examined to determine if fault detection problems appear. On Feeder 2 the same approach as in section 2.2 is applied to illustrate what effect the connection of a synchronous generator has on the grid contribution to the short-circuit current. This is done for different generator sizes in the range of 2-10 MVA. Via repetitive calculations the size and location of the generator is varied. For Feeder 2 the grid contribution as function of the generator size and location is given in figure 11.

The shape of the curves in figure 11 shows a great similarity with the curves in figure 5. The minima of these curves indicate the location of the largest generator impact. With the aid of the parameters of Feeder 2 and knowing the grid impedance this location can be calculated with equation (10). The local fault level of Node 12 is 180 MVA which corresponds with a grid impedance of $j0,551\ \Omega$. The electric parameters of Feeder 2 are given in table 2.

According to table 2 the total feeder impedance Z_L, is $5,04 + j3,57\ \Omega$. With the aid of the grid and feeder impedance and equation (10) the relative worst case generator location can be obtained. The relative worst case generator location l, is calculated as 0,43. The total feeder

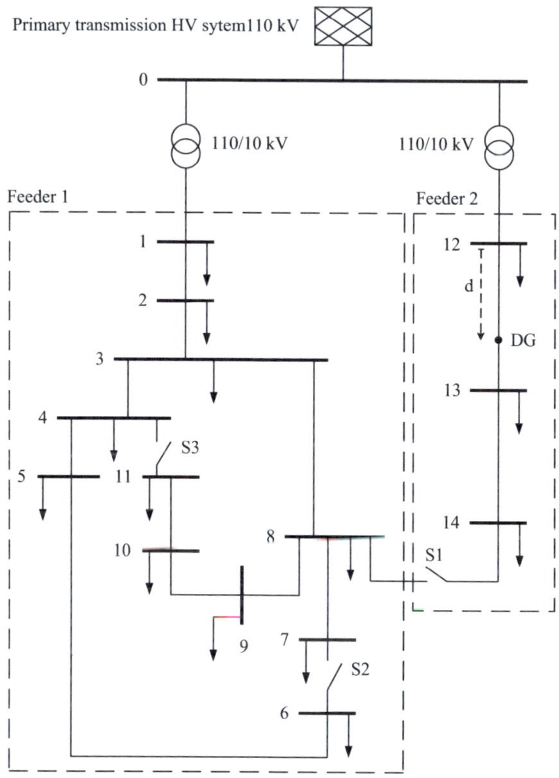

Fig. 10. Medium voltage benchmark network

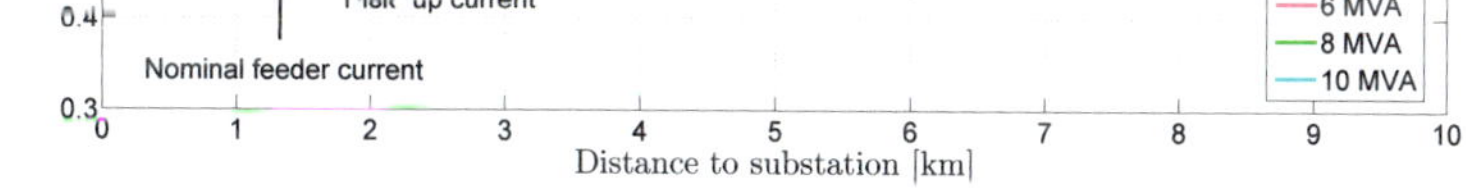

Fig. 11. Grid contribution as function of generator size and location

length d_{tot} is 9,88 km and with equation (1) the worst case generator location is 4,42 km which corresponds with the minima shown in figure 11.
Normally for benchmark networks the data of the protective system is not provided. Feeder 2 of the benchmark network is supposed here to be protected with a definite overcurrent protection which is located in Node 12. In (Anderson, 1999) general setting rules for overcurrent

Line	$R\ [\Omega/km]$	$X\ [\Omega/km]$	$d_{Line}\ [km]$	$Z_{tot}\ [\Omega]$	$I_{nom}\ [A]$
Line 12-13	0,51	0,361	4,89	$2,49 + j1,77$	500
Line 13-14	0,51	0,361	2,99	$1,52 + j1,08$	500
Line 14-8	0,51	0,361	2	$1,02 + j0,72$	500

Table 2. Line parameters of Feeder 2 of the benchmark network

protection are proposed. For instance, for the pick-up current I_m an approach of $\mathrm{I_m} = 2 \cdot \mathrm{I_{nom}}$ is proposed. Applying these rules on Feeder 2 do not lead to satisfactory relay settings. Due to the relative high feeder impedance the fault currents are such that $2 \cdot \mathrm{I_{nom}}$ will not be reached. Therefore the pick-up current of the protection is set to $1,3 \cdot \mathrm{I_{nom}}$ with a reaction time of 0,3 s. The pick-up current of the protection and the nominal feeder current is projected in figure 11. This figure indicates that for generators larger than 4 MVA the grid contribution declines under the pick-up current and causes blinding of protection. Strictly speaking generators larger than 4 MVA cannot be connected to Feeder 2 for locations where the grid contribution is smaller than the pick-up current. Connection of this size of generators is only possible when the protective system is modified and it is guaranteed that all possible faults can be detected. In (Deuse et al., 2007; Mäki et al., 2004) reduction of the pick-up current is proposed which makes the protective system more sensitive. For Feeder 2 reduction of the pick-up current reduces the reliability of the protection because the pick-up current approximates the nominal feeder current. In that case small switching transients can cause unwanted disconnection of the feeder.

For a more detailed analysis of the effect of DG on the protective system of Feeder 2, the DG-location is set at the worst case generator location. This is indicated in figure 10 with the parameter d. For a fixed DG-location the zone wherein fault detection problems occur, can be determined by calculating the grid contribution for all fault locations along Feeder 2. Per line segment for various fault locations a three-phase short-circuit calculation is performed. The fault location is adjusted with 1% of the length of the line segment. The results of these calculations are, including the pick-up current of the protection, depicted in figures 12(a) and 12(b). For a generator of 6 MVA the non-detection zone starts for faults at a distance $d_{Fault} > 9\ km$ while for a 10 MVA generator this zone already starts at $d_{Fault} > 8\ km$. In figure 12(b) it can be seen that for the mentioned zones the fault current will not be detected.

5.3 Dynamic simulations of fault detection problems

The calculations carried out so far do not incorporate the dynamics of the synchronous generator. As discussed in many textbooks (Grainger & Stevenson, 1994; Kundur, 1993; Machowski et al., 2008) the synchronous generator injects a time-varying short-circuit current which results in a time-varying grid contribution as well. Via dynamic simulations Feeder 2 is examined to determine how the protective system copes with the time-varying grid contribution. Therefore a three-phase fault at the end of Feeder 2 is simulated for generator sizes in the range of 2-10 MVA. The resulting grid contributions are given in figure 13.

The first conclusion that can be drawn from the simulations is that the fault stays undetected when the generator size is larger than 8 MVA. During the complete simulation period the grid contribution stays below the pick-up current of the overcurrent protection. The dynamic effect of the generator manifests itself in the increasing grid contribution to the fault current. This is mainly caused by the decaying DC-component in the generator contribution to the fault

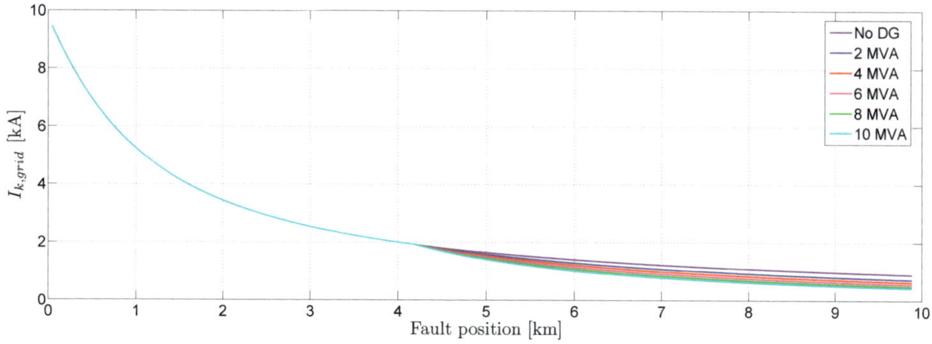

(a) Grid contribution for various fault locations in Feeder 2

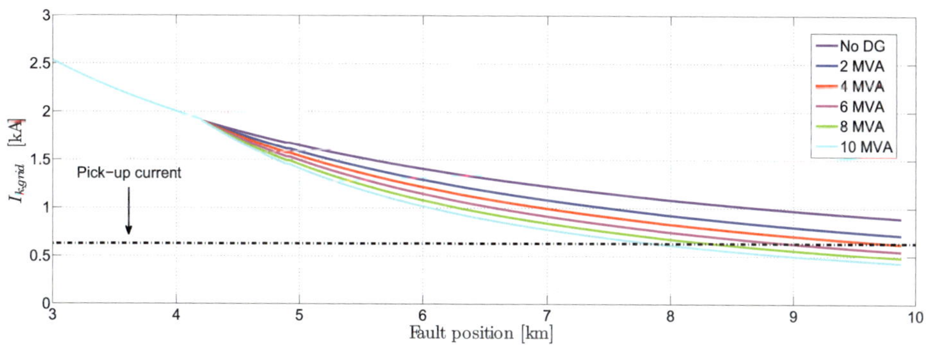

(b) Detail of figure 12(a)

(b) Detail of figure 12(a)

Fig. 12. Grid contribution as function of the fault location in Feeder 2

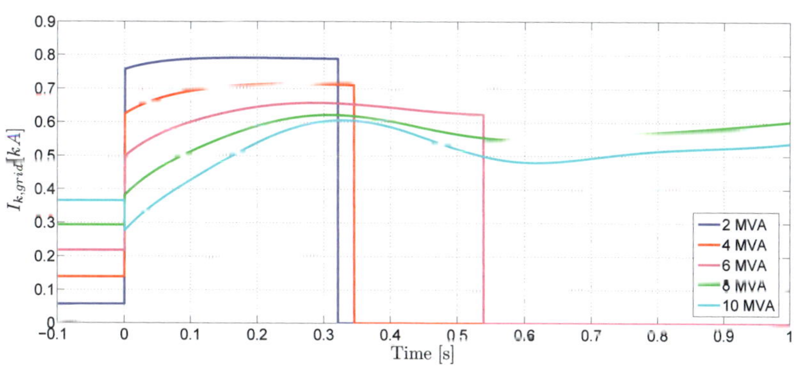

Fig. 13. Dynamic simulation of a three-phase fault at the end of Feeder 2 for various generator sizes

current. Initially the overcurrent protection is not triggered but after the DC-component is sufficiently damped for generator sizes smaller than 6 MVA the pick-up current is exceeded and the fault is cleared. However, for the case a 6 MVA generator is connected, the fault clearing is delayed with approximately 250 ms. This can cause serious coordination problems with upstream protective systems. A detailed look at the results in figure 13 shows that even connecting a 4 MVA generator leads to a delayed fault clearing.

Because of selectivity reasons and possible extra equipment damage, delayed fault clearing is unacceptable. Making in such situations the integration of DG possible then the protection settings or protective system has to be modified. The relatively weak feeder does not allow a reduction of the pick-up current of the overcurrent protection without the risk of unnecessary tripping of a healthy feeder.

The protection which has not been taken into account so far is the interconnect protection of the DG-unit itself. This type of protective system can affect the operation of the grid protection. Examples of interconnect protection relays are under- and overvoltage and under- and overfrequency protection. Interconnect protection differs from generator protection. The goal of generator protection is to protect the generator against internal short-circuits and abnormal operating conditions. These protection devices are connected at the terminals of the generator while the interconnect protection is connected at the point of common coupling (PCC).The major functions of an interconnect protection are (Mozina, 2006):

1. Disconnection of DG when it is no longer operating in parallel with the distribution grid
2. Protection of the grid system from damage caused by the connection of DG
3. Protection of the generator from damage from the grid system (e.g. auto reclosing)

The simulation model used in this section is extended with an interconnect protection of the DG-unit. The generator is equipped with a voltage transformer which measures the voltage of the PCC and an undervoltage protection. The settings of the undervoltage protection are a pick-up value of 0.8 p.u. and a clearing time of 200 ms. The pick-up time of the undervoltage protection is set at 20 ms. With the interconnect protection model the dynamic simulations are repeated to determine if the interconnect protection has a positive contribution to the fault detection problems. The results of the simulations are depicted in figure 14(a) and 14(b).

In figure 14(a) for generator sizes up to 6 MVA it can be seen that after 200 ms the generator is switched off by the undervoltage protection. After disconnecting the generator the grid contribution increases directly. For the generator sizes of 2 and 4 MVA the overcurrent protection was already triggered hence the fault-clearing time does not differ with the previous results. The disconnection of the 6 MVA generator results in an increase of the grid contribution in such a way that the pick-up current of the overcurrent relay is exceeded and the fault is cleared. However, the total fault clearing time is still approximately 550 ms. For these cases the undervoltage protection of the generator does not speed up the fault-clearing time of the protective system. For the generator sizes of 8 and 10 MVA the results significantly differ. In the previous simulations blinding of protection occurred and the overcurrent protection did not clear the fault. The addition of the undervoltage protection leads to the disconnection of the generator with the result that the overcurrent protection is triggered and the fault is cleared. However, the fault-clearing time is respectively 570 and 650 ms. These results can be explained in more detail with the aid of figure 14(b) where for all simulated generator sizes the voltage at the PCC is given. At the moment the fault occurs the voltage along the feeder

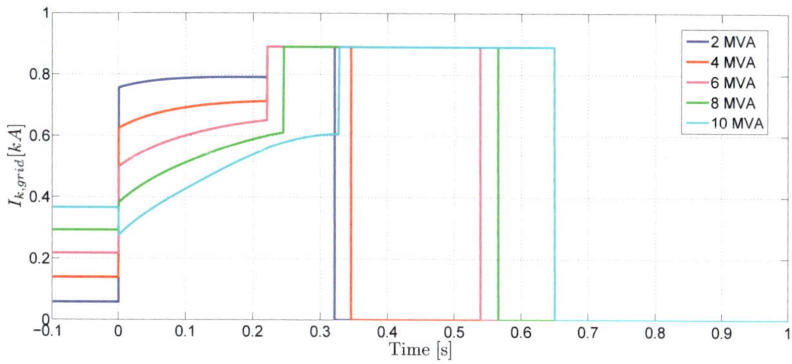

(a) Grid contribution to the fault current

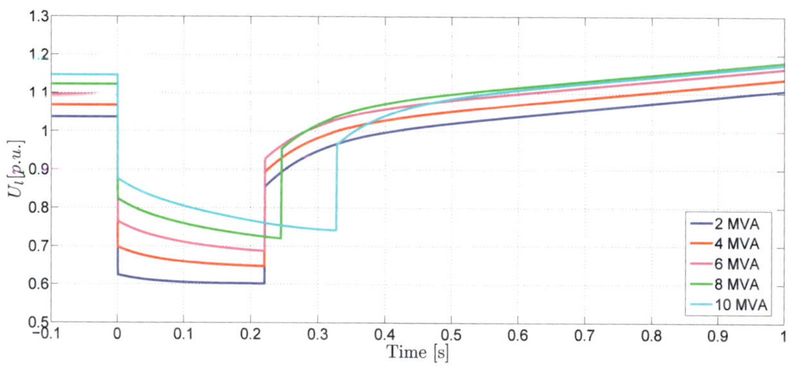

(b) Measured voltage by the undervoltage protection at the point of common coupling

Fig. 14. Dynamic simulation of a three-phase fault for various generator sizes including undervoltage protection

drops. In all simulations the generater tends to keep up the voltage at the PCC. For large generators this effect is stronger than for small generators. As discussed earlier the contribution of the generator consist of a decaying DC-component. Because of this declining DC-component the voltage at the PCC starts to drop as well, as can be seen in figure 14(b). For the cases till 6 MVA the voltage at the PCC drops below the pick-up value of the undervoltage protection immediately after the fault is applied. The 8 and 10 MVA generator keep up the voltage above the pick-up level of the undervoltage protection hence the undervoltage protection is not triggered. At a certain moment the voltage at the PCC exceeds the pick-up value of the undervoltage protection and the generator is disconnected from the feeder. This results directly in an increase of the grid contribution which triggers the overcurrent protection and removes the fault from the feeder. It can be concluded that the addition of the interconnect protection leads to fault clearing for all simulated cases. However, for larger generator sizes the fault clearing time is unacceptable.
To reduce the fault-clearing time the protective system has to be modified. The main goal of this modification is a guaranteed fault detection and a reduction of fault-clearing time. Gen-

erally, integration of a generator in a distribution grid increases the total fault current which can be used to improve the performance of the protective system. For faults after the PCC of the generator an overcurrent protection would sense the sum of the grid and generator contribution while the first upstream protection device before the PCC only senses the grid contribution. The difference in sensed fault currents can be used for coordinating the protection devices and reducing the fault-clearing time (Jäger et al., 2004; Keil & Jäger, January 2008). Therefore in Node 13 an overcurrent protection is modeled and coordinated with the overcurrent protection of Node 12. In figure 15 an overview of the location of all protection devices of Feeder 2 of the benchmark network is given.

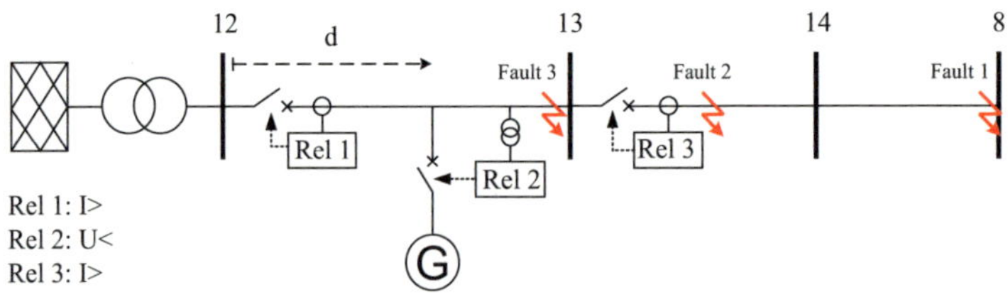

Fig. 15. Overview of all protection devices of Feeder 2

The protection devices Rel 1 and Rel 3 are coordinated such that faults between Node 13 and Node 8 are cleared by Rel 3 and faults between Node 12 and Node 13 by Rel 1. Protection device Rel 1 also serves as a backup protection for faults between Node 13 and Node 8 which are, for whatever reasons, not cleared by Rel 3. The relay settings are determined without the generator contribution. Hence the feeder protection operation is independent of the presence of the generator. In figure 18 the coordination of Rel 1 and Rel 3 is shown. In this case study a time grading of 200 ms is used.

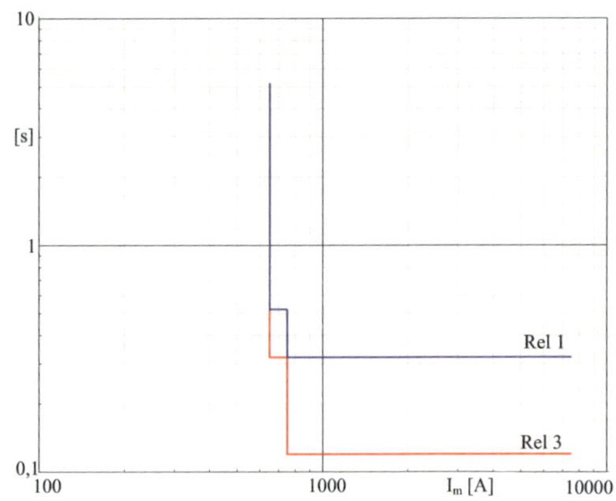

Fig. 16. Protection coordination graph of Feeder 2

The modified protection scheme is tested for three fault locations as indicated in figure 15. Fault location *Fault 1* is located at the end of Feeder 2 to check if Rel 3 can detect the fault current. Fault location *Fault 2* is near to Node 13 to determine selective fault clearing between

Rel 1 and Rel 3. The third fault location *Fault 3* is chosen just before Node 13 to study if the grid contribution to the fault current is large enough to trigger Rel 1. For these fault locations similar dynamic simulations as in the previous situations are preformed. The results of the first fault location are depicted in figure 17. In this figure the sensed fault currents by Rel 1 and 3 are given. It can be seen that for all generator sizes relay Rel 3 clears the fault in 100 ms. Relay Rel 1 does not react on the grid contribution to the fault current and the generator stays connected to the network. After fault-clearing the current swing sensed by Rel 1, as shown in 17(a) is caused by the dynamics of the generator.

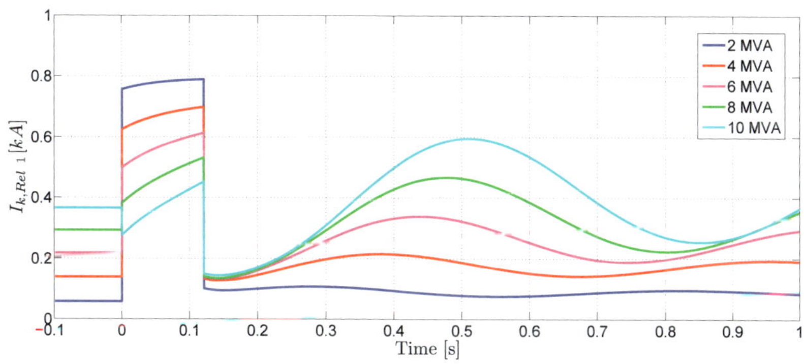

(a) Fault current at location Rel 1 for fault location 1

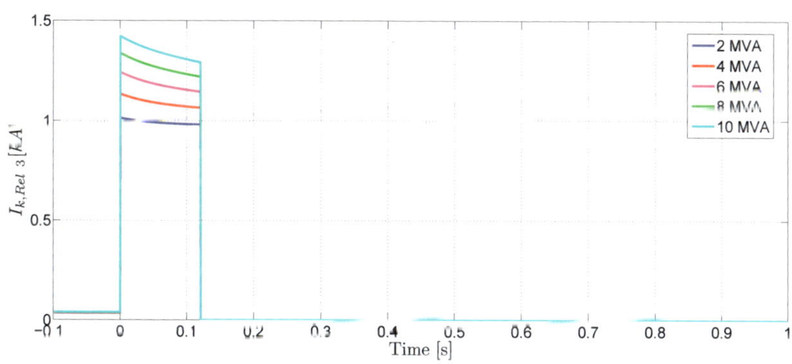

(b) Fault current at location Rel 3 for fault location 1

Fig. 17. Fault currents at relay location Rel 1 and Rel 3 for fault location 1

For fault location *Fault 2* the simulations indicate that this fault is also cleared in 100 ms by Rel 3. This is given in figure 18. The grid contribution is such that Rel 1 is triggered as well but the fault is cleared in a selective way. Because the fault is cleared in 100 ms the generator stays connected to the grid.

The results for the third fault location differ from the previous results. The clearing time of relay Rel 1 is set to 300 ms hence during the fault the generator is disconnected by the under-voltage protection. This can be seen in figure 19 which shows the results of the simulation of this fault location. In the first 200 ms of the fault the grid contribution is sufficient to trigger

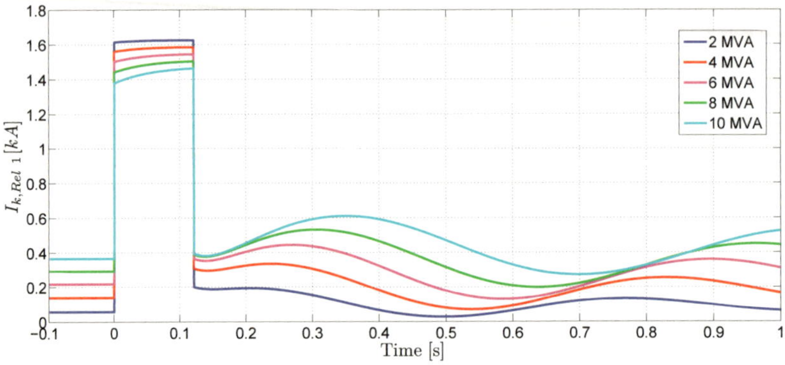

(a) Fault current at location Rel 1 for fault location 2

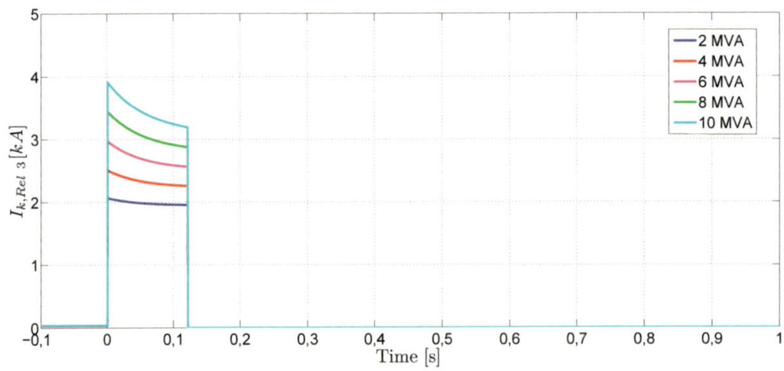

(b) Fault current at location Rel 3 for fault location 2

Fig. 18. Fault currents at relay location Rel 1 and Rel 3 for fault location 2

the overcurrent protection. For all generator sizes the generator is disconnected after 200 ms which causes a sudden increase in the grid contribution to the fault current. However, the overcurrent protection was triggered already and the fault is cleared after 300 ms.
It can be concluded that the addition of relay Rel 3 has resulted in a faster fault clearing for faults between Node 13 and Node 8 without disconnection of the generator. It is demonstrated that Feeder 2 is protected in a selective way. Faults between Node 12 and Node 13 are switched-off in 300 ms for all generator sizes. In comparison with the previous results the modification of the protective system has lead to a reduction in fault clearing time of 350 ms. Simulations have shown that all faults are cleared within 300 ms.

6. Conclusions

In this chapter the effect of distributed generation on the protection of distribution grids was treated. It was demonstrated that DG-units with a synchronous generator can have a strong influence on the grid contribution to the fault current. Analytical expressions were derived to

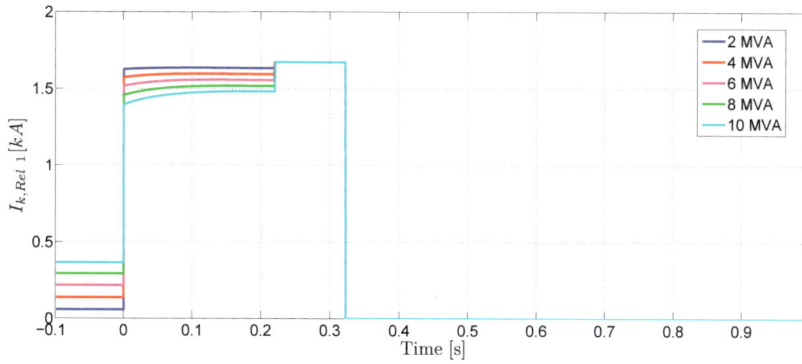

Fig. 19. Fault current at location Rel 1 for fault location 3

determine the key parameters which influence the grid contribution to the fault current. Important parameters which determine the effect of the generator on the grid contribution to the fault current are the total feeder impedance, the size and the location of the generator. Besides that, the local short-circuit power is also of importance. In a simulation of a 3-bus network it was shown that the generator has more effect on the grid contribution to the fault current for feeders consisting of overhead lines than for the same type of feeder built up of cables. This effect is caused by the fact that overhead lines have a significant larger impedance than an equivalent cable. Based on the results of the 3-bus network and the analytical description of the grid contribution to the fault current an equation was derived which can be used to calculate the worst-case generator location.

Then, an overview of all possible protection problems was given and categorized into fault detection and selectivity problems. This categorization shows that apparent different protection problems, such as blinding of protection and lost of fuse-recloser coordination, are related to each other. It was pointed out that solving the fault detection problem directly influence the reliability of a protective system while solving selectivity problems affect the security of a protective system. For both types of protection failure various solutions were discussed and an overview of new developments in protective systems to prevent these protection problems has been given.

The fault detection problem was demonstrated with a generic test feeder. The derived equations were applied on the test feeder to calculate the worst-case generator location. Stationary simulations were carried out for generator sizes of 2-10 MVA and showed that fault detection problems are expected for generator sizes > 4 MVA. Dynamic simulations gave a more accurate results and it could be concluded that for generator sizes of 8-10 MVA serious fault detection problems may occur while for generator sizes of 2-6 MVA a delayed fault clearing takes place. The different result can be explained by noticing that in the dynamic simulation the DC-component in the fault current of the generator is damped. Because of the declining generator contribution the grid contribution increases and triggers the overcurrent protection. Normally the DG-units are equipped with an undervoltage protection and with that the series of simulations were repeated. It resulted in guaranteed fault clearing for all generator sizes, however, for some sizes the fault clearing was still delayed. Hence, to improve the performance of the protective system it has to be modified. A simple modification is the addition of a protection device after the PCC of the generator which has to be coordinated with the

upstream protection device. This modified protective system was simulated and resulted for all generator sizes in a guaranteed and selective fault clearing.
In general the case study showed that dynamic simulations are necessary to evaluate the performance of a protective system of a feeder including DG.

7. References

Abarrategui, O., Zamora, I. & Larruskain, D. (2007). Comparative analysis of islanding detection methods in networks with DG, *CIRED 19th International conference on Electricity Distribution, May 21-24, Vienna.*

Anderson, P. (1999). *Power System Protection*, IEEE Press power engineering series, McGraw-Hill, New York.

Baran, M. & El-Markabi, I. (2004). Adaptive overcurrent protection for distribution feeder with distributed generation, october 10-13, new york ny, *IEEE Power Systems Conference and Exposition*, Vol. 2, pp. 715–719.

Baran, M. & El-Markabi, I. (2005). Fault analysis on distribution feeders with distributed generation, *IEEE transactions on power systems*, Vol. 20, pp. 1757–1764.

Benato, R., Caldon, R. & Cesena, F. (2003). Carrier signal-based protection to prevent dispersed generation islanding on mv systems, *CIRED 17th International conference on Electricity Distribution, May 12-15, Barcelona.*

Brahma, S. & Girgis, A. (2002). Microprocessor-based reclosing to coordinate fuse and recloser in a system with high penetration of distributed generation, *IEEE Power Engineering Society Winter Meeting, New York NY*, pp. 453–458.

Brahma, S. & Girgis, A. (2004). Development of adaptive protection scheme for distribution systems with high penetration of distributed generation, *IEEE Transactions On Power Delivery*, Vol. 19, pp. 56–63.

Chilvers, I., Jenkins, N. & Crossley, P. (2004). The use of 11 kV distance protection to increase generation connected to the distribution network, *Eight IEE international conference on developments in power system protection, Amsterdam, the Netherlands.*

Chilvers, I., Jenkins, N. & Crossley, P. (2005). Distance relaying of 11 kV circuits to increase the installed capacity of distributed generation, *IEE Proceedings Generation, Transmission and distribution*, Vol. 152, pp. 40–46.

Deuse, J., Grenard, S., Bollen, M., Häger, M. & Sollerkvist, F. (2007). Effective impact of der on distribution system protection, *CIRED 19th International conference on Electricity Distribution, May 21-24, Vienna.*

Doyle, M. (2002). Reviewing the impact of distributed generation on distribution system protection, *IEEE Power Engineering Society Summer Meeting, Chicago IL.*

Driesen, J. & Belmans, R. (2006). Distributed Generation: Challenges and possible solutions, *IEEE Power Engineering Society General Meeting, Tampa FL.*

Driesen, J., Vermeyen, P. & Belmans, R. (2007). Protection issues in microgrids with multiple distributed generation units, *Power Conversion Conference PCC'07, Nagoya.*

Dugan, R., Thomas, S. & Rizy, D. (1984). Integrating dispersed storage and generation with an automated distribution system, *IEEE transactions on power apparatus and systems*, Vol. PAS 103, pp. 1142–1146.

Funmilayo, H. & Butter-Purry, K. (2009). An approach to mitigate the impact of distributed generation on the overcurrent protection scheme for radial feeders, *IEEE Power Systems Conference and Exposition,Seattle WA*, pp. 1–11.

Grainger, J. & Stevenson, W. (1994). *Power System Analysis*, McGraw Hill Inc.

Hadjsaid, N., Canard, J.-F. & Dumas, F. (1999). Dispersed generation impact on distribution networks, *Computer applications in power*, Vol. 12, pp. 22–28.

IEEE1547, S. (July, 2003). IEEE 1547 standard for interconnecting distributed resources with electrical power systems, IEEE.

Jäger, J., Keil, T., L.Shang & Krebs, R. (2004). New protection co-ordination methods in the presence of distributed generation, *Eight IEE international conference on developments in power system protection, Amsterdam, the Netherlands*, pp. 319–322.

Jenkins, N., Allen, R., Crossley, P., Kirschen, D. & Strbac, G. (2000). *Embedded Generation*, Series 31, 2nd edn, IEE Power & Energy, London.

Kauhamieni, K. & Kumpulainen, L. (2004). Impact of distributed generation on the protection of distribution networks, *Eight IEE international conference on developments in power system protection, Amsterdam, the Netherlands*, Vol. 1, pp. 315–318.

Keil, T. & Jäger, J. (January 2008). Advanced coordination method for overcurrent protection relays using nonstandard tripping characteristics, *IEEE transactions on power delivery*, Vol. 23, pp. 52–57.

Kumpulainen, L. & Kauhaniemi, K. (2004). Analysis of the impact of distributed generation on automatic reclosing, *IEEE Power Systems Conference and Exposition, October 10-13, New York NY*, Vol. 1, pp. 603–608.

Kumpulainen, L., Kauhaniemi, K., Verho, P. & Vähämäki, O. (2005). New requirements for system protection caused by distributed generation, *CIRED 18th International Conference on Electricity Distribution, June 6-9, Turin.*

Kundur, P. (1993). *Power system Stability and Control*, McGraw Hill Inc., New York, USA.

Machowski, J., Bialek, J. & Bumby, J. (2008). *Power System Dynamics, Stability and Control*, 2nd edn, John Wiley & Sons, Chichester, United Kingdom.

Mahat, P., Chen, Z. & Bak-Jensen, B. (2008). Review of islanding detection methods for distributed generation, *Third international conference on Electric Utility Deregulation Restructuring and Power Technologies (DRPT2008), April 2008, Nanjing, China*, pp. 2743–2748.

Mäki, K., Repo, S. & Järventausta, P. (2004). Effect of wind power based distributed generation on protection of distribution networks, *Eight IEE international conference on developments in power system protection, Amsterdam, the Netherlands*, Vol. 1, pp. 327–330.

Morren, J. & de Haan, S. (2008). Impact of distributed generation units with power electronic converters on distribution network protection, *Ninth IET international conference on developments in power system protection, Glasgow, Scotland.*

Mozina, C. (2006). Distributed generator interconnect protection practices, *IEEE Power Engineering Society Power System Conference and Exposition, Dallas TX*, pp. 1164–1170.

Perera, N. & Rajapakse, A. (2006). Agent-based protection shceme for distribution networks with distributed generation, *IEEE Power Engineering Society General Meeting, Montreal, QC, Canada.*

Perera, N., Rajapakse, A. & Buchholzer, T. (October 2008). Isolation of faults in distribution networks with distributed generators, *IEEE transactions on power delivery*, Vol. 23, pp. 2347–2355.

Rizy, D., Jewell, W. & Stovall, J. (1985). Operational and design considerations for electric distribution systems with dispersed storage and generation, *IEEE transactions on power apparatus and systems*, Vol. PAS 104, pp. 2864–2871.

Ropp, M., Aaker, K., Haigh, J. & Sabbah, N. (2000). Using power line carrier communication to prevent islanding, *Twenty-eight IEEE Photovoltaic Specialists Conference, Anchorage, AK*, pp. 1675–1678.

Strunz, K. (2009). Benchmark systems for network integration of renewable and distributed energy resources, *Preliminary version, Cigre Task Force C6.04.02, July 2009.*

Tailor, J. & Osman, A. (2008). Restoration of fuse-recloser coordination in distribution system with high dg penetration, *IEEE Power Engineering Society General Meeting, Pittsburg PA.*

Vermeyen, P. (2008). *Effect of distributed generation on fault detection and ripple control*, Ph.D Thesis, Katholieke Universiteit Leuven.

Local and Remote Techniques for Islanding Detection in Distributed Generators

César Trujillo[1, 2], David Velasco[1], Emilio Figueres[1] and Gabriel Garcerá[1]
[1]*Universidad Politécnica de Valencia*
Spain
[2]*Universidad Distrital Francisco José de Caldas*
Colombia

1. Introduction

The condition of "Islanding" in Distributed Generators (DG) is an electrical phenomenon that occurs when the energy supplied by the power grid is interrupted due to various factors and DG continues energizing some or the entire load. Thus, the power grid stops controlling this isolated part of the distribution system, which contains both loads and generators. Therefore it may compromise some relevant issues like security, restoration of service and reliability of the equipment (Pietzsch, 2004) (Report IEA T5-09-2002). Figure 1 shows a diagram of a DG connected to the power grid.

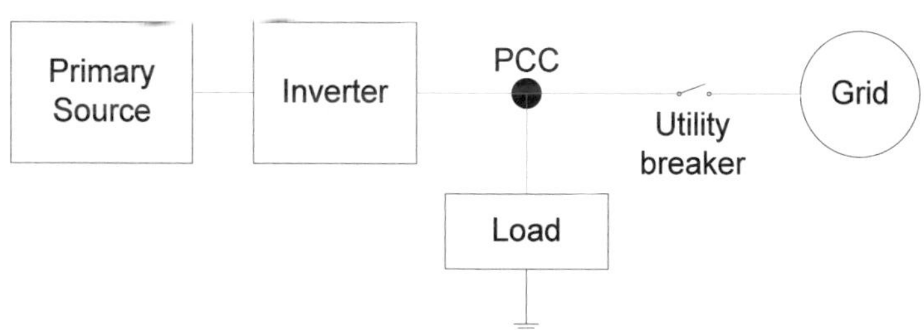

Fig. 1. Schematic configuration of the system

In the case of several Distributed Generation Systems connected to a low-voltage power grid, it is possible that the amount of energy generated by the distributed system matches the amount of energy consumed by the loads. Under this situation, there is no energy flow with the power grid and the distributed systems may fail to detect a possible power grid disconnection, so that the DG system might continue feeding the loads, which would entail an "Islanding" condition. In addition, when the islanding condition happens, a primary security condition is that the generator system disconnects from the de-energized grid without taking into account the connected loads.

Islanding may occur in inverters as a result of the following situations:

1. A failure detected by the grid, which results in a switch opening that is not detected by the inverter or the protection devices.
2. Accidental opening of the electrical supply after a system failure.
3. Sudden changes in the network distribution system and/or in the loads.
4. Intentional disconnection for maintenance services either on the grid or in the service entrance.
5. Human error or vandalism.
6. An act of nature.

There are many reasons why islanding should be anticipated in Distributed Generation systems connected to the grid. Some of the main reasons are safety, liability and maintenance of the quality of energy supplied to consumers. Consumers trust the quality of energy supplied by the grid, but they must also have anti-islanding inverters inside their distributed generation systems for the following reasons:

1. The grid can not control the voltage applied to the loads in islanding conditions, creating a possibility of user's equipment may be damaged.
2. The power grid, along with the owner of the generating system, may be responsible for damage to connected equipment, resulting from variations in voltage and frequency outside the allowed limits.
3. Islanding can be a hazard to workers or grid users, because a line that is supposedly disconnected from any power source can remain active.
4. Uncontrolled reconnection in an isolated DG can damage the generation equipment or any other connected equipment, even block the line, because voltages in both the grid and the inverter can be out of phase.
5. Islanding may interfere with manual or automatic restoration of normal grid service.

Anti-islanding requirements have been evolving over many years and in different ways, depending on the countries laws. For instance, in the Netherlands they only require changes in flow frequency. Other countries like Germany or Austria require specific methods based on sudden impedance changes, known as ENS o MSD. Other countries have adopted standards that require inverters which can detect fails and shut down within set time limits. United States, for instance, requires inverters for grid connection that are certified for that purpose and it also requires that the certification tests of the inverters use a standard test circuit and a methodology based on "worst case" among countries considered as members of the International Energy Agency (IEA). This test was also chosen to allow a single inverter to be tested instead of testing multiple inverters. The main idea to detect Islanding is to supervise the DG output parameters and/or system parameters in order to determine if there have been changes which may point out whether it exists. Islanding detection techniques can be divided into remote and local techniques. Local techniques can be subdivided in passive and active, as shown by Figure 2 (Mahat et al., 2008).

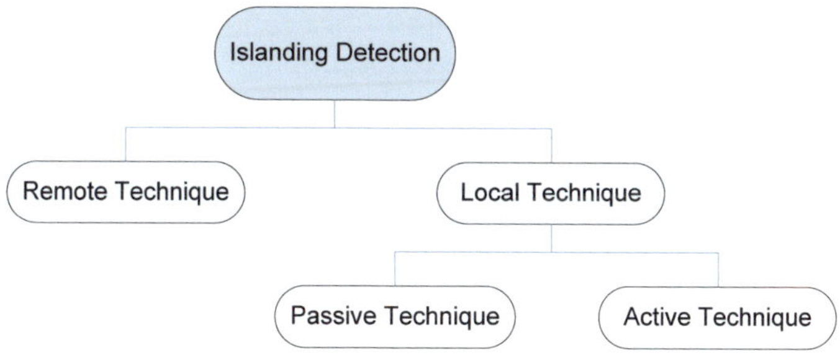

Fig. 2. Classification of Islanding detection techniques

Before defining the different methods of islanding detection, it is important to highlight two key features in order to understand the islanding phenomenon. The first one is associated with the so-called "Non-detection zone" (NDZ), which can be defined as the range (in terms of the difference between the power supplied by the DG inverter and that consumed by the load) in which an islanding detection scheme under test fails to detect this condition. The second one is associated with the nature of loads, which from a conservative point of view can be modelled as a parallel RLC circuit. The reason for using this model is primarily due to the difficulties that some detection techniques have to identify a islanding condition with such loads. Generally, non-linear loads, such loads that produce harmonics or constant power loads do not present difficulties in islanding detection. (Report IEA-T5-09, 2002).
In particular, RLC loads with high Q factor have problems with island detection. Quality Factor is defined as:

$$Q = R\sqrt{\frac{L}{C}} \tag{1}$$

This parameter describes the relationship between stored and dissipated energy in the RLC circuit. Loads with a high Q have large capacitance and small inductances and/or big parallel resistances.
Once they are defined both NDZ and the more critical loads connected to a DG inverter for islanding phenomenon detection, the most outstanding detection techniques will be described.

2. Local islanding detection techniques

These techniques are based on the measure of some parameters (voltage, current, frequency, among others) on the Distributed Generator side. They are classified as passive, based solely on the monitoring of these parameters, and active techniques, which intentionally introduce disturbances at the output of the inverter and observe whether the parameters outlined above are affected.

2.1 Passive techniques of islanding detection

These techniques are based on islanding detection through monitoring of parameters such as voltage, current, frequency and/or their characteristics. They interrupt the inverter energy conversion when there are changes beyond some limits previously established.

2.1.1 Over/under-voltage and over/under-frequency passive techniques of islanding detection

Before going into the method description, an analysis related to the power balance of the system will be done in order to identify how monitored parameters can be affected when the energy flow is interrupted by the grid. Figure 3 shows the power balance of the system.

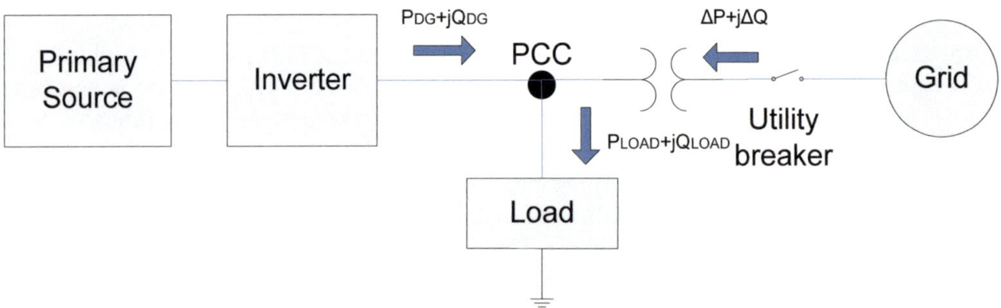

Fig. 3. Power balance in a local load supplied both by the grid and a distributed generator

Equations (2) and (3) describe the power balance of the system.

$$P_{LOAD} = P_{DG} + \Delta P \tag{2}$$

$$Q_{LOAD} = Q_{DG} + \Delta Q \tag{3}$$

If $P_{LOAD} = P_{DG}$ and/or if $Q_{LOAD} = Q_{DG}$, there is not active or reactive power mismatch between the DG and the power grid.

The behaviour of the system when the grid is disconnected depends on the previous values of ΔP y ΔQ (Liserre et al., 2006). It is worth to point out that the active power is directly proportional to the voltage. Therefore, if $\Delta P \neq 0$, the amplitude of the voltage will change. At first, when there is a disconnection of the grid, the power consumed by the load is forced to be the same as the one generated by the DG, so that the voltage value in the grid changes to:

$$V' = \sqrt{\frac{P_{DG}}{P_{LOAD}}} + V \tag{4}$$

In the case of $P_{DG} > P_{LOAD}$, the voltage increases, otherwise it decreases, which might indicate whether the islanding conditions appears.

Reactive power is a function of frequency and voltage width, so if $\Delta Q \neq 0$, the phase of the load voltage will present a sudden change and the control system will modify the signal frequency of the output current inverter to achieve $\Delta Q = 0$ (i.e. until it reaches the resonance frequency of the load). This change in the frequency may be detected to determine Islanding condition. The equation of reactive power in terms of frequency and voltage is presented by (5).

$$Q'_{LOAD} = Q_{DG} = \left(\frac{1}{\omega' \cdot L} - \omega' \cdot C \right) \cdot V' \tag{5}$$

Techniques of over/under voltage protection, OVP/UVP and over/under frequency protection, OFP/UFP, (Report IEA-T5-09, 2002), (Mahat et al., 2008) allow the detection of islanding phenomenon through the measure of voltage and/or frequency at the Point of Common Coupling (PCC), and subsequent comparison with the limits set for proper operation (IEEE 929, 2000). If the measured values are outside the established range, the inverter is stopped or disconnected. This method is not only a mechanism to detect islanding, it also protects the inverter. Figure 4 shows the mapping of the NDZ in the plane ΔP versus ΔQ for OUV and OUF.

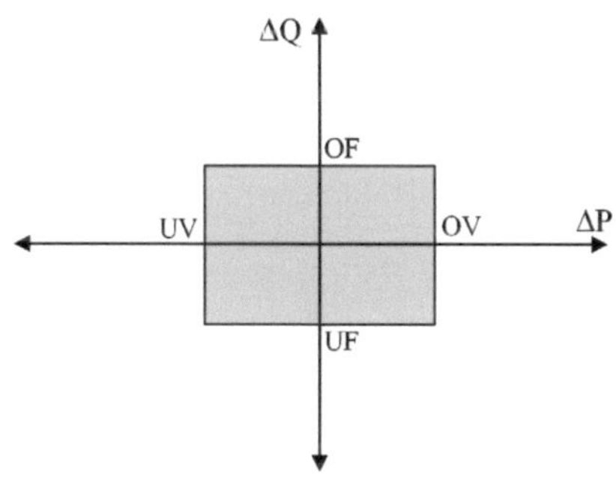

Fig. 4. Non-detection zone of OUV and OUF techniques

The methods outlined above present the advantage of being low-cost solutions but they have a large NDZ. Moreover, these methods are incapable of detecting the islanding condition when the power supplied by the DG matches the power consumed by the loads.

2.1.2 Phase Jump Detection

Phase Jump Detection technique (PJD), (Report IEA-T5 09, 2002), (Kobayashi et al., 1991) involves monitoring of sudden "jumps" of the inverter voltage as a consequence of differences between the voltage inverter and its output current.

During normal operation the current of the inverter is synchronized with the voltage of the power grid through a Phase Locked Loop (PLL).

Figure 5 shows the evolution of the voltage when it is disconnected from the power grid. This phenomenon occurs because only the output current is controlled by the inverter, so that the PCC voltage may be out of phase with regard the current in the case of islanding.

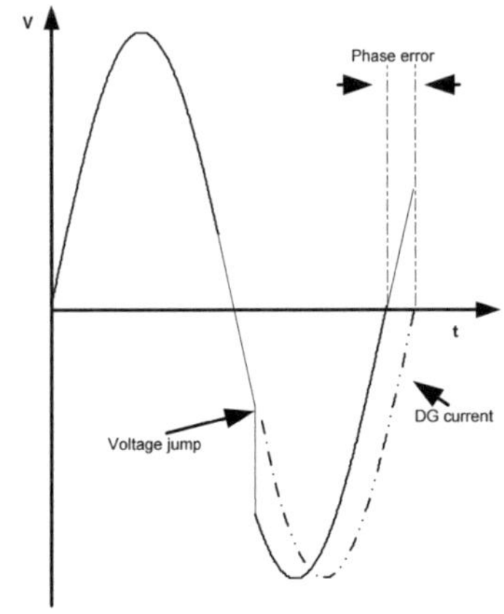

Fig. 5. Operation of PJD

The biggest advantage presented by PJD is its ease of implementation. As the inverter uses a PLL to synchronize with the grid, all that it is needed is the inverter capacity to be disconnected if the phase error between output current and voltage exceeds certain threshold. However, difficulty comes in the implementation of the threshold selection because the phase can be affected by the handling of certain loads such as motors or simply by being on presence of loads that can not produce phase error, which could induce an error in the detection of islanding.

2.1.3 Detection of voltage and current harmonics

This technique is based on the measurement of the voltage Total Harmonic Distortion (THD_v) at the PCC, the comparison of the measured value with a certain threshold and the inverter disconnection in case of this threshold is exceed (Report IEA-T5-09, 2002), (De Mango et al., 2006), (Jang & Kim, 2004). During normal operation, the voltage at the PCC is the grid voltage, so distortion can be considered as negligible ($THD_v \approx 0$) in most cases. However, when islanding condition happens, the current harmonics produced by the inverter are transmitted to the load, which usually presents higher impedance than the grid. The interaction of the harmonic currents and the grid impedance generates voltage harmonics which can be measured. Therefore, the THD_v variations beyond a certain threshold can be used to detect islanding.

This method has the advantage that its effectiveness does not change where there are multiple inverters. However, it is sensitive to grid perturbations, which makes the threshold establishment more difficult for islanding detection. For instance, with non-linear loads, the voltage distortion at the PCC can be so high that a fault may be erroneously detected even if the grid is present. Additionally, with linear loads the THD_v variation may be too low to be detected.

2.1.4 Detection based on state estimators (Liserre et al., 2006)

The basic idea of this technique is based on applying a voltage oriented control combined with the use of resonant controllers. As it is shown by Fig. 6, an algorithm based on Kalman filters is used to estimate the third and the fifth harmonic of the grid voltages. The correspondence between the energy of both the estimated values and the measured ones can be used to identify islanding condition.

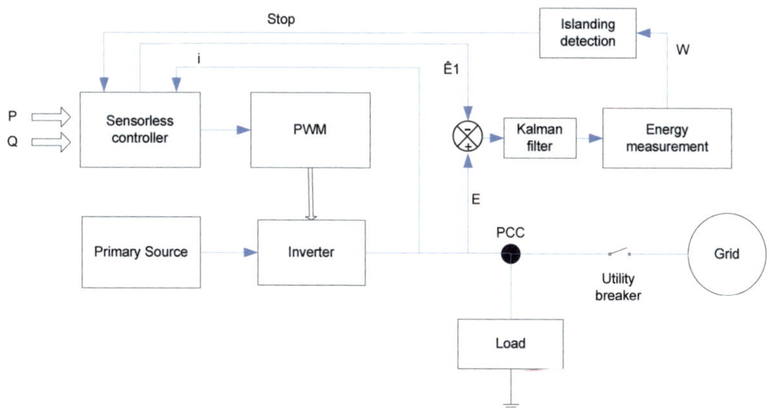

Fig. 6. Scheme of the islanding detection algorithm based on state estimators

Detection based on state estimators has the advantage of being a passive method which does not affect the system power quality. It presents a very low NDZ and an islanding detection rate very high, comfortably in line with IEEE. However, it demands more complexity from the programming point of view compared with other passive techniques of islanding detection.

2.2 Active techniques for islanding detection

These techniques intentionally introduce disturbances at the output of the inverter to determine if they affect voltage, frequency and impedance parameters, in which case it is assumed that the grid has been disconnected and the inverter is isolated from the load.

Active techniques have the advantage of remarkably reducing or even eliminating the NDZ. However, they may deteriorate the quality of the grid voltages or even they may cause instability.

2.2.1 Impedance measurement

Techniques based on impedance measurement seek to detect impedance changes at the output of the inverter, which is produced when the electric distribution grid, which is supposed to have low impedance, disconnects from the system. (Report IEA-T5-09, 2002), (Ropp et al., 2006), (Ciobotaru, 2007).

The inverter of the DG behaves as a current source which injects a current as follows:

$$i_{DG-inv} = I_{DG-inv} \sin(\omega_{DG} t + \phi_{DG}) \qquad (6)$$

Usually, a disturbance is added to the inverter output current, $I_{DG\text{-}inv}$, which causes the output voltage to suffer from changes when the grid is disconnected. This variation is monitored by calculating the *dv/di*, which represents the grid impedance that it is 'seen' by the inverter.

The main advantage of the impedance measurement method is its small NDZ. However, this method has many weaknesses. The first one is that its effectiveness decreases as the number of inverters connected to the grid increases unless all the inverters use this method and they all are synchronized. The second one is that it is necessary to establish an impedance threshold to identify when the grid is connected. This requires the exact value of the grid impedance which is a parameter initially unknown. It makes this method sometimes impractical.

2.2.2 Harmonic injection/detection of impedance

This method involves intentionally injecting a specific current harmonic at the PCC. (Chunjiang et al., 2006), (Ciobotaru et al., 2008). When the grid is connected, if the grid impedance is lower than the load impedance at the harmonic frequency, then the injected harmonic current will flow into the grid.

The size of the disturbance that appears at the voltage amplitude will depend on the nominal values of the grid impedance.

After the grid disconnection, the harmonic current will flow through the load, producing a specific harmonic voltage. The name of this method comes from the fact that the amplitude of the generated voltage harmonics will be proportional to the impedance of the load at the frequency of the injected harmonic current. Figure 7 shows a diagram of this method.

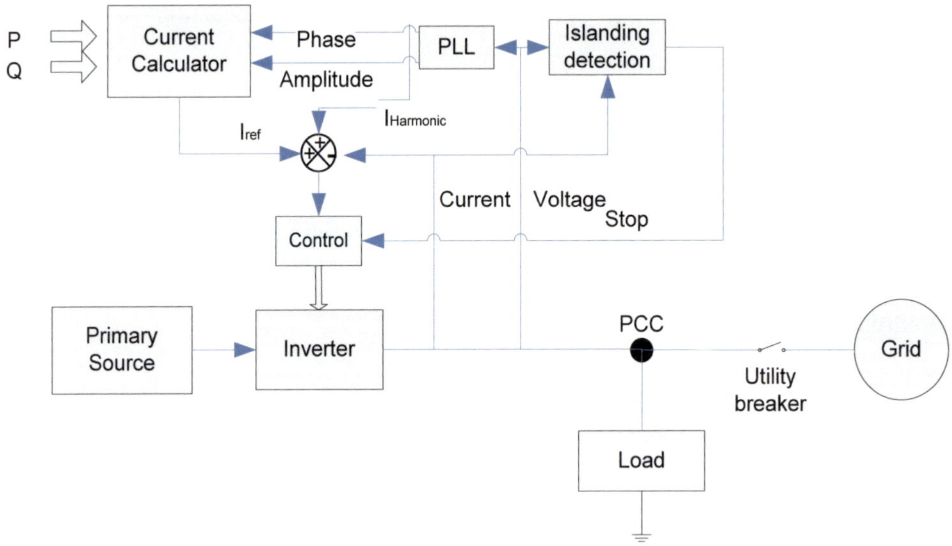

Fig. 7. Structure of islanding detection method based on harmonic injection

This method presents the same advantages and disadvantages of the harmonic detection technique. However, the disadvantages can be overcome if subharmonic signals are injected instead of high order harmonics. Unfortunately, problems are not definitively solved unless the amplitude of the injected harmonics is very small.

2.2.3 Sliding Mode Frequency Shift (SMS)

The operation principle of the SMS method is based on varying the inverter output frequency by controlling the phase of the inverter current (Sun et al., 2004), (Lopes & Sun, 2006). Usually, DG operates with unity power factor, so in normal operation the inverter output current-voltage phase angle of the inverter, instead to be controlled to be zero, is made to be a function of the frequency of the PCC voltage, as shown in Figure 8.

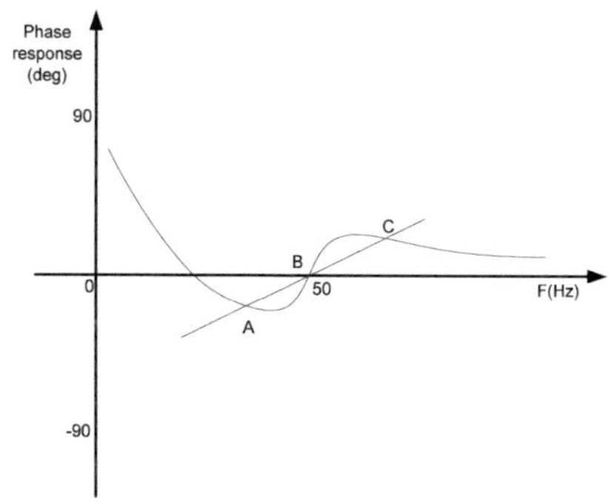

Fig. 8. Voltage-current phase Angle vs. Frequency controlled by SMS

The phase response curve of the inverter is designed such that the phase of the inverter increases faster than the phase of the (RLC) load with a unity-power factor in the region near the grid frequency. This makes the line frequency an unstable operating point for the inverter. While the grid is connected, it stabilizes the operating point at the line frequency by providing a solid phase and frequency reference. However, after the island is formed, the phase-frequency operating point of the load and inverter must be at an intersection of the load line and inverter phase response curve.

This method is relatively easy to implement because it is just a slight modification of a component which is already required, the PLL. Additionally, it has a small NDZ compared with other methods. It also has the advantage of being effective when dealing with multiple inverters and it offers a good compromise between islanding detection, the output power quality and transitory response. However, SMS method requires a decrease in the power quality of the DG inverter.

2.2.4 Active Frequency Drift (AFD)

The basis of AFD method is to vary the frequency of the output current by means of a positive feedback (Report IEA-T5-09, 2002), (Sun et al., 2004), (Lopes & Sun, 2006). The method is based on the injection of a current into the PCC slightly distorted in frequency as shown in Figure 9. When a grid disconnection occurs, a phase error appears between the inverter current and the voltage at the PCC. The inverter detects this error and tries to compensate it by increasing the frequency of the generated current. This process continues until the frequency exceeds the limits and is detected by the OFP / UFP.

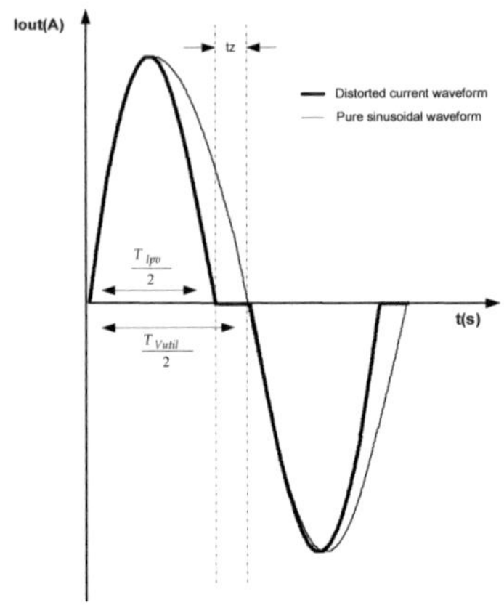

Fig. 9. Inverter current waveform distorted by using AFD

The relationship between t_Z in Figure 9 and half of the voltage period is called the chopping factor:

$$cf = \frac{2 \cdot t_Z}{T_{Vutil}} \tag{7}$$

This method can be easily implemented and applied to multiple inverters. However, the AFD method produces a small degradation in the quality of the DG output and the inverter has an NDZ that depends on the value of the chopping factor. There are similar techniques allowing to obtain better results by changing the chopping factor, with a significant reduction in the NDZ, like the following ones: Active Frequency Drift with Positive Feedback (AFDPF) (Jung et al., 2005), AFD with Pulsation of Chopping Fraction (AFDPCF) (Liu et al., 2007), among others.

2.2.5 Frequency jump

The Frequency Jump (FJ) method is a modification of AFD, and it is conceptually similar to the impedance estimation techniques. In the FJ method, dead zones are inserted into some cycles of the output current waveform. Instead, the frequency is "dithered" according to a pre-assigned pattern (Report IEA-T5-09, 2002), (Nehrir & Menon, 2007).
When the inverter is connected to the utility, the waveform of the voltage in the PCC is imposed by the grid. However, when the grid is disconnected, the Islanding situation is detected by forcing a deviation in frequency. The main advantage of this method is that if the pattern is sufficiently sophisticated, FJ may be relatively effective in Islanding detection when used with a single inverter. Furthermore, this method hardly presents NDZ in single-inverter case and it loses effectiveness when connecting multiple inverters, unless the frequencies dithering between inverters are synchronized.

2.2.6 Variation of active power and reactive power

This method is based on the ability of the inverter to generate independently both active and reactive power (Jeraputra & Enjeti., 2004), (Mango et al., 2006), (Ye et al., 2004), (Jeong et al., 2005). In islanding, the voltage variation with regard to the active power injected by the inverter may be obtained from the power flow shown by Figure 3. The power supplied by the inverter can be expressed as:

$$P_{DG} = P_{LOAD} = \frac{V^2}{R} \tag{8}$$

Differentiating P_{DG} with regard to voltage and expressing it in terms of power it results:

$$\frac{\partial P_{DG}}{\partial V} = 2 \cdot \frac{V}{R} = 2 \cdot \sqrt{\frac{P_{DG}}{R}} \tag{9}$$

Hence voltage variation in terms of power can be expressed as:

$$\Delta V = \frac{\Delta P_{DG}}{2} \cdot \sqrt{\frac{R}{P_{DG}}} \tag{10}$$

Since both R and P_{DG} are constant, voltage variation is directly proportional to the variation of active power. Hence, it is possible to vary the active power injected by the inverter in order to bring the amplitude of the voltage outside the normal operating range and be able to detect islanding.

It is necessary to choose carefully when the power is injected because continuous variations of the injected power can perturb the Maximum Power Point Tracking (MPPT) algorithms. For these reasons, this method involves the injection of active power only when the voltage measured at the PCC exceeds a certain threshold value (Vs).

The time needed by the algorithm to detect a fault can be adjusted with a Kv that increases o decreases dP proportionally to the voltage variation. This value should be chosen large enough to detect the islanding situation avoiding overcurrents which may damage the system elements. An initial value of the constant can be obtained from equation (10). The current reference for the inverter control can be calculated as follows:

$$I_{ref} = \frac{dP + P_{DG}}{V} \tag{11}$$

Where $dP=Kv(V-Vn)$, Vn being the amplitude of the nominal voltage and V the measurement of the feedback voltage amplitude.

Similarly to the relationship between voltage and active power, a strong dependence between frequency and reactive power exists, which may be used to develop another method of islanding, based on measuring the grid frequency (Sanchis et al., 2005). Since Std. 929-2000 recommends DG operation near to a unity power factor, the generated reactive power must be zero in normal operation. At the beginning of islanding, equation (12) remains valid, so that the frequency depends on the values of the inductive and capacitive components of the load. Therefore, the variation of Q with ω follows (13), where ω_0 is the resonance frequency of the equivalent RLC circuit (14).

$$Q_{DG} = Q_{LOAD} = 0 = V^2\left(\frac{1}{\omega L} - \omega C\right) \tag{12}$$

$$\frac{Q-0}{\omega - \omega_o} = \left.\frac{dQ}{d\omega}\right|_{\omega=\omega_o} \tag{13}$$

$$\omega_o = \frac{1}{\sqrt{LC}} \tag{14}$$

Solving equation (13), in order to determine the variation of the frequency as a function of Q and setting the frequency of the load at the resonance frequency, it is obtained:

$$\Delta\omega = -V^2 \cdot \left[\frac{\omega_o^2 L}{\omega_o^2 LC + 1}\right] \cdot Q \tag{15}$$

As observed in (15), the frequency variation is directly proportional to the changes of reactive power and it can be obtained the expression of Δf as a function of the quality factor Q_f, the resonance frequency f_o and the active power P_{DG}, following (16).

$$\Delta f = -\frac{f_o}{2P_{DG} \cdot Q_f} Q \tag{16}$$

The reference frequency for the inverter control can be calculated as follows:

$$f_{ref} = f_g + K_f \cdot (f - f_g) \tag{17}$$

Where f_g is the grid frequency, f is the measured frequency and K_f is a constant that allows to accelerate the islanding detection. K_f may be calculated taking into account the parameters of equation (16).

Figure 10 shows the diagram of the method.

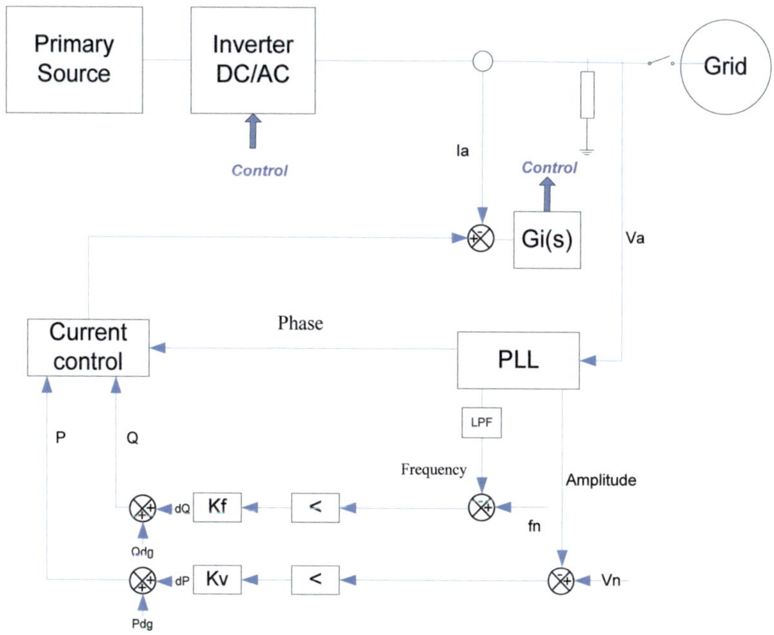

Fig. 10. Block diagram on P and Q injection method

The disadvantage of this method is that it can generate false detections of islanding when several inverters are connected to the same point of the grid. Moreover, instability problems may appear because the inverter is continuously injecting disturbances into the grid.

2.2.7 Sandia Frequency Shift (SFS)

This is an accelerated version of AFD and it is one of the positive feedback methods used to prevent the islanding operation (Report IEA-T5-09, 2002), (Lopes & Sun, 2006), (Wang et al. 2007) (John et al., 2004). With the grid connected, the method detects and tries to amplify small changes in frequency, but the presence of the grid avoid it. When the grid is disconnected, the frequency changes produce phase error and the positive feedback, in an iterative process, come the frequency beyond the threshold of OFP or UFP.
When the method is implemented, it is calculated the reference frequency for the inverter as a function of both the value at the iteration n and its variation Δf, following (18).

$$f_{n+1} = f_n + K_f \cdot \Delta f \qquad (18)$$

Where f_{n+1} is the reference frequency for the inverter in the $n+1$ cycle, f_n is the frequency in the n cycle and K_f is a constant that allows to accelerate the islanding detection. Finally Δf_n is the frequency variation in each cycle. K_f is designed to compensate the natural tendency of the system to move to the load resonance frequency when such a resonance frequency falls within the thresholds established to detect islanding.
In the $n+1$ cycle the inverter injects a current with a certain frequency. Therefore, the load introduces a sliding phase angle φ corresponding to an interval of time T_{ps} following (19).

$$T_{ps} = \frac{\varphi}{2 \cdot \pi} T_{n+1} = \frac{\varphi}{2 \cdot \pi \cdot f_{n+1}} \tag{19}$$

K_f is chosen to maintain the sliding frequency below the resonance frequency. In principle, this condition implies that the frequency in the $n+1$ cycle must be higher (for an increasing change) than the one of the n cycle, which yields in the following condition:

$$f_{n+1} = \left(\frac{1}{f_n + K_f \cdot \Delta f_n} - \frac{\varphi}{2 \cdot \pi \cdot (f_n + K_f \cdot \Delta f_n)} \right)^{-1} \geq \left(\frac{1}{f_n} \right)^{-1} \rightarrow K_f > \frac{-\varphi f_n}{2 \cdot \pi \cdot \Delta f_n} \tag{20}$$

Another issue for choosing the value of K_f is the desired time response. This can be accomplished by imposing a minimum time variation called shift time T_f. Therefore, it is obtained:

$$f_{n+1} = \left(\frac{1}{f_n + K_f \cdot \Delta f_n} - \frac{\varphi}{2 \cdot \pi \cdot (f_n + K_f \cdot \Delta f_n)} \right)^{-1} \geq \left(\frac{1}{f_n} - T_f \right)^{-1} \tag{21}$$

For frequencies close to the grid frequency the term $2\pi f_n T_f$ is appreciably greater that the phase-shift angle φ and the term $(1- f_n T_f)$ approaches 1. With these simplifications it is obtained (22):

$$K_f \geq f_n \frac{|\varphi| + 2 \cdot \pi \cdot f_n \cdot T_f}{(1 - f_n \cdot T_f) 2 \cdot \pi \cdot \Delta f_n} \rightarrow K_f \geq \frac{f_n^2 \cdot T_f}{\Delta f_n} \tag{22}$$

Figure 11 shows the block diagram of the method.

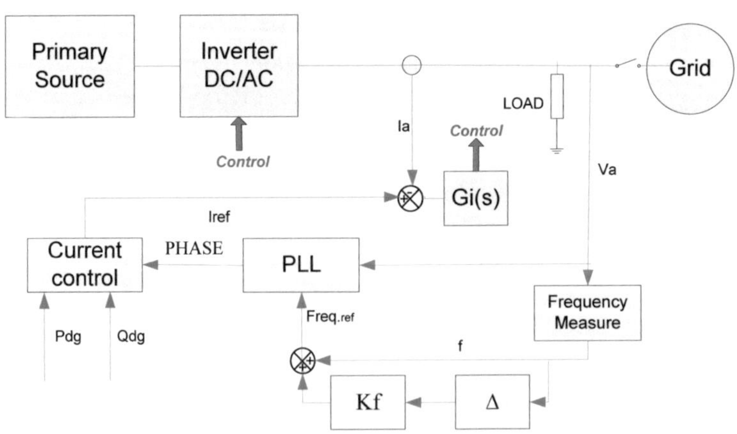

Fig. 11. Block diagram of the SFS method

2.2.8 Sandia Voltage Shift (SVS)

This method uses a positive feedback loop of the PCC voltage amplitude and it is similar to the active power variation technique. If the voltage amplitude (usually measured in RMS value) decreases, the inverter reduces its output current and thus the output power (Report IEA-T5-09, 2002), (John et al., 2004). The active power variation can be expressed following (23).

$$\Delta P_{DG} = 2 \cdot \Delta V \cdot \frac{V}{R} \tag{23}$$

ΔV can be calculated by comparing the voltage value with the one obtained after filtering. Therefore, with the grid connected, in steady state there is not voltage variation and the PCC won't be disturbed.

The response time of the algorithm can be adjusted by a factor *Kv* that increases or decreases the inverter current proportionally to the voltage variation. This value should be chosen following the same considerations that it was described in the active power variation method. The current reference for the inverter control can be calculated as follows:

$$I_{ref} = \frac{dP + P_{DG}}{V} = K_v \cdot \Delta V + \frac{P_{DG}}{V} \tag{24}$$

Finally, the method leads the voltage amplitude beyond the OVP/UVP limits allowing the islanding detection. To avoid any potential damage of the connected equipment, it is preferable to decrease the voltage amplitude instead increasing it. Figure 12 shows the block diagram of the method.

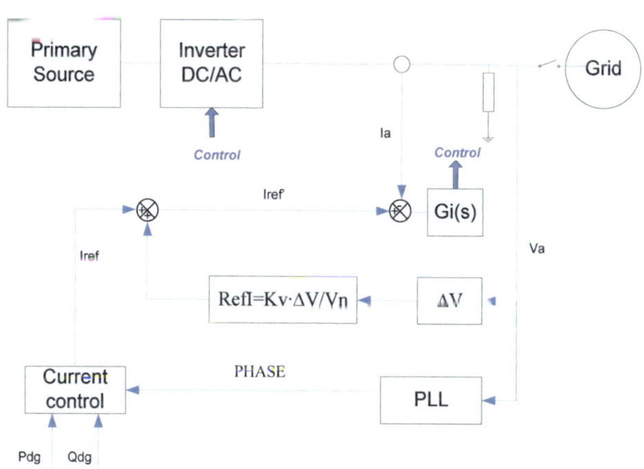

Fig. 12. Block diagram of the SVS method

SVS is easy to be implemented, and it is considered very effective among the methods that use positive feedback. Normally SVS & SFS are simultaneously implemented, improving the effectiveness of the method.

However, the SVS method has two disadvantages. On one hand it produces a small reduction in the power quality because the PCC voltage is continuously perturbed. On the other hand, efficiency of the MPPT algorithms may be affected.

2.2.9 General Electric Frequency Schemes – GEFS

This method injects a current disturbance into the system and evaluates the effects on the PCC (Ye et al., 2004), (Sun et al., 2004). The disturbance is added to the control signals in a Synchronous Reference Frame (SRF), usually known as DQ frame. The active power is proportional to the D axis component and the reactive power is proportional to the Q axis component.

This method is easy to be implemented and it has a reduced NDZ. In addition, it has a low impact on the power quality and it is very robust against grid disturbances. Nevertheless, the injection of disturbance signals (frequency and voltage) requires be as small as possible.

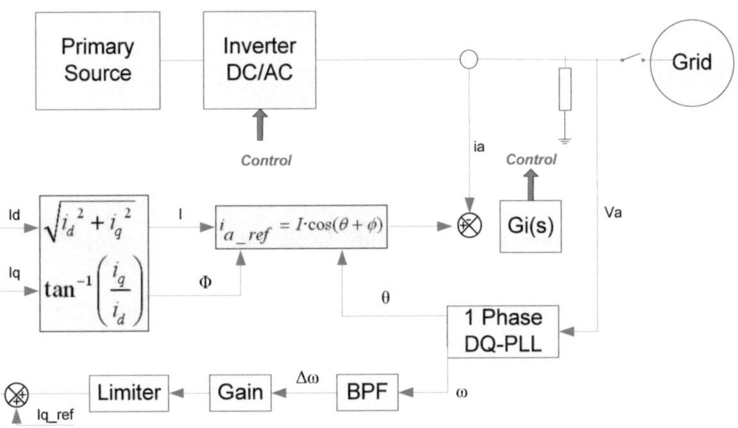

Fig. 13. Block diagram of the GEFS method

2.2.10 Mains Monitoring Units with Allocated All-pole Switching Devices Connected in Series (MSD)

This method relies on detection of the grid impedance. It uses two monitoring devices in parallel, connected to two series connected switch devices, which are independently controlled (Report IEA-T5-09, 2002). Each one of the units is continuously monitoring the voltage, the frequency and the impedance of the grid.

The advantages of this method are: small NDZ (very effective), redundancy monitoring and regular self-evaluation. However, the method has a high probability of interference with other devices including the grid itself.

3. Remotes techniques of islanding detection.

These detection techniques are based on some kind of communication between the grid and the DG. They are more reliable than the local techniques, but they are more expensive to implement. Some of these techniques are described in the following paragraphs.

3.1 Impedance Insertion

This method involves the insertion of a low impedance load, usually a bank of capacitors, which is connected to the PCC when the utility breaker opens (Report IEA-T5-09, 2002), (Hotta et al., 1993). As a result, the power balance between generation and load is modified. This disturbance causes a phase change and a sudden variation of the resonance frequency that can be detected by the OUF limits. A certain delay before connecting the additional impedance is mandatory to properly detect the frequency deviation. A scheme of the method is presented in Figure 14.

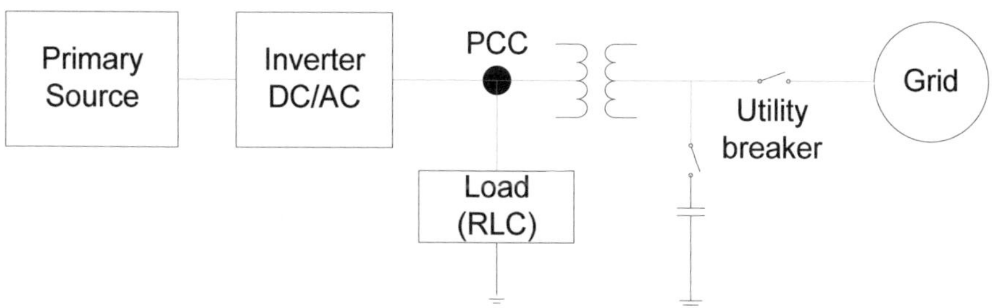

Fig. 14. Scheme of a method based on impedance insertion

This method has a low response time. In addition, the banks of capacitors can be used also for reactive compensation.
However, it is expensive to implement and the time needed to insert the capacitor bank after the grid disconnection could not meet certain standards. For this reason, the impedance value should be sized according to the minimum variation of phase (and therefore the frequency) that can be detected.

3.2 Power Line Carrier Communications (PLCC)

This method is a technique that relies on the use of the power line as a communication channel (Report IEA-T5-09, 2002), (Ropp et al., 2000), (Xu et al., 2007). The basic idea is to transmit a continuous low-energy signal between the transmitter (T) located on the side of the grid and receiver (R) located on the side of the DG. When this communication is disrupted, the receiver send a stopping signal to the inverter and/or a switch (included in the receiver) should be opened in order to isolate the load from the DG. The scheme of this method is shown in Figure 15.

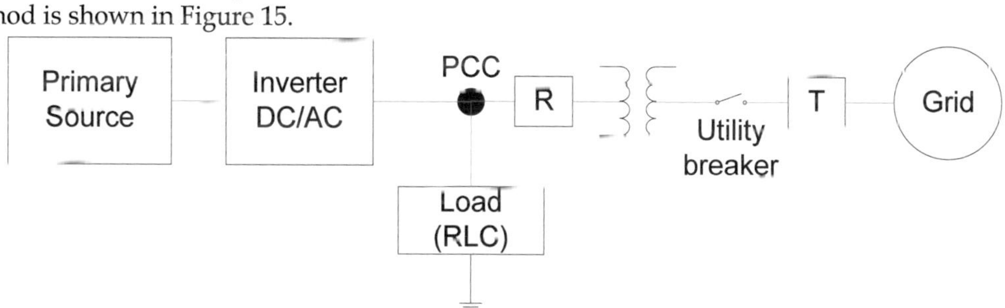

Fig. 15. PLCC system with both transmitter (T) and receiver (R) devices

Among others, the advantages of the method are: Ability to operate in areas with high density of DG. It does not have an NDZ. The inverter output power quality is not degraded. Its transient response depending on the type of signal to transmit, it is possible to use only one transmitter to cover a part of the grid.
Some of the weaknesses of this method are: the cost of the receiver and transmitter can be too high. Moreover certain charges under certain conditions highly abnormal, might replicate the emitted signal by the PLCC which would result in non-Islanding detection.

3.3 Signal Produced by Disconnect (SPD)

This method is similar to the PLCC-based method. The SPD method is based on communication between the network and the inverters to avoid Islanding (Report IEA-T5-09, 2002). SPD differs from the PLCC-based method in the type of transmission used (microwave link, telephone link or others). In this way the state of the switch is continuously known by the inverter. Figure 16 shows the schematic of the SPD method.

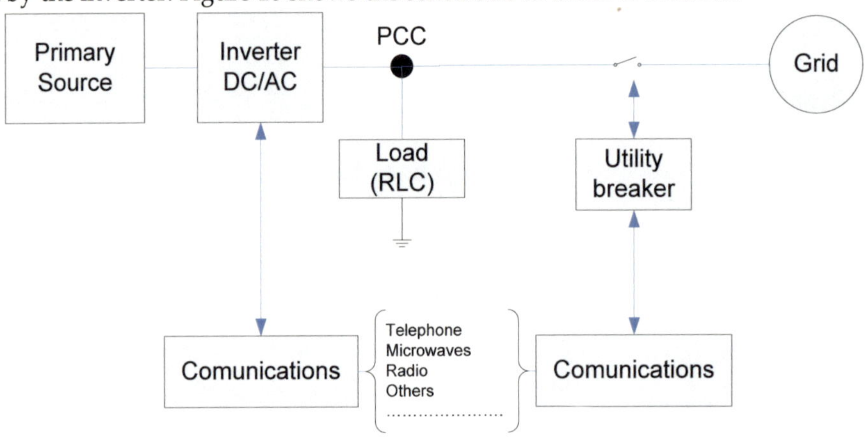

Fig. 16. Schematic of the SPD method

From the point of view of energy management, this method has the advantage of additional supervision and control of both the DG and the grid. Unfortunately, this method presents the great disadvantage of its high cost, which increases with every DG connected to the network. Besides, when the communication is via telephone, the communication wiring should be increased and communication protocols should be set up. This problem can be solved with the use of radio-frequency communications, but to cover up huge distances repeaters are needed, whereas a range of working frequencies should be established which might require licensing.

3.4 Supervisory Control and Data Acquisition (SCADA)

The inclusion of inverters in a SCADA system is a logical choice for Islanding prevention. SCADA systems use a wide communications network and sensors to control and monitor the grid connected equipment, allowing a fast response to contingencies that may arise in the grid, easing Islanding detection (Report IEA-T5-09, 2002), (Funabashi et al., 2003). When the grid is disconnected, a series of alarms are activated for disconnection of the DGs. This method is highly effective to detect islanding, eliminating the NDZ. However, this method presents the disadvantage of being too expensive and requiring a large number of sensors and additional features. Furthermore, it is not feasible in small installations.

4. Standards, codes and international recommendations about anti-islanding

Methods, circuits and procedures for anti-Islanding test have been documented in various international and national standards and recommendations. There are a great variety of approaches to verify anti-Islanding detection. Each country will choose its own methods to prevent islanding depending on its requirements. The test methods are generally defined for single and three phase inverters, as well as the requirements of the load to perform the tests. Following, a list of some standards, codes and recommendations concerning anti-Islanding is presented.

- IEEE Std. 929-2000, IEEE Recommended Practice for Utility Interface of Photovoltaic (PV) Systems, Sponsored by IEEE Standards Coordinating Committee 21 on Photovoltaics, Published by the IEEE, New York, NY, Apr 2000.

- UL1741, UL Standard for Safety for Static Converters and Charge Controllers for Use in Photovoltaic Power Systems, Underwriters Laboratories, First Edition, May 7, 1999, Revised Jan 2001.

- International Standard IEC 62116 DRAFT, Testing Procedure of Islanding Prevention Measures for Grid Connected Photovoltaic Power Generation Systems, International Electrotechnical Commission.

- Guidelines for the Electrical Installation of Grid-connected Photovoltaic (PV) Systems, Dutch guidelines to comply with NEN1010 (Safety provisions for low voltage installations), EnergieNed and NOVEM, Dec 1998.

- JIS C 8962:1997, Testing Procedure of Power Conditioners for Small Photovoltaic Power Generating Systems, Japanese Industrial Standard, 1997.EN61277, Terrestrial Photovoltaic (PV) Power Generating Systems -General and Guide.

- DIN VDE 0126:1999, Automatic Disconnection Facility for Photovoltaic Installations With a Nominal Output < 4.6 kVA and a Single-phase Parallel Feed by Means of an Inverter Into the Public Grid, (German National Standard for Utility Interconnection of Photovoltaic Systems).

- G77, UK Standard for Interconnection of PV and Other Distributed Energy, Generation, Standard Under Development, Expected Completion, Jan 2002.

5. Conclusion

In this chapter several techniques for Islanding detection have been presented. These techniques can be classified into two groups depending on their location in the DG system: remote techniques and local techniques. In the first group the detection algorithm is located at the grid side, whereas in the second group the detection method is located at the inverter side. Additionally, the local techniques can be divided into passive techniques, which are based on parameter measurement, and active techniques, which generate disturbances of the inverter output. Additionally, the passive methods do not affect the system THD, but present a NDZ which can be high enough to avoid islanding detection. On the contrary, active methods have the advantage that it is possible to detect Islanding in almost all situations, but they need to introduce a disturbance that could cause instability or distortion in the grid during normal operation, degrading power quality. Finally, the remote or 'non-resident in the inverter' methods, which do not affect power quality and have no NDZ, require extra hardware to be implemented, which means higher costs and may not be acceptable in many situations.

6. References

Ciobotaru, M., Agelidis, V. & Teodorescu, R. (2008). Accurate and less-disturbing active anti-islanding method based on PLL for grid-connected PV Inverters, *Proc. of IEEE Power Electronics Specialists Conference PESC 2008*, pp.4569-4576, Greece.

Ciobotaru, M., Teodorescu, R. & Blaabjerg, F. (2007). On-line grid impedance estimation based on harmonic injection for grid connected PV inverter, *Proc. of IEEE International Symposium on Industrial Electronics*, pp. 2437–2442, Spain.

Ciobotaru, M., Teodorescu, R., Rodríguez, P., Timbus, A. & Blaabjerg, F. (2007). "Online grid impedance estimation for single-phase grid-connected systems using PQ variations, *Proc. of the 35th Annual IEEE Power Electronics Specialists Conference PESC 2007*, pp. 2306-2312, U.S.A.

De Mango, F., Liserre, M., Aquila, A.D. & Pigazo, A. (2006). Overview of Anti-Islanding Algorithms for PV Systems. Part I: Passive Methods, *Proc. of the 12th International Conference on Power Electronics and Motion Control EPE-PEMC 2006.*, pp.1878-1883, Slovenia.

De Mango, F., Liserre, M. & Aquila, A.D. (2006). Overview of Anti-Islanding Algorithms for PV Systems. Part II: ActiveMethods, *Proc. of 12th International Conference on Power Electronics and Motion Control EPE-PEMC 2006*, pp.1884-1889, Slovenia.

Funabashi, T.; Koyanagi, K. & Yokoyama, R. (2003). A review of islanding detection methods for distributed resources, *Proc. of IEEE Power Tech 2003*, Vol.2, pp. 6, Italy.

Hotta, K., Kitamura, A., Okamoto, M., Takigawa, K., Kobayashi, H. & Ariga, Y. (1993) Islanding prevention measures: demonstration testing at Rokko test center for advanced energy systems, *Proc. of the 23rd Annual IEEE Power Electronics Specialists Conference PESC 1993*, pp. 1063-1067, Spain.

IEEE 929. (2000). *IEEE recommended practice for grid interface of photovoltaic (PV) systems*, Institute of Electrical and Electronics Engineers, U.S.A.

Jang, S. & Kim, K. (2004). An islanding detection method for distributed generations using voltage unbalance and total harmonic distortion of current, *IEEE Trans. on Power Delivery*, Vol. 19, No. 2, pp. 745-752.

Jeong, J.B., Kim, H.J., Back, S.H. & Ahn, K.S. (2005). An improved method for anti-islanding by reactive power control, *Proceedings of the 8th International Conference on Electrical Machines and Systems ICEMS 2005*, Vol.2, pp. 965-970, China.

Jeraputra, C. & Enjeti, P.N. (2004). Development of a robust anti-islanding algorithm for grid interconnection of distributed fuel cell powered generation, *IEEE Trans. on Power Electronics*, Vol.19, No.5, pp. 1163 -1170.

John, V., Ye, Z. & Kolwalkar, A. (2004). Investigation of anti-islanding protection of power converter based distributed generators using frequency domain analysis, *IEEE Trans. on Power Electronics*, Vol. 19, No. 5, pp. 1177 - 1183.

Jung, Y., Choi, J., Yu, B., So, J. & Yu, G. (2005). A Novel Active Frequency Drift Method of Islanding Prevention for the grid-connected Photovoltaic Inverter, *Proc. of IEEE Power Electronics Specialists Conference PESC 2005*, pp. 1915 - 1921, Germany.

Kitamura, A., Okamoto, M., Yamamoto, F., Nakaji, K., Matsuda, H. & Hotta, K. (1994) Islanding phenomenon elimination study at Rokko test center, *Proc. of the 1st IEEE World Conference on Photovoltaic Energy conversion*, pp. 759-762, U.S.A.

Kobayashi, H., Takigawa, K. & Hashimoto, E. (1991). Method for Preventing Islanding Phenomenon on Utility Grid with a Number of Small Scale PV Systems, *Proc. of the 21st IEEE Photovoltaic Specialists Conference*, pp. 695-700, U.S.A.

Liserre, M., Pigazo, A., Dell'Aquila, A. & Moreno, V.M. (2006). An Anti-Islanding Method for Single-Phase Inverters Based on a Grid Voltage Sensorless Control, *IEEE Trans. on Industrial Electronics*, Vol. 53, No. 5, pp. 1418-1426.

Liu, F., Kang, Y. & Duan, S. (2007). Analysis and optimization of active frequency drift islanding detection method, *Proc. of the 22nd Annual IEEE Applied Power Electronics Conference APEC 2007*, pp. 1379–1384, U.S.A.

Lopes, L.A.C. & Sun, H. (2006). Performance assessment of active frequency drifting islanding detection methods, *IEEE Trans. on Energy Conversion*, Vol. 21, No. 1, pp. 171 180.

Mahat, P., Zhe, C. & Bak-Jensen, B. (2008). Review of islanding detection methods for distributed generation. *Proc of the 3rd International Conference on Electric Utility Deregulation and Restructuring and Power Technologies*, pp. 2743-2748, China.

Menon, V. & Nehrir, M. H. (2007). A Hybrid Islanding Detection Technique Using Voltage Unbalance and Frequency Set Point, *IEEE Trans. on Power Systems*, Vol. 22, No. 1, pp. 442 - 448.

Pietzsch, M. (2004). *Convertidores CC/CA para la conexión directa a red de sistemas fotovoltaicos: comparación entre topologías de 2 y 3 niveles*, Bachelor thesis, U.P.C, Spain.

Report IEAPVPS T5-09. (2002). *Evaluation of islanding detection methods for photovoltaic utility interactive power systems*, International Energy Agency Implementing agreement on Photovoltaic Power Systems, U.S.A.

Ropp, M., Aaker, K., Haigh, J. & Sabhah N. (2000). Using Power Line Carrier Communications to Prevent Islanding, *Proc. of the 28th IEEE Photovoltaic Specialist Conference*, pp. 1675-1678, U.S.A.

Ropp, M., Ginn, J., Stevens, J., Bower, W. & Gonzalez, S. (2006). Simulation and Experimental Study of the Impedance Detection Anti-Islanding Method in the Single-Inverter Case, *Proc. of the 4th World Conference of the IEEE on Photovoltaic Energy Conversion*, Vol.2, pp.2379-2382, .

Sanchis, P., Marroyo, L. & Coloma, J. (2005). Design methodology for the frequency shift method of islanding prevention and analysis of its detection capability, *Progress in Photovoltaics: Research and Applications,* Vol. 13, No. 5, pp. 409-428, 2005.

Sun, H., Lopes, L.A.C. & Zhixiang, L. (2004). Analysis and comparison of islanding detection methods using a new load parameter space, *Proc. of the 30th Annual Conference of the IEEE Industrial Electronics Society IECON 2004,* Vol. 2, pp. 1172 - 1177, Korea.

Wang, X., Freitas, W., Xu, W. & Dinavahi, V. (2007). Impact of DG Interface Controls on the Sandia Frequency Shift Antiislanding Method, *IEEE Trans. on Energy Conversion,* Vol. 22, No. 3, pp. 792 - 794.

Xu, W., Zhang, G., Li, C.,Wang, W., Wang, G. & Kliber, J. (2007). A power line signaling based technique for anti-islanding protection of distributed generators—part I: scheme and analysis, *IEEE Trans. on Power Delivery,* Vol. 22, No. 3, pp. 1758-1766.

Ye, Z., Kolwalkar, A., Zhang, Y.,Du, P. & Walling, R. (2004). Evaluation of anti-islanding schemes based on non detection zone concept, *IEEE Trans. on Power Electronics,* Vol. 19, No. 5, pp. 1171 - 1176.

Ye, Z., Li, L., Garces, L., Wang, C., Zhang, R., Dame, M., Walling, R. & Miller, N. (2004). A new family of active antiislanding schemes based on DQ implementation for grid-connected inverters, *Proc. of the 35th Annual IEEE Power Electronics Specialists Conference PESC 2004,* pp. 235-241, Germany.

Zhang, C., Liu, W., San, G. & Wu, W. (2006). A Novel Active Islanding Detection Method of Grid-connected Photovoltaic Inverters Based on Current-Disturbing, *Proc. of the 5th International Power Electronics and Motion Control Conference IPEMC'06,* Vol. 3, pp. 1-4, China.

Single-Phase Photovoltaic-Inverter Operation Characteristic in Distributed Generation System

Muh. Imran Hamid and Makbul Anwari

Abstract

Single-phase grid tied inverter is one among types of inverters widely used in photovoltaic (PV) generation system due to the advantages they offer. This chapter describes model and simulation of such inverter in operation as distributed generation in electrical power system. Power characteristics including power quality, grid interaction behavior and load sharing that are important aspects in their operation as grid connected inverter will be simulated and analyzed. The role of current or voltage control and associated mechanism in photovoltaic inverter such as photovoltaic I-V characteristic, maximum power point tracker (MPPT), and other mechanism that involves in power flow and load sharing control are described.

Further, some observation and measurement from a 5-kWp laboratory scale grid interconnected photovoltaic plant that employ single phase photovoltaic inverter will be presented. The load sharing behavior between photovoltaic plant and utility grid during supplying both linear and non linear load that connected on their point of common coupling. In addition, observation and measurement results of power quality parameter behavior during photovoltaic inverter operation along extremely density variation of photovoltaic produced energy that comes from the atmospheric condition will be presented.

Keywords: single phase PV Inverter, distributed generation

1. Introduction

Application of photovoltaic (PV) as a source of electrical energy showed a tendency to increase in terms of generation capacity and in terms of its spread in large areas around the world. Many aspects trigger the trend; economic, technology and policy are some among many. The restricted reserve of fossil fuel sources and followed by the increasing cost of fossil fuel based electricity generation has motivated the effort to exploit other alternative energy sources. In the other hand, the high price of equipment and system of photovoltaic generation as the main constraint on implementing this renewable generation system shows significant reduction during recent years; implicates to declination of production cost per kW electric from photovoltaic. IEA reported that over a decade (1996-2006), the price of photovoltaic system have decreased by probably more than 40% [1]. The maturity and continuously improved technology implemented on photovoltaic generation system that causes the photovoltaic power conversion more efficient, the typical advantages of PV generation compared with other electrical generation systems of renewable energy sources such as its flexibility and simplicity to build in any places, their dependency from transportation system are some technical factors causes the change to this type of renewable energy generation for electricity is preferred. In energy policy, environment issue and global warming as important consideration for the authority in deciding the choice of energy sources for their electricity make the photovoltaic generation system as the clean and convenient energy as one among priorities. Photovoltaic energy conversion becomes main focus of many researches due to its promising potential as source for future electricity.

The photovoltaic generation systems can either be operated as isolated system or be connected to the grid as a part of an integrated system, with other electrical generation, they form the distributed generation system. As renewable distributed generation, PV has some advantages if it is compared to other renewable energy generations. PV generation plant needs not a specific geographic or geo-morphological requirement such as on the wind and micro/small hydropower generation. In contrary, PV generation plant can be built in almost all area where the sun irradiation is available; allows the flexibility to determine the place of the plant according to its main allotment. The solar field and building integrated photovoltaic (BIPV) [2], [3] are some cases if the place- flexibility is taken as the advantage. In power system point of view, the place-flexibility of PV generation allows to employ the plant not only as power source for distributed generation, but can also as part of scheme for the transmission/distribution losses reduction [4], as compensator and power conditioner for the power system [5]. In addition, the module-based production of PV plant components that enables ones to build and adjust the size of PV plant from small capacity and then expand it to follow the demand growth is also one of advantages of this type of generation system. These entire make the PV generation becomes an interesting choice for development the distributed generation as direction of transition of the electrical power generation system [6].

However, besides the advantages, there are also several limitations of PV plant as electrical power generation for distributed generation. The reality that PV plant is built based on a number of power electronics equipment as the characteristic of renewable energy generation plant [7] causes the PV plant is considered as distortion and power quality problems source for the connected consumers or for electrical power component and system. Further, as the part of distributed generation, interaction with the power system component and load are also a factor influences the power quality resulted from PV generation plant operation. Regarding to these issues, some research has been conducted to investigate the effect PV generation connection to system, in [8] and [9], the harmonics effects of a large number PV inverter penetration in distribution network is analysed, the possibility of resonance event is found as the effect of interaction of harmonics produced from the inverters in the network. Injection of dc component in ac grid concerning to PV inverter integration have also reported in [10].

Other aspects that may influence the performance of PV generation for DG that should also be consider are: the effect of the extreme variation of power density variation in a range of hourly and daily as the nature of photovoltaic energy source, effect of PV plant configuration, type of PV inverter implemented, and interaction with the system in form of load flow and load sharing with grid. The influence of these aspects can be reviewed by analyzing the role of the PV inverter as the heart of a PV generation system. This chapter presents an analysis of performance and power quality aspect of operation a PV plant as distributed generation inverter concerning to their operation within dynamic atmospheric condition, the type of inverter used, their configuration with the photovoltaic array and their interaction with the grid parameters and loads. The single phase type PV inverter is stressed to be analyzed. A brief review of the PV generation systems and their setup components, and their potential to affect the quality of power output is firstly presented. Further, some measurement results from a laboratory scale generation system that show some power quality behaviors during plant operation are described and analyzed.

2. Photovoltaic Generation System

To analyze the power quality behaviors resulted from operation of a PV plant in distributed generation system, a review of the PV generation system and some aspects that involved in their operation as the part of the system must be done, they are: the PV module, PV inverter and the module-inverter configuration, the PV plant-grid interaction and the atmospheric condition. The main components of a PV generation plant are the PV modules and the PV inverters. The PV module is used as energy conversion equipment, converting the light energy to electrical form of the dc voltage and current. The conversion involves interaction process of the light, thermal and electrical parameters in a photovoltaic material. The PV inverter is then used to convert the dc to ac power to be used by consumer or to be connected to the grid. Some functions such as matching the array's dc voltage output with the inverter circuit voltage operation, matching the inverters' output voltage with the grid voltage and grid synchronizing are integrated in this compact equipment.

2.1 PV Module

Photovoltaic module is set up from formation of photovoltaic cells that convert the energy in the light to electric power. The modules are then arranged in both series and parallel configuration as a photovoltaic array to reach the voltage and current requirement. Photovoltaic cells are semiconductor devices that draws non-linear characteristics between output current *(I)* and voltage *(V)* on their terminal. In a condition when the light shapes their surface, the light generations current are produced in proportional to the light intensity, in the same time a dc voltage is generated. This condition presents the current source behavior of the cell. If the generated voltage is high enough, the solar cell current drops extremely, similar to the behavior of the diode as seen on the knee point of the diode's characteristic. From these behaviors, based on circuit perspective, photovoltaic cell can be modeled as configuration of current sources in parallel with some diodes, serial and parallel resistors are then added to present the voltage and current losses during cell operation. A photovoltaic cells model called the single diode model is widely used as shown in Fig.1. The model is a simplification of the two diode model by setting the value of ideality factor -a number that indicates the dominate region in a semiconductor material- of photovoltaic cell based on whether the diode's behavior of the cell is dominated by recombination or depletion region [11].

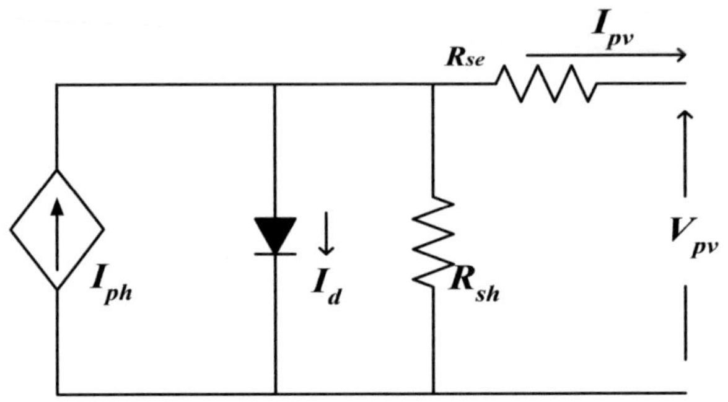

Fig. 1. Single diode model of a PV cell

This single diode model of PV cell can be described using the follows mathematical model to form the cell's *V-I* characteristic as [12]:

$$I_{pv} = I_{ph} - I_d - I_{sh} \tag{1}$$

$$I_{pv} = I_{ph} - I_o\left[e^{\frac{q(V_{pv}+I_{pv}.R_{se})}{A_o kT}} - 1\right] - \frac{(V_{pv} + I_{pv}.R_{se})}{R_{sh}} \tag{2}$$

Where

$$I_{ph} = [I_{SCR} + K_1\,(T - 25)] - \frac{G_a}{100} \tag{3}$$

$$I_o = I_{or}\left[\frac{T}{T_r}\right]^3 exp\left[\frac{qE_{Go}}{A_o k}\left(\frac{1}{T_r} - \frac{1}{T}\right)\right] \tag{4}$$

I_{pv} and V_{pv} are the current and voltage of the PV cell. I_{ph} is the light generating current, its value depends on irradiance and the physical dimension of photovoltaic cell, I_o is the diode dark saturation current. The Boltzmann constant = 1.3807 x 10^{-23} JK^{-1} and electric charge = 1.6022 x 10^{-19} C are presented by k and q. R_{se} and R_{sh} are the representation of the parasitic series and shunt resistances that associated with real solar cells in operation condition. A_o is the diode ideality (quality) factor; its value is taken between 1 and 2, A_o = 1 indicates that diode behaviour of cell is dominated by recombination in the quasi-neutral regions and A_o = 2 indicates that recombination in the depletion region dominates. I_{SCR} is short circuit current of the cell on 1000 W/m^2 and 25 ^oC of temperature. K_1 is the short circuit temperature coefficient at I_{SCR}. T and G_a are cell temperature and irradiance on cell surface. I_{or} is the cell saturation current at reference temperature T_r. E_{GO} is band gap energy.

Equation (1)-(4) shows the dependency of I-V characteristics of a photovoltaic cell to irradiance and temperature condition. The irradiance contributes to the cells current; the higher irradiance the higher current generates by the photovoltaic cell, while the temperature makes effect to the cell's voltage; the higher temperature the lower voltage appears on the cell's terminal. Figure 2(a) shows a set of I-V characteristics of a photovoltaic cell under varying irradiance, but at a constant temperature, while Fig. 2(b) shows the one at the same irradiance value, but under varying temperature. Both figures also show the points where the multiplication of voltage and current of PV cell reaches the maximum value;

maximum power point (MPP). At these points, the PV cell operates in maximum efficiency and produces maximum output power for the related irradiances and temperatures.

2.2 PV Inverter

The second main components of the PV generation system are the PV inverters. This component interfaces the various power density generated by the photovoltaic array to the utility level of electrical power. As this role, some functions are employed: adjusting the voltage level of photovoltaic array output to meet the voltage operation of the inverter circuit; tracking the voltage/current to a point where maximum power can be extracted and creating sinusoidal ac power. If the PV inverter is used as grid tied PV inverter, it must be completed with a synchronizing and power flow control mechanism, a reliable protection, such as anti islanding to protect the inverter from over load, must be added. In addition, the modern PV inverter is also completed by an advanced data communication and monitoring system.

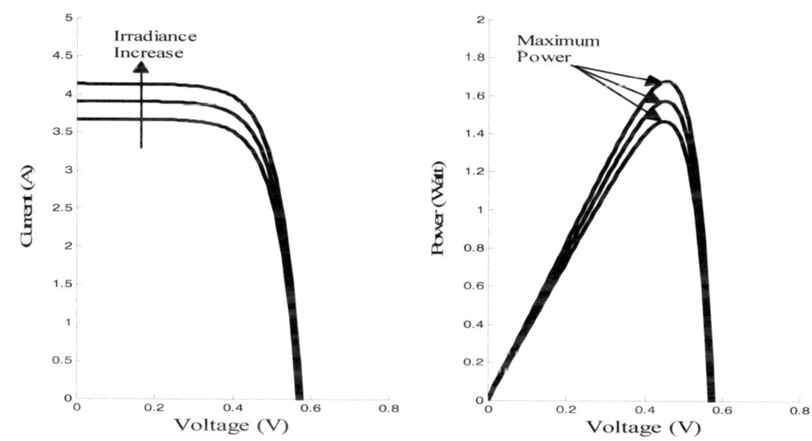

a. V-I characteristic on various irradiance and constant temperature

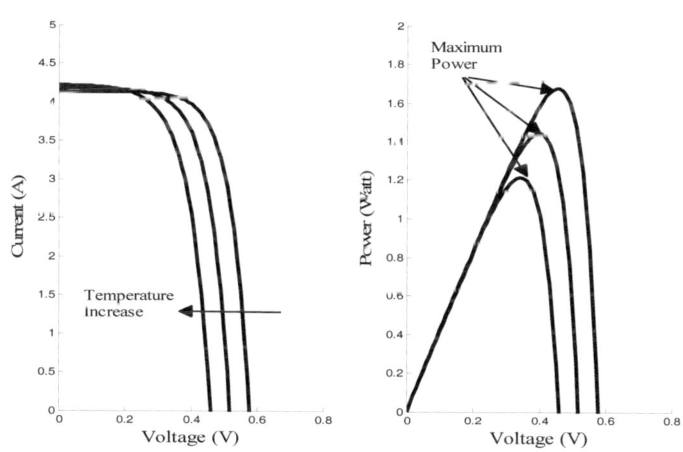

b. V-I characteristic on various temperature and constant irradiance

Fig. 2. *I-V* characteristics of a photovoltaic cell under varying irradiance and temperature

Maximum power point tracker (MPPT) is an important mechanism in a PV inverter; it is embedded to ensure that PV inverter extracts the optimum power generated by photovoltaic module or array. Cause of the unique properties of the PV array, when it is connected directly to a load, the operating point of the system lies on the crossing point of the *I–V* characteristics curve of the PV array and *I-V* characteristic curve of the load. Generally, this point is not at the PV array's maximum power point (Fig. 3), thus optimum extracted power cannot be achieved. In this case, to ensure that the load's power requirement can be supplied, a larger capacity of array is needed; leads to an expensive system. MPPT overcomes this problem, using certain algorithm, the mechanism forces the converter's voltage or current to operate on the value where the multiplication of array's voltage and current lie on the maximum condition, thus optimum power extraction can be achieved. The important role of the MPPT in PV inverter make this component becomes depth attention to the researchers, various techniques and development algorithms have been developed as in reference [13] and according to reference [14], the techniques can be categorized to three main categories: Look-up table, perturbation and observation (P&O) and computational methods.

In a photovoltaic generation where the dc voltage output of PV array is not match with the associated ac voltage must be generated by the inverter circuit, DC-DC converter is implemented on the input side of the inverters' power circuit and the inverter system is called the multi stage PV inverter. Boost, buck and buck-boost or other types of DC-DC converter are used for this need. The MPPT mechanism in this case is implemented in the DC stage as direct duty cycle control for the DC-DC converter. Some types of multi stage PV inverter, for safety consideration, employ a transformer placed on the line side (LF transformer) or between power stages (HF transformer) of the inverter system.

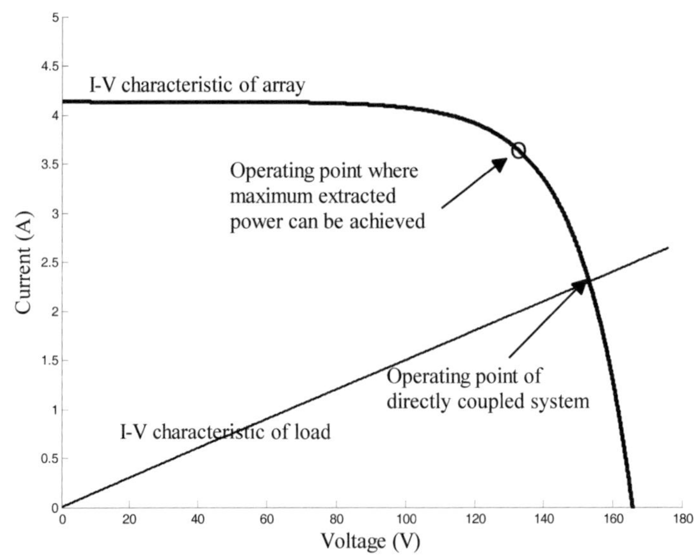

Fig. 3. Operating point of directly coupled system

For the grid connected inverter (grid tied inverter), one among functions of control mechanism of PV inverter is aimed to ensure that all the generated power can be sent to the grid optimally, voltage and current control methods together with the MPPT are used for this requirement [15]. Protection mechanism such as anti islanding is added to avoid the PV inverter from overloaded condition when a fault exists on the grid. Synchronization with the grid is performed by various methods such as: by filtering the grid voltage, using PLL method, and using zero crossing detector based synchronization method.

Array-Inverters Configuration

From the overall requirements, the PV inverters are hoped to operate in optimum conversion efficiency. Photovoltaic material exploration, converter topologies and control mechanism have and being continuously developed. Development of power conversion efficiency can also be reached by optimizing the configuration between PV module/array and their associated PV inverter. Array-PV inverter configuration implicates to technical characteristic of plant such as voltage level, losses, reliability and power quality. In the real application, according to connection and configuration between PV module and PV inverter, photovoltaic generation can be identified as [16]:

Photovoltaic generation using central PV inverter, this configuration is reached by paralleling some PV module strings (some PV modules in serial connection) and connected to dc side of a relatively large capacity PV inverter. High conversion efficiency can be reached by using this configuration but faces a weakness along with diversity of PV module types and diversity of irradiance or partial shading. This configuration is also susceptible to reliability problem because the plant depends on single equipment only. Central inverter configuration is often implemented in a high capacity generation plant to optimize the cost, even though tend to limit the flexibility for adjusting plant capacity.

Photovoltaic generation using string PV inverter, to overcome the weakness of central inverter configuration, the string PV inverter configuration was introduced. Each module string is connected to one inverter with own MPPT. Optimum power extracting from each string can be achieved, overall efficiency is better and PV generation reliability is increased because the system is not depend to only one equipment anymore. Plant capacity is also easy to adjust to follow demand growth.

Photovoltaic generation using multi-string PV inverter, this configuration is developed to adopt the cost and technical advantageous of both previous configurations. Some module strings with dc-dc converter and their own MPPT mechanism are connected to one inverter. Optimum power extracting from each module string can be achieved and in other side reduced cost for dc-ac converter can be done. This configuration allows using various types of inverter, various modules with different electrical characteristic and configuration.

Photovoltaic generation using team system, even though optimum power extraction can be achieved using multi-string PV inverter configuration, but implementation of single dc-ac converter in second stage of electrical conversion in other hand will decrease the reliability of generation system. Team system configuration was introduced to solve this problem, this configuration contains several module strings and its own PV inverters set up in parallel. When irradiance high enough, each module strings and its own inverter work independently for injecting power to ac side, on the contrary, if the irradiance goes down, module string configuration is changed in such a way so that not all of dc-ac converters operate. This scheme ensures that dc-ac converter always operates around their rated power.

Figure 4 illustrates the above configurations of PV generation system. In addition, module integrated PV inverter; a system in which a PV inverter is designed with its associated module specifically is also being developed. The string PV inverter, influenced by cost consideration, plant capacity, and flexibility to enlarge the plant capacity have widely implemented and become standard PV system technology for grid connected PV generation plant [17].

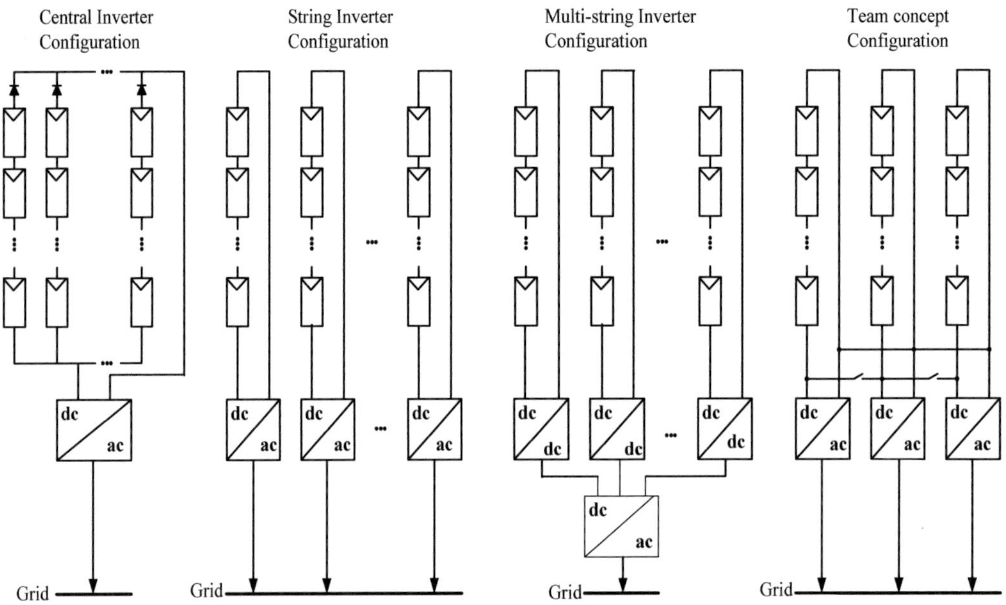

Fig. 4. Some PV module configurations in a PV plant [16, 17]

3. Single Phase Photovoltaic Inverter

Along with the development and diversity application of photovoltaic as electrical source cause the PV inverters also spread to various types with the trend of single phase application. Single phase PV inverter has advantageous in flexibility for connecting to both single phase and three phase grid (three phase connection is formed by connecting at least three unit single phase inverters); this trend is also influenced by trend of PV generation plant configuration, as has been be discussed. For single phase application, PV inverter can be categorized in many point of views, in [18] single phase inverter is categorized based on the number of power switch that used and the number of power conversion sequences, while in [19] categorizing is done according to the number of power stages in cascade, type of power decoupling between the PV modules and the single-phase grid, is it utilizes a transformer or not and the types of grid-connected power stage. Figure 5 shows the diagram of some functions in a single phase PV inverter.

In the markets, PV inverters are found as a three-phase or single-phase unit, both in isolated and grid tied connection type. The three phase PV inverters are produced in relatively high capacity and used for large power application such as central inverter in a PV generation plant. Whereas, the single phase inverters are generally produced as string inverter in lower capacity, used in small power application such as in building photovoltaic system or in individually residential photovoltaic electricity. In a PV generation plant using single phase inverters, enlarging the plant capacity is done by implementing a number of PV inverters and connecting them to form the three phase connection.

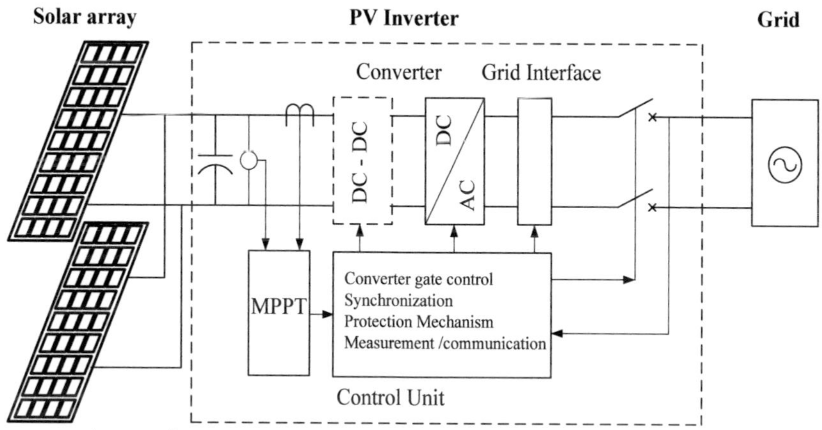

Fig. 5. Diagram of some functions in a single phase grid-tie PV inverter

4. Simulation of photovoltaic generation using single phase inverter

4.1 Photovoltaic characteristic generator model for simulation

One important aspect in photovoltaic generation study is to simulate the electrical parameter characteristic of photovoltaic module as the power source converter. The first step have to do is to model the characteristic relation of module/array output parameters (voltage and current) and input parameter (irradiance, temperature, humidity, etc) in form of mathematical model that can be use in digital simulation. Regarding to this, some approach can be done, mathematical model can be obtained from empirical characteristic of the array [20], here the model is built based on data from direct measurement of photovoltaic array's voltage/current in various inputs condition, model is then derived from plotted curve that describes voltage/current relation with the input parameter. If photovoltaic modules come with the *I-V* characteristic curve, mathematics model can also be derived from that curves. Other approach is by modeling the module/array characteristic based on the equations that present the physical behavior of the photovoltaic cell; involves the relation between light, thermal and electrical parameters [21]-[23], and such approach is indicated by Eq. (1)-(4).

Generally, as shown by Eq. (1)–(4), mathematic model obtained from described methods are in form of implicit and exponential equations that difficult to be solved and simulated. Therefore using Matlab/Simulink, to solve and simulate such model, the photovoltaic array equations are presented using several blocks, arranged such way so that implicit equation can be solved. An S-Function file is built in a block to solve numerically the photovoltaic cell characteristic equation and to synthesize the cell equation as array characteristic equation that contains of photovoltaic cell and module. If photovoltaic module contains N_s cells in series and N_p cell in parallel, thus (2) become:

$$I_{pv} = N_P \left\{ I_{ph} - I_o \left[e^{\frac{q\left(\frac{V_{pv}}{N_S} + \frac{I_{pv}.R_{se}}{N_P}\right)}{A_O kT}} - 1 \right] \right\} - \frac{\left(\frac{V_{pv}}{N_S} + \frac{I_{pv}.R_{se}}{N_P}\right)}{R_{sh}} \tag{5}$$

Simulation block of the PV array characteristic generator can be seen on Fig. 6. It is drawn as an S-Function block connected to a controlled current source block. As the input of S-Function blocks are the irradiation, temperature, and a loop of one-sample delayed voltage data from the output of whole PV array characteristic block generator. The loop is used to solve the implicit function in the *V-I* characteristic equation of the array. Output of the S-Function block is used to drive the controlled current block so that the entire blocks configuration present a controlled current source characteristic of the PV array for used in time domain simulation.

To evaluate the performance of Simulink block of PV array model, the block model is simulated as follow:
two *I-V* curves; *I(V)*, in different irradiance value according to Eq. (5) are plotted on an *i-v plane*. With the same temperature value, the Simulink block of array model and connected load are then run, firstly with the irradiance value is similar to one used in plotting the curve on *i-v* plane, voltage level and generated current as the array operating point is marked. Then, during simulation is running, the value of irradiance is changed to the value used to plot the second curve in same *i-v* plane, operating point is marked once again. For both conditions, the operation marked points should be lie on a point along the corresponding *I-V* curve.

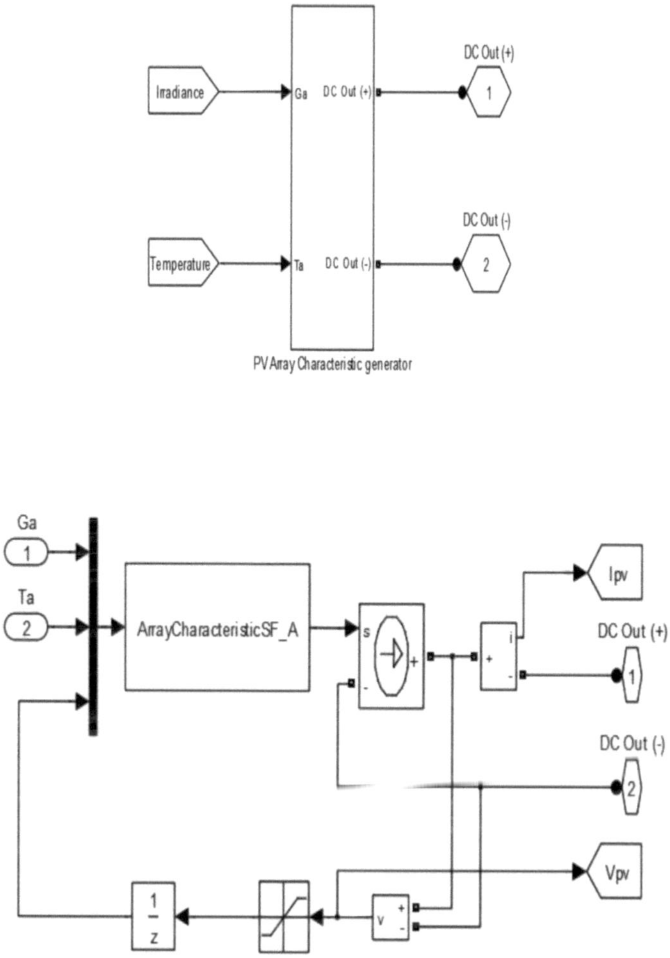

Fig. 6. Simulation block for the PV array characteristic generator in Simulink

Figure 7 shows the result of the treatment; *I-V* characteristic curves are plotted with 500 and $1000 W/m^2$ of irradiances and in same temperature of 25^oC. The PV array block model then is run with connected load of 10-Ohm impedance. Figure 7(b) and 7(c) are taken from running the array block in Simulink, for irradiance as $500 W/m^2$ on temperature 25 ^oC, it shows that the connected load causes photovoltaic array operates on 31.36 Volt and supplying 3.129 A of current. If irradiance is increased to 1000 W/m^2 on same temperature, the array operating point will shift to 32.01 Volt and 3.2 A. As supposed, these values are precisely located at a point along *I-V* characteristic curve of photovoltaic array as shown on Fig. 7(a) as detail portion of Fig. 10(d).

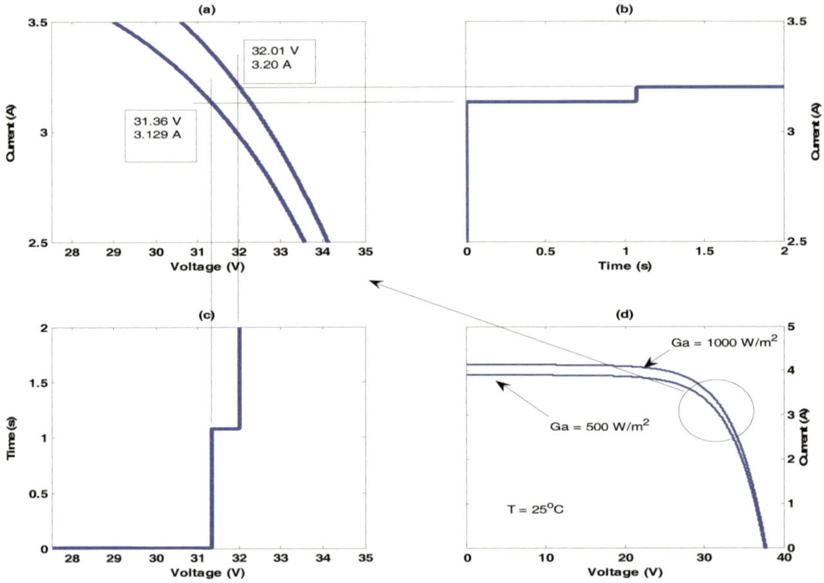

Fig. 7. PV array operation point obtained from plotting PV equation characteristic and from simulation in Simulink

4.2 Power and control circuit

Power converter block contains of the inverter power and control circuit. Inverter power circuit is set up from a coupling capacitor as power balancing for the instantaneous generated and delivered power to the grid, a single-phase bridge inverter circuit block controlled by PWM mode, an output filter, and a grid connection trough a step up transformer for matching the photovoltaic plant and grid voltage. Control circuit contains maximum power point (MPPT), current control and synchronizing block. In this case, the incremental conductance algorithm is used for the MPPT; as illustrated in Fig. 8, this algorithm works based on the condition that at the point of maximum power, the rate of change of output power to the array voltage is zero:

$$\frac{dP}{dV} = 0 \tag{6}$$

$$\frac{dP}{dV} = \frac{d(VI)}{dV} = I + V\frac{dI}{dV} = 0 \tag{7}$$

$$I + V\frac{dI}{dV} \cong I + V\frac{\Delta I}{\Delta V} = 0 \tag{8}$$

$$\frac{\Delta I}{\Delta V} = -\frac{I}{V} \tag{9}$$

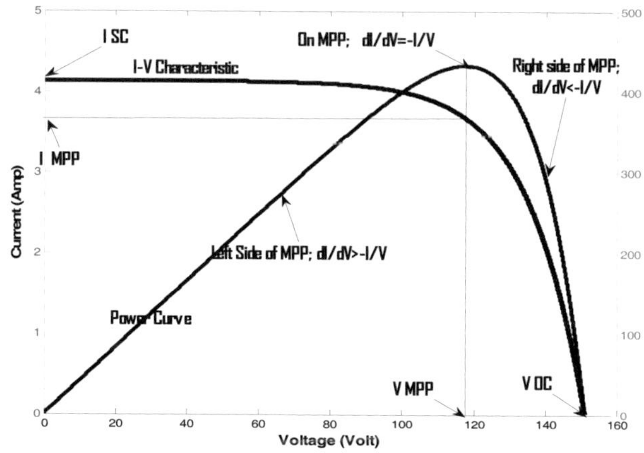

Fig. 8. Increment conductance ΔI/ΔV around characteristic curve of photovoltaic array

On the maximum power point, the value of incremental conductance *ΔI/ΔV is* *-I/V* and varies around it as see on array characteristic curve, briefly:

$$\frac{\Delta I}{\Delta V} = -\frac{I}{V} \qquad \text{on MPP}$$

$$\frac{\Delta I}{\Delta V} > -\frac{I}{V} \qquad \text{on the left of MPP and}$$

$$\frac{\Delta I}{\Delta V} < -\frac{I}{V} \quad \text{on the right of MPP and} \qquad (10)$$

Flowchart of the incremental conductance algorithm is shown on Fig.9.

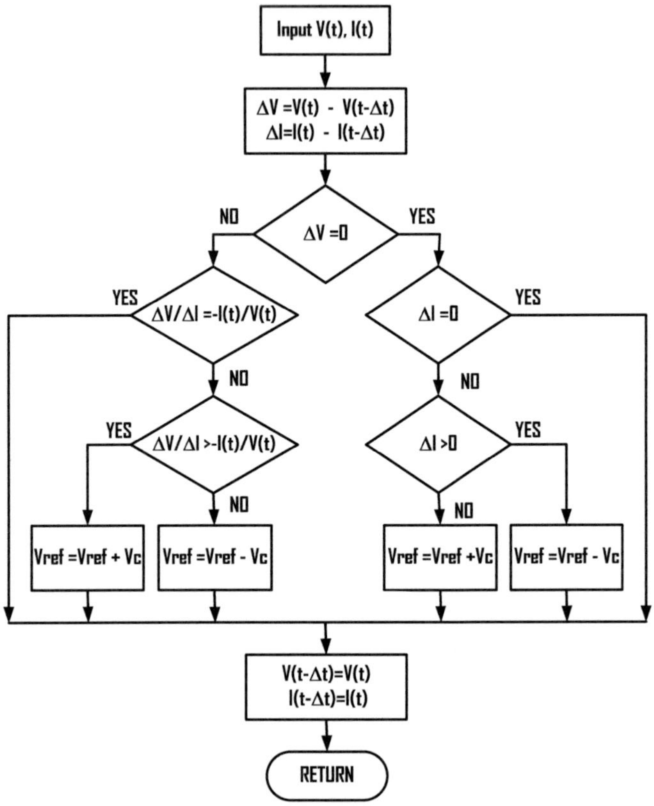

Fig. 9. Incremental conductance algorithm flowchart

In simulation, the algorithm is implemented in a S-Function block with the instantaneous current *I(t)* and voltage *V(t)*, and each of their previous value in one sample time Δt are used as inputs signal. The algorithm than calculates the increment conductance and evaluates it according to Eq. 7, the result is used to determine if the correction voltage V_C for reference voltage V_{ref} become positive or negative. An initial array voltage is required for this algorithm; values of 0.7 or 0.8 V_{NL} (No load voltage) are preferred values. The error between V_{ref} and actual PV array voltage is then used by controller to form reference current for inverters' current controller after added by actual current reference generated from power calculator block. Power calculator is a block where the instantaneous generated power in the array is calculated. Implementation of current control has an advantageous that current can be controlled independently from grid voltage, thus active and reactive power and certain power factor level is allowed in all range of load [15]. Reference current as the output of MPPT block is fed to current control block where the switching instant of the inverter switch is formed. In current control block, reference current determines the magnitude of injected current, while angle and current shape is determined by generated sinusoidal wave from the phase lockup loop (PLL) block. PLL operates by continuously observe the phase angle of grid voltage and compare it with an internally generated angle, a loop control is employed so that a reference phase angle can be determined according to grid phase voltage.

To perform the switching instant for current control, some modulating techniques can be used; here the hysteresis and ramp comparison current control [24], is taken as the case, Fig.10. By using hysteresis current control, inverter output current is forced to follow the current reference. Deviation between these two quantities is limited by upper and lower band in a hysteresis loop. If actual current reaches the upper limit of hysteresis band, the inverter leg is switched off so that the current decrease untill reach the lower band of hysteresis loop. In this point, the inverter leg is switch on again and actual current back to increase to the upper band, the process repeat continuously. If ramp comparison current controller is used, a sinusoidal-wave signal is added to a triangle signal for creating a sinusoidal-triangle reference. This reference is then compared with the actual current. The instant when the sinusoidal-triangle wave and the actual current crosses becomes time when the inverter legs are switched. If the current error greater than sinusoidal-triangle wave, the inverter leg is switch off and in contrary if the current error less than sinusoidal-triangle then the inverter leg is switch on.

As the end of power circuit, a step up power transformer is used to connect the PV inverter circuit to the grid. Three-phase connection is formed by connecting at least three unit single-phase inverters that connected to each phase of the grid. Inverters in this configuration operate independently with their own control. Control signal for each PLL block in each inverter are picked from each grid phase. Complete simulation block diagram is shown in Fig. 11.

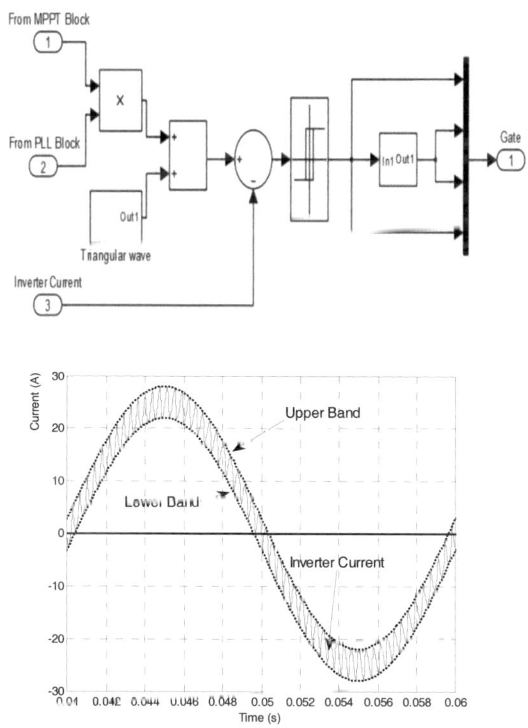

a. Hysteresis current control

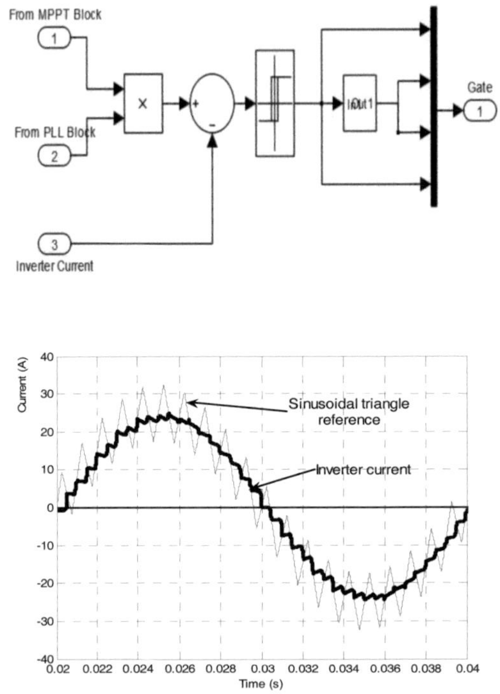

b. Ramp comparison current control

Fig. 10. Modulating techniques for current controller used in the simulation

In order to evaluate the operation of power and control circuits, the complete simulation diagram is run under variation of irradiance and temperature. Photovoltaic inverter is connected to the grid and is assumed in parallel with other generation equipment in the grid for supplying the connected grid load. Grid load is set up such a way so that enable to accept all generated power from photovoltaic plant -or in other word- load capacity is set up more than the photovoltaic plant capacity. By this way, whatever power can be generated by photovoltaic plant, it is able to be absorbed by the grid load.

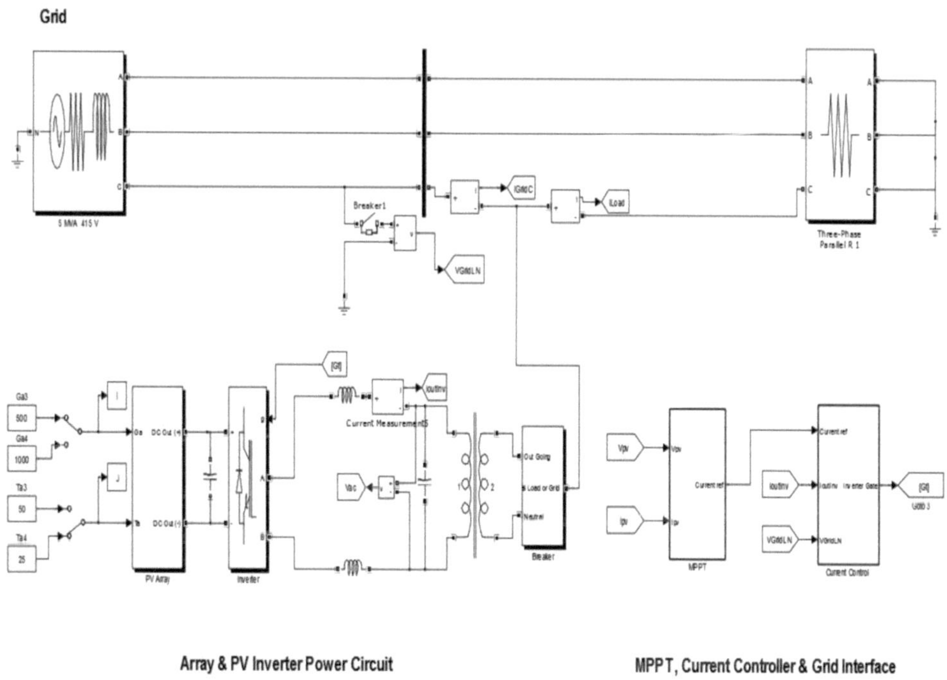

Fig. 11. Complete single-phase grid-tie PV inverter simulation diagram

Operation of MPPT as optimizing mechanism of delivered and injected power to the grid can be seen by measuring the generated power of PV array (dc side) and output of inverter (ac side) in a condition when the inputs of photovoltaic array are varied. MPPT operation indication will appear when the measured value of output power inverter always follows the value of dc power as the output of photovoltaic array. This condition is shown in Fig.12, the values of generated power by PV array correspond to irradiance and temperature level, each combination of their values will give different generated power on array output. It is also shown the value of power measured on output terminal of inverter that always run before the value of generated power that measured on array terminal. For G_a = 500 W/m^2 and T_a = 25°C that equivalent with dc output power of array P_{dc} around 475 Watt, injected power on output terminal of PV inverter P_{ac} shows the same value. When the irradiance is changed to G_a = 1000 W/m^2 on the same temperature with equivalent to P_{dc} = 500 Watt, the trend of P_{ac} curve tend to this value. These behaviors indicate that the MPPT mechanism works properly.

The current drawn by connected load is supplied from both grid and PV plant. It is different with paralleling two sources such as transformers or generators in supplying a connected load, where capacity and internal impedance are parameter that determine load sharing between both equipments, load current drawn by PV inverter in a grid is not depend on its capacity. Optimum power can be sent to grid as long as there is sufficient power generated by the plant. Figure 13 shows simulation result how a connected load current is supplied from both grid and PV inverter. Initially, whole load drawn by the constant load is supplied from the grid. At point when the PV plants switched on, the PV plant current increase and the current from grid decrease and then go to steady condition untill the PV plant is switched off.

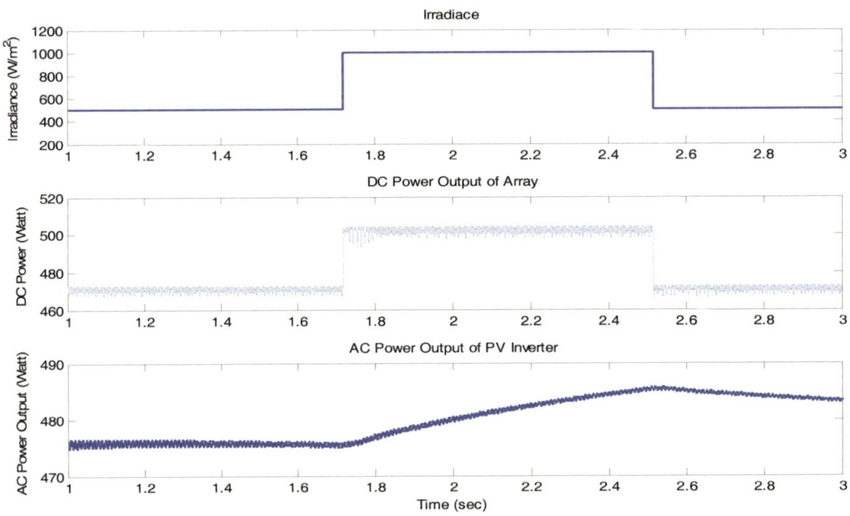

Fig. 12. MPPT operation is indicated by the trend of ac power output that is always tend to dc power value

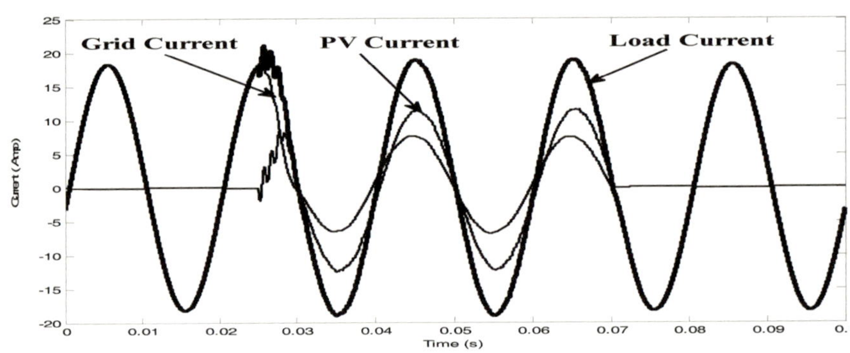

Fig. 13. Load current sharing between grid and plant on phase *a* when plant switched

5. Observation and measurement results of a laboratory scale PV distributed generation operation using single phase inverter

In this section, some power behaviors from the operation of inverters in a PV plant such as power quality, voltage stability, harmonics, wave form distortion and load sharing characteristic will described. The schematic diagram of the plant is shown in Fig. 13.

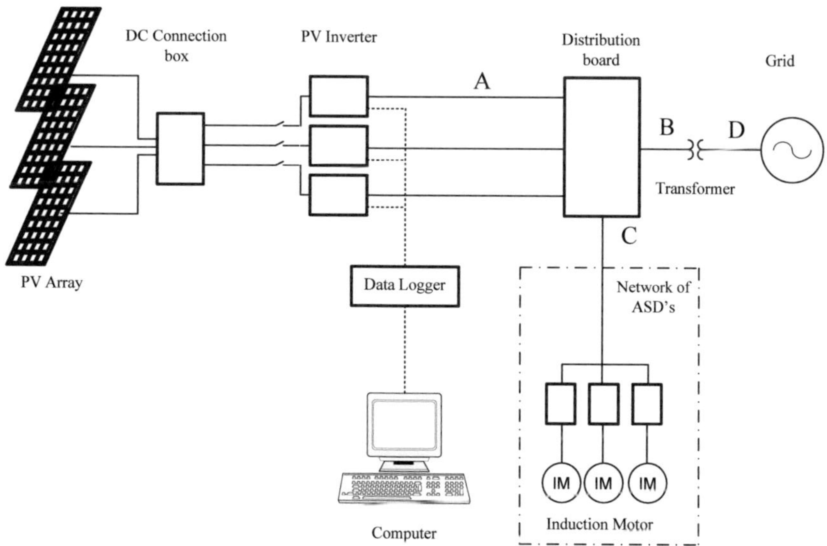

Fig. 13. Diagram of laboratory scale PV generation system

5.1 System Description

The PV plant is located at the position of 1° 33′37.52″ N and 103° 38′ 30.87″ E, consists of a PV array of 24 modules with totally 5 *kWp* installed capacity. The PV array is then divided into three sub arrays in order to configure the plant type as a string-inverter configuration. Three unit PV inverters of 700 Watt each are then fed by these array outputs. The sub arrays are connected to PV inverters by a DC connection box where the modules are configured in serial to obtain the DC voltage requirement for the PV inverter. The 70-meter long and 10 mm2 DC cable links DC connection box and PV inverters. Each single-phase inverter is connected to each phase of the grid to form the three-phase connection. Output side of PV inverters are fed to an interconnection switchboard where the grid integration, measurement, and local load supply of resistive and adjustable speed drives are connected.

The three units string PV inverters role as distributed generator (grid tied inverter). The inverters are from single stage type with H-bridge topology; operate in carrier based pulse-width modulation of 14 k-Hz carrier frequency. MPPT mechanism, current control, synchronization, and grid impedance monitoring complete each PV inverter. Connecting or disconnecting the PV inverter to the grid is a part of the control system; it is accomplished based on the availability of power from PV array and in coordination with the internal grid monitoring system. Some parameters of the PV inverter are described in Table 1.

Input		Output	
V_{DC} Max, Volts	250	V_{AC} nom, Volts	250
$V_{DC\ MPP}$, Volts	119 - 200	f_{AC} nom, Hz	50 / 60
I_{dc} Max, Amps	7	P_{AC} nom, Watts	700
		I_{AC} nom, Amps	3
		Cos Φ	1

Table 1. PV Inverter parameters for the observation

5.2 Voltage Stability

Figure 14.a shows the power output, the line to neutral voltage of an inverter and the total power output of the PV plant during a whole day operation that measured at the point of common coupling of the PV plant and the grid. The inverter starts to operate and connect to the grid when the generated powers reach the minimum predetermined value. The injected power to the grid is begin to increase along with the increasing of the solar power density in the morning, tend to constant in the middle of the day, and then decrease in the afternoon. The maximum injected power from the phase A-connected inverter is 503.90 Watt, phase B inverter 481.6 Watt and phase C inverter 453.10 Watt; coincide with the maximum power output of the PV plant of 1439.0 Watt. With the installed capacity of each inverters and plant are 700 and 5000 Watt, the values indicate the maximum capacity factors of the PV inverter and the plant on that day measurement are 71.98% and 28.78% respectively.

Figure 14.b is a portion of Fig.14.a is a 180 minutes measurement during peak power time, the figure shows the phase-A line voltage and the injected power from the connected inverter. It is shown that variation of injected power to the grid caused by variation of extracted power density from about 100 Watt to 400 Watt is followed by variation of voltage level on the plant terminal as about 3 V. The shape of the curves shows that during the measurement interval, the variation of the plant parameters occur very often along with the atmospheric dynamic condition. This condition, further implicates to voltage stability problem. Figure 15 is a histogram obtained from measurement data of Fig. 14.a; it is performed to show the frequency of occurrence of various power levels (in range of 50 Watt each) of the PV plant, data is collected for every 7 second. The number of power levels and the frequency of occurrence on each power levels in the histogram indicate that the grid is susceptible to voltage stability problem concerning to operation of a PV plant. The instability of voltage in this case is more important along with the increasing capacity of the plant and the fact that the atmospheric dynamic condition is occurring along the whole day.

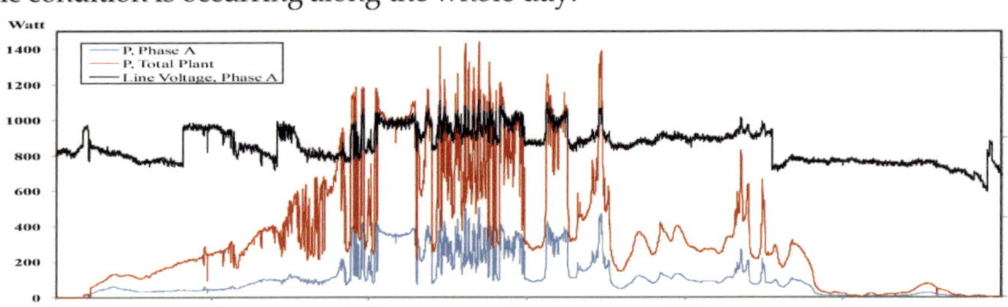

a. Line voltage, injected power on a phase of the grid and total injected power of the PV plant

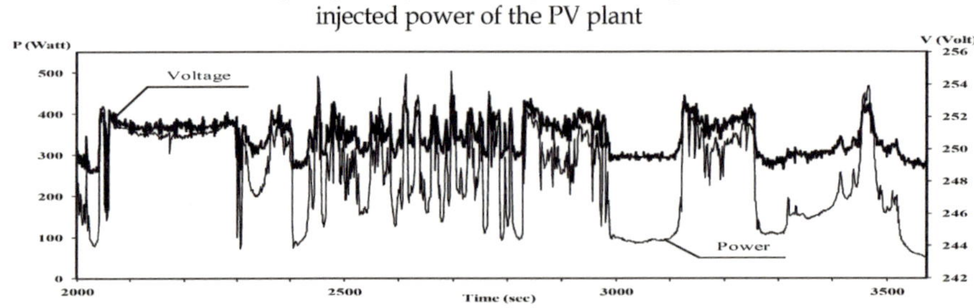

b. Line voltage, injected power on a phase A for 180 minutes measurement during the peak time

Fig. 14. P and V_{LN} on the point of common coupling between the PV Inverter and Grid

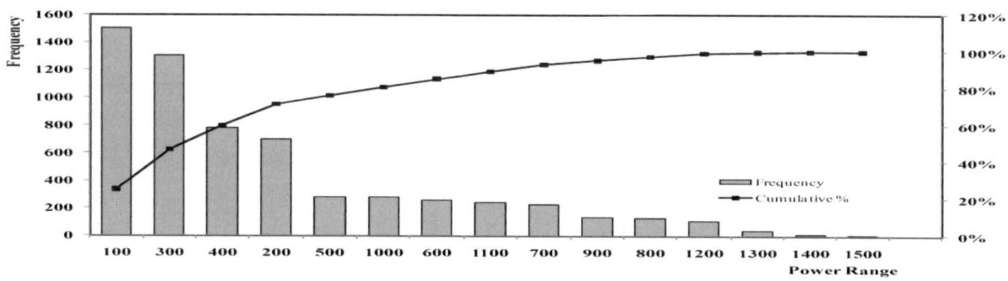

Fig. 15. Pareto diagram of various power levels generated from PV Inverter

Variation of injected power is also followed by variation of other power parameters such as *Cos θ* and power factor (PF). Their values according to magnitude of power level that injected to the grid are shown in Fig. 16. During injecting relative low power, *Cos θ* and power factor tend to decrease and then increase along with the increasing of generated power. Optimum *Cos θ* and power factor are reached when generated power is more than 50% of PV inverter rated. The measurement result indicates the importance for employing the PV inverter near to its rated power.

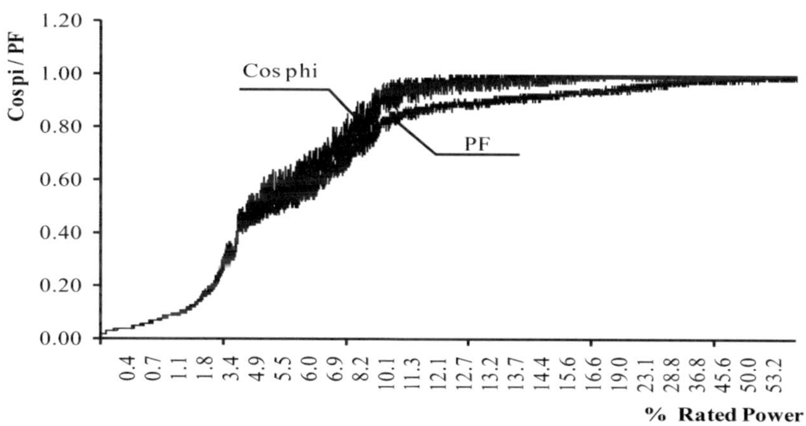

Fig. 16. *Cos θ* and power factor as function of generated and injected power

5.3 Waveform and harmonics distortion

Other aspect in PV plant operation using grid tied inverters is the appearance of distortion on the wave shape of power parameters where the plant is connected. In injecting the generated power to the grid, the control of the PV inverter forces the current to flow with the wave shape is fashioned to follow and in phase with the grid voltage. Nevertheless, depend on the design component of the inverter and the injected control type that used; the resulted current may draw a distorted wave that indicates the existence of harmonics, the voltage and current shape of a single phase PV inverter during operate is shown in Fig.17.

The level of the harmonics distortion -presented as the total harmonics distortion (THD) - on the inverters current is varying during operation; its value shows a relation with the level of generated and converted power from the array. Low harmonics content appears during high power conversion operation and in contrary high harmonics content during low power operation. However, an anomaly appears during low power operation due to nonlinearity of the inverter components, in a certain condition, harmonics tend to increase while in the other one, tend to decrease. A measurement result that describes the plant behaviors is shown in Fig. 18. The figure presents the harmonics measurement of the inverter's current and connected grid voltage as function of time during a day operation.

During connected to the grid, harmonics contents on the produced current of the PV inverter interact with the grid impedance causes the voltage where the inverter is connected is distorted. Further, the interaction between these two distorted waves causes the power factor on the connection point is influenced. Effect of the voltage and current distortion to the power factor can be obtained from the following relations:

$$PF = \frac{P_{avg}}{V_{1rms}.I_{1rms}} \frac{1}{\sqrt{1+THD_v^2}.\sqrt{1+THD_i^2}} \tag{11}$$

where:

$$D_f = \frac{P_{avg}}{V_{1rms}.I_{1rms}} = Cos\ \theta \tag{12}$$

and

$$F_f = \frac{1}{\sqrt{1+THD_v^2}.\sqrt{1+THD_i^2}} \tag{13}$$

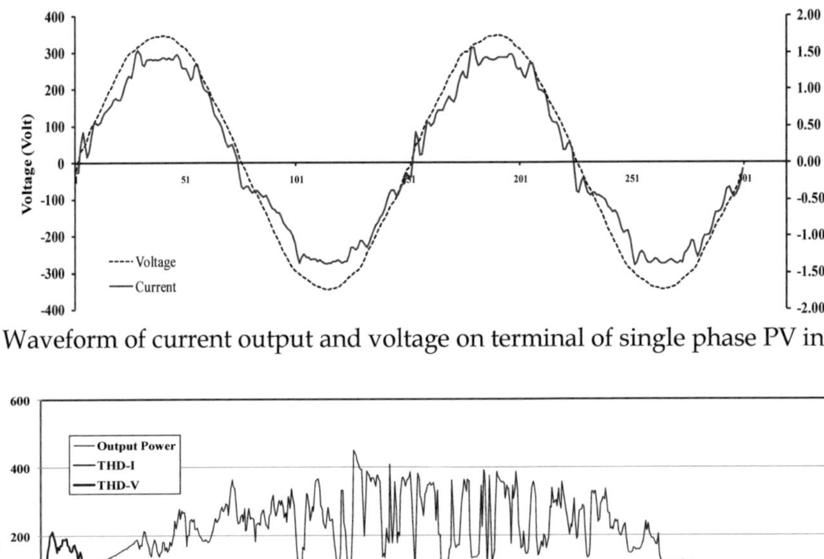

Fig. 17. Waveform of current output and voltage on terminal of single phase PV inverter

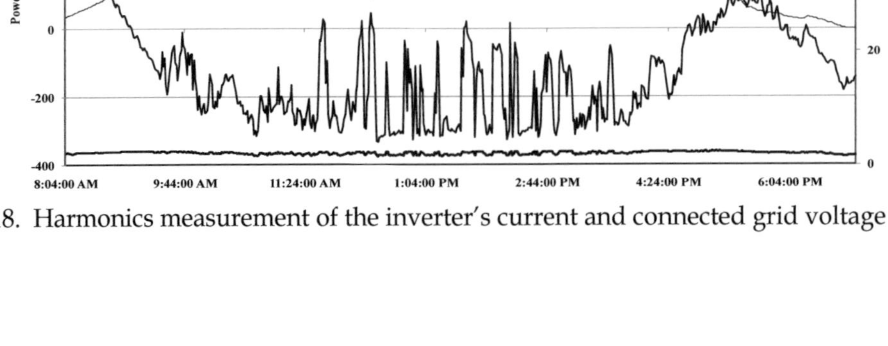

Fig. 18. Harmonics measurement of the inverter's current and connected grid voltage

The equations show that the power factor PF consists of two components; the displacement factor D_f that depends on the phase shift between voltage and current, and the form factor F_f that depends on the wave shape of voltage and current. According to Fig. 17, it is obvious that the power factor of the PV inverter is dominantly influenced by form factor. Low level of displacement factor of the PV inverter is caused by the reality that the control current of PV inverter is designed so that the current and voltage of the inverter are always in phase.

As the part of the distributed generation, PV plant operation must comply with the power quality requirement as described in various power quality standards. On the subject of harmonics contents, the standards generally present an acceptable maximum value of each harmonics order of injected current and voltage in the grid. For example, IEEE standard 519-1992 [25] and 1547-2003 [26] limits the individual harmonics in some groups of range for the first 40 harmonics and give a maximum level of the total demand harmonics distortion of the injected current. Regarding to the distortion and converted power level relation of PV inverter as described, it is obvious that operation of inverters in the PV plant causes the grid is distorted, especially by the current harmonics in low power operation. In other words, refers to capacity factor of inverter, we can say that the PV inverters will draw high contents of harmonics if they are employed in low capacity factor. Further, in order to comply with the limitation of injected current harmonics, it is important to perform the suitability combination between the capacity of both inverter and PV array. Minimum output of PV array during operation cycle should not be smaller than a level in which PV inverters operate in low capacity factor and produce harmonics that excess the accepted standard. Besides the conversion efficiency, harmonics and distortion factor should be considered in designing the array installed-capacity for a certain capacity of PV inverter.

5.4 PV plant and grid interaction

Once a PV inverter is connected to the grid, it become the part of the entire power system and associated load. Further, both system influence and interact each other, including power sharing in supplying a connected load. Figure 19 and 20 shows the measurement results of current on phase R at point A, B and C -as shown in Fig. 13 -when the system is supplying the local load. Two types of variable load are taken as test case: a network of adjustable speed drives (ASDs) and a resistive load. For both types of load, load current is set up so that their magnitudes are larger than the magnitude of current from the PV plant, direction of current is than observed in order to show load sharing between PV plant and grid.

During steady state operation of the ASD, measurement in point C shows the typical current shape of DC converter in input side of ASD (three-phase diode bridge rectifier). At the instant when the ASD draws input current, injected current from PV plant to the grid is reduced; and becomes supply current for the ASD. The wave shape of the injected current to the grid is in form of a distorted sinusoidal with a portion is cut by the three phase bridge rectifier current shape. In this condition, harmonics spectrum of current measured at the grid are different with the usual ASD current shape; a sinusoidal spectrum in with the spectrum of the three phase diode bridge rectifier is cut out. If the amount of the current for the ASD cannot be fulfilled fully by current from the PV plant, thus a portion of current flows from the grid, harmonics spectrum of current in the grid are same as the ASD current spectrum. This condition can be seen by the appearances of negative portion on the shape of the grid current in each half of fundamental frequency, that mean that the current flow in opposite direction with the reference current flow. The same case is happen when the resistive load is connected.

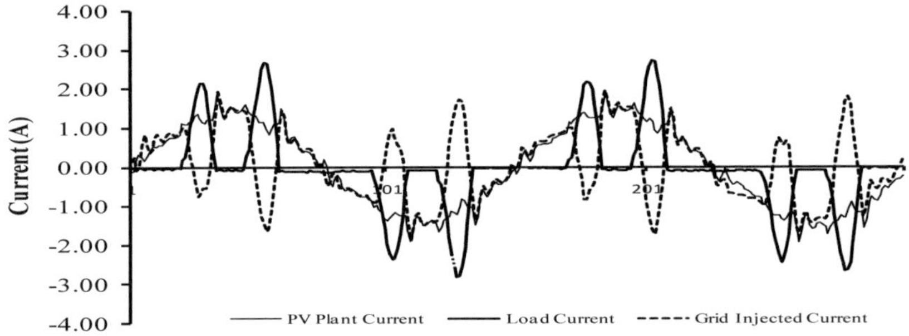

Fig. 19. PV plant, load and grid current with ASD type load is operated

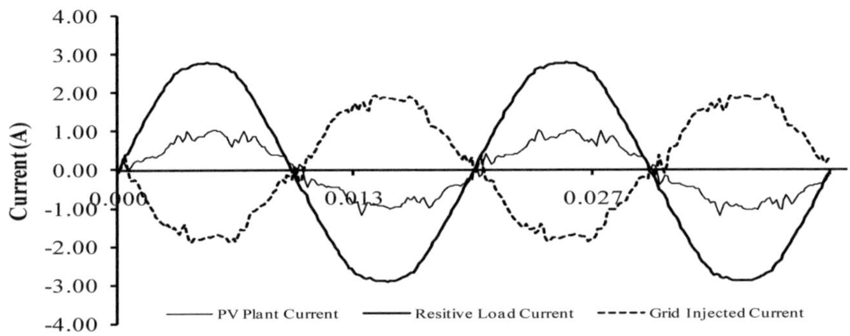

Fig. 20. PV plant, load and grid current with resistive type load is operated

6. Conclusion

In this chapter, some aspects in photovoltaic conversion using single-phase PV inverters, their involvement as a part of the distributed generation and the role of some embedded parts and functional devices that supporting this role has been described. Modeling of such parts and functions for digital simulation has also been presented. Through the simulation, some aspects of the PV inverters operating behavior -for certain reasons cannot be obtained from direct observation- can be evaluated.

This chapter has also presented some of the characteristics of the single-phase PV inverters in real operation, primarily in the characteristics of the power generated and the interactions with the distribution network where they are connected. Observation and measurement show that the voltage stability and waveform distortion are some problems that appear in the operating a PV plant as distributed generation. The distortion influences the other power output parameters of the plant such as the grid voltage and power factor. The power factor of the inverter output is dominantly caused by form factor rather than displacement factor and the current harmonics highly involved in the condition. It has been shown that capacity of PV inverters should be chosen so that they operate in a plant with high capacity factor in order to minimize the appearance of harmonics. Load sharing between the PV plant and the grid for supplying a connected load have also been presented. Finally, the relationship between the compliance capacity of the inverter with the quality of output power has been clarified and the importance for involving the distortion factor in sizing and planning the capacity of photovoltaic plant components have been stressed.

7. References

[1] Trends in Photovoltaic Applications. Survey Report of Selected IEA Countries Between 1992 and 2006. International Energy Agency Photovoltaic Power Systems, IEA-PVPS-T1-16:2007. [Online]. Available: www.iea-pvps.org

[2] M. Oliver, T. Jackson, "Energy and Economic Evaluation of Building-integrated Photovoltaic," *Energy* vol. 26. pp.431-439. 2001.

[3] V. Acquaviva. "Grid-connected Rooftop PV Systems for Reducing Voltage Drop at the End of the Feeder—A Case Study in Corsica Island," Energy, vol. 25. pp. 741-756. 2000.

[4] A.S. Jhutty, "System implication of embedded generation and its protection and control," IEE Colloquium Birmingham, Feb 1998

[5] L. Borle, M. Dymond, and C.V. Nayar, "Development and testing of a 20 kW grid interactive photovoltaic power conditioning system in Western Australia", 1996.

[6] G.M. Masters, "*Renewable and Efficient Electric Power Systems*". Hoboken New Jersey: John Wiley & Sons Inc., 2004.

[7] F. Blaabjerg, Z. Chen and S. B. Kjaer, "Power Electronics as Efficient Interface in Dispersed Power Generation Systems." *IEEE Trans. on Power Electronics* 2004; vol.19 no. 5. Pp. 1184-1194. 2000.

[8] J.H.R. Enslin, Peter J. M. Heskes, "Harmonic Interaction Between a Large Number of Distributed Power Inverters and the Distribution Network," *IEEE Trans. On Power Electronics*, vol. 19, no. 6, November 2004.

[9] P.J.M. Heskes, "Power Quality Behaviour of Different PV Inverter Topologies" PCIM-2003 The 24th International Conference, Nierenberg Germany.

[10] A. Woyte, V.V. Thong VV, R. Belmans, and J. Nijs J., "Voltage Fluctuations on Distribution Level Introduced by Photovoltaic Systems," *IEEE Trans. on Energy Conversion*; vol. 21 no.1. pp. 202-209. 2006.

[11] Luque A. *Handbook of Photovoltaic Science and Engineering*. West Sussex England: John Wiley and Sons Inc., 2003.

[12] E. Koutroulis, K. Kalaitzakis, and N. C. Voulgaris, "Development of a Microcontroller-Based, Photovoltaic Maximum Power Point Tracking Control System," *IEEE Trans. On Power Electronics*, vol. 16, No. 1, Jan 2001.

[13] T. Esram, "Comparison of Photovoltaic Array Maximum Power Point Tracking Techniques," *IEEE Transaction on Energy Conversion*, vol. 22 no.2. pp. 439-449. 2007

[14] M.A.S.Masoum, H.Dehbonei, "Theoretical and Experimental Analyses of Photovoltaic Systems With Voltage- and Current-Based Maximum Power-Point Tracking," *IEEE Trans. on Energy Conversion*, vol.17, no.4. pp. 514-522. 2002.

[15] S.H. Ko, S.R. Lee, and H. Dehbonei, "Application of Voltage- and Current-Controlled Voltage Source Inverters for Distributed Generation System," *IEEE Trans. Energy Conversion*, vol. 21, no.3, pp. 782-792, 2006.

[16] M. Calais, J. Myrzik, T. Spooner, V.G. Agelidis, "Inverters for Single-phase Grid Connected Photovoltaic Systems - An Overview," *Proc. IEEE PESC'02, vol. 2. 2002. Pp 1995-2000.*

[17] F. Blaabjerg, Z. Chen and S.B. Kjaer, "Power Electronics as Efficient Interface in Dispersed Power Generation Systems," *IEEE Trans. on Power Electronics,* vol. 19, no.5. pp. 1184-1194, 2004.

[18] Y. Xue, L.Chang, S. B. Kjær, J. Bordonau, and T. Shimizu, "Topologies of Single-Phase Inverters for Small Distributed Power Generators: An Overview," *IEEE Trans. on Power Electronics,* vol. 19, no. 5, pp. 1305-1314, 2004.

[19] S.B. Kjær, J.K. Pedersen, and F. Blaabjerg, "A Review of Single-Phase Grid-Connected Inverters for Photovoltaic Modules," *IEEE Trans. On Industry Application*, vol.41, no.5, pp. 1292-1306, 2005.

[20] H. Yamashati, K. Tamahashi, M. Michihira, A. Tsuyoshi, K. Amako, M. Park, "A Novel Simulation Technique of the PV Generation Systems using Real Weather Conditions," *IEEE 2002, PCC-Osaka 2002.*

[21] S. Liu and R. A. Dougal, "Dynamic multi-physics model for solar array," *IEEE Trans. on Energv Conversion,* Vo1.17, No.2, pp.285-294, June 2002

[22] A. Cow and C.D. Manning, "Development of a photovoltaic array model for use in power-electronics simulation studies", *IEE Proc. Elecrr. Power Appl.,* Vol. 146, Na.2, pp.193-200, March 1999

[23] W. Xiao, W.G. Dunford, A. Capel, "A Novel Modelling Method for Photovoltaic Cells" *35th Annual IEEE Power Electronics Specialists Conference Aachen,* Germany. 2004

[24] D. M. Brod and D. W. Novotny, "Current Control of VSI-PWM Inverters," *IEEE Trans. on Industry App.* vol. IA-21, no. 4, pp. 562-570, May/Jun 1985.

[25] *IEEE Standard for Interconnecting Distributed Resources with Electric Power Systems,* IEEE Standard 1547, 2003.

[26] *IEEE Standard for Interconnecting Distributed Resources with Electric Power Systems,* IEEE Standard 519, 2003.

[27] The Mathwork, accelerating the pace of engineering and science, http://www.mathworks.com

8

Single-Phase Distributed Generation System Based on Asymmetrical Cascaded Multilevel Inverter

Sergio P. Pimentel
Federal Institute of Education, Science and Technology of Goias
Campus Itumbiara
Brazil

Rodolfo M. M. Martinez and Jose A. Pomilio
University of Campinas
Brazil

1. Introduction

The attention and researches on distributed generation systems have been increasing lately and this situation tends to increase more and more. The main reasons for it are the growth of the discussions about environmental issues, the development of renewable energy sources technologies and the imminent possibility of an energy crisis. Power electronics devices are an useful alternative in this process. Power electronics can improve the dynamic response of the distributed generation systems, increase its global efficiency and permit the operation of a large number of systems based on photovoltaic cells, fuel cells or wind turbines. Besides these systems can be also an alternative UPS system if they were not connected to the grid utility. Generally low power distributed generation systems are connected to the grid utility at low voltage levels ($\leq$13.8kV), which are considerably affected by the presence of nonlinear loads. It means that these systems must be controlled considering harmonics, resonances and other power quality issues. In case of single-phase systems (residential power generation) the attention on those disturbances must be bigger. According to (Jung, Bae, Choi & Kim, 2006), a lot of researches about three-phase systems have been presented but a few number of single-phase systems that operates connected to grid utility or not have been studied. There are five possible obstacles for it:

- The main control concepts of three-phase systems are based on $dq0$ transformation and they can not be applied directly in single-phase systems.
- If they could be applied after some modifications, the $dq0$ components would not be equal to constant values for situations involving harmonics, resonances or unbalance.
- The classical control of exchange power by changes on angle phase and on amplitude voltage is based on sinusoidal steady-state equations.
- The use of reactive power control systems assumes that the reactive power concept has already been defined.

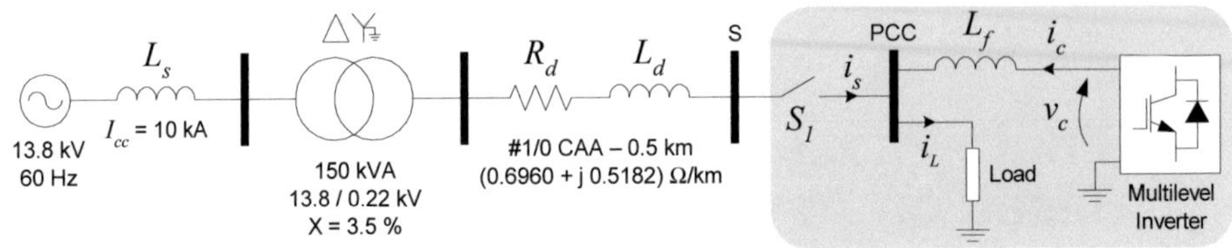

Fig. 1. Grid connection diagram of the DGS.

- High costs of inverters and renewable energy technologies disable a expansion of small distributed generation systems, mainly on single-phase applications.

The proposed distributed generation system (DGS) is connected to a single-phase grid utility and it is supplied by DC voltage sources. Details about it and its operation modes are presented. Simulation and experimental results are also presented. Finally the conclusions about the effectiveness of the proposed DGS are drawn.

2. Distributed Generation System (DGS)

The structure of the proposed DGS consists of an inverter connected to the PCC (Point of Common Coupling) by an inductance (see Figure 1). The inductance L_f could represent a power transformer element or just a first order low pass filter element. Besides, this inductance is responsible for injecting current into PCC throw it and also for attenuating the impact of high frequency components from the inverter output voltage (v_c) on the inverter current (i_c).
The main purpose of the DGS is to supply load continuously even if an abnormal condition (a grid failure or an islanding situation) has occurred. At this point, the DGS has similarities with on-line UPS systems. When there is not an abnormal condition occurring, the DGS operates at the grid-connected mode. The voltage at PCC is held by the grid utility and variations on it could change the operation mode of the DGS. During the grid-connected mode the DGS could supply all the load power (active and nonactive), only a part of it (in addiction to the grid utility supply) or supply all the nonactive power of load (like an active power filter would do). However, this last possibility does not represent exactly a distributed generation scheme. Considering that all the load power is supplied by the DGS, the inverter current (i_c) should be equal to the load current (i_L) and the source current (i_s) should be null. Except when a high di_L/dt occurs and it demands energy rapidly from the grid utility. It happens mainly for nonlinear loads due to the inductance L_f and to the dynamic response limitations of the inverter. If the inverter maximum power is higher than the load power, the DGS could use this spare power for delivering active power to the grid utility. This situation represent a distributed generation scheme and that is why it is not an UPS system.
The DC-links of the inverter could be supplied by an amount of fuel cells, photovoltaic (or solar) cells or both. There is also the possibility of wind turbines supplying them by energy storage in batteries. The energy from those elements should be conditioned by an isolated DC/DC converter before being stored in batteries, capacitors or another storage element. As the primary energy of the DGS comes from DC voltage sources or DC switching power supplies, the proposed DGS could be characterized as an asynchronous (or nonrotational) system.

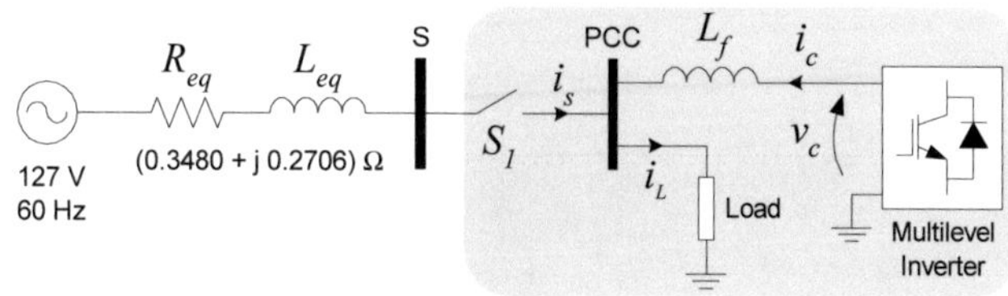

Fig. 2. Grid connection diagram of the DGS with an equivalent circuit of the Area EPS.

2.1 Grid Connection

Figure 1 shows the connection of the multilevel inverter to the PCC by an inductance. The load (or an amount of loads) is also connected to the PCC. According to (*IEEE Standard for Interconnecting Distributed Resources with Electric Power Systems*, 2003), the complex formed by the load, the inverter and the PCC is called Local EPS (from Electric Power System). The Local EPS stood out in Figure 1 and the switch S_1 (closed or open) determines if the Local EPS is connected, respectively, to the utility grid or not. The utility grid away from switch S_1 is called Area EPS, as in (*IEEE Standard for Interconnecting Distributed Resources with Electric Power Systems*, 2003). During the grid-connected mode (switch S_1 closed), the DGS operates as a controlled current source connected in parallel to the load. And during the stand-alone mode (switch S_1 open), as a controlled voltage source. Additional information about these modes is available in subsections 3.1 and 3.3.

The considered Area EPS is a common radial power system and consists of a low voltage bus (127/220V) connected to the high voltage bus (13.8kV) by a power transformer in series and a typical distribution line. The grid frequency is 60Hz. The DGS is single-phase and it is connected at the low voltage for representing a possible power exchange among utility grid and residential consumers. At the low voltage bus, the effect of harmonics from nonlinear loads increases and the DC system should attempt this and others disadvantages. The Area EPS presented in Figure 1 could be replaced by its equivalent circuit per phase. Considering a short circuit current equals to 10kA at the high voltage bus (13.8kV) and data available in Figure 1, the calculated equivalent circuit of Area EPS is presented in Figure 2.

2.2 Multilevel Inverter

The utility-interactive inverter of the DGS consists of a multilevel structure with 19 levels produced by an asymmetrical cascaded topology (Daher et al., 2005; Pimentel, 2006; Rodriguez et al., 2002; Silva et al., 2005b). Figure 3 shows the structure of the used inverter. Instead of the symmetrical, the asymmetrical topology produces a high number of levels using a few numbers of switching devices and H-bridge cells. The asymmetrical and symmetrical topologies should be formed by 3 and 9 H-bridge cells, respectively, to allow the same 19 possible levels in the output voltage. For single-phase or low voltage applications, this feature of the asymmetrical topology could be useful due the number and the type of available primary energy sources. There may be different types of sources naturally isolated from each other. For example, the distinct sources could be a group of solar cells, another of fuel cells and other of wind turbines interfaced by converters. That was one of the reasons for choosing the asymmetrical topology. Others are related to power quality and conducted EMI.

The main disadvantages of the used multilevel inverter are: many switching devices; the DC voltages should be isolated from each other; the DC voltage value of the highest voltage H-bridge cell rises rapidly if the numbers of levels increases fairly; and its control system and

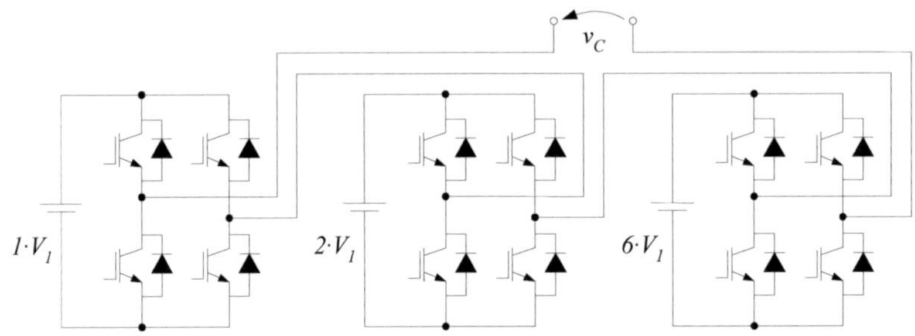

Fig. 3. Asymmetrical cascaded multilevel inverter with 19 levels.

modulation strategy are not simple (Rech & Pinheiro, 2005; Silva et al., 2005a). However the asymmetrical topology produces an output voltage v_c waveform with lower THD than the standard PWM 3-level inverter. Besides, it can operates at low switching frequency. Both advantages can eliminate the second order (or higher) low pass filters (LC filters) that are responsible for reducing the harmonics from the inverter output voltage before connecting it to the PCC. The low Total Harmonic Distortion (THD) of v_c decreases the impact of the switching components on the power quality at PCC and on the current i_c supplied by the multilevel inverter. Reducing the impact on i_c, the magnitude of conducted EMI is also attenuated. Even if a PWM modulation were used by only the lowest voltage H-bridge cell, the conducted EMI would be attenuated due to the lower voltage involved.
The multilevel inverter can achieve high voltage levels using a lot of H-bridge cells that operates at low voltage levels. This characteristic could permit a directly connection of the inverter to high voltage buses without using power transformers (Rodriguez et al., 2002). The configuration of the cascaded topology should be "1:2:6" to produce 19 levels on the inverter output voltage. According to this configuration, the DC voltages of the lowest voltage, of the middle voltage and of the highest voltage cells should be equal to V_1, $2V_1$ and $6V_1$, respectively. The lowest voltage cell (DC voltage equals to V_1) operates with a 3-level PWM modulation strategy to cover all the possible voltage range among levels and to reduce the voltage error. The others cells operate at low switching frequency with a special modulation strategy presented in (Lipo & Manjrekar, 1999; Pimentel, 2006; Rech et al., 2002; Silva et al., 2005a).

3. Operation Modes and Transitions

The state of switch S_1 determines which mode of operation is responsible for controlling the proposed DGS: grid-connected mode or stand-alone mode. However it is necessary two auxiliary control systems for determining the state of this switch and its transitions. These auxiliary systems should determine that the switch must be open if it were closed, and vice versa. The auxiliary control systems of the proposed DGS: an islanding detection scheme; and a procedure for grid synchronization before reconnection. These operation modes and their transitions are represented as a state machine in Figure 4.

3.1 Grid-Connected Mode

During the grid-connected mode, the DGS is able to deliver active power from the Local EPS to the Area EPS and supply all the active and the nonactive power of the load. A suitable inverter current control is very important because the DGS operates as a controlled current

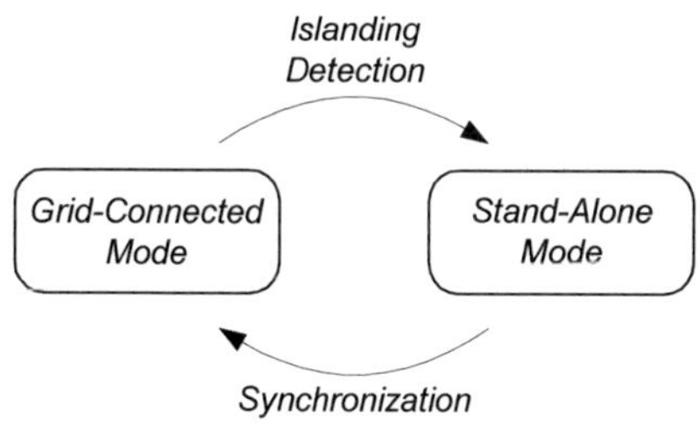

Fig. 4. Operation modes and transitions of the DGS.

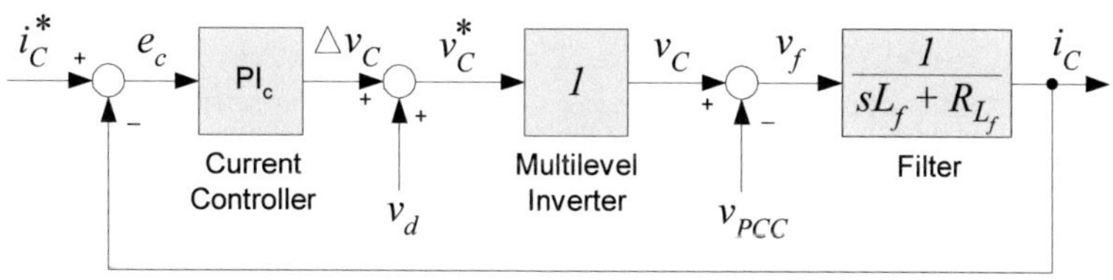

Fig. 5. Inverter current control diagram during grid-connected mode.

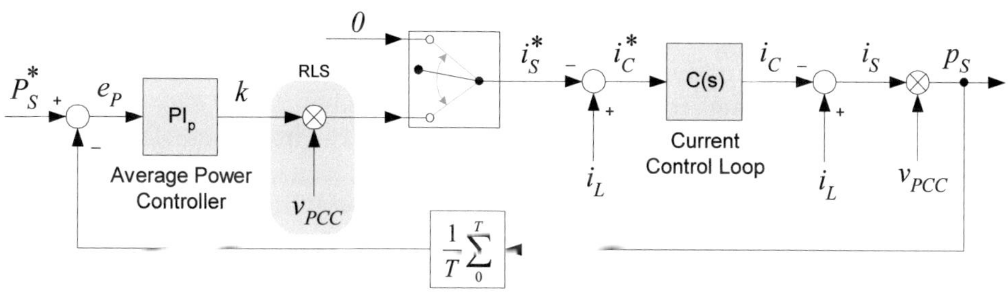

Fig. 6. Source average power control diagram during grid-connected mode.

source. Besides the exchange of active power between Local EPS and Area EPS should result in a source power factor closed to unity. This is achieved by the source average power control.

3.1.1 Current Control

The inverter current control diagram is shown in Figure 5. It is very similar with those used in active power filters. A PI controller with a high time constant is used to guarantee that current error e_c be null at the grid frequency (60Hz). The output of the PI controller (Δv_c) is added to the voltage v_d (a feed-forward action) to elaborate the reference signal of the inverter output voltage $v_c{}^*$. This signal is used by the modulation strategy to create the real inverter output voltage v_c. From variations on v_c it is possible to control the inverter current i_c by the inductance L_f. The voltage v_d could be equal to the PCC voltage or to a sinusoidal signal waveform which frequency is determined by the islanding detection scheme. The reference signal waveform to the inverter current $i_c{}^*$ comes from the source average power control.

3.1.2 Average Power Control

The source average power control diagram is shown in Figure 6. The average value of the source instantaneous power p_s is equal to source active power (i.e., $\overline{p_s} = P_s$). That is why the control is based on $\overline{p_s}$. The source average power control is responsible for regulating P_s and for defining the reference signal waveform to the inverter current $i_c{}^*$. If none active power is delivered to the Area EPS (i.e., $i_s{}^* = 0$), the signal $i_c{}^*$ is equal to the load current i_L and the DGS supplies all the active and nonactive load power. If an amount of extra active power is available for delivering to the Area EPS, the signal $i_c{}^*$ is equal to the sum of the load current i_L with the reference signal waveform of the source current $i_s{}^*$. The reference signal $i_s{}^*$ is obtained from the multiplication of the PCC voltage v_{PCC} with the output average power PI controller k. This multiplication represents a technique called "Resistive Load Synthesis" (RLS) (Nunez-Zuniga et al., 2000; Silva et al., 2005b). The delivering of only active power to the Area EPS is possible because current and voltage have the same waveform in the RLS technique, just like in a resistor. A PI controller with a very low time constant and a low proportional gain is used to guarantee that average power error e_p be null at 0Hz. It is important to notice that the reference source average power P_s^* must assume negative values to permit a correct delivering of active power to the Area EPS (e.g., $P_s^* = -500W$). A procedure to calculate the parameters of the source average power PI controller is also given in (Pimentel, 2006).

3.2 Islanding Detection

An abnormal condition of grid utility should determine that the Local EPS operates disconnected ("islanded") from the Area EPS. That is called islanding and the main islanding situations are grid failure, under/over voltage and under/over frequency. During the grid-connected mode the voltage at PCC is held by the grid utility. If an islanding situation happens the PCC voltage would be uncontrolled, the load current would follow these unintentional changes, the Local EPS could supply wrongly some loads from Area EPS and the inverter current control may be unstable. Besides the DGS should not be controlled at the grid-connected mode anymore, it should be controlled at the stand-alone mode thereafter. An islanding detection scheme is necessary to select the right operation mode and to change the state of switch S_1 for avoiding those related problems with the DGS. The proposed DGS uses an active islanding detection based on an active technique called "Active Frequency Drift with Positive Feedback" (AFDPF) (Jung, Choi, Yu & Yu, 2006). The positive feedback action is responsible for forcing an unstable condition of the PCC voltage.

The basic concept of active islanding detection techniques is to modulate a power system parameter and to measure its corresponding system response (Jung, Choi, Yu & Yu, 2006). Passive islanding detection techniques could failure if there is no power exchange between Local EPS and Area EPS or could generate unintentional islanding situations, which must be avoided. Originally the AFDPF technique presented in (Jung, Choi, Yu & Yu, 2006) introduces a small perturbation at the inverter output current modulating its frequency. The proposed islanding detection scheme modulates also the frequency but introduces these perturbation at the inverter output voltage by the voltage signal v_d. The signal v_d is produced from some data collected by a PLL and a FWD (Fundamental Wave Detector) that are pursuing the PCC voltage (see Figure 7). The PLL and the FWD are presented in (Marafao et al., 2005). The logical signal ISD is high when an islanding situation has been detected.

The frequency of the PCC voltage (f_{PCC}) is used to calculate the chopping factor (cf) as in (1). The initial chopping factor cf_0 is equal to 0.0083264 and it represents the chopping factor when there is no frequency error. The initial frequency f_{PCC_0} is equal to 60Hz and the gain k_f is equal to 1.0.

$$cf = cf_0 + k_f \cdot \left(f_{PCC} - f_{PCC_0}\right) \tag{1}$$

The chopping factor (cf) is used to determine the distortion factor (d) as in (2). The factor d describes how many higher the frequency distortion (f_d) is than f_{PCC} and it is very important to the positive feedback. If there is a frequency error, the factor d would tend to increase that frequency error and to destabilize the frequency at PCC. It happens continuously and the islanding situation could be detected if the values of under/over voltage and under/over frequency were reached. When there is no frequency error, f_d is equal to 60.05Hz.

$$d = \frac{f_d}{f_{PCC_0}} = \frac{1}{(1 - cf)} \tag{2}$$

The factor d is used to produce the distortion angle (θ_d) from the phase angle of the voltage at PCC (θ_{PCC}) as in (3). The angle θ_{PCC} is manipulated by scaling its range during each half period of f_{PCC}. It results in a change of the rise time of the angle θ_{PCC} as is shown in Figure 8(a). If f_d is higher than f_{PCC}, θ_d would rise faster (a frequency drift up situation) than θ_{PCC} during each half period. Otherwise, it rises slower (a frequency drift down situation).

$$\theta_d = \begin{cases} d \cdot \theta_{PCC}, & \text{if } (0 < \theta_{PCC} \leq \pi) \\ \pi + \left[d \cdot (\theta_{PCC} - \pi)\right], & \text{if } (\pi < \theta_{PCC} \leq 2\pi) \end{cases} \tag{3}$$

The distortion angle (θ_d) and the peak value of the fundamental voltage at PCC (V_{PCC_1}) are used to compose the distortion voltage (v_d) as in (4). Therefore, v_d receives all the modifications produced by the d at θ_{PCC}. It changes the frequency of the distortion voltage waveform on each half-cycle as is shown in Figure 8(b). The distortion voltage (v_d) is added to the output current controller (Δv_c) to produce the reference signal of the inverter output voltage. This action qualifies the DGS for detecting islanding situations even when there is a minimal power exchange between Local EPS and Area EPS.

$$v_d = V_{PCC_1} \cdot \sin\left(\theta_d\right) \tag{4}$$

3.3 Stand-Alone Mode

During the stand-alone mode the Local EPS is disconnected from the Area EPS and the DGS becomes a controlled voltage source that supplies the load. Therefore the main purpose of this operation mode is to keep voltage at the PCC closed to a sinusoidal waveform with a peak voltage equals to $127\sqrt{2}$V (=179.605V), a frequency equals to 60Hz and an initial phase angle equals to value of the grid phase angle just before islanding. The inverter must provide an output voltage that is composed by the desired voltage to PCC and by a compensation for the voltage drop across the inductance L_f caused due the load current.

3.3.1 Voltage Control

It is possible to achieve the purpose of the stand-alone operation mode using a control system that regulates the inverter output voltage based on a feedback of the PCC voltage. The voltage control system diagram is shown in Figure 9. It uses a proportional plus resonant controller (P+Ress) instead of the traditional PI compensator for annulling the voltage error e_v at 60Hz. At this frequency, both controllers have different frequency responses. The reason for it is the

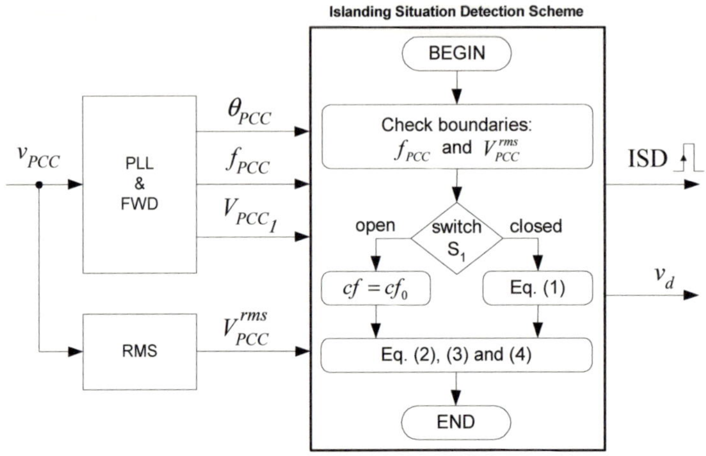

Fig. 7. Data flux of the islanding detection scheme.

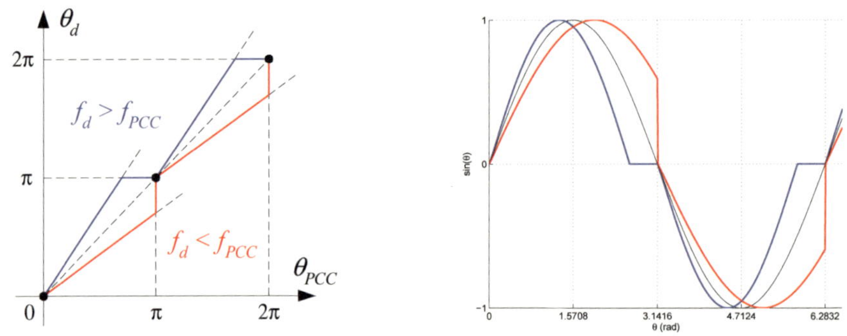

(a) (b)

Fig. 8. AFDPF angle (a) and waveform (b) modifications with frequency drift up (blue) and frequency drift down (red).

frequency response of both controllers at 60Hz. The P+Ress controller has a high gain at 0Hz and at 60Hz, while the PI controller has only at 0Hz (Zmood & Holmes, 2003). It means that voltage error e_v at 60Hz could be null if a P+Ress controller were used. This is an attractive characteristic because the reference signal of PCC voltage v_{PCC}^* is a sinusoidal waveform with frequency equals to 60Hz. Others control schemes have been tested to regulate the peak value or the RMS value of the PCC voltage using PI controllers, but they were not so suitable as the control scheme shown in Figure 9 was.

The proportional part of the P+Ress is represented by the gain α_1 and it is responsible for the transient response of the controller. The resonant part is represented by the resonant frequency ω_1 ($2\pi 60$rad/s) and by the gain β_1. At steady-state the output of the resonant part is a sinusoidal waveform at 60Hz with a stable amplitude, that could be faster achieved if the gain β_1 were increased. The P+Ress controller is also known as synchronous controller or

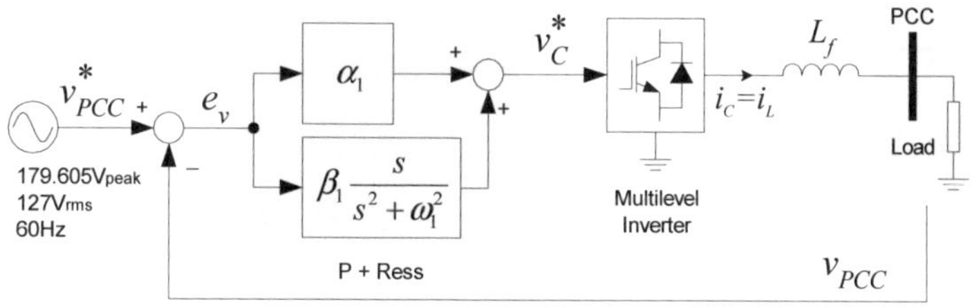

Fig. 9. PCC voltage control diagram during stand-alone mode.

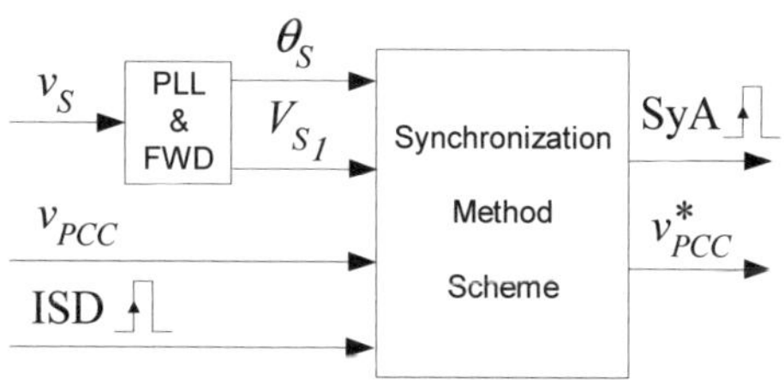

Fig. 10. Data flux of the SM scheme.

stationary frame controller because its resonant frequency is at fundamental (or synchronous) frequency.

3.4 Synchronization

During the stand-alone mode, the voltage at the PCC bar is controlled by the DGS and the voltage at the "S" bar is unknown and independent just because the switch S_1 is open (see Figure 2). The S voltage (v_S) is controlled by the Area EPS operator and it could not be exactly equal to the PCC voltage. If they were not equal in amplitude, in phase angle or in frequency, the switch S_1 should not be closed. That is why a synchronization method (SM) is necessary to adjust the PCC voltage with the purpose of decreasing these differences and reconnecting the Local EPS to Area EPS.

The scheme of the used synchronization method is shown in Figure 10. It is noticed that the SM receives data from the voltage v_S by a PLL (θ_S and V_{S_1}), from the PCC voltage (v_{PCC}) and from the islanding detection method (ISD). The SM process all of these collected data and provides two output signals: the logical signal (SyA) that means "Synchronism Achieved"; and the reference signal of the PCC voltage (v^*_{PCC}). The signal SyA is high if the error voltage were below an acceptable level for some cycles of the PCC voltage and if the signal ISD were high. An association of both logical signals, ISD and Sya, could be used to determine the state of switch S_1.

The elaboration of the reference signal v^*_{PCC} is based on a PLL, which is similar to that used by the islanding detection scheme. When the logical signal ISD is low, the phase angle of v^*_{PCC} (θ^*_{PCC}) is equal to θ_S and the peak voltage of v^*_{PCC} ($V^*_{PCC_1}$) is equal to V_{S_1}. After ISD has became high, θ^*_{PCC} varies cyclically (60Hz) from 0 to 2π with an initial value based on θ_S just before ISD had become high, and $V^*_{PCC_1}$ is equal to $127\sqrt{2}$V. If a stable and nominal voltage has been checked at the "S" bar during the stand-alone mode, the SM starts up. The SM adjusts slowly (for a few cycles of 60Hz) the peak voltage $V^*_{PCC_1}$ and the phase angle θ^*_{PCC} for equalling them to V_{S_1} and θ_S, respectively.

4. Simulation Results

A model of the proposed DGS has been simulated using the software PSIM. An uncontrolled bridge rectifier with an DC output capacitor supplying a resistor was considered as the load connected to PCC. By it, the performance of the DG system has been tested considering non-linear loads. Table 1 presents the main parameters used by the simulation model.

Parameter	Value	Units
Maximum Output Inverter Voltage $v_{C_{max}}$	$1.15 \times 127\sqrt{2}$	V
Filter Inductance L_f	3.85	mH
Filter Resistance R_f	0.01	Ω
Grid Equivalent Inductance L_{eq}	0.70	mH
Grid Equivalent Resistance R_{eq}	0.35	Ω
AC Input Rectifier Inductance L_{ac}	4	mH
AC Input Rectifier Resistance R_{ac}	0.01	Ω
DC Output Rectifier Capacitance C_{dc}	470	μF
DC Output Rectifier Resistance R_{dc}	$100 - 200$	Ω

Table 1. Parameters of the simulation model

The normal operation (normal condition, without islanding situations) of the DGS during the grid-connected is presented in Figure 11. The DGS is connected to PCC at t=0.3s and it starts to supply immediately all the load current i_L. At t=0.4s the DGS starts to deliver active power to the Area EPS in addition with the load current supply. The reference source active power P^*_s is equals to -500W and it is achieved after t=0.45s. The phase angle of the source current i_s (lagging by 180ř) proves the injecting of that active power into the Area EPS.
The islanding detection scheme was also tested by the simulation model. Figure 12 shows its test results. Considering that there was no power exchange between Local EPS and Area EPS (i.e., $\overline{p_s} \rightarrow 0$), an islanding situation (an abnormal condition) has occurred at t=0.5s. From this point, the PCC voltage waveform v_{PCC} contains all the output inverter voltage harmonics attenuated by the filter inductance. The islanding is detected at t=0.5316s by an under frequency protection. The under frequency level was reached by the variation of f_d on v_d and on v_{PCC} as a result. It validates the islanding detection scheme. After detecting the islanding, the switch S_1 is open at t=0.5316s and the DGS starts to operate at the stand-alone mode. Therefore v_{PCC} is controlled and held closed to its reference signal, a sinusoidal waveform. This procedure guarantees a uninterrupted load supply even during abnormal conditions.
The grid utility supply could be available a few cycles after the failure or it could take a long time. While it not occurs, the DGS operates at the stand-alone mode continuously. During the simulation, the availability of grid utility at the "S" bar was reestablished at t=0.8s but with another phase angle, i.e., its value is different from the phase angle of the voltage at PCC. At t=1.0s the SM starts to run and adjusts the frequency of the PCC voltage for equalling both phase angles θ_{PCC} and θ_S, as is shown in Figure 13. It is noticed that the f_{PCC} and V_{PCC_1} varies slowly along the time due the synchronization adjustments. The signal Δv represents the difference between v_S and v^*_{PCC}.

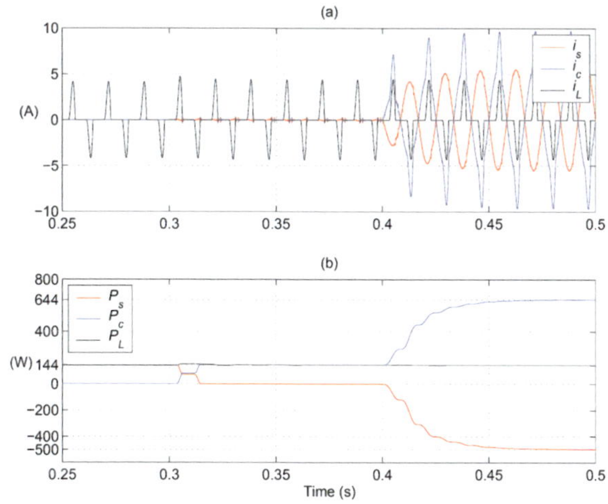

Fig. 11. Current waveforms (a) and active power waveforms (b) during a normal operation of the grid-connected mode.

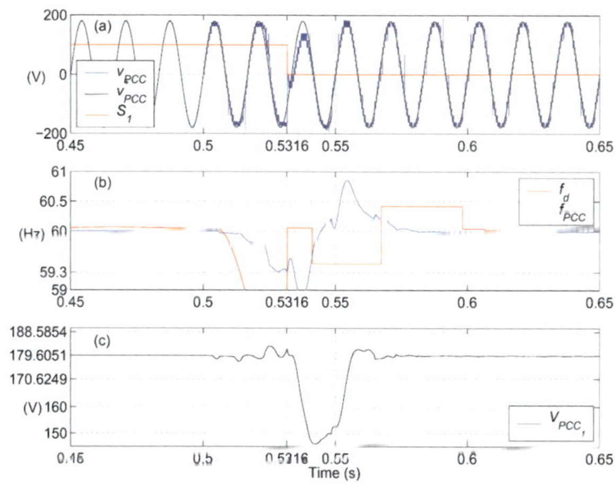

Fig. 12. Islanding detection and stand-alone mode: (a) v_{PCC} (blue), v^*_{pcc} (black) and S_1 gate signal scaled by 100 (red); (b) f_d (red) and f_{PCC} (blue); (c) V_{PCC_1}.

Figure 14 shows the behavior of the Δv and frequency f_{PCC} during the SM adjustments. If they were lower than predetermined levels (e.g., $\pm 1\%$) for five consecutive cycles of 60Hz, the SM allows the reconnection of Local EPS and Area EPS by setting the SyA signal. It happens at t=1.559s and the switch S_1 is closed at this time, as it is shown in Figure 15. After the reconnection, the operation of the DGS returns to the grid-connected mode and the islanding detection scheme is reactivated.

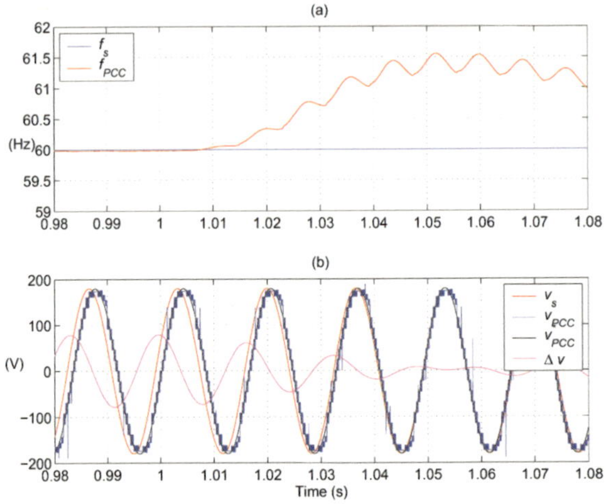

Fig. 13. After $t = 1.0$s, the SM adjusts f_{PCC} (a) for equalling θ_{PCC} and θ_S (b).

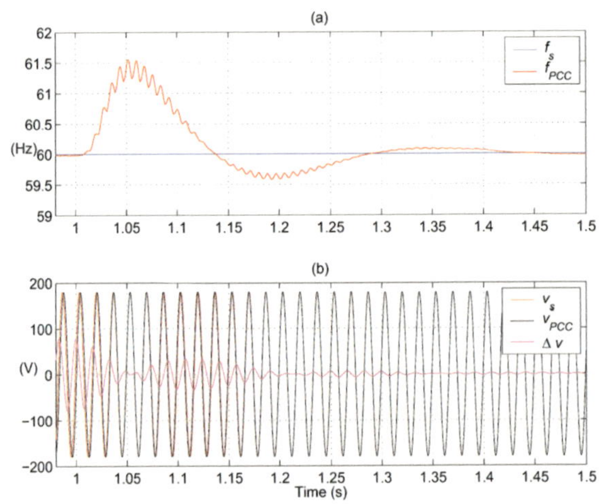

Fig. 14. Behavior of f_{PCC} (a) and Δv (b) during the synchronization adjustments after $t = 1.0$s

5. Experimental Results

A multilevel inverter supplied by isolated DC voltage sources was used to compose a prototype of 1kW for validating the proposed DGS. As the configuration of the asymmetrical cascaded multilevel inverter should be "1:2:6", these DC voltage sources were equal to 22.95V, 45.90V and 137.70V. The lowest voltage cell (22.95V) operates with a 3-level PWM modulation strategy to cover all the possible voltage range among levels and to reduce the voltage error. The highest and middle voltage cells operate at low switching frequency with a special modulation strategy presented in (Lipo & Manjrekar, 1999; Pimentel, 2006; Rech & Pinheiro, 2005; Silva et al., 2005a).

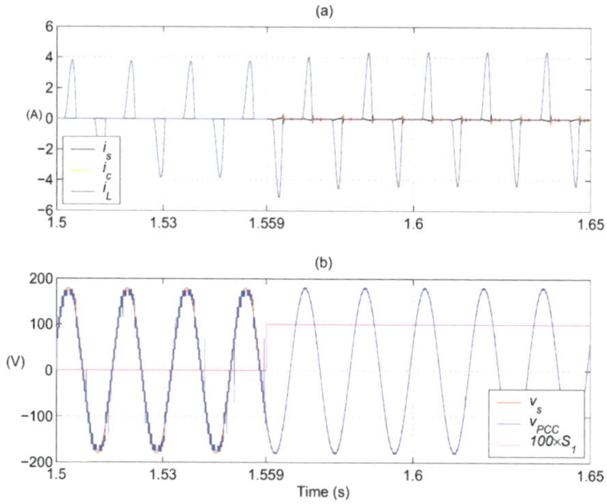

Fig. 15. Reconnection at $t = 1.559$s: (a) current waveforms; (b) voltage waveforms and S_1 gate signal scaled by 100 (magenta).

The DGS prototype is implemented with integrated power modules IRAMX16UP60A as the semiconductor devices and is controlled by a fixed-point DSP TMS320F2812. The prototype parameters are similar to those simulation model parameters presented in Table 1. Additional parameters are: sampling frequency equals to 36kHz; and switching frequency equals to 18kHz. The considered load is also similar to that from the simulation model, i.e., a nonlinear load.

Figure 16 shows the connection of the DGS to PCC and its performance for supplying all the load current during the grid-connected mode. Differently from the simulation model, the DGS prototype does not supply immediately all the load power. It happens gradually for one cycle of 60Hz and facilitates the stability of the prototype operation. However it happens just at the initialization of the prototype. Future load changes must be supplied immediately by the DGS.

After the stabilization of the DGS operation and the current controller, the prototype is able to provide an extra active power to the Area EPS. Figure 17 shows the behavior of the current and voltage waveforms after the active power injection has been started. After that the DGS supplies the load power and the source active power. The source average power controller achieves a stable operation point after two cycles of 60Hz and the source active power is similar with its predetermined reference. Besides the load power supply has not been interrupted and the load current has not been changed. It could be noticed from Figure 17 that the source current has the same waveform of the PCC voltage. This similarity validates the RLS technique and the exchange of only active power between Local EPS and Area EPS.

Figure 18 shows the operation of the DGS during the stand-alone mode. The nonlinear load are still powered by the DGS and the load current i_L waveform is similar to that load current during the grid-connected mode shown in Figures 16 and 17. It is noticed from Figure 18 that the load voltage (at PCC bar, v_{PCC}) is similar to its reference signal v^*_{PCC}, which is a sinusoidal waveform with a frequency of 60Hz and a peak voltage of $127\sqrt{2}$V. This is possible because the voltage control compensates the voltage drop at inductance L_f caused by the load current

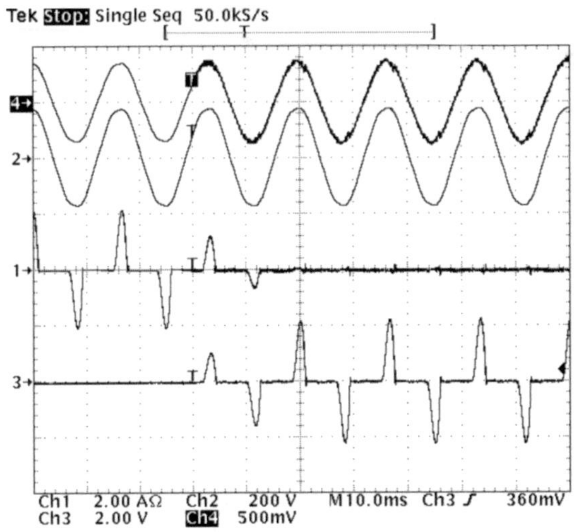

Fig. 16. From top to bottom: v_c (Ch4, 250V/div), v_{PCC} (Ch2, 200V/div), i_s (Ch1, 2A/div) and i_c (Ch3, 2A/div).

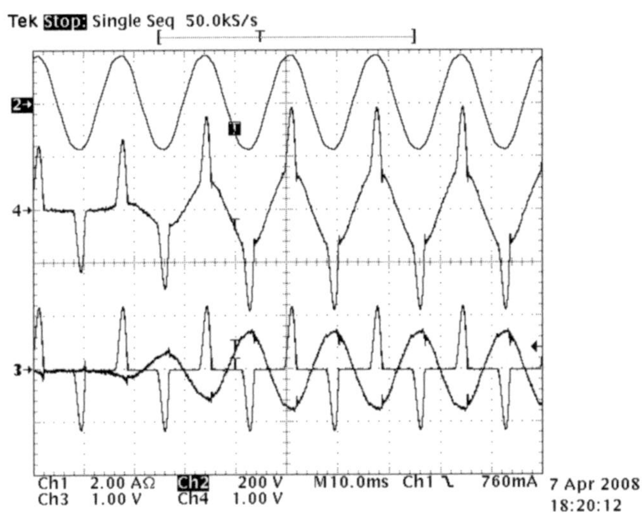

Fig. 17. From top to bottom: v_{PCC} (Ch2, 200V/div), i_c (Ch4, 2A/div), i_L (Ch4, 2A/div) and i_s (Ch1, 2A/div).

i_L. The drop voltage is added to the reference signal v^*_{PCC} by the gain α_1 from the P+Ress controller (see Figure 9). Voltage distortions are also added by α_1 and it allows the DGS to handle with linear and nonlinear loads. All those adjustments are made in the reference signal of the inverter output voltage v^*_c. In Figure 18 the signal v^*_c was adjusted mainly at the peak voltage due to the nonlinear load characteristics. The inverter output voltage v_c follows those adjustments.

High frequency components of the load voltage v_{PCC} in Figure 18 are due to the switching frequency (18kHz) from the lowest voltage cell (V_1) of the multilevel inverter that operates ac-

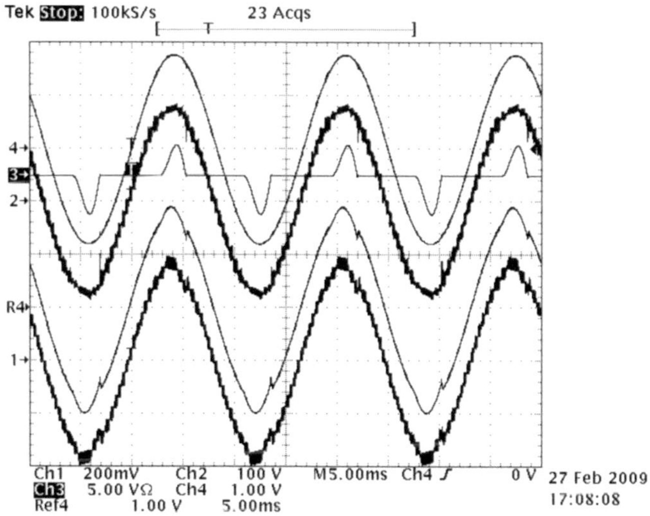

Fig. 18. From top to bottom: v_{PCC}^{*} (Ch4, 100V/div), i_L (Ch3, 5A/div), v_{PCC} (Ch2, 100V/div); v_c^{*} (Ref4, 100V/div) and v_c (Ch1, 100V/div).

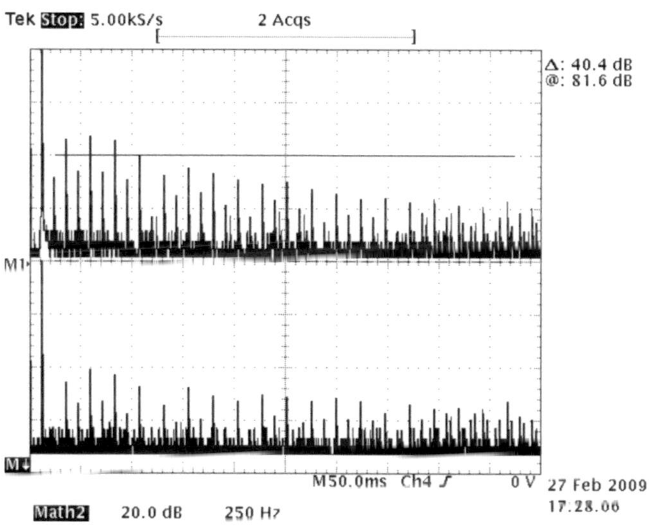

Fig. 19. From top to bottom: FFT of v_c (M1, 20dB/div) and FFT of v_{PCC} (M2, 20dB/div), both voltages from Figure 18.

cording to PWM. Those high frequency components can be also noticed in the output voltage v_c of multilevel inverter. Although both voltages (v_{PCC} and v_c) suggest a poor power quality in a time domain analysis, they are well from the frequency domain point of view. Figure 19 shows the harmonic spectrum of both voltages during the situation presented in Figure 18. The switching frequency carrier (18kHz) and its multiple frequencies are not shown because Figure 19 shows frequency response until 2.5kHz. It is noticed that harmonic amplitudes of voltages v_c and v_{PCC} are, respectively, 35dB lower ($\leq$1.78%) and 40dB lower ($\leq$1%) than fundamental amplitude. The measured THD values were 118% to i_L, 4.1% to v_c and 1.7% to

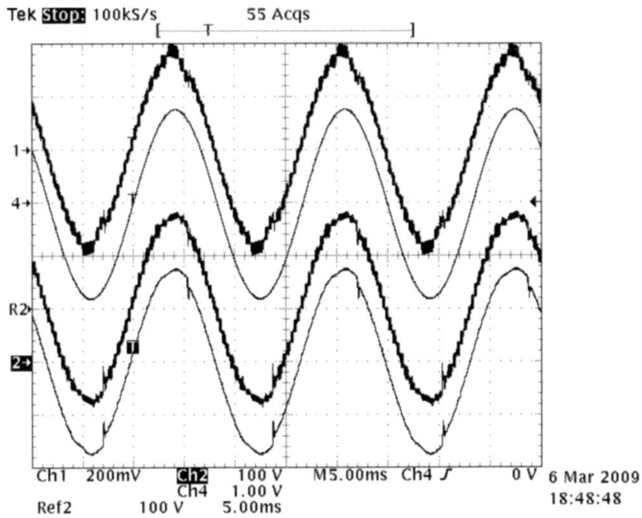

Fig. 20. From top to bottom: v_c (Ch1, 100V/div), v^*_{PCC} (Ch4, 100V/div), v_{PCC} without (Ref2, 100V/div) and with LC filter (Ch2, 100V/div).

v_{PCC}. The low THD of voltage v_{PCC} confirms that the load voltage is similar to a sinusoidal waveform. Besides, it is lower than the maximum THD requirement (5%) presented in (*IEEE Standard for Interconnecting Distributed Resources with Electric Power Systems*, 2003). In addition, each individual harmonics should be also evaluated for attending (*IEEE Standard for Interconnecting Distributed Resources with Electric Power Systems*, 2003) in all harmonic requirements. Considering that a filtered sinusoidal voltage were required at the PCC bar, a LC filter could be used. The design procedure of the elements to that LC filter are relaxed due the low THD of voltage produced by the multilevel inverter. The inductance of the LC filter will be lower in size than that designed for 3-level PWM inverters considering the same capacitance on both cases. Besides a specific modulation strategy that decreases the amplitude of some harmonics (e.g., Selective Harmonic Elimination) may be also used. As the switching frequency is 18kHz, a cut-off frequency closed to 5kHz could be chosen. Considering an inductance L_f equals to 3.85mH (same value of Table 1) and a capacitance C_f equals to 330nF, the natural resonance frequency of LC filter (without load) is equal to 4.46kHz. Figure 20 shows the filtered load voltage for these design values. It is noticed that the high frequency components have been attenuated and the load voltage is more similar to its reference signal.

Figure 21 shows the frequency response of voltage v_{PCC} using and not the designed LC filter. With only the L filter, the switching frequency carrier can be noticed at 18kHz and its multiple at 36kHz. Using the LC filter, they have been attenuated by 20dB at least. By the way, frequency components closed to resonance frequency of LC filter have been fairly increased. It has indicated that the damped resonance frequency was approximately 3.75kHz. The THD of load voltage was equivalent on both cases ($\approx$2%) because low frequency harmonics are more significant to THD evaluation than high frequency harmonics.

6. Conclusion

A proposal of a single-phase distributed generation system with the ability to handle linear and nonlinear loads is presented. Details about the utility-interactive multilevel inverter, grid

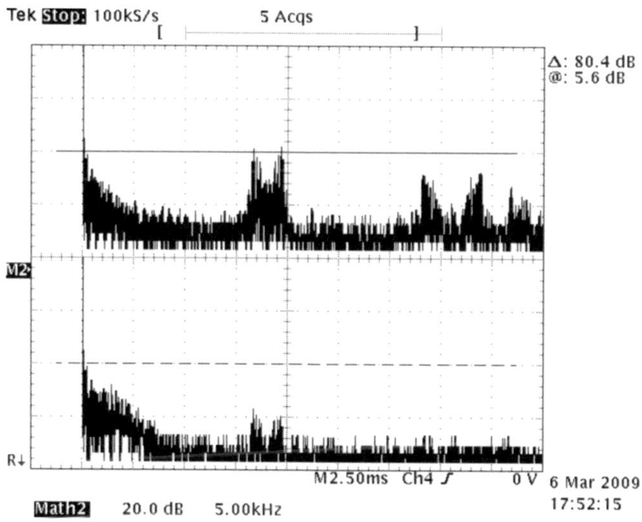

Fig. 21. From top to bottom: FFT of v_{PCC} without (M2, 20dB/div) and with LC filter (Ref2, 20dB/div), both voltages from Figure 20.

connection, voltage levels and operation modes have been described. Concepts of the control schemes used by the DGS have been discussed. Simulation results have been presented for illustrating those concepts and facilitating a prototype model design.

A prototype of the proposed DGS has been implemented and it operates based on a fixed-point DSP TMS320F2812. The validity of the proposed DGS during the grid-connected mode and the stand-alone mode has been proved by the experimental results. The current controller of the grid-connected mode is based on a PI compensator. During the experimental tests, it was noticed that this type of controller has not a wide range of possible operation points and it could not be efficient even if it were stable. The initialization issue could be solved using another type of current controller, probably a nonlinear type. The authors are still working on this issue and on the attempt to get additional experimental results from the prototype.

The similarity between voltage v_{PCC} and its reference signal v^*_{PCC} turns easier to validate the effectiveness of the proposed synchronization method. All the adjustments for decreasing the difference between voltages v_{PCC} and v_S are made by the reference signal v^*_{PCC}. This procedure seems to be less difficult to test than that used by the proposed islanding detection scheme.

7. References

Daher, S., Schmid, J. & Antunes, F. (2005). Design and implementation of an asymmetrical multilevel inverter for renewable energy systems, *COBEP 2005* pp. 199–204.

IEEE Standard for Interconnecting Distributed Resources with Electric Power Systems (2003).

Jung, S., Bae, Y., Choi, S. & Kim, H. (2006). A low cost utility interactive inverter for residential power generation, *IEEE Power Electronics Specialists Conf. (PESC'06)* pp. 1–6.

Jung, Y., Choi, J., Yu, B. & Yu, G. (2006). Optimal design of active anti-islanding method using digital pll for grid-connected inverters, *IEEE Power Electronics Specialists Conf. (PESC'06)* pp. 1–6.

Lipo, T. A. & Manjrekar, M. D. (1999). Hybrid topology for multilevel power conversion.

Marafao, F., Deckmann, S., Pomilio, J. & Machado, R. (2005). Metodologia de projeto e analise de algoritmos de sincronismo pll, *Revista Eletronica de Potencia* **10**(1): 7–14.

Nunez-Zuniga, T., Ataide, M. & Pomilio, J. (2000). Filtro ativo de potencia sintetizando cargas resistivas, *Revista Eletrônica de Potência* **5**(1): 35–42.

Pimentel, S. P. (2006). *Application of multilevel inverter for active power filters*, Master's thesis, School of Electrical and Computer Engineering (FEEC), University of Campinas, UNICAMP, Brazil.

Rech, C., Grundling, H., Hey, H., Pinheiro, H. & Pinheiro, J. (2002). A generalized design methodology for hybrid multilevel inverters, *IEEE Annual Conf. of the Industrial Electronics Society (IECON'02)* **1**: 834–839.

Rech, C. & Pinheiro, J. (2005). Qualitative analysis of hybrid multilevel inverters in closed-loop systems, *COBEP 2005* pp. 223–229.

Rodriguez, J., Lai, J.-S. & Peng, F. Z. (2002). Multilevel inverters: a survey of topologies, controls, and applications, *IEEE Trans. on Industrial Electronics* **49**(4): 724–738.

Silva, L. A., Pimentel, S. P. & Pomilio, J. A. (2005a). Analysis and proposal of capacitor voltage control for an asymmetric cascaded inverter, *IEEE Power Electronics Specialists Conf. (PESC'05)* pp. 809–815.

Silva, L. A., Pimentel, S. P. & Pomilio, J. A. (2005b). Nineteen-level active filter system using asymmetrical cascaded converter with dc voltages control, *IEEE Power Electronics Specialists Conf. (PESC'05)* pp. 303–308.

Zmood, D. & Holmes, D. (2003). Stationary frame current regulation of pwm inverters with zero steady-state error, *IEEE Trans. on Power Electronics* **18**(3): 814–822.

Performance Of Microturbine Generation System in Grid Connected and Islanding Modes of Operation

Dattatraya N. Gaonkar
National Institute of Technology Karnataka Surathkal
India

1. Introduction

The distributed generation system based on microturbine technology is becoming more potential and viable distributed energy source in the recent years. This is due to their salient features such as high operating efficiency, ultra low emission levels, low initial cost and small size (Scott, 1998 and Gaonkar, 2006). The microturbine generation system (MTG) produce power in the range of 25-500 kW and can be operated in stationary or mobile, remote or interconnected with the utility grid. Once connected to power distribution system, these generators will affects the dynamics of the system. Hence dynamic models are necessary to deal with issues in system planning, interconnected operation and management. There is lack of adequate information on the performance of MTG system when connected to distribution network, even though microturbine is based on gas turbine technology, which is well established (Scott, 1998 and Al-Hinai, 2002). In particular, little development has been made on the effective modeling of these DG systems. The design of power electronic converter interface, needed for single shaft MTG system presents a significant challenge (Nikkhajoei, 2005 and Illanda, 2002). The dynamic models of MTG system for isolated operations are reported in (Al-Hinai, 2002, Gaonkar, 2006 and Guda, 2004). But in these works only isolated operation of MTG system is considered. The grid connected model of MTG system with various power electronic topologies are reported in (Azmy, 2003, Fethi, 2004 and Nikkhajoei 2005). These topologies have there own disadvantages which over come by the back to back converter topologies.

One of the important technical issues created by DG interconnection to the utility is the islanding of DG units (Villeneuve, 2004). This mode of DG operation is encouraged by the ongoing rapid development of new generation technologies such as micro turbines, wind generators, fuel cells, photovoltaic plants and their increased penetration in the utility network. Furthermore, the remarkable improvements in the power electronic technology make it possible to design seamless scheme for transferring DG operation between grid connected and islanding modes without interruption in the power supply [Barsali, 2002]. The intentional islanding mode operation is utmost important in the case of sensitive and mission-critical industrial loads as there is a need to maintain a continuous and uninterrupted AC power during planned or unplanned grid outage conditions (Tirumala ,

2002 and Illanda, 2002). The IEEE Std. 1547-2003 states the need for implementing intentional islanding operation of DG systems (IEEE, 2003). As a consequence, research in the last few years has been motivated to study the possibility of intentional islanding of various DG systems.

In the above context this chapter presents the technology, application and performance study of the micro turbine generation system. The detailed modeling of a single-shaft MTG system suitable for grid connection in Simulink of the Matlab is given in this chapter. The algorithm for seamless transfer of microturbne generation system operation between grid connected and islanding mode is presented in this chapter.

2. Types of Microturbine Design

The basic components of a microturbine are the compressor, combustor, turbine generator and recuperator. It operates on the same principles as traditional gas turbines. Air is drawn into the compressor, where it is pressurized and forced into the cold side of the recuperator. Here, exhaust heat is used to preheat the air before it enters the combustion chamber. The combustion chamber then mixes the heated air with fuel and burns it. This mixture expands through the turbine, which drives the compressor and generator.

There are two types of micro turbine designs available, based on position of compressor turbine and generator. Figure 1 (a) shows a high speed single shaft design with the compressor and turbine mounted on the same shaft along with the permanent magnet synchronous generator. The generator generates power at very high frequency ranging from 1500 to 4000 Hz. The high frequency voltage is first rectified and then inverted to a normal AC power at 50 or 60 Hz. Another design is shown in Fig. 1 (b) in which the turbine on the first shaft directly drives the compressor while a power turbine on the second shaft drives the gearbox and conventional electrical generator (usually induction generator) producing 60 Hz power. The two-shaft design features more moving parts but does not require complicated power electronics to convert high frequency AC power output to 60 Hz.

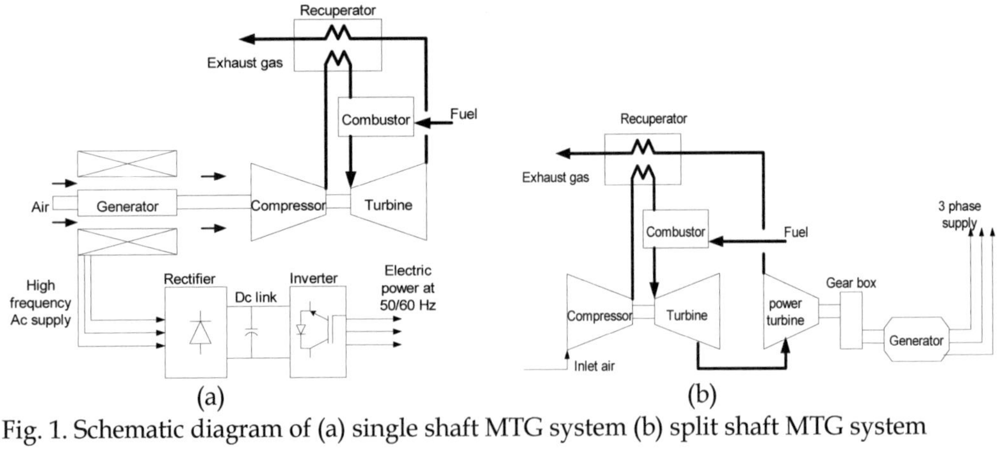

Fig. 1. Schematic diagram of (a) single shaft MTG system (b) split shaft MTG system

Microturbine turbo-machinery is based on single-stage radial flow compressors and turbines. This offers highest efficiency in various size ranges; where as moderate to large size gas turbines use multi-stage axial flow turbines and compressors. In a microturbine, the turbo-compressor shaft generally turns at high rotational speed as high as 1, 20,000 rpm based on the power rating. The microturbine utilizes gas foil bearings (air bearings) for high reliability, low maintenance and safe operation. This needs minimum components and no liquid lubrication is necessary to support the rotating group. Recuperators are heat exchangers that use the hot exhaust gas of the turbine (typically around 1,200°F) to preheat the compressed air (typically around 300°F) going into the combustor. This reduces the fuel needed to heat the compressed air to turbine inlet temperature. With recuperator microturbine efficiency can go above 80%.

3. Power Electronic Interface Topologies

The single shaft MTG system is widely used for power generation applications rather than split shaft due its advantages. As discussed, single shaft microturbine design requires power electronic converter interface to convert the high frequency AC power produced by the generator into usable electricity. This is a critical component in the single shaft microturbine design. It represents significant design challenges, specifically in matching turbine output to the required load. Generally, power electronic interface circuits are designed to with stand transients and voltage spikes up to seven times the nominal voltage.

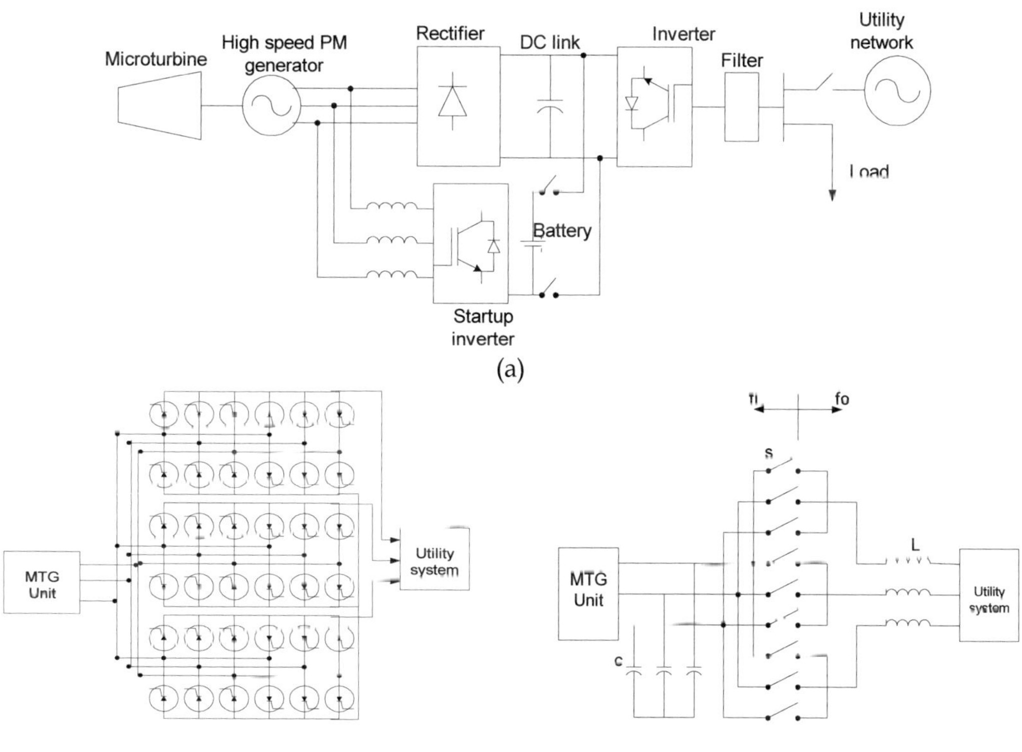

Fig. 2. (a) Passive rectifier and inverter combination (b) Cycloconverter interface model (c) Matrix converter interface model.

There are different interface topologies available for connecting single shaft MTG systems to grid. Figure 2 (a) shows the passive rectifier and inverter combination with DC link. This design needs separate startup inverter for motoring operation of generator to launch the microturbine and also during cool-down process, to remove heat stored in the recuperator and microturbine engine, in order to protect the system components. A cycloconverter and matrix converter shown in Fig. 2 (b) and Fig. 2 (c) can be used to interface the microturbine generator to the grid (Azmy, 2003 and Nikkhajoei 2005). These converters directly convert AC voltages at one frequency to AC voltages at another frequency with variable magnitude. For this reason, they are also called frequency changers. The disadvantages of these converters are that they have double the number of switches compared to the DC link approach and energy storage is not possible. If there is no DC link, any fluctuation on either side of the converter will directly influence the other side.
The matrix converter shown in Fig.2 (c) can be used at lower frequency compared to PWM based converters. Some advantages of the matrix converters are less thermal stress on the semiconductors during low output frequency and absence of the DC link capacitors which increases the efficiency and life time. The drawbacks of this topology are the intrinsic limitation of the output voltage, the unavailability of a true bi-directional switch; absence of decoupling between the input and the output of the converter. This may lead to some instability issues (Teodorescu, 2004). The back to back voltage source converters (VSC) interface topology is shown in Fig. 3. This topology allows bi-directional power flow between the converter and the grid and hence no separate starting arrangement is required. At the time of starting, permanent magnet synchronous machine acts as motor and draws power from the grid to bring the turbine to certain speed. During the generating mode PMSM acts as generator and power flows from MTG system to grid (Fethi, 2004 and Gaonkar, 2008).

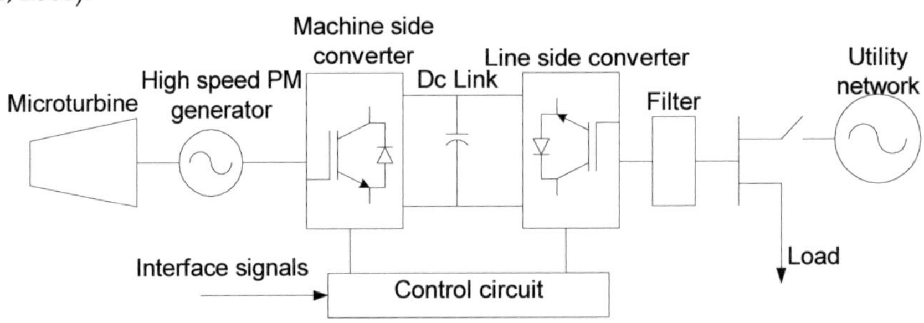

Fig. 3. MTG system with back to back converter interface

4. Applications of MTG System

Microturbine generation system can be used for a wide range of applications. Some of the applications are discussed in this section (Scott, 1998).

4.1 Base load, peak shaving and stand-alone power

Microturbine based DG system can operate in parallel with grid or with any other generation source. The microturbine can augment utility supply during peak load periods, thus increasing power reliability and reducing or eliminating peak demand charges. Shaving peaks will increase overall system efficiency which will reduce investments in traditional generation, bulk transmission, and distribution facilities. Shaving peaks will also enable the utility to serve incremental load growth in areas where there is a shortage of substation and/or distribution feeder capacity. The Microturbine can provide prime power generation where the electric utility grid is not readily available or where service is unreliable.

4.2 Combined Heat and Power

Cogeneration, or combined heat and power (CHP) generation refers to the process of utilizing the heat produced by a combustion engine as energy output. During normal operation, microturbine produces significant quantities of high-temperature exhaust that can be easily integrated with a heat exchanger and a hot water loop to produce valuable energy output. From a cost-benefit perspective, this yields a significant saving compared to heating fuel and purchasing power. When the heat energy is utilized, overall system fuel efficiency can range between 70 and 90+%.

4.3 Resource Recovery

Flared gases often have low-energy yield or high contents of corrosive "sour" (hydrogen sulphide, or H_2S) gas, making them an infeasible fuel source for conventional generators. Microturbine, on the other hand has no problem operating exclusively on low-energy gases. Microturbine can also be used in oil and gas recovery applications. With the ability to convert unprocessed casing gas that contains up to 7% corrosive H_2S gas, microturbine can eliminate the need for flaring. At the same time, their power output reduces or eliminates the need for additional electricity sources.

4.4 UPS and Stand by Services

The microturbine technology can be integrated into a wide variety of products and systems. Uninterruptible power supplies, all-in-one combined heat and power systems, and welding machines are just a few examples for original equipment manufacturing applications.

5. Modeling of MTG System Components

The integrated model of MTG system consists of microturbine, permanent magnet synchronous machine, machine and grid side converters with control and filter. The modeling of individual components is described in this section (Rowen, 1983).

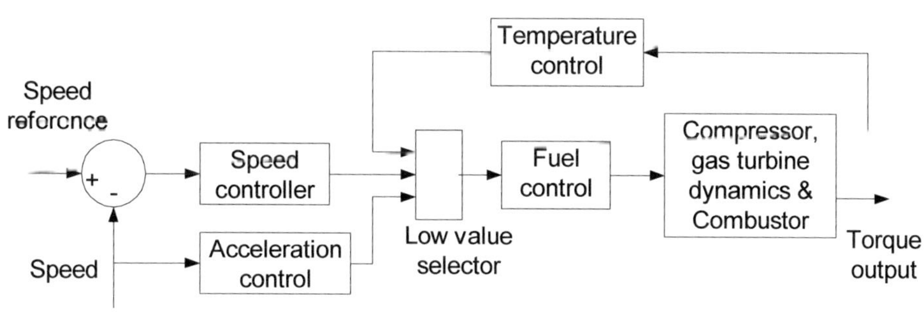

Fig. 4. Block diagram of microturbine system with controls

5.1 Microturbine

The block diagram of microturbine along with its control is shown in Fig. 4. This consists of fuel, speed, acceleration and temperature control along with the combustor and turbine dynamics. The implementation of microturbine model using Simulink of the Matlab is shown in Fig 5.

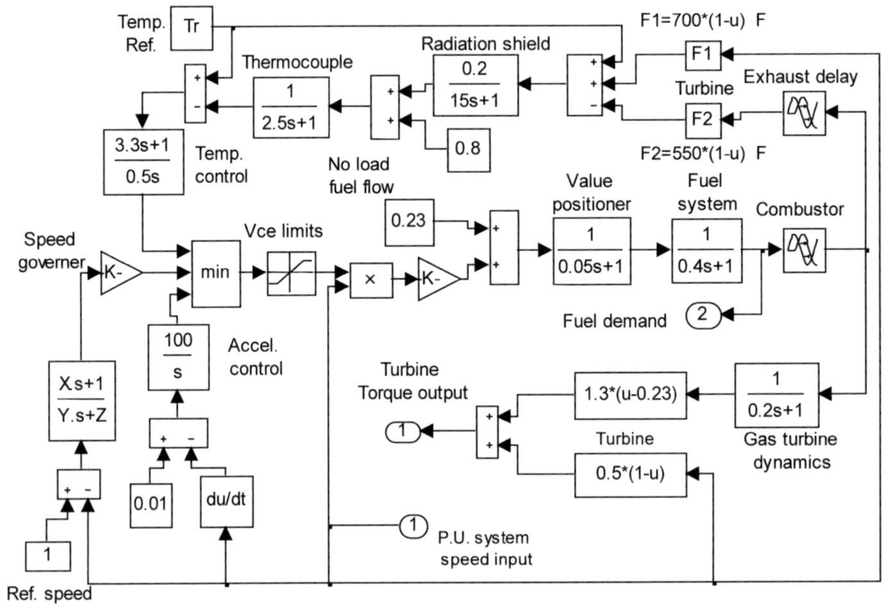

Fig. 5. Simulink model of the microturbine

Speed and acceleration control: The speed control operates on the speed error formed between a reference (one per-unit) speed and the rotor speed of the MTG system. It is the primary means of control for the microturbine under different load conditions. Speed control is usually modeled by using a lead-lag transfer function or by a PID controller (Hajagos, 2001).Acceleration control is used primarily during gas turbine startup to limit the rate of rotor acceleration prior to reaching governor speed. This ameliorates the thermal stress encountered during startup. Acceleration controller is an integrator as shown in Fig. 5 and acts on the error between the derivative of p. u. speed of generator and constant reference signal.

Temperature control: The temperature control is the common method of limiting gas turbine output at a predetermined firing temperature, independent of variation in ambient temperature or fuel characteristics. The fuel burned in the combustor results in turbine torque and in exhaust gas temperature. The exhaust gas temperature is measured using a series of thermocouples incorporating radiation shields. The thermocouples and radiation shields are represented by transfer functions as shown in Fig. 5.The output from the thermocouple is compared with a temperature reference value. Normally the reference value is higher than the thermocouple output. This forces the output from the temperature control to stay on the maximum limit permitting uninhibited governor/speed control. When the thermocouple output exceeds the reference temperature, the difference becomes negative and it starts lowering the temperature control output. When the temperature control output becomes lower than the speed governor output, the former value will pass through the low value selector to limit the output and the unit will operate on temperature control (Rowen, 1983).

Fuel control: The outputs of the speed governor, acceleration controller and temperature control system go as input to a minimum value selector, which selects the lowest value among three inputs. The output of low value selector represents the least amount of fuel required for that particular operating point and is denoted as 'Vce'. The per unit value for 'Vce', corresponds directly to the per unit value of mechanical power on turbine base in steady state. The output of the low value selector is compared with maximum and minimum limits. The maximum limit acts as back up to temperature control and is not encountered in normal operation and minimum limit is more important dynamically. The out put of the 'Vce' limiter is multiplied by 0.77 and offset by no load fuel flow value to ensure the continuous combustion process. The fuel flow controls are represented by series of blocks including the valve position and flow dynamics.

Compressor, combustor and turbine: Turbine is a linear, non dynamic device with the exception of the rotor time constant. There is a small transport delay, which is the time lag associated with the compressor discharge volume. Another transport delay is due to the transport of gas from the combustion system through the turbine. The values of these delays are given in (Rowen, 1983 and Hajagos, 2001]. Both the torque and exhaust temperature characteristics of the single shaft gas turbines are essentially linear with respect to fuel flow and turbine speed. They are given by following equations.

$$\text{Turbine torque} = T = 1.3\,(w_{f2} - 0.23) + 0.5\,(1 - N) \quad (1)$$

$$\text{Exhaust gas temperature} = T_{ex} = T_R - 700\,(1 - w_{f1}) + 550\,(1 - N) \quad (2)$$

Where T_R is the reference temperature, N is per unit speed and W_f is per unit fuel demand signal. The constant value 1.3 in the turbine torque expression depends on the enthalpy or higher heating value of the gas stream in the combustion chamber. In this chapter microturbine is assumed to operate under normal operating conditions. Temperature control and acceleration control are of no significance under normal system conditions. They can be omitted in the turbine model. In this chapter electro-mechanical behavior of microturbine is our main interest. The recuperator is not included in the model as it is only a heat exchanger to raise the energy efficiency. Also due to the very slow response time of the recuperator, it has little influence on the timescale of dynamic simulations presented in this chapter. Thus microturbine model shown in Fig. 5 is simplified by using the above assumptions and resulted model is used in this investigation.

5.2 Permanent Magnet Synchronous Machine (PMSM)

The development of advanced magnetic materials, power electronics and digital control systems are making permanent magnet (PM) machine as an interesting solution for a wide range of applications. The advantages of PMSM compared to other AC machines are its simple structure, high-energy efficiency, reliable operation, high power density and possibility of super high speed operation. Recent important applications of permanent magnet synchronous machine are in the area of distributed generation, mainly in wind and microturbine generation systems. An advantage of a high speed generator is that the size of the machine decreases almost in directly in proportion to the increase in speed, leading to a very small unit. Super high speed PMSM is an important component of single shaft MTG system. The mathematical model of a PMSM is similar to that of the wound rotor synchronous machine.

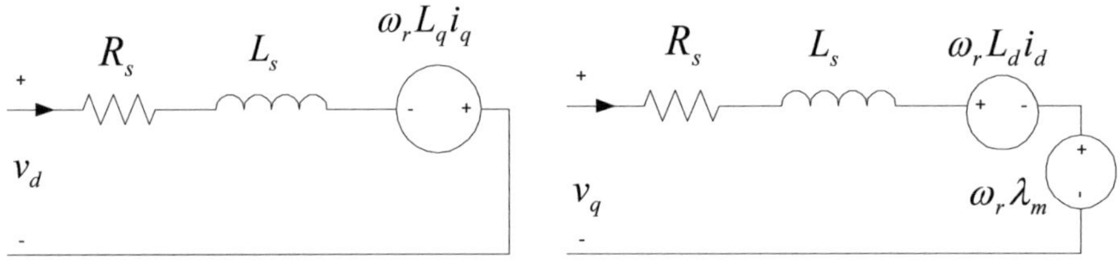

Fig. 6. *dq*-axis equivalent circuit model of the PMSM a) *d*-axis b) *q*-axis

The *dq*-axis equivalent circuit model of PMSM is shown in Fig. 6
The PMSM drive modeling is done with the assumption of sinusoidal distributed windings, saturation is neglected, eddy currents and hysteresis losses are negligible (Pillai, 1989). With these assumptions the stator *dq* equations of the PMSM in the rotor reference frame are:

$$v_d = R_s i_d + L_d \frac{di_d}{dt} - p\omega_r L_q i_q \tag{3}$$

$$v_q = R_s i_q + L_q \frac{di_q}{dt} + p\omega_r L_d i_d + pw_r \Phi_m \tag{4}$$

Where, the stator resistance is denoted by $R_{s,}$, the *d*-axis and *q*-axis inductances are L_d and L_q respectively, Φ_m is the flux linkage due to the permanent magnets, v_d and v_q are *dq* axis voltages. In the *dq*-frame, the expression for electro-dynamic torque becomes:

$$T_e = 1.5p\,(\Phi_m i_q + (L_d - L_q) i_q i_d) \tag{5}$$

The equation for motor dynamics can be given as:

$$\frac{d}{dt}\omega_r = \frac{1}{J}(T_e - F\omega_r - T_M) \tag{6}$$

$$\frac{d}{dt}\theta_r = \omega_r \tag{7}$$

$$\begin{bmatrix} v_q \\ v_d \\ v_0 \end{bmatrix} = \frac{2}{3}\begin{bmatrix} \cos\theta_r & \cos(\theta_r - 120) & \cos(\theta_r + 120) \\ \sin\theta_r & \sin(\theta_r - 120) & \sin(\theta_r + 120) \\ 1/2 & 1/2 & 1/2 \end{bmatrix}\begin{bmatrix} v_a \\ v_b \\ v_c \end{bmatrix} \tag{8}$$

$$\begin{bmatrix} v_a \\ v_b \\ v_c \end{bmatrix} = \begin{bmatrix} \cos\theta_r & \sin\theta_r & 1 \\ \cos(\theta_r - 120) & \sin(\theta_r - 120) & 1 \\ \cos(\theta_r + 120) & \sin(\theta_r + 120) & 1 \end{bmatrix}\begin{bmatrix} v_q \\ v_d \\ v_0 \end{bmatrix} \tag{9}$$

Where p is the number of pole pairs, T_e is the electromagnetic torque, F is combined viscous friction of rotor and load, ω_r is the rotor speed, and J is the moment of inertia, θ_r is rotor angular position and T_m is shaft mechanical torque. The d, q variables are obtained from a, b, c variables through the Park transform given in (8) and a, b, c variables are obtained from the d, q variables through the inverse of the Park transform given in (9).

5.3 Machine Side Converter Control

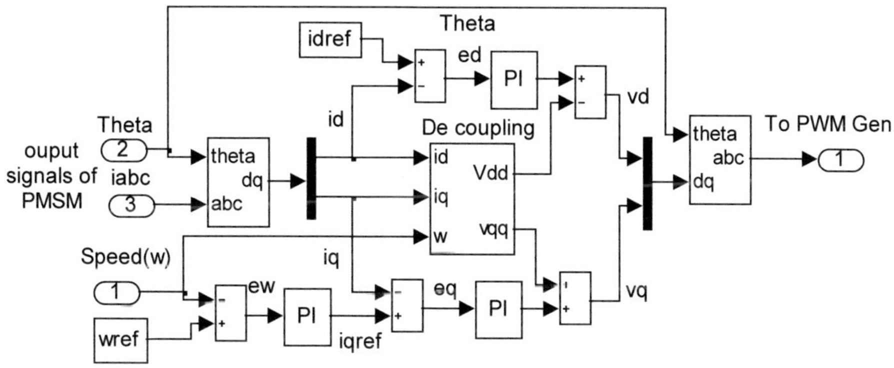

Fig. 7. Machine side converter controller implemented in Simulink

Figure 7 shows the Machine side converter controller implemented in Simulink of the Matlab. The commanded speed ω_{ref} is pre-calculated according to the turbine output power and set to the optimum speed (Morimoto, 1994 and Fethi, 2004). Based on the speed error the commanded q axis reference current iq_{ref} is determined through the speed controller. In this system the following PI controller is employed as the speed controller.

$$i_{qref} = K_{p\omega} e_{\omega} + K_{i\omega} \int e_{\omega} dt \tag{10}$$

Where, $K_{P\omega}$ and $K_{I\omega}$ are the proportional and integral gains of the speed controller respectively while, e_{ω} is the error between the reference speed and measured speed. The commanded optimal d-axis current i_{dref} is obtained from the maximum allowed phase voltage and phase current constraints of the drive, which are given in (11) and (12). These constraints depend upon the machine rating and DC link voltage.

$$v_d^2 + v_q^2 \leq V_{\max}^2 \tag{11}$$

$$i_d^2 + i_q^2 \leq I_{\max}^2 \tag{12}$$

Using the above constraints and neglecting the voltage drop due to the stator resistance, the optimal d-axis current for a non salient PMSM (L_d=L_q) can be obtained as:

$$I_d = \frac{\frac{V_{\max}^2}{\omega^2} - L_q^2 I_{\max}^2 - \lambda_m^2}{2L_d \lambda_m} \tag{13}$$

Considering the relationship $I_{\max}^2 = i_d^2 + i_q^2$ the optimal d-axis current can be given as a function of the q-axis current i_q, as :

$$i_d = -\frac{\lambda_m}{L_d} + \sqrt{\left(\frac{V_{\max}}{\omega L_d}\right)^2 - \left(\frac{L_q}{L_d} i_q\right)^2} \quad (14)$$

Based on the current errors the d-q axis reference voltages are determined by PI controllers, as given in (15) and (16).

$$v_d = K_{Pi} e_{id} + K_{Ii} \int e_{id} dt - \omega_r L_q i_q \quad (15)$$

$$v_q = K_{Pi} e_{iq} + K_{Ii} \int e_{iq} dt + \omega_r (L_d i_d + \lambda_m) \quad (16)$$

Where, K_{Pi} and K_{Ii} are the proportional and integral gains of the controller respectively. $e_{id} = i_{dref} - i_d$ is the d-axis current error and $e_{iq} = i_{qref} - i_q$ is the q-axis current error. The decoupling terms $(-\omega_r L_q i_q)$ and $(\omega_r (L_d i_d + \lambda_m))$ are used in (15) and (16) respectively for the independent control of d and q-axis currents. The commanded dq-axis voltages (v_d, v_q) are transformed into a, b, c variables (v_a, v_b, v_c) and given to the PWM generator to generate the gate pulse for machine side converter.

5.4 Line Side Converter Control

The objective of the supply-side converter is to keep the DC-link voltage constant, regardless of the magnitude and direction of the rotor power. A vector control approach is used here, with the reference frame oriented along the stator (or supply) voltage vector position (Pena, 2001).

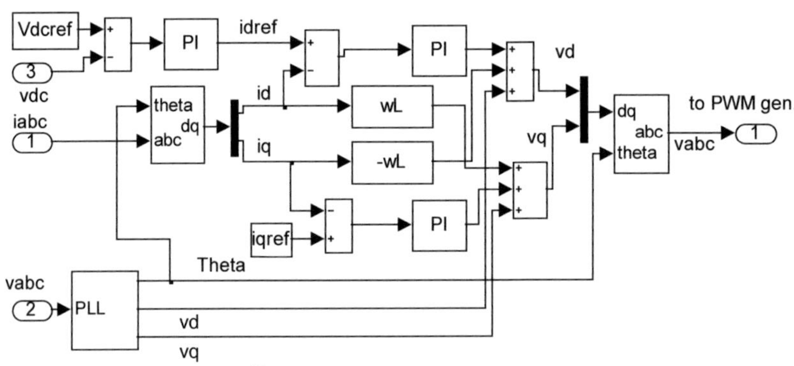

Fig. 8. Grid side converter controller.

Grid connected mode: The PQ control strategy with DC link voltage control is employed for grid connected operation of MTG system. In this scheme the power injected to the grid is regulated by controlling the injected current. The control structure for grid-connected mode operation of MTG system is shown in Figure 8. The standard PI-controllers are used to regulate the currents in the dq synchronous frame in the inner control loops as they have satisfactory behavior in regulating DC variables, as well as filtering and controlling can be easily achieved. Another PI controller is used in the outer loop to regulate the capacitor voltage in accordance with the current injected in to the grid. Its output is the reference for the active current PI controller. In order to obtain only a transfer of active power, the i_q current reference is set to zero. And also to have independent control of the current components i_d and i_q the decoupling voltage components are added to the output of current PI controllers.

Converter voltages can be given as:

$$v_a(t) = \sqrt{\frac{2}{3}}\, V \cos(\omega t),\quad v_b(t) = \sqrt{\frac{2}{3}}\, V \cos(\omega t - \frac{2\pi}{3}),\quad v_c(t) = \sqrt{\frac{2}{3}}\, V \cos(\omega t - \frac{4\pi}{3}) \tag{17}$$

dq component in synchronous reference frame with decoupling terms can be represented as:

$$u_d = Ri_d + L_d \frac{di_d}{dt} + v_d - \omega L\, i_q \tag{18}$$

$$u_q = R\, i_q + L_q \frac{di_q}{dt} + \omega L_d i_d + v_q \tag{19}$$

Where, the dq currents are controlled by means of the right choice of the dq converter side voltages. Two PI regulators are command a PWM modulator to generate the voltage that should control the current.

Islanding mode: The islanding mode operation of single shaft MTG system requires a control different from that of grid connected mode. In this mode, the system has already been disconnected from the utility. Therefore the voltage and frequency is no longer regulated by it. Thus the output voltages are to be controlled in terms of amplitude and frequency which leads to control of the reactive and active power flow.

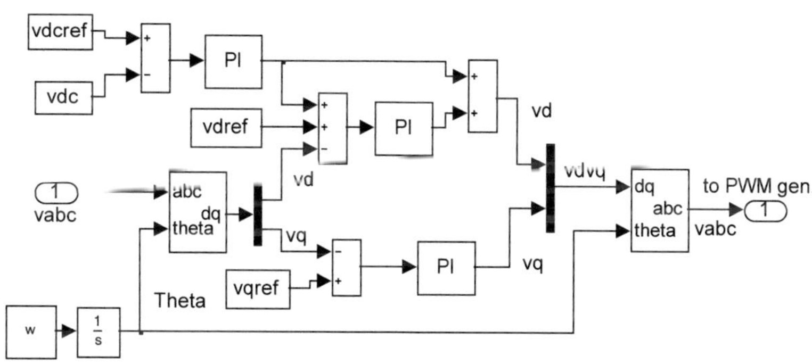

Fig. 9. Control structure for islanding mode operation

This is done by controlling the amplitude and frequency of the modulating input-signal to the PWM inverter. The control structure for islanding mode is depicted in Fig. 9. It consists of output voltage and DC link voltage PI controllers. The output voltage controllers control the output voltage with a minimal influence from nature of the load currents or load transients. A standard PI controller operating in the synchronously rotating coordinate system where, v_q is kept to zero is used. The DC voltage PI controller controls the DC voltage level based on the reference. For fast response the output of the DC voltage controller is feed forwarded to the voltage controller output. The DC link voltage controller acts only when the DC link voltage is below the reference and it lowers the voltage reference of the main voltage controller in order to avoid inverter saturation. The frequency control is done by integrating the constant reference frequency ω and using it for coordinate transfer of the voltage components from *abc* to *dq* and vice versa. In islanding mode, when there is more generation than the load demand the DC link voltage becomes higher than the reference. .For this case damping chopper control is implemented. By activating this control excess energy stored in the Dc -link is dissipated in the damping resistor and hence maintain the Dc link voltage constant (Teodorescu, 2004). A proportional controller is used to control the duty cycle of the chopper. In practice, batteries are used to sore the excess energy.

One of the important requirements in the interconnection design of the power electronic converter interfaced DG system is that of synchronization to the utility system. (Chung, 2000) Synchronism of converter control with the grid is achieved by using a PLL. The Simulink block diagram of the PLL used in this work is shown in Fig. 10, where γ is the grid phase angle, v_x and v_y are the grid voltage components in the stationary reference frame. The philosophy of the PLL is that the sine of the difference between grid phase angle γ and inverter phase angle θ can be reduced to zero using a PI-controller, thus locking the inverter phase to the grid with small arguments ($\sin(\gamma-\theta) \cong (\gamma-\theta) = \Delta\theta$). The output of the PI controller is the inverter output frequency that is integrated to obtain the inverter phase angle θ. In order to improve the dynamic response at startup, the nominal frequency of the grid, ω_o is feed forwarded to the output of the PI controller. The major disadvantage of the three phase PLL is its sensitivity to grid voltage unbalance. Some attempts are made to extend this method for unbalanced voltages based on Symmetrical components (Ghartemani, 2004).

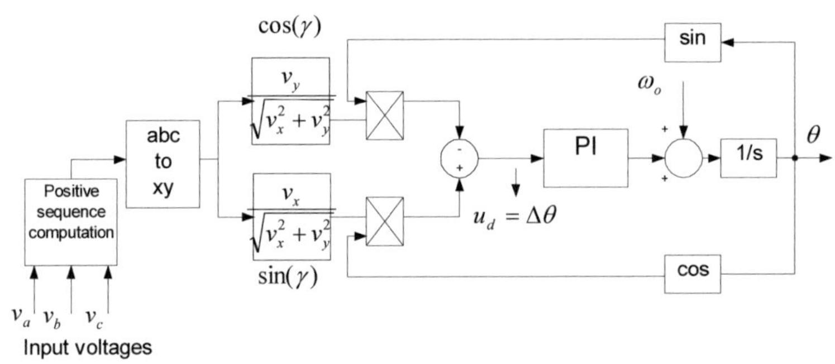

Fig. 10. PLL implemented in the Simulink

6. Simulation and Results

Figure 11 shows the simulation model implemented in the SimPowerSystems of the MATLAB to study the performance of the MTG system operation in grid connected mode (Gaonkar, 2008). The utility network, to which the MTG system is connected, is represented by a 3 phase sinusoidal source with its impedance. The series RL filter is used at the grid side of the MTG system. The simulation parameters of the model are given in Table 1. The micro turbine generation system takes per unit speed of the PMSM as input. The torque output of the microturbine is given as an input mechanical torque (T_m) to the PMSM. The direction of the torque T_m, is positive during motoring mode and made negative during generating mode of the PMSM. The machine side converter controller takes the rotor angle speed and 3 phase stator current signals of the PMSM as inputs. In all the presented cases the voltage across the capacitor is zero, at the starting of simulation. During the start up, the PMSM operates as a motor to bring the turbine to a speed of 30,000 rpm. In this case power flows from the grid to MTG system.

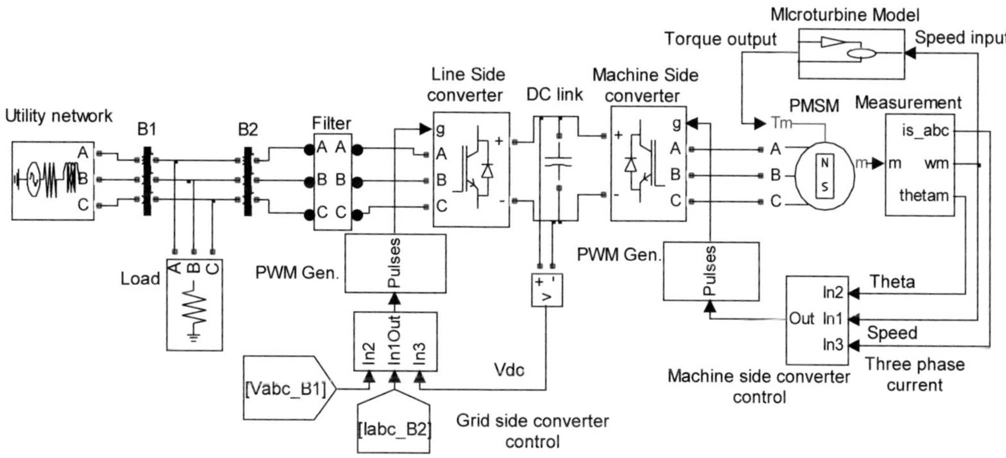

Fig. 11. Matlab/ SimPowerSystems implementation of MTG system connected to grid.

Grid parameters	480V, 60Hz, R_s=0.4Ωand L_s=2mH
Filter Parameters	L=0.97mH, R=0.21Ω
Switching Frequency	Grid side converter = 8 Khz Machine side Converter = 20 KHz
DC link capacitance	5000 μF
PI controllers sampling time	100 μsec
PMSM parameters	480 V, 30 kW ,1.6 KHz, 96000 rpm R_s=0.25 Ω, L_q=L_d=0.0006875 H
Microturbine parameters	Gain(K)=25,X=0.4,Y=0.05 and Z=1

Table 1.Simulation parameters for the model shown in Fig. 11

Figure 12 shows that the microturbine reaches the set value of speed in 0.4 sec. At this speed, the MTG system absorbs power of 5.4 kW as shown in Fig. 12 (b). The PMSM terminal voltage reaches 192 V at a frequency of 500 Hz at this speed. To ensure this operating condition at an unity displacement factor, the pre-calculated reference speed and direct current component i_d are set to 3142 rad/s and -5.36 A respectively (Morimoto, 1994). The speed regulator provides the reference for the i_q current component. At t=0.4 sec, the sign of the PMSM input torque is changed to operate it in generating mode. The power starts flowing from the MTG system to grid as shown in Fig. 12 (b).

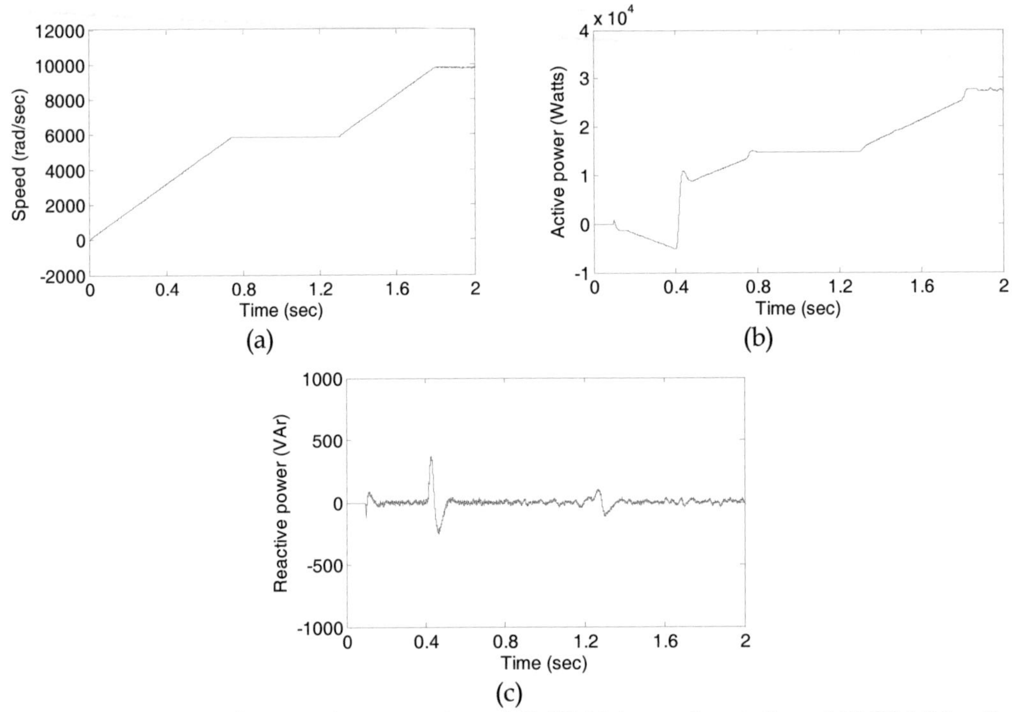

Fig. 12. Motoring and generating operation of PMSM (a) speed variation of PMSM (b) active power variation at the grid side of the MTG system (c) reactive power variations at the grid side of the MTG system.

At t=0.4 sec, the reference speed and i_d current are set to the pre-calculated values of 5849 rad/sec and -15.89 amps in order to generate power of 14 kW. In order to study the performance of the MTG system model for the change in power, the reference values of speed and i_d current component are again changed at t=1.3 sec to generate the rated power of 28 kW. When PMSM generates 28 kW, its line to line voltage and fundamental RMS output current reach the value of 480 V and 33.84 A respectively. Fig. 12 (c) shows that the reactive power injected to the grid during the simulation period is zero.

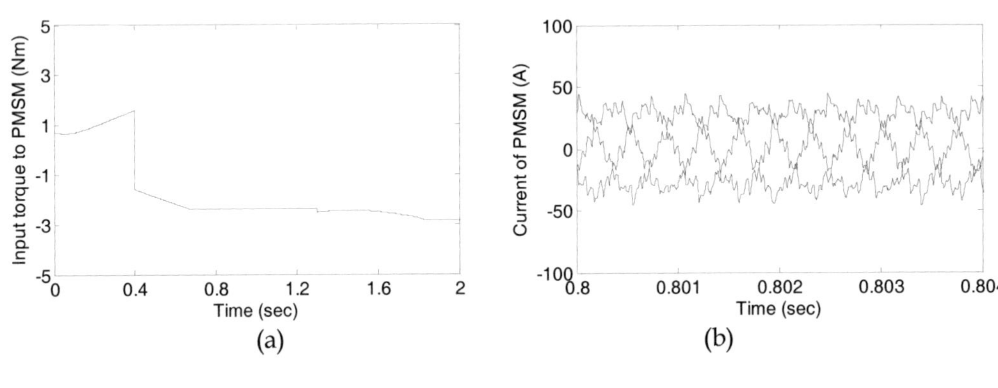

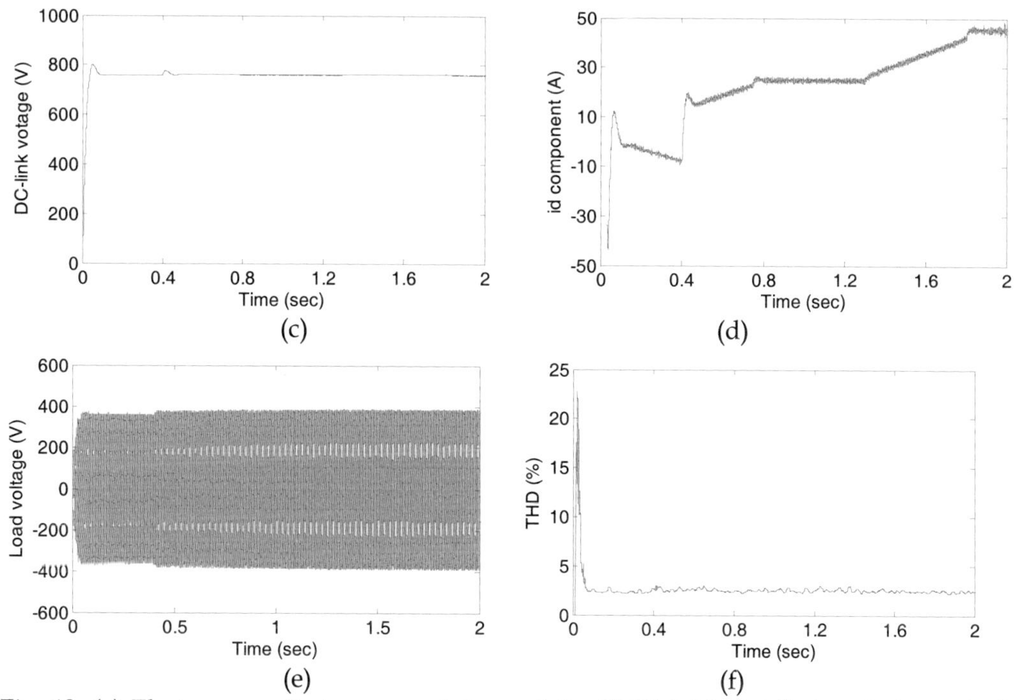

Fig. 13. (a) Electromagnetic torque variations of the PMSM (b) Detailed variations of the stator current of PMSM (c) DC link voltage variation (d) i_d component of the injected grid current (e) Line to line voltage at the load terminals (f) %THD variation at the load terminals.

Figure 13 (a) shows the variation of electromagnetic torque of the PMSM. In this it can be observed that, the change in the operation mode of PMSM in simulation is instantaneous. But this may not be the same in practical because of the inertia of the machine. Figure 13 (b) shows the nature of the stator current waveform of the PMSM. It can be observed from Fig. 13 (c) that the DC link voltage is regulated to 760 V by the grid side converter. Figures 13 (d) and (e) show the variation of i_d component of the injected grid current and the voltage across the terminals of the load. There is a small decrease in the voltage for t< 0.4 sec, as shown in Fig. 13 (e). This is due to the increasing power drawn by the MTG system during motoring mode operation as shown in Fig. 12 (b). In motoring mode both MTG system and load draw power from the grid. The total harmonic distortion (THD) of the voltage is about 2.3% during the entire simulation time as shown in Fig. 13 (f).

6.1 Response of the MTG model for various disturbances in the grid

Simulations are carried out to study the performance of developed the model of the MTG system under various disturbances originating from the grid. Three grid disturbance conditions are considered for study. They are balanced voltage dip, unbalanced voltage and harmonic distortion in the grid voltage. In all these conditions the MTG system is operated to deliver 28 kW.

Balanced voltage dips: The simulation results in Fig. 14 show the performance of the MTG system under balanced grid voltage dip. At t =1.4 to 1.5 sec (10 cycle), 20% balanced voltage dip is introduced in the grid as seen from Fig. 14 (a). The variation of the injected reactive power due to the balanced dip is regulated to zero, by the grid side converter controller as shown in Fig. 14 (b). Figure(s) 14 (c), (d) and (e) show the variation in the active power output of the MTG system and i_q, i_d component of the injected current to the grid respectively.

Voltage unbalance: At t=1.4 sec the voltage unbalance in the grid is introduced by creating short circuit fault between phase A and ground. Thus the phase A voltage is reduced by 20% of its nominal value. In this case it is assumed that unbalance exist till the end of the simulation time. The simulated phase voltage of the grid is shown in Fig. 15 (a).

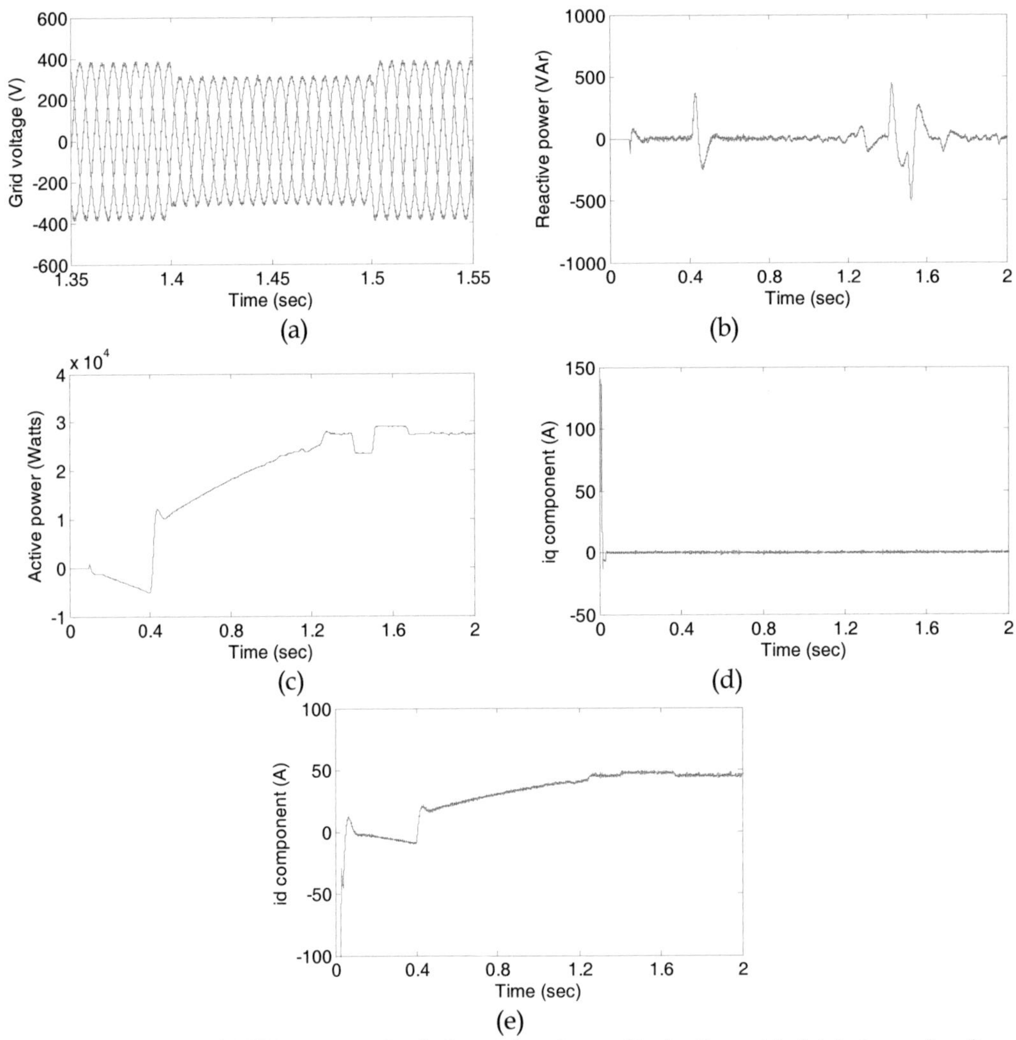

Fig. 14. Response of MTG system for balanced voltage dip in the grid. (a) balanced voltage dip in line to line grid voltage. (b) reactive power exchanged with grid (c) the injected active power to the grid. (d) i_q component of the injected current (e) i_d component of the injected current to the grid.

At the point of voltage unbalance, reactive power injected to the grid suddenly increases and then is regulated to zero as seen in Fig. 15 (b). The active power injected to the grid is not affected by the unbalance in the grid voltage as observed in Fig. 15 (c). There is no variation in the i_d and i_q components of the injected grid current during unbalance in the grid voltage as shown in Fig. 15 (d) and (e).During this unbalance phase A current increases to keep the active power output of the MTG system to constant.

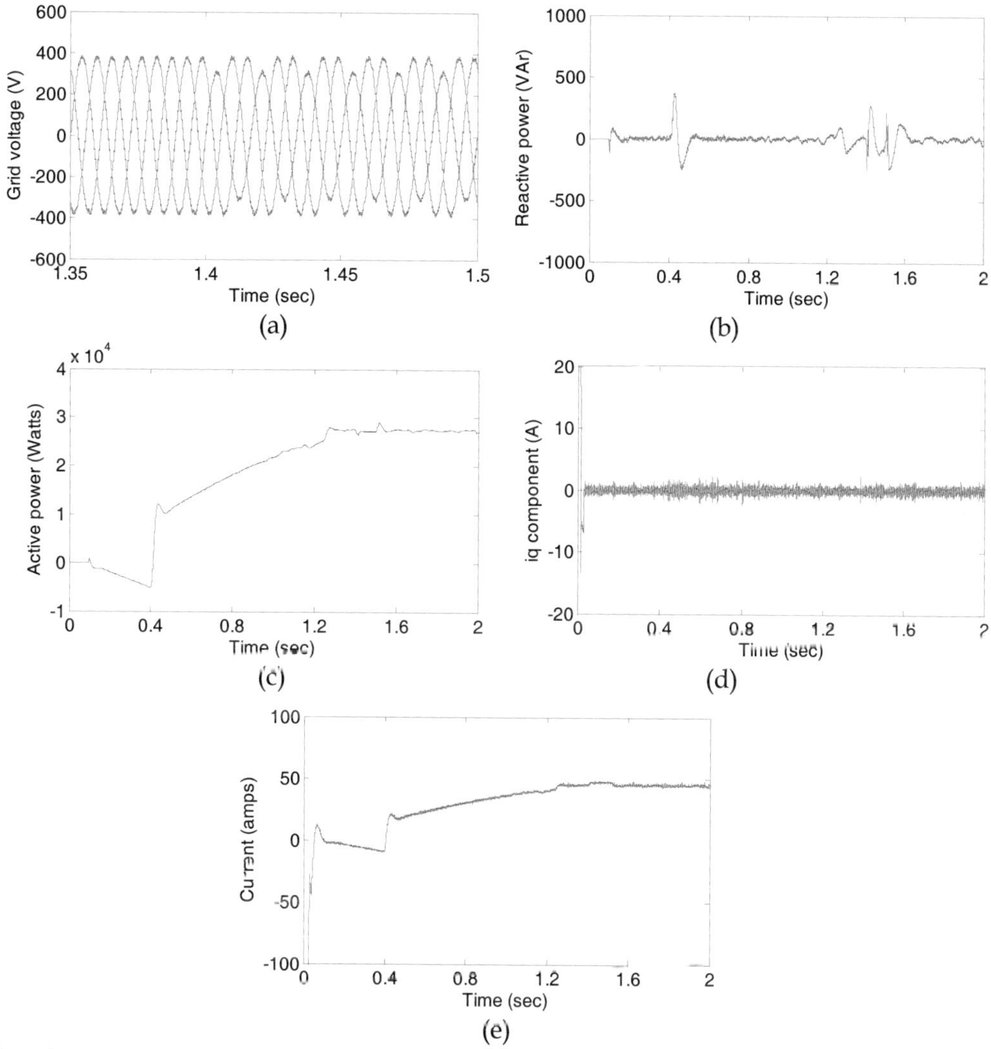

Fig. 15. Response of MTG system for voltage unbalances in the grid. (a)Unbalanced phase voltage of the grid (b) variation of injected reactive power to the grid (c) variations of injected active power to the grid (d) i_q component of the injected current to the grid (e) i_d component of the injected current to the grid.

Polluted grid voltage: The simulation results in Fig. 16 show the performance of the MTG system under harmonic polluted grid voltages. The various non linear loads connected to the grid are the main cause for the harmonic pollution in the grid voltage. In order to study the performance of the model under polluted grid voltage, along with the fundamental component 10% of 5th and 6% of 7th harmonics are injected to the grid. The polluted grid voltages under this condition are shown in Fig. 16 (a). As per IEEE Standard 1547-2003 (IEEE, 2003), the THD at the point of common coupling to the grid should be kept below 5% but in this case it varies from 10% to 8% as shown in Fig. 16 (b) Fig. (s) 16 (c) and (d) show the variation of reactive power exchanged with the grid and injected active power respectively. The variation of the *dq* component of the injected line currents is shown in Fig. 16 (e) and (f) respectively.

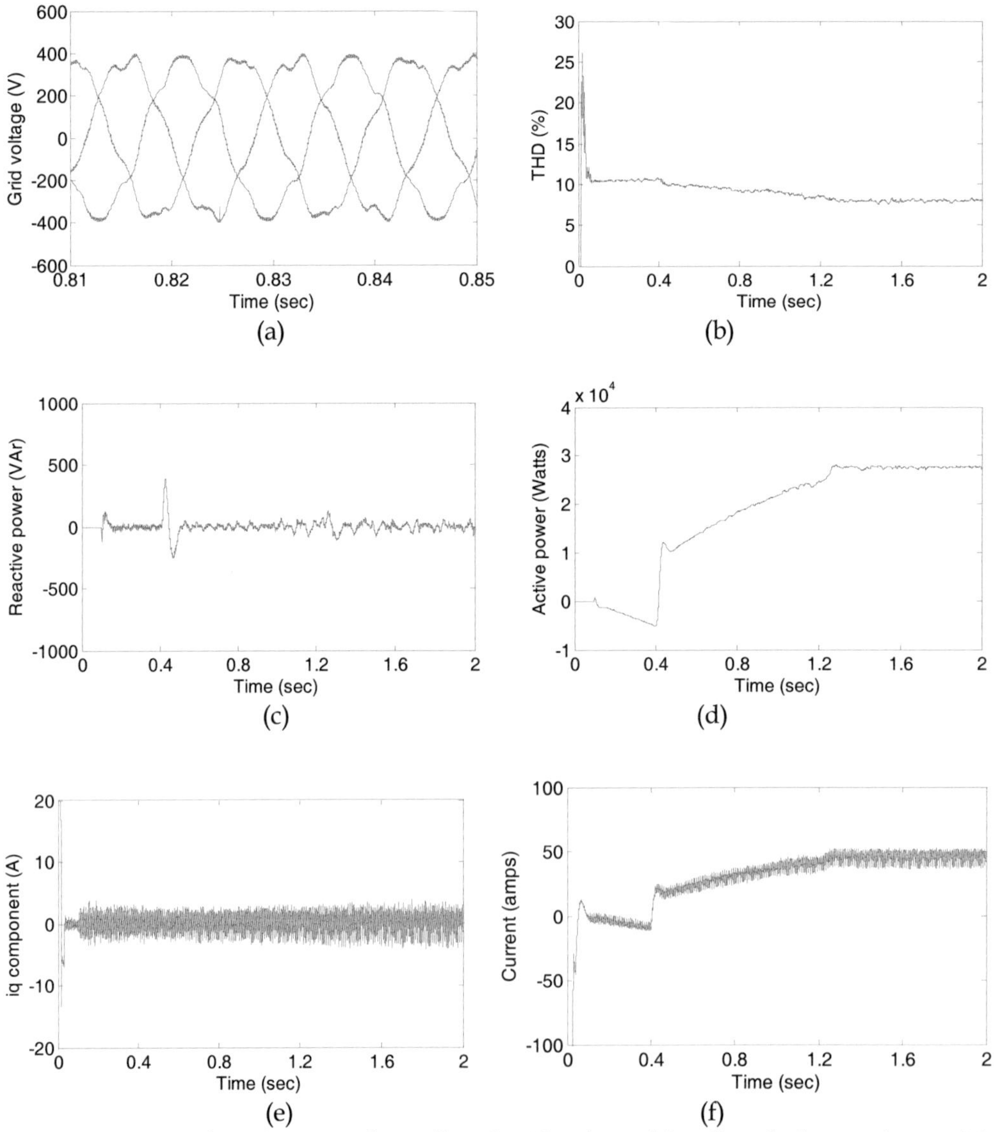

Fig. 16. Response of MTG system for polluted grid voltage (a) polluted phase voltage of the grid (b) variation of the THD in percentage (c) variation of injected reactive power to the

grid (d) variations in injected active power to the grid (e) i_q component of the injected current to the grid (f) i_d component of the injected current to the grid.

6.2 Seamless Transfer Scheme

The scheme consists of passive islanding detection and re-closure method. The presented islanding detection method uses the phase angle estimated by PLL to detect the islanding condition (Gaonkar, 2009). The re-closure scheme continuously monitors the phase angle and terminal voltage magnitude to determine whether disturbance in the grid is over or not. This is necessary in order to synchronize the MTG system and to connect back to grid, without any down time. The PLL block used in inverter control is employed for this scheme, hence no additional hardware needed. The structure of the PLL used to determine the phase angle error required by the proposed scheme has been described in the section 5.4.

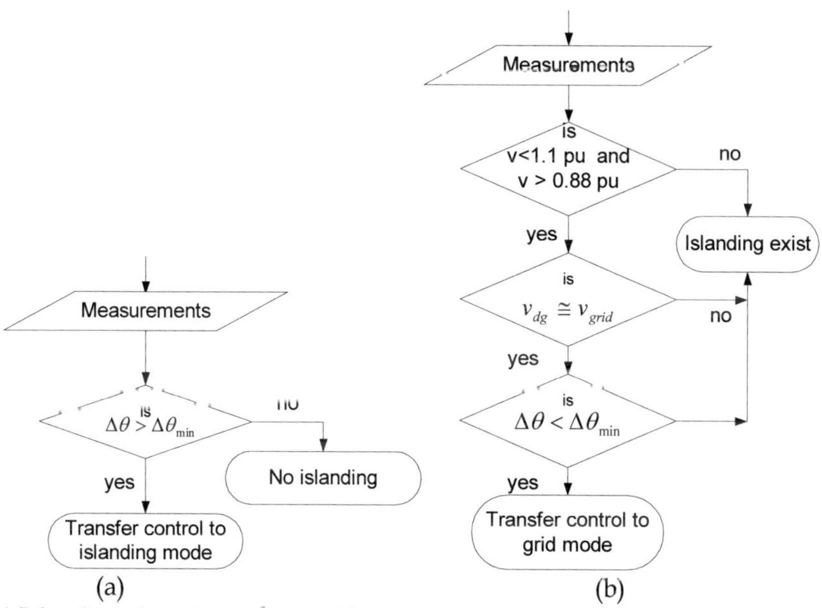

Fig. 17. (a) Islanding detection scheme (b) Re-closure scheme

Once the voltages are approximately equal then algorithm compares the $\Delta\theta$ value obtained from PLL with the set threshold limit. As long as these minimum requirements are met (voltage and phase), there are no major issues hindering the reconnection of islanded systems to the utility without de-energizing and causing downtimes.

Islanding Detection and Re-closure Method: The algorithm devised for the detection scheme is shown in Fig. 17 (Gaonkar 2009). The algorithm compares the phase angle error ($\Delta\theta$) obtained from the PLL, with that of the threshold value. If $\Delta\theta$ exceeds the set threshold limit, then islanding is confirmed. At the same time control of the inverter is transferred from the current control mode (grid connected) to voltage control mode (islanding mode). During the grid connected mode $\Delta\theta$ is approximately zero. Once the grid fails, phase angle error starts to increase. The $\Delta\theta$ value is obtained by PLL as explained in section 5.4. This algorithm gives accurate results even under matching load conditions. The re-closure algorithm for connecting MTG system to grid, when utility recovers from the disturbance is shown in Fig. 17. The two main obstacles in implementing this type of re-closure are matching voltage magnitude and phase angle between converter and the grid. If this is not done, large transient will take place which will damage the MTG system. The re-closure algorithm continuously monitors terminal voltage of the grid and MTG system. Both voltage magnitudes are compared. As per IEEE std.1547-2003 DG terminal voltage should be in between 1.1 and 0.88 per unit of the nominal voltage. Both the voltage should be approximately equal, to avoid large transients during re-connection to grid.

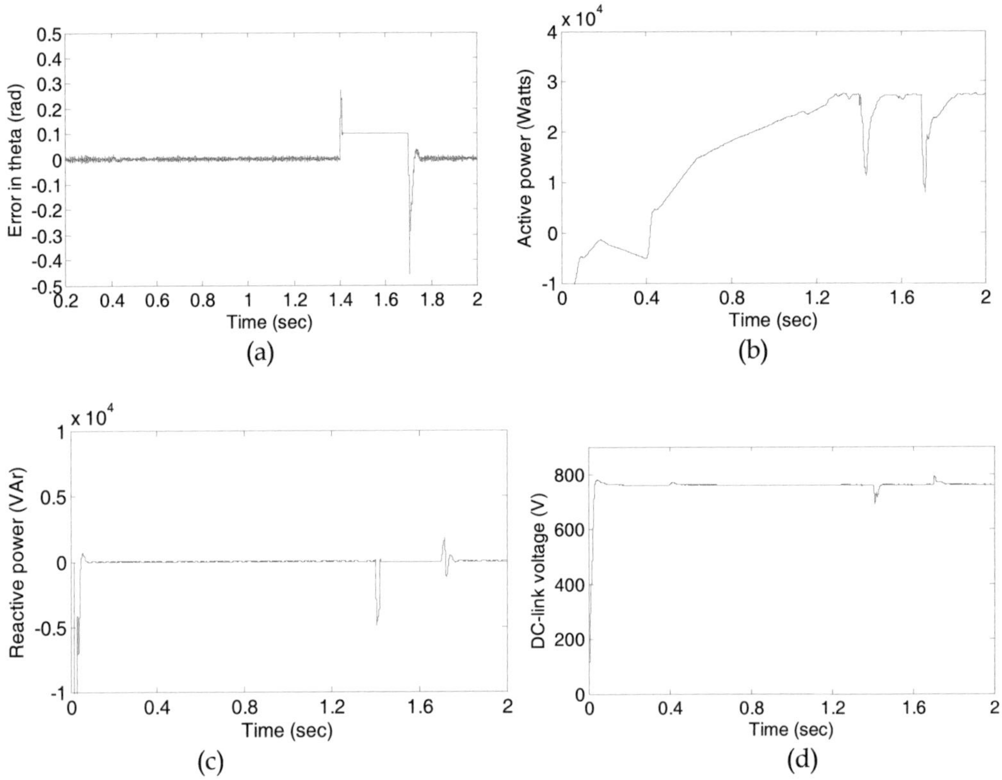

Fig. 18. (a) Phase angle error ($\Delta\theta$) variation (b) active power injected to the grid by MTG system (c) reactive power exchanged with the grid (d) DC link voltage

Utility Interactive to Islanding Mode: The MTG system acts as a motor during starting to launch the microturbine. Once microturbine reaches the speed of 3142 rad/sec at t=0.4 sec, the PMSM is operated as generator by changing the direction of its input torque to negative. During generating mode power flows from MTG system to grid. The phase angle error between the grid voltage and converter voltage is approximately zero as shown in Fig. 18 (a). In grid connected mode MTG system is operated to inject 28 kW as shown in Fig. 18 (b). The other parameters variations during this mode of operation are given in Fig(s). 18 (c) and (d). At t=1.4 sec, utility connection to the DG is disconnected by opening the circuit breaker.

This results in formation of a planned islanding situation comprising local load and MTG system. Opening of the circuit breaker can occur due to the faults and other grid disturbances. When the circuit breaker is opened, the phase angle difference ($\Delta\theta$) between grid voltage and the inverter voltage obtained from PLL starts to increase. The islanding detection algorithm given in Fig. 17 continuously monitors the $\Delta\theta$ variation. Once $\Delta\theta$ crosses the threshold limit of 0.08 *rad* then islanding is detected and converter control switches to islanding mode. The algorithms detect the islanding condition and transfer the control almost instantaneously. The variation of $\Delta\theta$, active power, reactive power and i_d component of the injected grid current are shown respectively in Fig (s). 18 (a), (b), (c) and (d). During this transition dip in DC link voltage can be observed from Fig. 18 (d).This is due to the variation in active power output of the MTG system. During the islanding mode MTG system supplies the power requirement of the load. Hence current supplied by the utility is zero as observed from Fig. 19 (a)

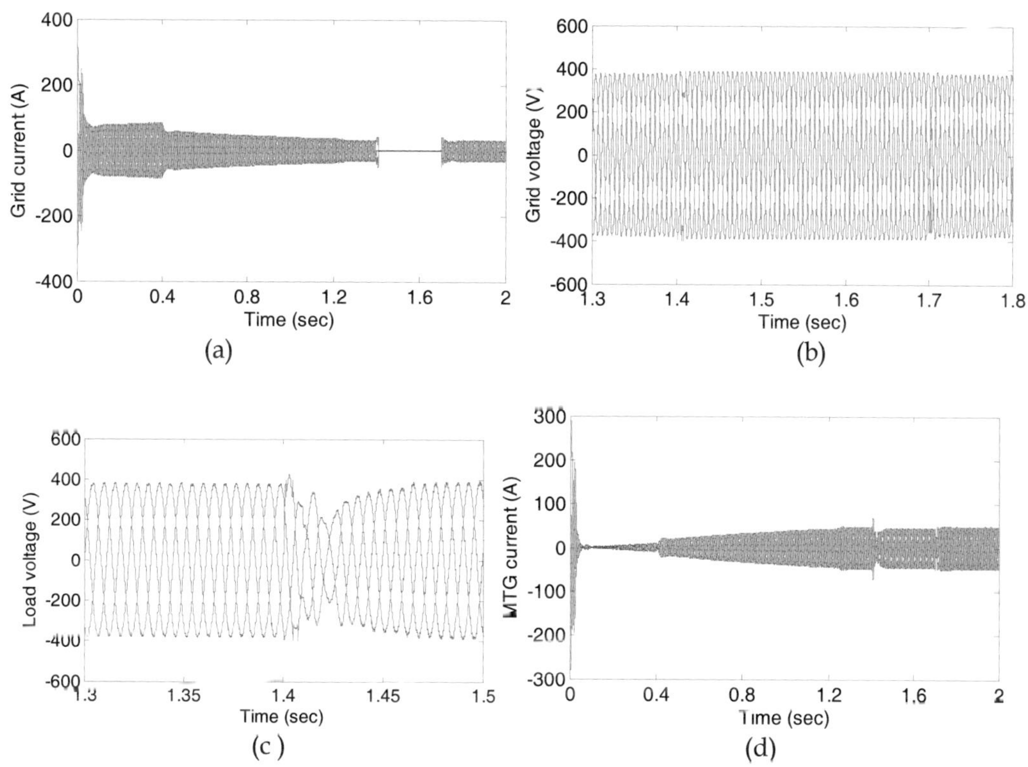

Fig. 19. (a) Line current of the grid (b)phase voltage of the grid (c)detailed variation of load phase voltage (d)line current at the grid side of MTG system

Islanding To Utility Interactive Mode: At t= 1.7 sec, the synchronization process starts to reconnect the DG to utility. The re-closure algorithm compares the grid voltage magnitude and phase angle with the MTG systems terminal voltage and phase angle. When the utility voltage magnitude approximately matches with that of MTG system and $\Delta\theta$ is less than the threshold then DG is connected to the utility. During this transition variation of $\Delta\theta$, active power, reactive power, i_d component of the injected grid current and DC link voltage are shown respectively in Fig(s). 4.8 (a), (b), (c), (d) and (e). The variation of the line current of the grid, phase voltage and current variation of the MTG System terminals are shown in Fig.19.

7. Conclusion

The Distributed generation based on microturbine technology is new and a fast growing business. These DG systems are quickly becoming an energy management solution that saves money, resources, and environment in one compact and scalable package- be it stationary or mobile, remote or interconnected with the utility grid. In this thesis the MTG system model suitable for grid connected and islanding operation has been presented.The detailed modeling of a single-shaft MTG system suitable for grid connection and islanding operation has been developed in Simulink of the Matlab and is described in this chapter. The developed model allows the bidirectional power flow between grid and MTG system, hence need of extra converter for starting and shut down operation of MTG system can be avoided. The multi-loop control structure with decoupling terms used for grid and machine side converters has better performance. The three phase PLL structure described in this chapter uses only positive sequence component of the voltage. Hence gives accurate estimation of the phase angle even under grid disturbance conditions. A seamless transfer scheme for MTG system operation between grid connected and intentional islanding mode has been proposed in this chapter. The presented automatic mode switching scheme helps in providing continuous power supply to the customer even during outages of the utility and. is simple to implement without any additional hardware. The model shows good performance in both grid connected and islanding mode of operations. Thus the developed model of MTG system can be used as a tool suitable for studying and for performing accurate analysis of most electrical phenomena that occur when a microturbine is connected to the grid. As the issues are new and are the key for sustainable future power supply, lot of research is required to study their impacts and exploit them to the full extent.

8. References

Al-Hinai, A., and Feliachi, A., (2002), Dynamic model of microturbine used as a distributed generator, *Proceedings of the 34th South Eastern Symposium on System Theory*, ,pp. 209-213, 18-19 March 2002, Huntsville, AL.

Azmy, A. M., and Erlich, I., (2003), Dynamic simulation of fuel cells and microturbines integrated with a multi-machine network," *Proceedings of the IEEE Bologna Power Tech Conference*, Vol. 2, pp. 2212-2219, Bologna, Italy.

Barsali, S., Ceraolo, M., Pelacchi, P., and Poli, D., (2002), Control techniques of dispersed generators to improve the continuity of electricity supply, *Proceedings of the IEEE PES Winter Meeting*,Vol. 2, pp. 27-31, 2002.

Capstone Microturbine Documentation, Available at : www.capstoneturbine.com

Chung, S.-K., (2000), Phase-locked loop for grid-connected three-phase power conversion systems, *Proc. Inst. Elect. Eng.*, Vol. 147, No. 3, pp. 213-219, 2000.

Fethi, O., Dessaint, L. A., and Al-Haddad, K., (2004), Modeling and simulation of the electric part of a grid connected micro turbine, *Proceedings of the IEEE PES General Meeting*, Vol. 2, pp. 2212–2219, 2004.

Gaonkar, D. N., and Patel, R. N., (2006), Modeling and simulation of microturbine based distributed generation system, *Proceedings of the IEEE Power India Conference*, pp. 256–260, 10–12 April 2006. New Delhi, India,

Gaonkar, D. N., Pillai, G. N., and Patel, R. N., (2008), Dynamic performance of microturbine generation system connected to grid, *Journal of Elect. Power Compon. Syst.*, Vol. 36, No. 10, pp. 1031–1047, 2008.

Gaonkar, D. N., Pillai, G. N., and Patel, R. N., (2009), Seamless transfer of microturbine generation system operation between grid connected and islanding modes, *Journal of Elect. Power Compon. Syst.*, Vol. 37, No. 10, pp. 174–188, 2009.

Ghartemani, M. K., and Iravani, M. R., (2004), A method for synchronization of power electronic converters in polluted and variable frequency environment, IEEE Trans. Power Syst., Vol. 19,No. 3, pp. 1263–1270, 2004.

Guda S. R., Wang C. and Nehrir M. H., (2004), Modeling of microturbine power generation systems, *Electrical Power System and Components*, vol. 34, no. 9, pp. 1027–1041.

Hajagos, L. M., and Berube, G. R., (2001), Utility experience with gas turbine testing and modelling, *Proc. IEEE PES Winter Mtg.*, Vol. 2, pp. 671–677, 2001.

IEEE, (2003), IEEE standard for interconnecting distributed resources with electric power systems, IEEE Std. 1547-2003, 2003.

Illinda, M., and Venkatramanan, G. (2002), Control of distributed generation systems to mitigate load and line imbalances, *Proceedings of the IEEE Electronics and Specialists Conference*, Vol. 4, pp. 2013–2201, 2002.

Morimoto, S., Sanada,M., and Takeda,Y., (1994), Wide speed operation of interior permanent magnet synchronousmotors with high performance current regulator, *IEEE Trans. Indl. Appl.*, Vol. 30, pp. 920–926, 1994.

Nikkhajoei, H., and Iravani, R., (2005), A matrix converter based microturbine distributed generation system, *IEEE Trans. Power Delivery*, Vol. 20, No. 3, pp. 2182–2192, 2005.

Pillai, P., and Krishnan, R., (1989), Modeling, simulation, and analysis of permanent-magnet motor drives, Part I: The permanent-magnet synchronous motor drive, *IEEE Trans. Indl. Appl.*,Vol. 25, No. 2, pp. 265–273, 1989.

Pena, R. Cardenas, R. Blasco R., G. Asher G. and J. Clare J., (2001), A cage induction generator using back-to-back PWM converters for variable speed grid connected wind energy system," in *Proc. IECON'01 Conf.*, vol. 2, 2001, pp. 1376–1381,2001.

Rowen, W. I., (1983), Simplified mathematical representations of heavy duty gas turbines, *ASME Trans. J. Eng. Power*, Vol. 105, No. 4, pp. 865–869, 1983.

Scott, W. G., (1998) , Microturbine generators for distribution systems, *IEEE Industry Appl. Mag.*,Vol. 4, No. 3, pp. 57–62,1998.

Teodorescu, R., and Blaabjerg, F., (2004), Flexible control of small wind turbines with grid failure detection operating in stand-alone and grid-connected mode, *IEEE Trans. Power Electronics.*,Vol. 19, No. 5, pp. 1323–1332, 2004.

Tirumala, R., Mohan, N., and Henze, C., 2002, Seamless transfer of grid-connected PWM inverters between utility-interactive and stand-alone modes, *Proc. 17th Ann. IEEE Appl. Power Electron. Conf. Exposit.*, Vol. 2, pp. 1081–1086, 2002.

Villeneuve, P. L., 2004, Concern generated by islanding, *IEEE Power Energy Mag.*, Vol. 2, No. 3, pp. 49-53, 2004.

10

Distributed Generation and Islanding – Study on Converter Modeling of PV Grid-Connected Systems under Islanding Phenomena

N. Chayawatto[1], N.Patcharaprakiti[2], V. Monyakul[1], K.Kirtikara[1] and K. Tunlasakun[1]

Clean Energy Systems Group (CES),
King Mongkut's University of Technology Thonburi (KMUTT) [1],
Bangkok 10140, Thailand
Rajamangala University of Technology Lanna[2] *, Chiang Mai, Thailand*
Corresponding author: V. Monyakul <v_monyakul@yahoo.com>

Abstract

Thailand government has launched a 15-year (2008-2022) strategic plan on new and renewable energy. Possible electricity generated from solar photovoltaic has been estimated with a potential of 50,000 MW, whereas at present the cumulative installed wattage is only 32 MW. Under the Plan, numbers of measures and incentives are provided for participation of private very small power producers (VSPP) generating and selling the electricity into the utilities. Most VSPPs generate electricity from renewable sources such as mini-hydro, biogas and biomass, wind and solar. Examples of measures and incentives are the Renewable Portfolio Standard (RPS) for the generating utility and independent power producers (IPP), a feed in tariff with an extra adder, soft loans and tax reduction.

The past decade in Thailand has seen shifts from PV used in the public market through government demonstration projects to the consumer market, installations of PV VSPPs and domestic roof-top grid connected PV units gain momentum. With the government incentive more households will be attracted to produce electricity from solar PV and wind energy. As domestic roof sizes are limited, PV roof-top grid-connected units will be of small capacity, less than 10 kW. It is this possible large expansion of market for thousands of small PV roof-top grid-connected units or wind systems in Thailand, and eastern Asia, that draws our attention to the study of single phase distributed generator grid-connected systems. Our focus will be on the anti-islanding protection, which is of concerns to Thai electrical utilities. In order to know the behavior and the effect of anti-islanding techniques, the converter modeling of PV grid-connected systems under islanding phenomena is studied. The approach of modeling is to model a dc-ac full bridge switching converter PV grid-connected system under islanding phenomena using two mathematical modeling techniques. One corresponds to a state-space averaging technique (no linearization) and the other a piecewise technique. The former technique applies a state-space averaging technique (no-

linearization) and voltage source inverter with current control as "large signal modeling". The latter one employs piecewise functions. Each piecewise function ON and OFF interval are derived by using a state-space equation and solved by the Laplace Transform technique. Both are used to analyze the dynamic response of load voltages under 3 different resistive loads, i.e, 125%, 100% and 25% of inverter output and a RLC load when the grid system is disconnected. An experiment on islanding detections is set up. The equations from two proposed modeling technique are handled by MATLAB/SIMULINK software. The results of the proposed models are compared with experiments and the PSpice simulation showing good agreement. It is found that the proposed models consumes much less computation time than the PSpice Program and does not encounter any convergence problem. For islanding trip times, it is found that in all cases tested they are about 2-3 cycles or 40-60 ms, passing the criteria of regulations and standards of the electric utilities of Thailand.
This technique can potentially be further developed for implementation in larger systems consisting of a large array of grid-connected PV modules or other distributed generators.

1.Introduction

1.1 Renewable Energy and Electricity Generation in Thailand

In 2007, the Thailand government launched a strategic 15- year plan for promotion of the renewable energy technology with the goal of increase uses of renewable energy from 5.8% at present to 20% of final energy consumption [1]. The renewable energy targets, covering solar, wind, hydro, wastes, hydrogen, various types of biomasses and biofuels, are outlined in Table 1, expressed as electricity power generation and ktoe. In terms of electrical power generated, the electrical power from solar PV, wind, biomass, biogas, garbage and hydrogen will increase from the existing 1,714 MW to 5,508 MW equivalent to 2,990 ktoe, the thermal energy from renewable energy sources from 3,007 ktoe to 7,433 ktoe, and the biomass energy from less than 10 ktoe to 3,986 ktoe..

Fuel	Potential	Existing	2007-2011		2012-2016		2017-2022	
Electricity	**MW**	**MW**	**MW**	**ktoe**	**MW**	**ktoe**	**MW**	**ktoe**
Solar PV	**50,000**	**32**	**55**	**6**	**95**	**11**	**500**	**55**
Wind	1,600	1	115	13	375	42	800	89
Hydro	700	56	165	43	281	73	324	85
Biomass	400	1,610	2,800	1,463	3,220	1,682	3,700	1,933
Biogas	190	46	60	27	90	40	120	54
Garbage	400	5	78	35	130	58	160	72
Hydrogen	-		0	0	0	0	3.5	1
Total Electricity		1,750	3,273	1,587	4,191	1,907	5,608	2,990
Thermal Energy	Ktoe	ktoe		ktoe		ktoe		ktoe
Solar Thermal	154	1		5		17.5		38
Biomass	7,400	2,781		3,660		5,000		6,760
Biogas	600	224		470		540		600
Garbage		1		15		24		35

Total Thermal		3,007		4,150		5,528		7,433
BioEnergy	Ml/day	ktoe	Ml/day	ktoe	Ml/day	ktoe	Ml/day	ktoe
Ethanol	3.00	1.24	3.00	805	6.20	1,686	9.00	2,447
Biodiesel	4.20	1.56	3.00	950	3.64	1,145	4.50	1,415
Hydrogen			0	0	0	0	$0.1x10^6$ kg	124
Total Bioenergy			6.00	1,755	9.84	2,831	13.50	3,986

Table 1 Potential Renewable Energy and 15 –Year Plan of Thailand

The Plan is divided into 3 periods, the Short Term, the Mid Term and the Long Term. Its essences are

1) The Short Term (2007-2011) : The first stage will concentrate on promoting proven technologies and high potential renewable energy such as bio-energy biomass, biogas and compressed natural gas.

2) The Mid Term (2012-2017) : Renewable energy industries will be promoted with supports on research and development on ethanol production, biodiesel from algae, oil production from biomass, and hydrogen generation. It is envisaged that a "green city" demonstration will be carried out and applications of renewable energy to community levels strengthened.

3) The Long Term (2017-2022) : Hydrogen economy, the green city and renewable energy in community will be expanded. Thailand would lead as an energy leader in bioenergy and renewable energy of ASEAN.

Among many renewable energy sources of significant, solar PV has a very high potential, estimated to about 50,000 MW. The daily average of solar radiation in Thailand is about 5.0 to 5.3 kWh/m^2-day, corresponding to 18 to19 MJ/m^2-day. High values 20-24 MJ/m^2-day are recorded during April and May. The north eastern and northern regions receive about 2,200 to 2,900 hours of sunshine on annual basis, 6-8 sunshine-hours per day. However, the exiting PV power generation is only 32 MW.

Thailand electricity sector of the country can be divided into generation, transmission and distribution sectors. In the past, it was dominated by 3 state enterprise utilities. They are one generation and transmission utility, the Electricity Generating Authority of Thailand-EGAT, ,and two distribution utilities, the Provincial Electricity Authority-PEA and the Metropolitan Electricity Authority-MEA. The industry was liberalized in early 1990s allowing participation of the private sector in power generation. Presently in the generation sector, two major power generators are EGAT 48% and 39 % from private independent power producers- IPPs, with generating capacity more than 90 MW, initiated in early 1990s. 3% are imported from neighboring Laos. The remaining are from small power producers- SPP, generating capacity 10-90 MW, and very small power producers – VSPP, less than 10 MW. SPPs use mainly biomass whereas VSPPs are mostly mini hydroelectric, wind and solar PV. Electricity generated by EGAT, and IPPs are transmitted through the network of EGAT to the 2 distributors, PEA and MEA. SPPs directly sell power to users, mainly large industries, some are connected to PEA and MEA. The structure of electricity industry of Thailand is shown as Fig.1.

It should be noted that electricity generated by renewable energy sources will be SPPs and VSPPs. They will become more common as renewable energy promotion expanded, hundreds or thousands of them can be connected to the distribution network in the future. The 3 utilities, EGAT, PEA and MEA are concerned on power electrical quality resulting from connecting of large numbers of grid-connected SPPs and VSPPs, being distributed in nature, to the network. Therefore, they have come up with guidelines on such distributed generators.

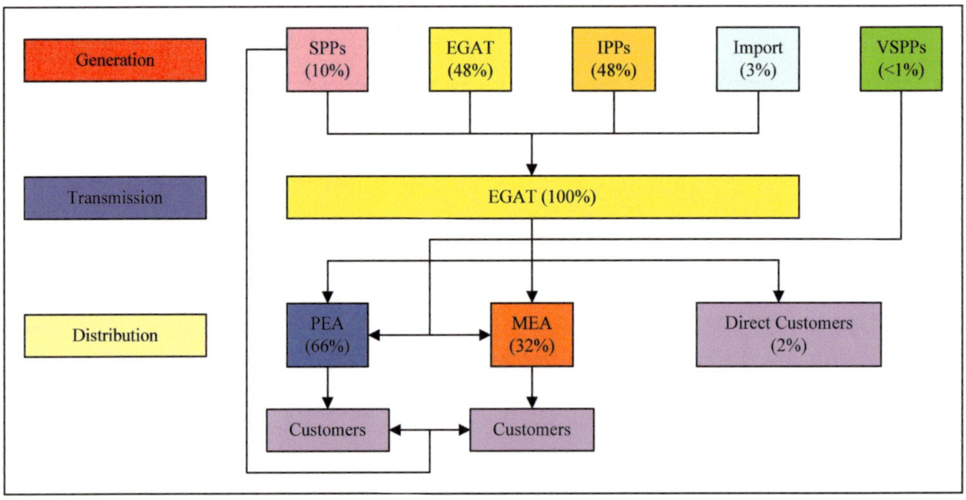

Fig 1. Structure of electricity industry of Thailand

The Power Development Plan - PDP of Thailand sets targets for electricity production from various sources [2]. The present one is PDP 2007. In 2008 148,197 million units were generated. Natural gas share is about 70% while renewable energy, excluding hydroelectric, is only 1.4%. From electricity forecasting of the national energy policy committee, Thailand has the peak power demand and energy demand approximately 44,281 MW and 288,920 million units in 2021. Natural gas would still the major source with 60% share and renewable energy, excluding hydroelectric, will increase to 2%. In order to promote electricity generation from renewable energy, SPPs and VSPPs will play a major role. To achieve this, economic and technology obstructions need to be removed. As a consequence, an extensive promotion program has been initiated through various mechanism such as the Renewable Portfolio Standard-RPS mandatory for the power producers, financial incentives, feed-in-tariff, adders, soft loans, and measures of the Board of Investment- BOI and tax reductions schemes [3].

The government establishes the RPS for power plants, mainly for EGAT. Consequently, new EGAT plants have to invest so that 5% of their electricity production comes from renewable energy sources such as solar cells, wind energy and waste to energy. However, power plants can either buy RPS credits from SPP operating on renewable energy. Furthermore, to motivate more electricity generation from renewable energy, adders for VSPPs are introduced, shown in Table 2.

Type of VSPPs	Capacity	New adder (Bath/kWh)	Duration(years)
Biomass	Capacity ≤ 1MW Capacity > 1MW	0.50 0.30	7 7
Biogas	Capacity ≤ 1MW Capacity > 1MW	0.50 0.30	7 7
Municipal waste	Digestion system Thermal system	2.50 3.50	7 7
Wind energy	Capacity ≤ 50 kW Capacity > 50 kW	4.50 3.50	10 10
Mini Hydro	Capacity 50-200 kW Capacity ≤ 50 kW	0.80 1.50	7 7
Solar PV		8.00	10

Table 2. Adders for VSPPs using renewable energy

1.2 Thai Utilities Guidelines on Distributed Generation

To assure the quality of power, stability, reliability and safety of operation, Thai utilities consider the effects of distributed generator-DG connecting to the power systems in many aspects and conditions. IPPs and SPPs are mainly involved with and, therefore, follow regulations of EGAT, the nature of guidelines are not covered in this paper. PEA and MEA will increasingly deal more with VSPPs and have come up with numbers of regulations on purchasing of power from VSPPs or DGs.

PEA and MEA regulations can be categorized into 3 groups, namely, first, the terms for connection of VSPPs to the electrical network, secondly, operation of grid-connected VSPPs, and thirdly, specifications of VSPPs connecting to the network [4] [5] [6][7].

On specifications of VSPPs, the followings are covered in guidelines : (i) Power capacity and voltage levels for connection, shown in Table 3, (ii) Voltage regulations, (iii) Short circuit levels, (iv) Power quality such as voltage fluctuations, frequencies, power factor and harmonics, (v) Protection systems, and (vi) Connection configurations of VSPPs.

On the connection configurations, MEA and PEA have identified types of VSPPs based on voltage levels, generator types, and power capacity. Examples are (i) synchronous generator rated not exceed to 1.25 MVA and rating of transformer not exceed 2.5 MVA for multi generators, (ii) single synchronous generator capacity more than 1 MW, (iii) multi synchronous generators with capacity more than 1 MW, (iv) single induction generator with capacity less than 5 MVA, (v) multi induction generators with capacity more than 5 MVA, (vi) single induction generator at low voltage, (vii) multi induction generators at low voltage, (viii) three-phase inverter at medium voltage, (ix) three- phase inverter at low voltage and (x) single phase inverter at low voltage.

Utility	Voltage Level	Normal Circumstances		Exceptional Circumstances		Short Circuit Level
		Min (kV)	Max (kV)	Min (kV)	Max (kV)	kA
MEA	115 kV	109.2	120.7	103.5	126.5	31.5-40
	69 kV	65.5	72.4	62.1	75.9	40

	33 kV	31.3	34.7	29.7	36.3	25
	22 kV	20.9	23.1	19.8	24.2	25
	380 V	342 V	418 V	342 V	418 V	10
	220 V	200 V	240 V	200 V	240 V	10
PEA	115 kV	106.4	117.6	96.0	123.0	31.5-40
	69 kV	63.6	70.4	57.3	72.5	40
	24 kV	21.8	23.6	21.6	24.0	25
	12 kV	10.9	11.8	10.8	12.0	25
	400 V	371 V	410 V	362 V	416 V	10
	230 V	214 V	237 V	209 V	240 V	10

Table 3. Maximum and minimum voltage level standard of MEA and PEA

The voltage levels of the networks at which DGs are connected are determined by the power capacity of DGs. For the DG with power capacity of more than 250 kW, the voltage level at medium level, i.e. 22-33 kV for PEA and 12-24 kV for MEA. For, DGs with capacity between 56 -250 kW, PEA and MEA will consider voltage level on a case by case basis. For DGs less than 56 kW, the voltage level is 380 V three-phase for PEA and 400V three-phase for MEA. For DGs less than 10 kW, connection is made at 220 V, single phase for PEA and 230 V single phase for MEA.

In relation to solar PV applications in Thailand, the past decade has seen shifts from PV used in the public market through government demonstration projects to the consumer market, with installations of numbers of PV VSPPs and domestic roof-top grid connected PV units gaining momentum. The latter approach has been adopted in Japan, Germany, Switzerland and, to a certain extent, in the U.S. to promote PV applications in a consumer market. It is envisaged that with the government incentive on adders on electricity generated from renewable energy, more households will be attracted to produce electricity from solar PV and wind energy. As domestic roof sizes are limited, PV roof-top grid-connected units will be of small capacity, less than 10 kW. It is this possible large expansion of market for small PV roof-top grid-connected units in Thailand, and eastern Asia, that draws our attention to the study of single phase PV grid-connected systems. Our focus will be on the anti-islanding protection study.

In connecting a single phase PV-based inverter to the power system, a basic configuration consists of PV generators, an inverter, a circuit breaker, and local loads, shown schematically in Fig 2 . Major features of the inverter used must comply with abovementioned regulations of utilities on

protections of undervoltage and overvoltage, phase and ground over-current, underfrequency and overfrequency, and anti-islanding protection. For other types of distributed generators, similar considerations hold.

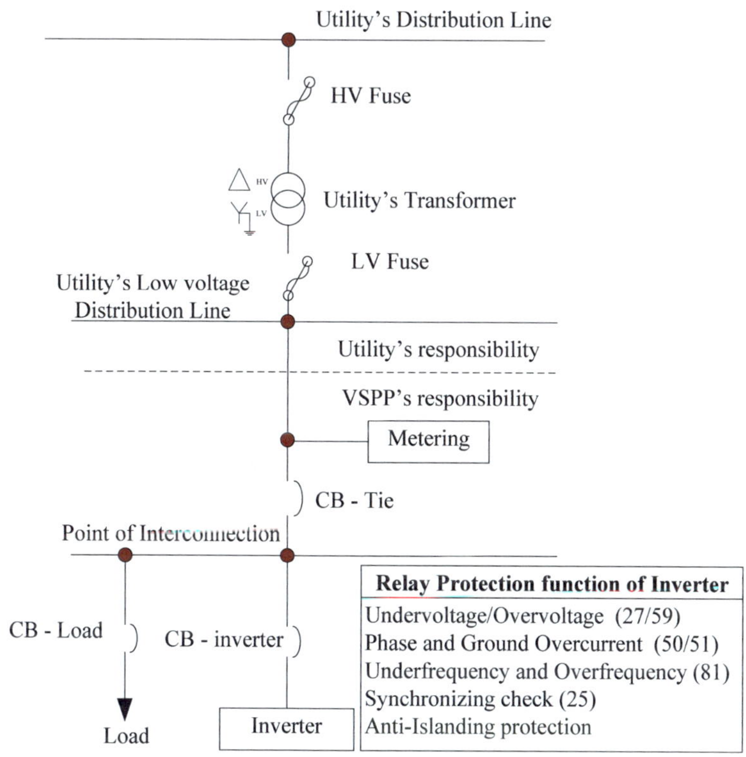

Fig. 2. Single phase inverter connection to the utility

Islanding Protection

The anti islanding function is one of requirement protection function of VSPP inverter based systems. Islanding phenomena is a condition in which the utility grid is disconnected from the distributed generation which still supplies to any section of local loads. Normally, the distributed generation is required to sense the absence of utility-controlled generation and cease energizing the grid. Otherwise, damages can occur to VSPP equipment if the generation in the islanding area, no longer under utility control, operates outside of normal voltage and frequency conditions. Besides, customer and utility equipment can be damaged if the main grid recloses into the island out of synchronization. Energized lines within the island present a shock hazard to unsuspecting utility line workers who think the lines have no electricity.

The anti-islanding protection must follow the regulations of PEA and MEA, requiring that after tripping an interconnection breaker or a fuse, the generator should separate from the system within 0.1 seconds. In addition, the inverter need to pass either

(i)the IEEE 929-2000 Recommended Practice for Utility Interface of Photovoltaic Systems [8] and IEEE 1547.1-2005 Standard Conformance Test Procedures for Equipment Interconnecting Distributed Resources with Electric Power Systems [9], or

(ii) the IEC 62116 Test Procedure of Islanding Prevention Measures for Utility-Interconnected Photovoltaic Inverters [10] and the IEC 61727 Photovoltaic Systems-Characteristics of the Utility Interface [11].

In this study, the goal is to develop a mathematical model of a dc-ac full bridge switching converter voltage source with current control of a PV grid-connected system under islanding phenomena with the state-space averaging and piecewise technique. To simplify the mathematical models and equivalent circuits, some basic assumptions have been neglected such as the exclusion of parasitic elements effects (equivalent series inductance, ESL, of inductor-winding resistance and core loss or equivalent series resistance, ESR, of filter capacitors). The two proposed modelings, implemented by MATLAB/SIMULINK, are verified with the PSpice and experiments.

2. Modeling Single Phase PV-Based Inverter under Islanding Phenomena

2.1 Components of a PV-Grid Connected System under Study

In this study, the single phase photovoltaic system has been proposed. Fig 3 shows a block diagram of a voltage source inverter using a current control technique, applied for a PV grid-connected system in this study. This configuration is common for most available commercial units. The main components are a PV Panel, a DC-DC converter with an isolated transformer, a DC-AC converter, and a AC filter, a Phase Lock Loop (PLL), a Maximum Power Point Tracking (MPPT) unit, a PI controller and a Pulse Width Modulation (PWM) for switching scheme.

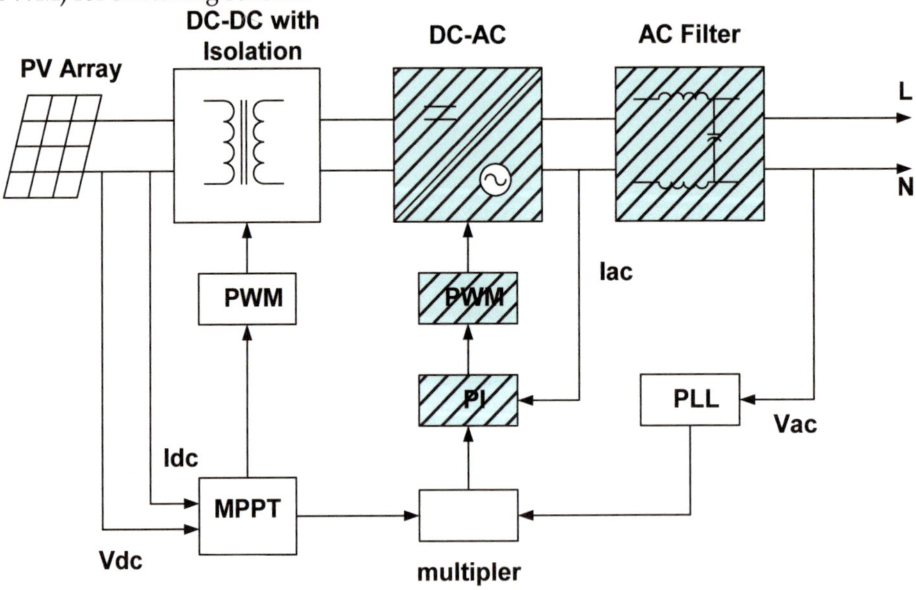

Fig. 3. Block diagram of a voltage source inverter with a current control.

PV Panel: generates direct current from sunlight. In this study, the PV panel is assumed to be a fixed DC voltage source neglecting the variation of sunlight.

DC-DC switching converter with isolated transformer [12][13]: There are several topologies for converting DC to DC voltage with desired values, for example, Push-Pull, Flyback, Forward, Half Bridge and Full Bridge. The choice for a specific application is often based on many considerations such as size, weight of switching converter, generation interference and economic evaluation. In this study, the full bridge configuration is designed for achieving the maximum power from PV application which is controlled with MPPT command. Also included is an isolated transformer commonly used to provide isolation between both sides of the DC input and output of AC systems. The main purpose of this is to eliminate the ground loops between circuitries.

DC-AC switching converter or inverter [14]: It is used to convert from DC to AC. Commonly, the inverter can be classified into two types (a) Voltage Source Inverter (VSI) if input voltage remains constant and (b) Current Source Inverter (CSI) if input current remains constant. The CSI is mostly used in large motor applications, whereas the VSI is adopted for and alone systems. The CSI is a dual of a VSI. In a VSI, the load current depends on the load impedance, whereas the load voltage in a CSI depends on the load impedance. The advantages of the CSI are (i) as the input dc current is controlled and limited, thus misfiring of switching device or short circuit would not be a serious problem (ii) peak current of the power device is limited (iii) the commutation circuit for thyristors are simple and (iv) it can handle reactive or regenerative load without freewheeling diodes. Both can use controlled turn-on and turn-off power electronics devices, for example, metal oxide semiconductor field effect transistor (MOSFET), bipolar junction transistors (BJT) and insulated gate bipolar transistors (IGBTs). Generally, the inverter uses pulse width modulation (PWM) control signals for generating the desired value.

A control technique for voltage source inverters consists of two types, a Voltage Control Inverter as shown in Fig.4 (a) and a Current Control Inverter as shown in Fig. 4 (b). The voltage control inverter is designed to control the angle between the inverter voltage and the grid voltage before synchronizing. In the case of the current control inverter, the output current from the inverter is controlled in phase with the grid voltage. The advantages of this are, first, peak current protection , secondly, extremely good dynamics, thirdly, control of instantaneous current waveform and high accuracy [15]. The disadvantages of current control are that it is costly due to current sensor device, and the difficulty to eliminate the noise signal from the sensed current especially with small current.

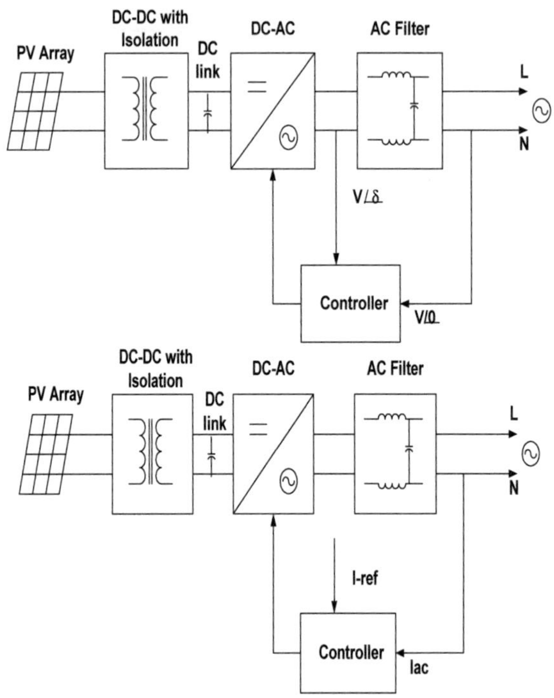

(a) Voltage Control Inverter (b) Current Control Inverter

Fig. 4. Control techniques for an inverter.

2.2 The Operation of Block Diagram of PV Grid Connected System with Current Control

The direct current and voltage from the PV panel are measured and formed as input for an MPPT block to generate the gate signal for a dc-dc converter in order to operate in maximum power generation mode. The current amplitude at maximum operation from MPPT block is multiplied with a unity sinusoidal waveform which is produced from Phase Locked Loop (PLL) block. Its result is designated as the current reference signal. At the output of a dc-ac converter stage, the actual current is sensed and compared with the current reference, then its error is compensated with PI controller which is called an error amplify stage. Finally, this output is compared with the saw-tooth signal to generate a PWM signal for gate drive of a dc-ac converter in a comparator stage.

In this study PV panel, a dc-dc converter with an isolated transformer, an MPPT and a PLL are neglected, therefore the shaded area in Fig.3 is focused and re-illustrated in Fig.5. The input source of a dc-ac switching converter is replaced with a constant DC source instead of a dc-dc switching converter. The switching converter configuration is a full bridge or H-bridge switching converter with a LC filter. The actual current inductor flowing through the filter is sensed for the feedback control loop. This current feedback is compared with constant current reference. The result of this is compensated with PI controller. A simple circuit of dc-ac switching converter is shown in Fig.6.

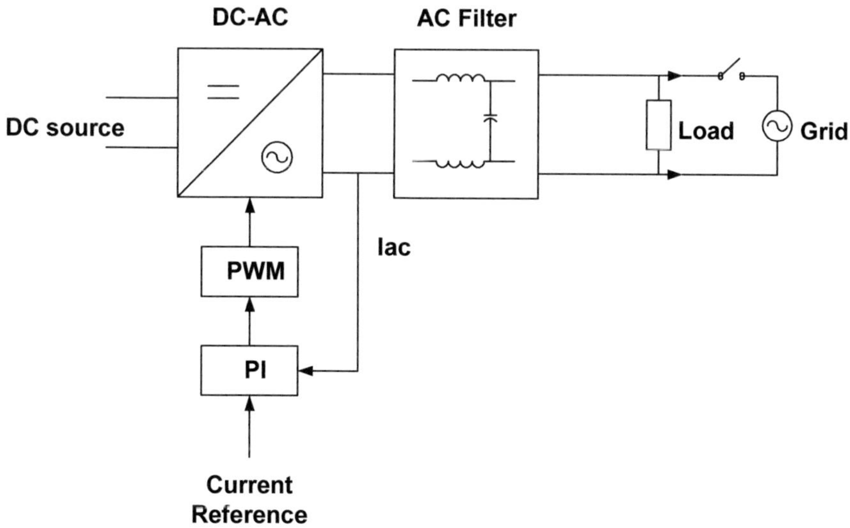

Fig. 5. Block diagram of dc-ac switching converter.

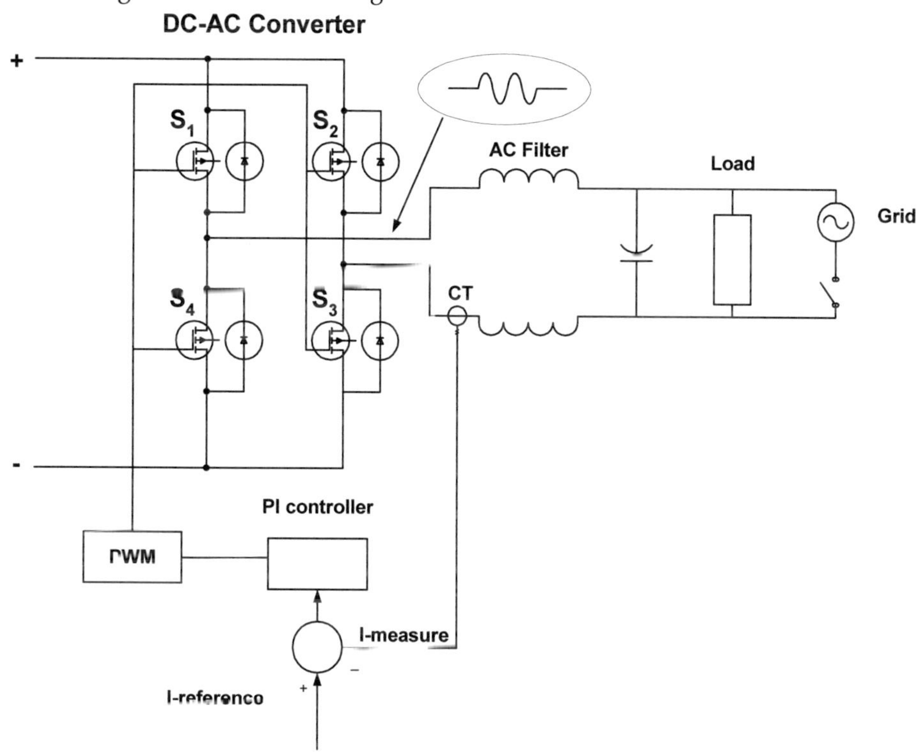

Fig. 6. A simple circuit of dc-ac full-bridge switching converter.

As shown in Fig.6, a pair of the switching converters S1- S3 and S2-S4 are operated alternately of a switching period with its duty cycle (d). The duty cycle (d) is the ratio of the

ON time (t_{on}) to the switching period (T), $d = \frac{t_{on}}{T} = t_{on} f_s$ and 1-d for OFF time as plotted in Fig 7. During one switching cycle, two circuits for the switch states of a dc-ac full bridge switching converter are shown in Fig. 8 (a) and (b) for S1-S3 ON state and S2-S4 ON state respectively. It is noted that the differences of both circuits are only the reversed polarity of input voltage (v_s).

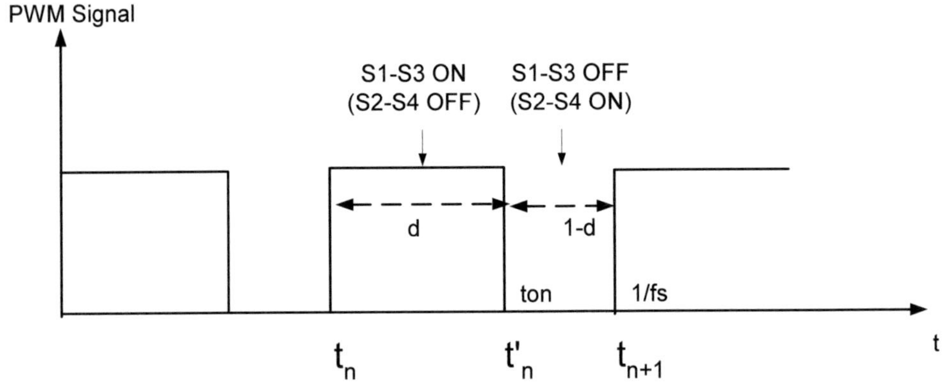

Fig. 7. PWM Signal for d and 1-d interval.

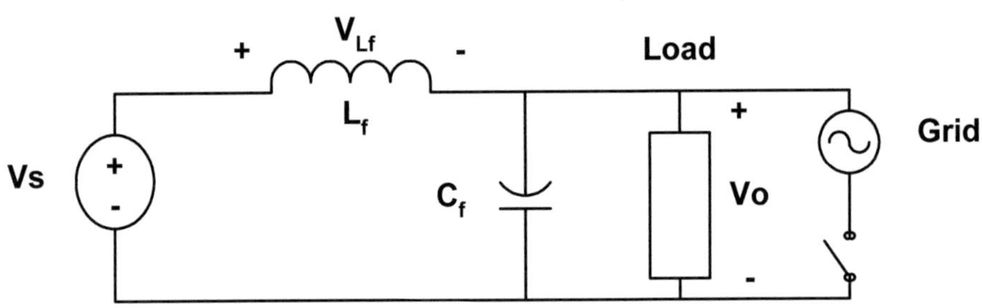

(a) S1-S3 ON state and S2-S4 OFF state.

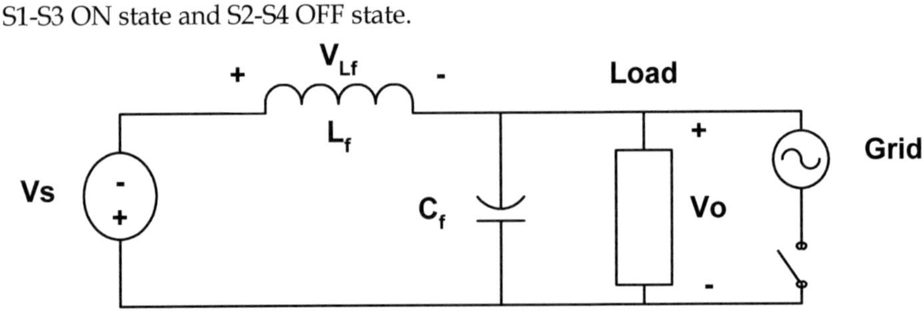

(b) S1-S3 OFF state and S2-S4 ON state.
Fig. 8. Circuit diagram during one switching cycle.

2.3 Voltage Conversion Ratio of a DC-AC Full-Bridge Switching Converter

To calculate the voltage conversion ratio, we use the fact that the averaging of inductor voltage $v_{L(avg)}$ per cycle must be zero in the steady stage. Therefore, we can identify two linear switched circuits modeling as follows:

- For S1-S3 ON state and S2-S4 OFF state, V_{Lf} is

$$V_{Lf} = V_S - V_O$$

The value of V_{Lf} with multiplied by D is given

$$(V_S - V_O)D \tag{1}$$

Where D = $t_{ON} f_s$

f_s = switching frequency

- For S1-S3 OFF state and S2-S4 ON state, V_{Lf} is

$$V_{Lf} = -(V_S - V_O)$$

The value of V_{Lf} with multiplied by (1-D) is equal to

$$-(V_S - V_O)(1-D) \tag{2}$$

Combine Eq.(2.1) and Eq.(2.2) , we have the voltage conversion ratio as below;

$$(V_S - V_O)D - (V_S - V_O)(1-D) = 0$$

$$\frac{V_O}{V_S} = 2D - 1 \tag{3}$$

Where D = duty cycle in steady state term

From the voltage conversion ratio in Eq.(3), it indicates that if the duty cycle (D) is set equal to 0.5, then the output voltage is zero value. In this study, the need for a dc-ac switching converter is to convert the DC signal into AC signal. That means that if we set the duty cycle (D) in terms of 0.5 + sinusoidal function, as a result, the output signal will be sinusoidal function across the x-axis. On the other hand, if we vary the DC term not to be 0.5 value, the output signal will be shifted above or below the x-axis with the DC term.

3. Analytical Methods

The switching converter can be represented by two basic methods, (i) a state-space averaging(in this study, non-linear behavior is focused, thus linearization is not performed) and (ii) a piecewise method [16] .

3.1 State-Space Averaging Modeling

Dynamic analysis is developed by using the state-space averaging technique developed by Middlebrook and Cuk. The main objective is to eliminate the time-varying parameters and it is well suited for characterizing the switching converter in the low-frequency domain. Generally, it is performed in linearization terms for linear time-invariant models which can be analyzed in the standard frequency domain. However, this study aims to evaluate the islanding phenomena of a PV grid connected system which is nonlinear behavior, thus any linearization is not implemented.

Procedures for state-space averaging (no linearization) are as follows [12]:

identify a switched model over a switching cycle and draw a linear switched circuit for each state,

identify state variables and write state an equation in each state,

the d interval

$$\dot{\mathbf{x}} = \mathbf{A}_1\mathbf{x} + \mathbf{B}_1\mathbf{v} \tag{4}$$

the 1-d interval

$$\dot{\mathbf{x}} = \mathbf{A}_2\mathbf{x} + \mathbf{B}_2\mathbf{v} \tag{5}$$

Where $\mathbf{x}$ = state variable vector

$\mathbf{A}$ = state coefficient matrix

$\mathbf{v}$ = source vector

$\mathbf{B}$ = source coefficient matrix

perform state space averaging using the duty cycle as a weighting factor and combine state equations into a single averaged state equation as shown below:

$$\dot{\mathbf{x}} = \left[\overline{\mathbf{A}_1 d} + \overline{\mathbf{A}_2 (1-d)}\right]x + \left[\overline{\mathbf{B}_1 d} + \overline{\mathbf{B}_2 (1-d)}\right]\mathbf{v} \tag{6}$$

Where $\mathbf{x}$ = state variable vector

$\mathbf{A}$ = state coefficient matrix

$\mathbf{v}$ = source vector

$\mathbf{B}$ = source coefficient matrix

We note that any nonlinear continuous systems can be approximated as a linear system with a small neighborhood about its DC operating point. For example, duty cycle (d) comprises a steady state (DC) term designated by capital letter and a dynamic (AC) term by the "hat", as shown:

$$d = D + \hat{d}$$

To implement the state space averaging of a dc-ac full bridge switching converter with current control scheme, the converter can be divided into three sections, i.e. (a) power stage which comprises full-bridge switching configuration, a filter and resistive load R , (b) combination of resistive, inductive and capacitive loads, RLC and (c) a control stage which is a feedback current loop (an error amplify and comparator part), as shown in Fig.9.

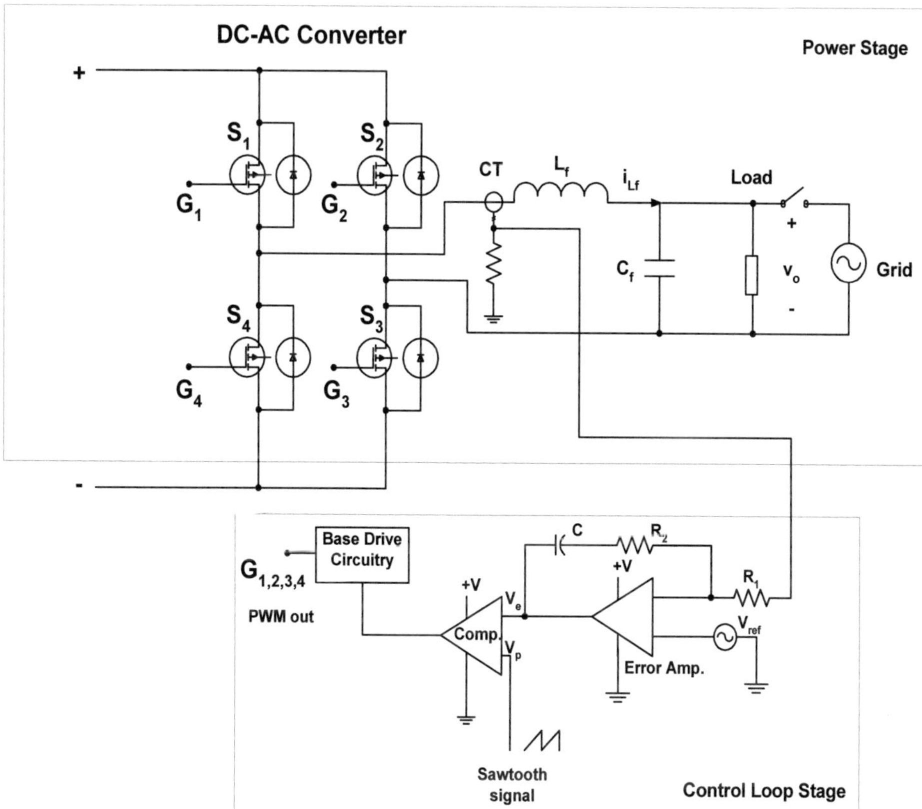

Fig. 9. Schematic diagram of a dc-ac full-bridge switching converter with a current control scheme for a resistive load (R)

Power stage for resistive load

The power stage is drawn in Fig.10 where the resistor is connected in parallel as load.

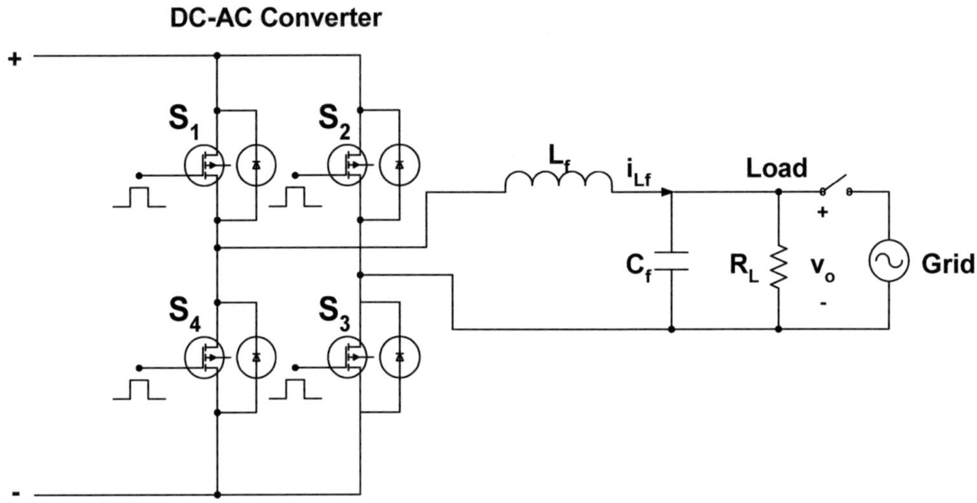

Fig. 10. Power stage for resistive load.

The equivalent circuit during S1and S3 ON, d interval (or S2 and S4 OFF) is illustrated in Fig. 11.

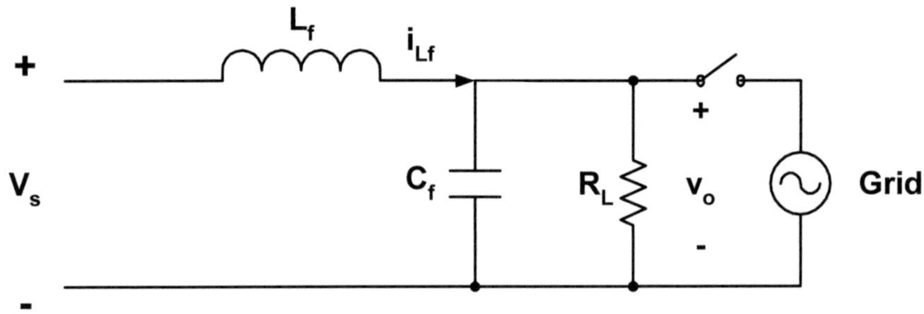

Fig. 11. Equivalent circuit during S1-S3 ON.

From Fig.11, the two state variables for this converter are considered as an inductor current flowing through the filter, i_{Lf} and a load voltage, v_o. Thus two state equations are

$$v_s - v_o = L_f \frac{di_{Lf}}{dt}$$

$$C_f \frac{dv_o}{dt} = i_{Lf} - \frac{v_o}{R_L}$$

Written in matrix, we have

$$\begin{bmatrix} \dfrac{di_{Lf}}{dt} \\ \dfrac{dv_o}{dt} \end{bmatrix} = \begin{bmatrix} 0 & \dfrac{-1}{L_f} \\ \dfrac{1}{C_f} & \dfrac{-1}{R_L C_f} \end{bmatrix} \begin{bmatrix} i_{Lf} \\ v_o \end{bmatrix} + \begin{bmatrix} \dfrac{1}{L_f} \\ 0 \end{bmatrix} [v_s] \tag{7}$$

For 1-d interval, the equivalent circuit during S1and S3 OFF, 1-d interval (or S2 and S4 ON) is illustrated in Fig. 12.

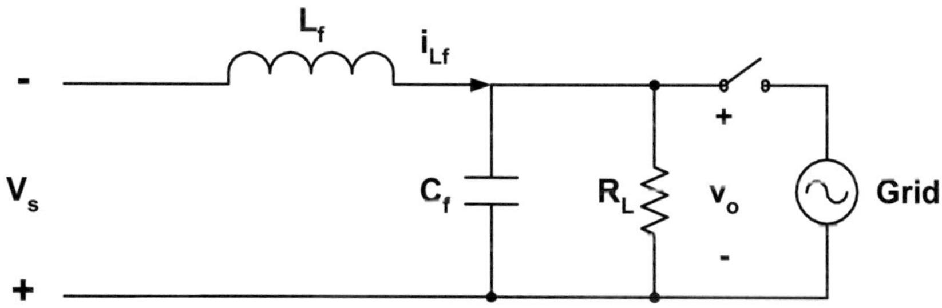

Fig. 12. Equivalent circuit during S1-S3 OFF.

Two linear state equations can be expressed by applying Kirchoff's voltage and current law, from Fig.12, as shown below:

$$v_s + v_o = -L_f \frac{di_{Lf}}{dt}$$

$$C_f \frac{dv_o}{dt} = i_{Lf} - \frac{v_o}{R_L}$$

Written in matrix, it becomes

$$\begin{bmatrix} \dfrac{di_{Lf}}{dt} \\ \dfrac{dv_o}{dt} \end{bmatrix} = \begin{bmatrix} 0 & \dfrac{-1}{L_f} \\ \dfrac{1}{C_f} & \dfrac{-1}{R_L C_f} \end{bmatrix} \begin{bmatrix} i_{Lf} \\ v_o \end{bmatrix} + \begin{bmatrix} \dfrac{-1}{L_f} \\ 0 \end{bmatrix} [v_s] \tag{8}$$

From Eq.(7) and (8), we can calculate the state space averaged state coefficient matrix as expressed in Eq.(6). The result is

$$\mathbf{A} = \begin{bmatrix} 0 & \dfrac{-1}{L_f} \\ \dfrac{1}{C_f} & \dfrac{-1}{R_L C_f} \end{bmatrix} d + \begin{bmatrix} 0 & \dfrac{-1}{L_f} \\ \dfrac{1}{C_f} & \dfrac{-1}{R_L C_f} \end{bmatrix} (1-d)$$

$$\mathbf{A} = \begin{bmatrix} 0 & \dfrac{-1}{L_f} \\ \dfrac{1}{C_f} & \dfrac{-1}{R_L C_f} \end{bmatrix}$$

$$\mathbf{B} = \begin{bmatrix} \dfrac{1}{L_f} \\ 0 \end{bmatrix} d + \begin{bmatrix} \dfrac{-1}{L_f} \\ 0 \end{bmatrix} (1-d) = \begin{bmatrix} \dfrac{2d-1}{L_f} \\ 0 \end{bmatrix}$$

Therefore, we obtain the stage space average matrix as follows:

$$\begin{bmatrix} \dfrac{di_{Lf}}{dt} \\ \dfrac{dv_o}{dt} \end{bmatrix} = \begin{bmatrix} 0 & \dfrac{-1}{L_f} \\ \dfrac{1}{C_f} & \dfrac{-1}{R_L C_f} \end{bmatrix} \begin{bmatrix} i_{Lf} \\ v_o \end{bmatrix} + \begin{bmatrix} \dfrac{2d-1}{L_f} \\ 0 \end{bmatrix} [v_s] \tag{9}$$

Power stage for RLC load

For the RLC load, the power stage which is connected with the combination of R, L and C in parallel as load is shown in Fig. 13. To simplify the state equation, the combination between the filter capacitor and the load capacitor is designated as C_{new} term. Therefore, an inductor current flowing through the filter i_{Lf}, an inductor current flowing through load i_{LL} and a load voltage v_O are chosen as state variables.

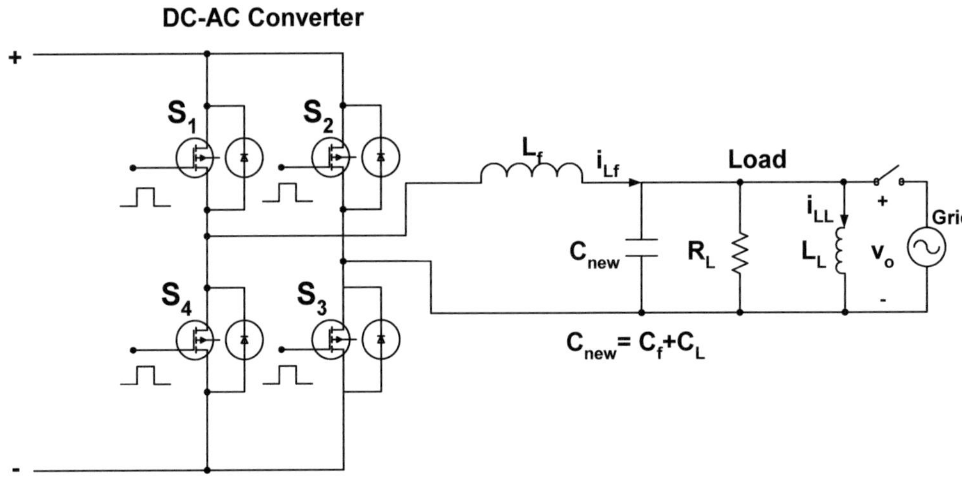

Fig. 13. Power stage for RLC load.

The equivalent circuit during S1and S3 ON, d interval (or S2 and S4 OFF) is illustrated in Fig. 14.

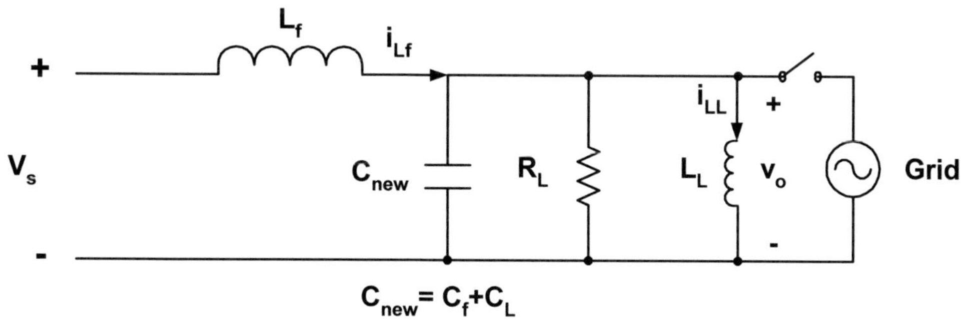

Fig. 14. Equivalent circuit during S1-S3 ON.

We can derive three linear state equations in d interval, from Fig.14, as follows:

$$v_s - v_o = L_f \frac{di_{L_f}}{dt}$$

$$v_o = L_L \frac{di_{LL}}{dt}$$

$$C_{new} \frac{dv_o}{dt} = i_{Lf} - i_{LL} - \frac{v_o}{R}$$

Written in a matrix form, it gives

$$\begin{bmatrix} \frac{di_{Lf}}{dt} \\ \frac{di_{LL}}{dt} \\ \frac{dv_o}{dt} \end{bmatrix} = \begin{bmatrix} 0 & 0 & \frac{-1}{L_f} \\ 0 & 0 & \frac{1}{L_L} \\ \frac{1}{C_{new}} & \frac{-1}{C_{new}} & \frac{-1}{R_L C_{new}} \end{bmatrix} \begin{bmatrix} i_{Lf} \\ i_{LL} \\ v_O \end{bmatrix} + \begin{bmatrix} \frac{1}{L_f} \\ 0 \\ 0 \end{bmatrix} [v_s] \tag{10}$$

Also, for 1-d interval, the equivalent circuit can be drawn as in Fig. 15.

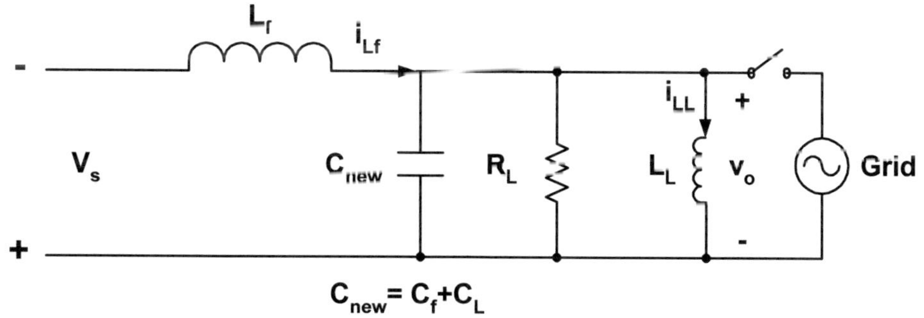

Fig. 15. Equivalent circuit during S1-S3 OFF.

From Fig. 15, the state equation is obtained by applying the Kirchoff's voltage and current law as follows:

$$-v_s - v_o = L_f \frac{di_{Lf}}{dt}$$

$$v_o = L_L \frac{di_{LL}}{dt}$$

$$C_{new} \frac{dv_o}{dt} = i_{Lf} - i_{LL} - \frac{v_o}{R_L}$$

Written in matrix form, we have

$$\begin{bmatrix} \frac{di_{Lf}}{dt} \\ \frac{di_{LL}}{dt} \\ \frac{dv_o}{dt} \end{bmatrix} = \begin{bmatrix} 0 & 0 & \frac{-1}{L_f} \\ 0 & 0 & \frac{1}{L_L} \\ \frac{1}{C_{new}} & \frac{-1}{C_{new}} & \frac{-1}{R_L C_{new}} \end{bmatrix} \begin{bmatrix} i_{Lf} \\ i_{LL} \\ v_o \end{bmatrix} + \begin{bmatrix} \frac{-1}{L_f} \\ 0 \\ 0 \end{bmatrix} [v_s] \tag{11}$$

Similarly, from Eq.(10) and (11), we apply the averaging technique from Eq.(6), the nonlinear modeling can be represented as follows:

$$\begin{bmatrix} \frac{di_{Lf}}{dt} \\ \frac{di_{LL}}{dt} \\ \frac{dv_o}{dt} \end{bmatrix} = \begin{bmatrix} 0 & 0 & \frac{-1}{L_f} \\ 0 & 0 & \frac{1}{L_L} \\ \frac{1}{C_{new}} & \frac{-1}{C_{new}} & \frac{-1}{R_L C_{new}} \end{bmatrix} \begin{bmatrix} i_{Lf} \\ i_{LL} \\ v_o \end{bmatrix} + \begin{bmatrix} \frac{2d-1}{L_f} \\ 0 \\ 0 \end{bmatrix} [v_s] \tag{12}$$

The stage space averaged equations from Eq.(9) and (12) for R and RLC load respectively are nonlinear equations because duty cycle (d) is a function of an inductor current flow through the filter, i_{Lf}.

Feedback current control stage

A feedback current control technique is implemented by controlling inductor currents flowing through filters corresponding to a reference current. As a result, the output current is in phase with grid voltages and produces a good power factor which controlled by a feedback current control loop stage consisting of sensing of inductor current flowing through the filter (i_{Lf}). The current is converted to a voltage form by multiplying with a resistor. Subsequently, this voltage is compared with a sinusoidal reference value. This stage is called "error amplification". Then the error amplification is compensated with the PI

controller. The result is designated as the error voltage (v_e). Finally, the error voltage is compared with the saw tooth signal to generate the PWM signal.
To model the dc-ac full-bridge switching converter and produce the output voltage in a sinusoidal waveform, we have to set up a duty cycle (d) with a variation in terms of a sinusoidal waveform around an average level of 0.5. This then provides [17]:

$$d(t) = 0.5 + m(\sin(\omega t)) \tag{13}$$

where m is a modulated duty cycle variation. The factor of 0.5 is determined from the relationship of voltage conversion ratio of the dc-ac full bridge switching converter, which is $v_o / v_s = 2d - 1$. Therefore, if v_s is dc source and d is represented from Eq.(13), then the output voltage (v_o) can produce the sinusoidal waveform across the x-axis. The term $m(\sin(\omega t))$ can be represented as $\frac{v_e(t)}{V_P}$. Thus, we can substitute it into Eq.(13) and obtain

$$d(t) = 0.5 + v_e(t)/V_p \tag{14}$$

As shown in Fig. 16, we can calculate $v_e(t)$ with a basic Op-amp circuit calculation as shown below:
The current flowing through R_1 is

$$i_{R1} = \frac{v_{iLf} - v_{ref}}{R_1}$$

The current flowing through R_2 is

$$v_{ref} - v_e = i_{R2} R_2 + v_C$$

Applying the Kirrchoff's current law, that is $i_{R1} = i_{R2}$ and $v_C = \frac{1}{C}\int i_{R2}dt$, thus we have

$$v_e(t) = v_{ref}\left(1 + \frac{R_2}{R_1}\right) - v_{iLf}\left(\frac{R_2}{R_1}\right) - \frac{1}{C\,R_1}\int v_{iLf}dt + \frac{1}{C\,R_1}\int v_{ref}dt \tag{15}$$

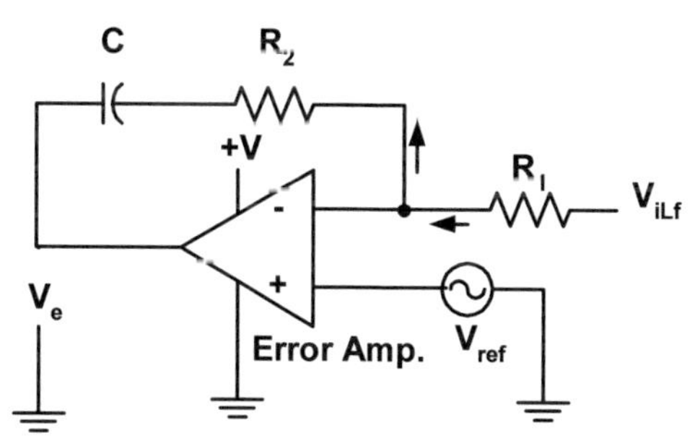

Fig. 16. Error amplification circuit with a PI controller.

We obtain the duty cycle (d) in a time domain by substituting $v_e(t)$ from Eq.(15) into Eq.(14). We construct a nonlinear differential equation model by using Eq.(9), (12) and (14) for power stages of R loads, RLC loads, and duty cycle (d) from a feedback current control stage, respectively.

The solution of a nonlinear differential equation can easily be handled by MATLAB computation as well as passive detection method as shown in Fig. 17.

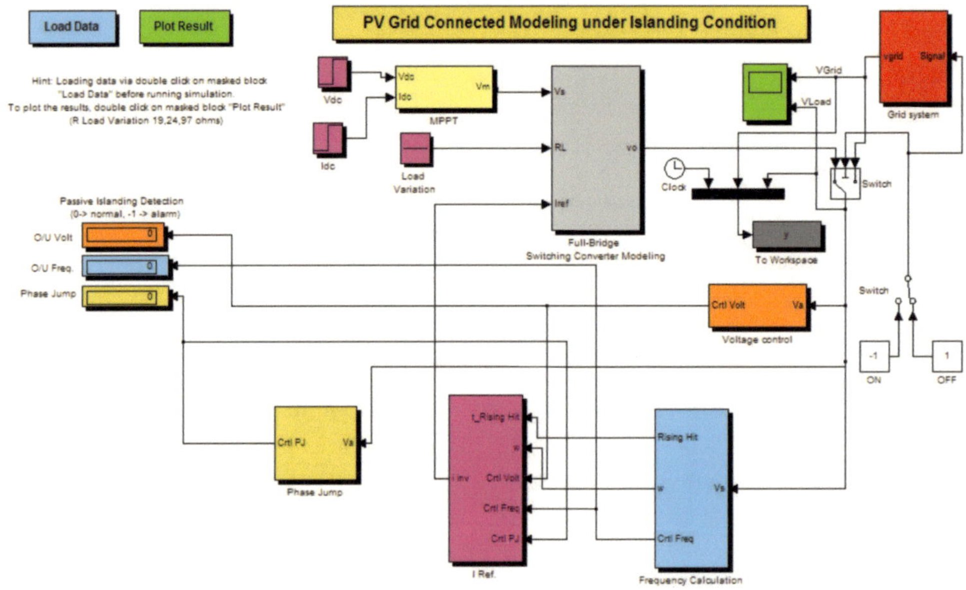

Fig. 17. Proposed modeling of PV grid connected on MATLAB/SIMULINK.

3. 2 Piecewise Technique Modeling

The basic idea of the piecewise technique is the point by point conversion of a continuous time signal (A/D) with a sampling period (T) as illustrated in Fig. 18

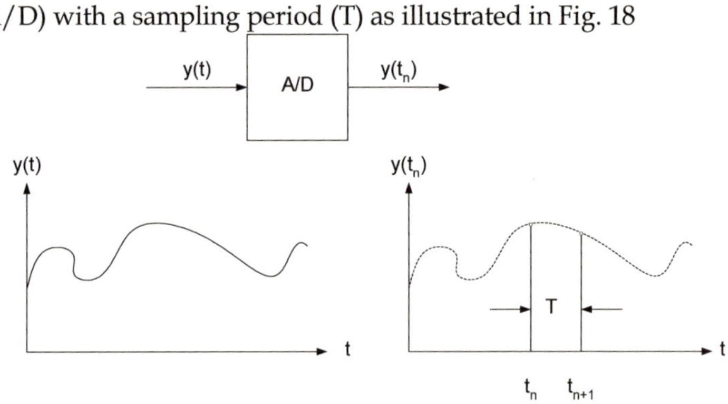

Fig. 18. Basic conversions from continuous time signal into piecewise signal.

During one switching cycle of a switching converter, the operation consists of each piece at a particular interval of time and then they repeat themselves periodically. Thus, it is straightforward to model such kind of operation by splitting the system into several sub-system topologies corresponding to time sub-intervals. For the solution at a particular time, this consists of taking an initial value at specific sub-circuit and solving it. Then, we continue to solve the next sub-circuit with the previous solution as the initial value. The main concept of piecewise technique model involves the substitution of the solution of previous sub-interval as the initial condition of the sub-interval under consideration The flow chart is shown in Fig.19.

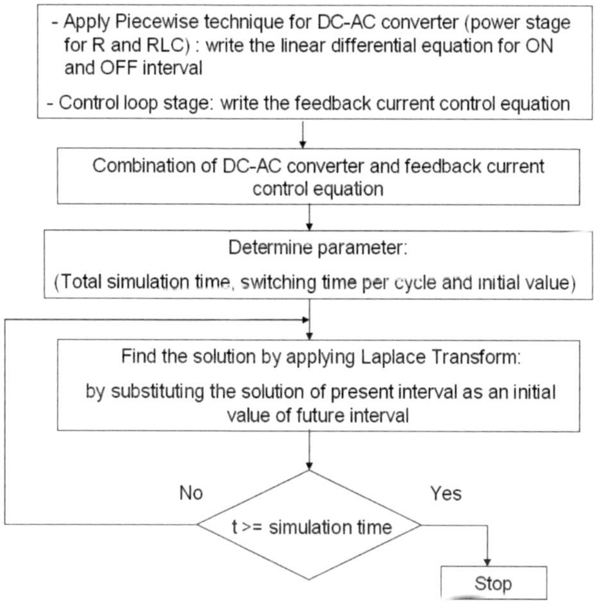

Fig. 19. Flow chart for a proposed model.

In doing so, firstly we write down the state equations which describe the individual switched circuit of a multi-topological circuit. The inverter operates in continuous conduction mode. Thus, two switched circuits can be identified, one for the 'switch-on' interval (S1 and S3 ON) and the other to 'switch-off' interval (S1 and S3 OFF). Two state equations can be derived by applying the Kirchoff's voltage and current laws. We solve the state-space equation of the switch-on and switch- off intervals from Eq.(4) and (5) by using the Laplace transformation which is obtained in s-domain as shown below:

$$\mathbf{x}(s) = \left[s\mathbf{I} - \mathbf{A}_{(1)\,\text{or}(2)}\right]^{-1}\left[x(t_{initial}) + \mathbf{B}_{(1)\,\text{or}\,(2)}V_s(s)\right] \tag{16}$$

Following this, the partial fraction technique is applied for splitting up a ratio of large polynomials into a sum of ratios of small polynomials, for example:

$$Y(s) = \frac{s+1}{(s+2)^2} = \frac{A}{s+2} + \frac{B}{(s+2)^2}$$

Then, this result with the partial fraction can be easily taken using an inverse Laplace transformation into the time domain with standard inverse Laplace forms. Finally, the piecewise method is implemented in each interval as explained above.
To analyze the behavior of a DC-AC full-bridge switching converter PV grid-connected system under islanding phenomena, similarly to the state-space averaging method, we have to split this circuit into three sections, for power stage; (a) a resistive load, R; (b) a combination of resistive, inductive and capacitive load, RLC and control stage and (c) feedback current control.

Power stage for resistive load
The power stage is illustrated Fig. 10 where the resistor is connected in parallel as a load. The equivalent circuit during S1 and S3 ON, d interval or $t_n \le t < t_n'$ (or S2 and S4 OFF) is illustrated in Fig. 11 and the state equation in Eq.(7). For the 1-d interval, the equivalent circuit during S1 and S3 OFF is illustrated in Fig. 12 and Eq.(8).
The state-space equation of the switch-on and-off interval is solved by using the Laplace transformation, then the partial fraction technique is applied , and taking the inverse Laplace transform. The solution is

$$i_{Lf}(t) = K_{1R-on} + \frac{K_{3R-on} - K_{2R-on}\sigma}{\omega} e^{-\sigma(t-t_n)} \sin\omega(t-t_n) + K_{2R-on} e^{-\sigma(t-t_n)} \cos\omega(t-t_n) \tag{17}$$

$$v_o(t) = K_{4R-on} + \frac{K_{6R-on} - K_{5R-on}\sigma}{\omega} e^{-\sigma(t-t_n)} \sin\omega(t-t_n) + K_{5R-on} e^{-\sigma(t-t_n)} \cos\omega(t-t_n) \tag{18}$$

where $\sigma = \frac{1}{2C_f R_L}$, $\omega = \sqrt{\frac{1}{L_f C_f} - \sigma^2}$

$$K_{1R-on} = \frac{v_s}{R_L}, \qquad K_{2R-on} = i_{Lf}(t_n) - \frac{v_s}{R_L}$$

$$K_{3R-on} = \frac{i_{Lf}(t_n)}{R_L C_f} - \frac{v_o(t_n)}{L_f} + \frac{v_s}{R_L}\left(\frac{R_L}{L_f} - 2\sigma\right), \qquad K_{4R-on} = v_s$$

$$K_{5R-on} = v_o(t_n) - v_s, \qquad K_{6R-on} = \frac{i_{Lf}(t_n)}{C_f} - 2v_s\sigma$$

Following this, inserting $t = t_n'$ and $t_n' - t_n = dT = t_{on}$ in switch-on interval into Eq. (17) and (18), we can obtain the value of i_{Lf} and v_o at the end of switch-on interval as expressed:

$$i_{Lf}(t_n') = K_{1R-on} + \frac{K_{3R-on} - K_{2R-on}\sigma}{\omega} e^{-\sigma(dT)} \sin\omega(dT) + K_{2R-on} e^{-\sigma(dT)} \cos\omega(dT) \tag{19}$$

$$v_o(t_n') = K_{4R-on} + \frac{K_{6R-on} - K_{5R-on}\sigma}{\omega} e^{-\sigma(dT)} \sin\omega(dT) + K_{5R-on} e^{-\sigma(dT)} \cos\omega(dT) \tag{20}$$

Since K-values are a function of $\mathbf{x}(t_n)$, we can write the general form of the difference equation in the switch-on interval, which involves the value of $\mathbf{x}(t_n)$ as initial value and d. The equation is

$$\mathbf{x}(t_n^{'}) = \mathbf{f}(\mathbf{x}(t_n), d) \tag{21}$$

For switch (S1,S3)-off interval $t_n^{'} \le t < t_{n+1}$, in this interval, $\mathbf{x}(t_n^{'})$ value, which is the solution value of the previous interval or switch-on interval, is determined as the initial value. Similar to the switch-on interval, the expression of the solution from Eq. (8) can be obtained:

$$i_{Lf}(t) = K_{1R-off} + \frac{K_{3R-off} - K_{2R-off}\sigma}{\omega} e^{-\sigma(t-t_n^{'})} \sin\omega(t-t_n^{'}) + K_{2R-off} e^{-\sigma(t-t_n^{'})} \cos\omega(t-t_n^{'}) \tag{22}$$

$$v_o(t) = K_{4R-off} + \frac{K_{6R-off} - K_{5R-off}\sigma}{\omega} e^{-\sigma(t-t_n^{'})} \sin\omega(t-t_n^{'}) + K_{5R-off} e^{-\sigma(t-t_n^{'})} \cos\omega(t-t_n^{'}) \tag{23}$$

where

$$K_{1R-off} = -\frac{v_s}{R_L}, \qquad K_{2R-off} = i_{Lf}(t_n^{'}) - \frac{v_s}{R_L}$$

$$K_{3R-off} = \frac{i_{Lf}(t_n^{'})}{R_L C_f} - \frac{v_o(t_n^{'})}{L_f} + \frac{v_s}{R_L}\left(2\sigma - \frac{R_L}{L_f}\right), \quad K_{4R-off} = -v_s$$

$$K_{5R-off} = v_o(t_n^{'}) + v_s, \qquad K_{6R-off} = \frac{i_{Lf}(t_n^{'})}{C_f} + 2v_s\sigma$$

In this interval, we substitute $t - t_{n+1}$ then, $t_{n+1} - t_n^{'} = (1-d)T = t_{off}$ into Eq. (22) and (23), then we can obtain the value of i_{lf} and v_o at the end of switch-off interval as

$$i_{Lf}(t_{n+1}) = K_{1R-off} + \frac{K_{3R-off} - K_{2R-off}\sigma}{\omega} e^{-\sigma(1-d)T} \sin\omega(1-d)T + K_{2R-off} e^{-\sigma(1-d)T} \cos\omega(1-d)T \tag{24}$$

$$v_o(t_{n+1}) = K_{4R-off} + \frac{K_{6R-off} - K_{5R-off}\sigma}{\omega} e^{-\sigma(1-d)T} \sin\omega(1-d)T + K_{5R-off} e^{-\sigma(1-d)T} \cos\omega(1-d)T \tag{25}$$

From Eq. (24) and (25), the general form of the difference equation in the switch-off interval can be expressed

$$\mathbf{x}(t_{n+1}) = \mathbf{f}(\mathbf{x}(t_n^{'}), d) \tag{26}$$

To determine the general form of difference equation in one switching period, the $\mathbf{x}(t_n^{'})$ value of switch-on interval and the $\mathbf{x}(t_{n+1})$ value of switch-off interval are combined by substituting $\mathbf{x}(t_n^{'})$ in Eq. (21) with Eq. (26). Thus, the general form of difference equation is

$$\mathbf{x}(t_{n+1}) = \mathbf{f}(\mathbf{x}(t_n), d) \tag{27}$$

3.3 Power stage for RLC loads

For RLC loads, the combination of R, L and C are connected in parallel as load, as in Fig.12. The state equations of d and 1-d interval are expressed in Eq.(10) and (11) respectively: Similar to R load, Eq. (10) and (11) are solved by taking the Laplace transformation and re-arranging it to a partial fraction, then taking the inverse Laplace. The solution of S1and S3 ON, *d* interval is

$$i_{Lf}(t_n^{'}) = K_{1RLC-on} + K_{2RLC-on}dT + \frac{K_{4RLC-on} - K_{3RLC-on}\sigma_1}{\omega_{RLC}} e^{-\sigma_1(dT)} \sin \omega_{RLC}(dT) + K_{3RLC-on}e^{-\sigma_1(dT)} \cos \omega_{RLC}(dT) \quad (28)$$

$$i_{LL}(t_n^{'}) = K_{5RLC-on} + K_{6RLC-on}dT + \frac{K_{8RLC-on} - K_{7RLC-on}\sigma_1}{\omega_{RLC}} e^{-\sigma_1(dT)} \sin \omega_{RLC}(dT) + K_{7RLC-on}e^{-\sigma_1(dT)} \cos \omega_{RLC}(dT) \quad (29)$$

$$v_o(t_n^{'}) = K_{9RLC-on} + \frac{K_{11RLC-on} - K_{10RLC-on}\sigma_1}{\omega_{RLC}} e^{-\sigma_1(dT)} \sin \omega(dT) + K_{10RLC-on}e^{-\sigma_1(dT)} \cos \omega_{RLC}(dT) \quad (30)$$

Where $\sigma_1 = \dfrac{1}{2C_{new}R_L}$, $\omega_{RLC} = \sqrt{\dfrac{L_L + L_f}{L_L L_f C_{new}} - \sigma_1^2}$

$$K_{1RLC-on} = \left(\frac{L_f L_L C_{new}}{L_f + L_L}\right)\left(\frac{i_{Lf}(t_n)}{C_{new}L_L} + \frac{v_s}{R_L C_{new} L_f} + \frac{i_{LL}(t_n)}{C_{new}L_f} - \frac{K_{2RLC-on}}{R_L C_{new}}\right)$$

$$K_{2RLC-on} = \frac{v_s}{L_f + L_L}$$

$$K_{3RLC-on} = i_{Lf}(t_n) - K_{1RLC-on}$$

$$K_{4RLC-on} = \frac{i_{Lf}(t_n)}{R_L C_{new}} + \frac{v_s}{L_f} - \frac{v_o(t_n)}{L_f} - \frac{K_{1RLC-on}}{R_L C_{new}} + K_{2RLC-on}$$

$$K_{5RLC-on} = \left(\frac{L_f L_L C_{new}}{L_f + L_L}\right)\left(\frac{i_{Lf}(t_n)}{C_{new}L_L} + \frac{i_{LL}(t_n)}{C_{new}L_f} - \frac{K_{6RLC-on}}{R_L C_{new}}\right)$$

$$K_{6RLC-on} = \frac{v_s}{L_f + L_L}$$

$$K_{7RLC-on} = i_{LL}(t_n) - K_{5RLC-on}$$

$$K_{8RLC-on} = \frac{i_{LL}(t_n)}{R_L C_{new}} + \frac{v_o(t_n)}{L_L} - \frac{K_{5RLC-on}}{R_L C_{new}} - K_{6RLC-on}$$

$$K_{9RLC-on} = \frac{v_s L_L}{L_f + L_L}$$

$$K_{10RLC-on} = v_o(t_n) - K_{9RLC-on}$$

$$K_{11RLC-on} = \frac{i_{LL}(t_n) - i_{Lf}(t_n)}{C_{new}} - \frac{K_{9RLC-on}}{R_L C_{new}}$$

From Eq. (28), (29) and (30) we can write the general form of the difference equation in the switch-on interval as follows:

$$\mathbf{x}(t_n^{'}) = \mathbf{f}(\mathbf{x}(t_n), d) \quad (31)$$

The solution of S1and S3 OFF, 1-d interval is

$$i_{Lf}(t_{n+1}) = K_{1RLC-off} + K_{2RLC-off}(1-d)T + \frac{K_{4RLC-off} - K_{3RLC-off}\sigma_1}{\omega_{RLC}} e^{-\sigma_1(1-d)T} \sin\omega_{RLC}(1-d)T + K_{3RLC-off}e^{-\sigma_1(1-d)T}\cos\omega_{RLC}(1-d)T \quad (32)$$

$$i_{LL}(t_{n+1}) = K_{5RLC-off} + K_{6RLC-off}(1-d)T + \frac{K_{8RLC-off} - K_{7RLC-off}\sigma_1}{\omega_{RLC}} e^{-\sigma_1(1-d)T} \sin\omega_{RLC}(1-d)T + K_{7RLC-off}e^{-\sigma_1(1-d)T}\cos\omega_{RLC}(1-d)T \quad (33)$$

$$v_o(t_{n+1}) = K_{9RLC-off} + \frac{K_{11RLC-off} - K_{10RLC-off}\sigma_1}{\omega_{RLC}} e^{-\sigma_1(1-d)T} \sin\omega_{RLC}(1-d)T + K_{10RLC-off}e^{-\sigma_1(1-d)T}\cos\omega_{RLC}(1-d)T \quad (34)$$

where

$$K_{1RLC-off} = \left(\frac{L_f L_L C_{new}}{L_f + L_L}\right)\left(\frac{i_{Lf}(t'_n)}{C_{new}L_L} - \frac{v_s}{R_L C_{new} L_f} + \frac{i_{LL}(t'_n)}{C_{new}L_f} - \frac{K_{2RLC-off}}{R_L C_{new}}\right)$$

$$K_{2RLC-off} = \frac{-v_s}{L_f + L_L}$$

$$K_{3RLC-off} = i_{Lf}(t'_n) - K_{1RLC-off}$$

$$K_{4RLC-off} = \frac{i_{Lf}(t'_n)}{R_L C_{new}} - \frac{v_s}{L_f} - \frac{v_o(t'_n)}{L_f} - \frac{K_{1RLC-off}}{R_L C_{new}} + K_{2RLC-off}$$

$$K_{5RLC-off} = \left(\frac{L_f L_L C_{new}}{L_f + L_L}\right)\left(\frac{i_{Lf}(t'_n)}{C_{new}L_L} + \frac{i_{LL}(t'_n)}{C_{new}L_f} - \frac{K_{6RLC-off}}{R_L C_{new}}\right)$$

$$K_{6RLC-off} = \frac{-v_s}{L_f + L_L}$$

$$K_{7RLC-off} = i_{LL}(t'_n) - K_{5RLC-off}$$

$$K_{8RLC-off} = \frac{i_{LL}(t'_n)}{R_L C_{new}} + \frac{v_o(t'_n)}{L_L} - \frac{K_{5RLC-off}}{R_L C_{new}} - K_{6RLC-off}$$

$$K_{9RLC-off} = \frac{-v_s L_L}{L_f + L_L}$$

$$K_{10RLC-off} = v_o(t'_n) - K_{9RLC-off}$$

$$K_{11RLC-off} = \frac{i_{LL}(t'_n) - i_{Lf}(t'_n)}{C_{new}} - \frac{K_{9RLC-off}}{R_L C_{new}}$$

From Eq. (32), (33) and (34) we can write the general form of the difference equation in the switch-off interval. It is

$$\mathbf{x}(t_{n+1}) = \mathbf{f}(\mathbf{x}(t'_n), d) \tag{35}$$

To obtain the general form of difference equation in one switching period, the $\mathbf{x}(t'_n)$ value of switch-on interval and the $\mathbf{x}(t_{n+1})$ value of switch-off interval are combined by substituting $\mathbf{x}(t'_n)$ into Eq. (31) with Eq. (35). Thus, the general form of difference equation is

$$\mathbf{x}(t_{n+1}) = \mathbf{f}(\mathbf{x}(t_n), d) \tag{36}$$

Now, we can construct a completed differential equation model by using Eq.(27),(36) and (14) for the power stage with R load, RLC load and the duty cycle (d) from a feedback current control stage respectively.

The solution for such a model can readily be handled using simple numerical iteration methods with sampling time by using embedded MATLAB function on MATLAB/SIMULINK with the passive islanding method and PLL block [18] as shown in Fig. 20.

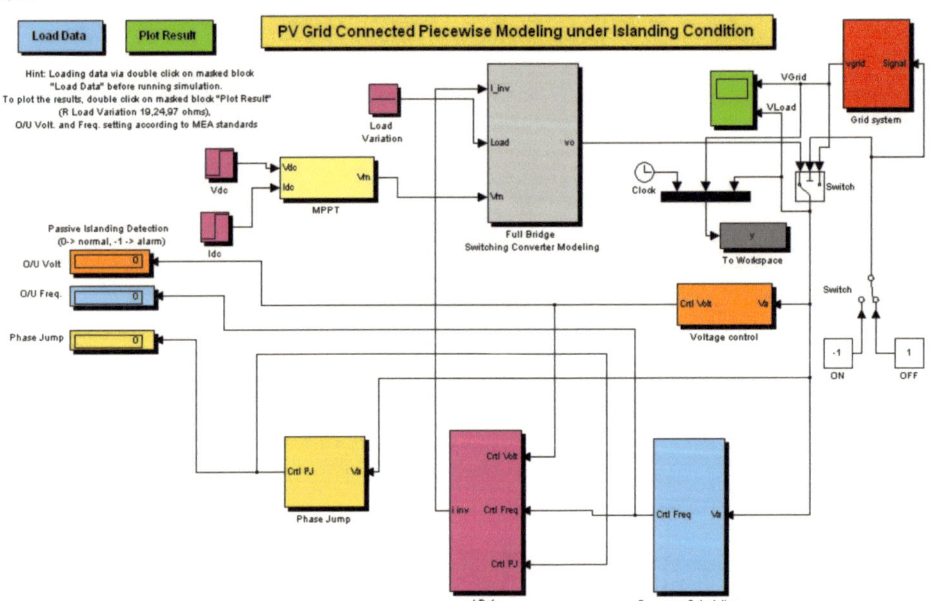

Fig. 20. The proposed model on MATLAB/SIMULINK.

4. Experimental Set Up and Results

The islanding test circuit, as shown in Fig.20, is set up to validate the two proposed modelings under islanding condition. The grid voltage reference and load voltage waveform are recorded from the oscilloscope for analysis with various load types. The various load types and adjustment of the ratios of the real load to an inverter output are undertaken and connected in parallel with the inverter. The input source of the inverter is a fixed a dc supply instead of a PV array to avoid the uncertainty of sunlight intensity. A 2 kW inverter purchased locally is used in this experiment.

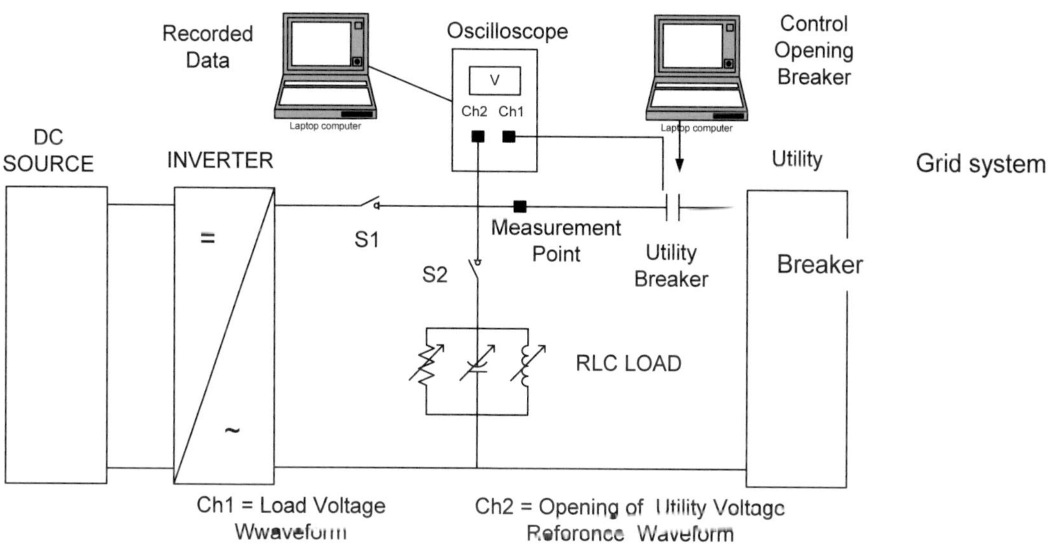

Fig. 20. Islanding test circuit.

Components used in the study

L_f	=	2mH
C_f	=	6.8 μ F
R_L	=	19, 24 and 97 ohms (125%, 100% and 25% of the inverter output)
C_L	=	2.66 μ F
L_L	=	3.55H
R_1, R_2	=	10k, 25k ohms
C	=	470 μ F
V_p	=	6V
f_s	=	10kHz
v_S	=	450Vdc
I_{ref}	=	1.5Aac

R_{ref} = 1.0 ohm
r = 0.1

Initially, the grid system is connected to the inverter, and then removed intentionally in order to form an islanding condition. For the experiment, the inverter is automatically stopped with its own islanding detection. The islanding detection of the proposed model and PSpice have been implemented with the passive method such as O/U voltage and O/U frequency as well as phase jump methods by sensing the output voltage. To avoid the malfunction of islanding detection, the inverter will be delayed 2-3 cycles before shutdown. The results between experiments, PSpice and the proposed model for various resistive load types and RLC load are compared under 3 different resistive loads: 125%, 100% and 25% and RLC. The results are plotted in Fig.21,22,23 and 24. The grid voltage is plotted as reference in the upper trace, superimposing values from measurement, the PSpice Program, the state space method and the piecewise method and the lower trace is the load voltage with similar superposition.

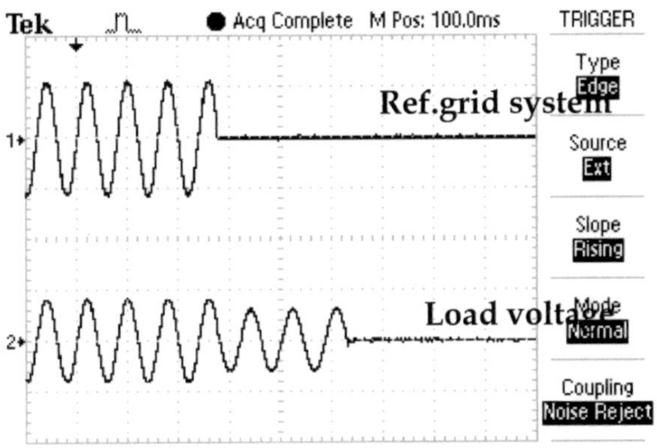

(a) Experiment result

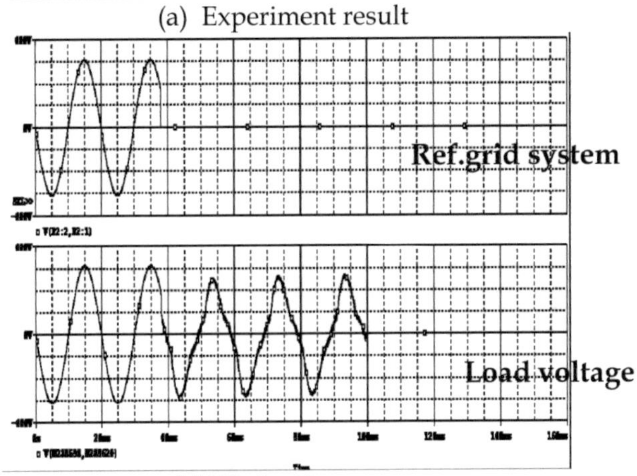

(b) PSpice Program

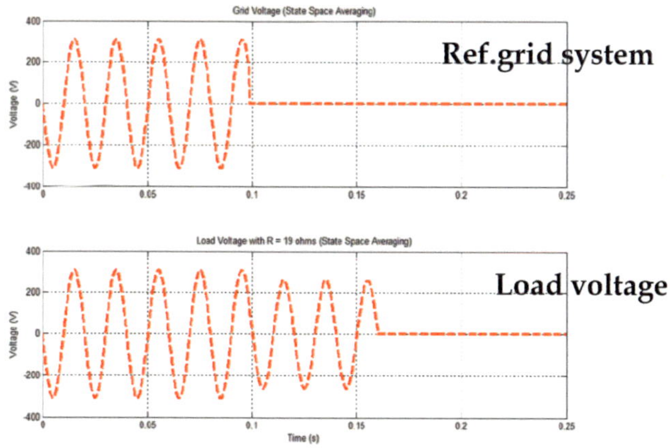

(c) State space averaging modeling

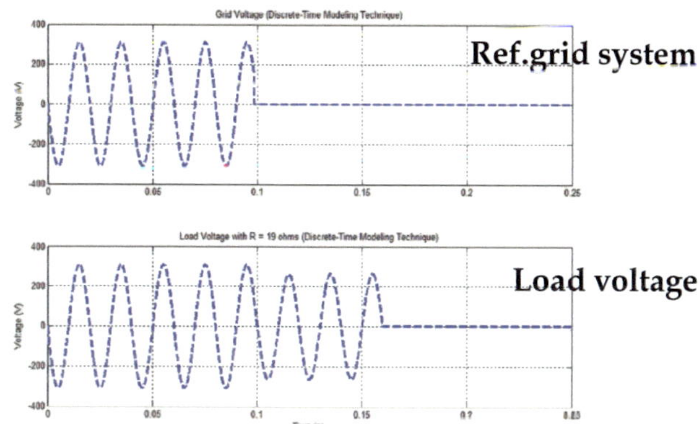

(d) Piecewise techniques modeling

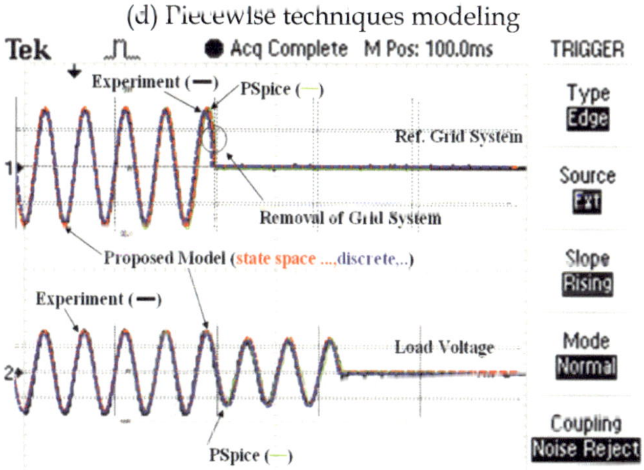

(e) Comparison of grid voltage (top) and load voltage (bottom) from experiment results, the PSpice Program, the state space method and the piecewise method

Fig. 21. Comparison of results for resistive load of 125% inverter output.

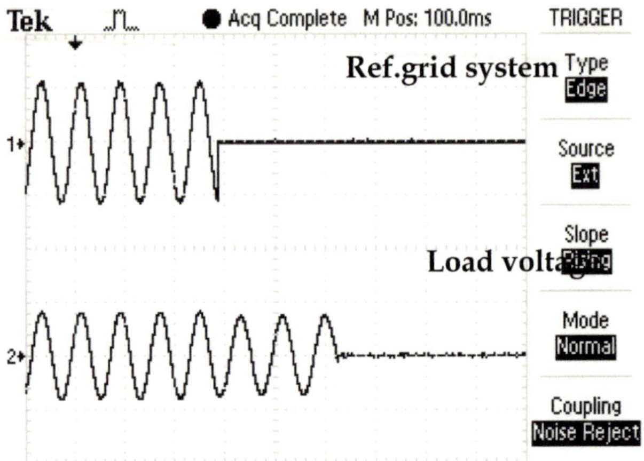

(a) Experiment result

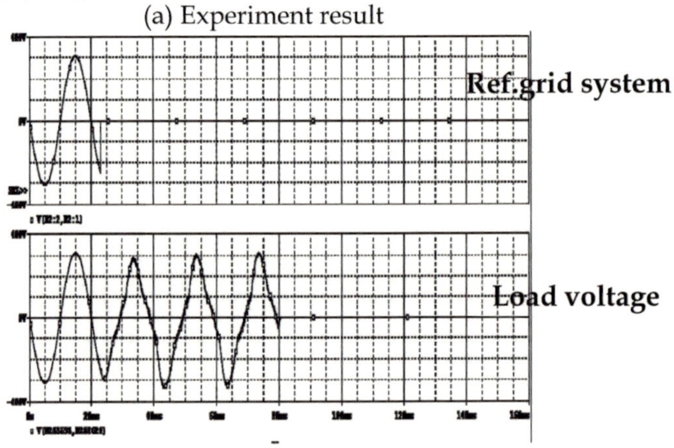

(b) PSpice Program

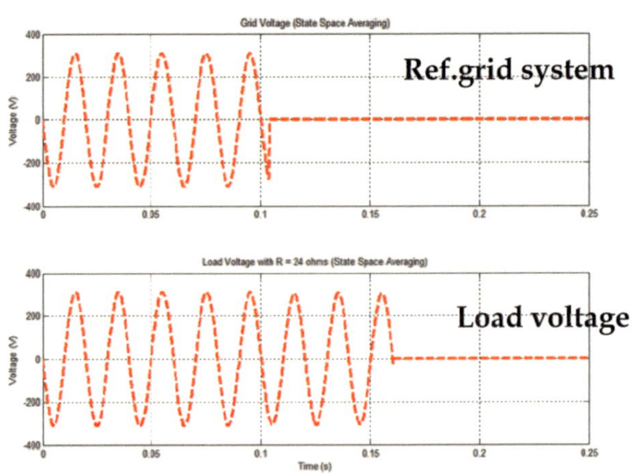

(c) State space averaging modeling

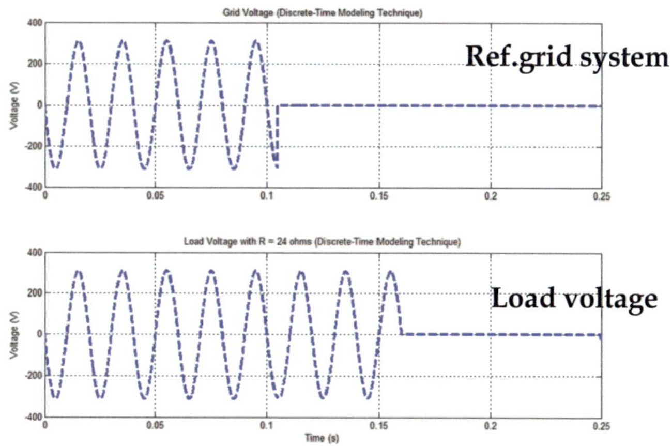

(d) Piecewise techniques modeling

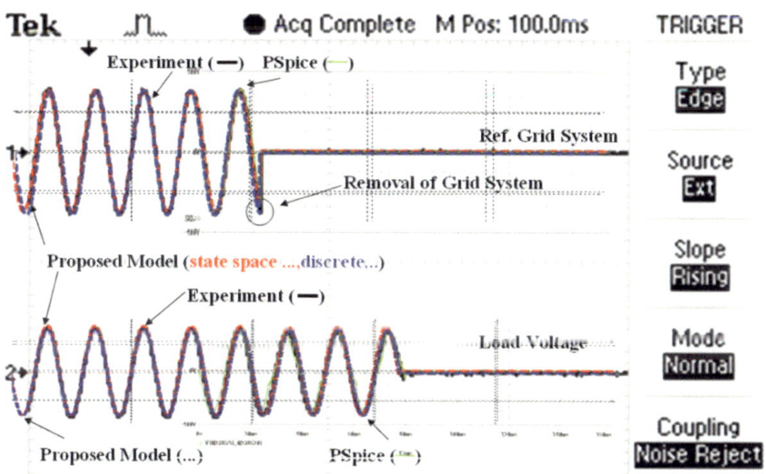

(e) Comparison of grid voltage (top) and load voltage (bottom) from experiment results, the PSpice Program, the state space method and the piecewise method

Fig. 22. Comparison of results for resistive load of 100% inverter output.

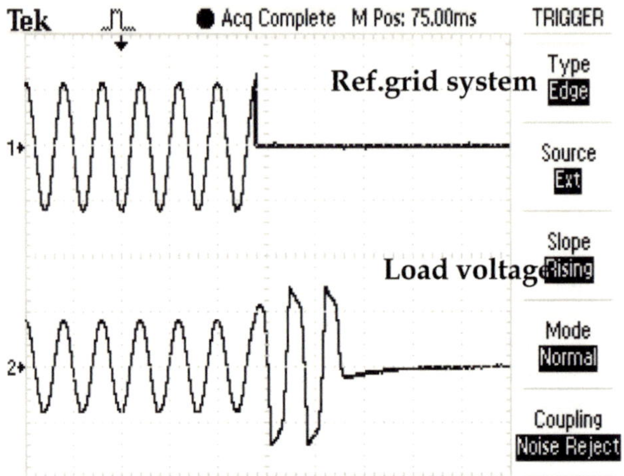

(a) Experiment result

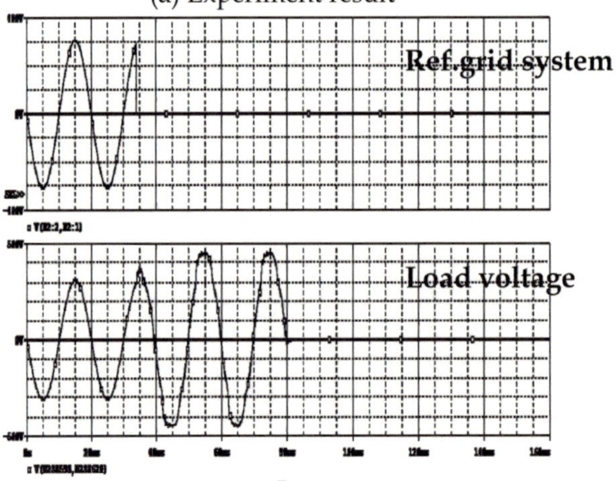

(b) PSpice Program

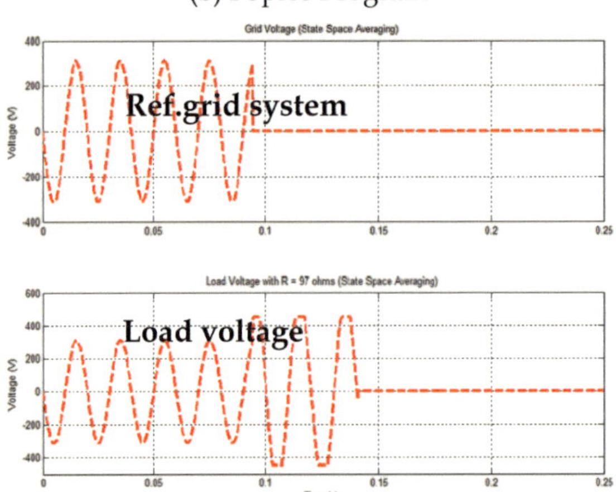

(c) State space averaging modeling

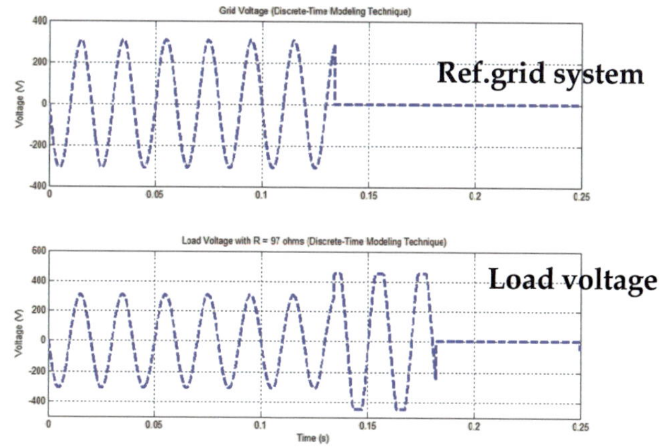

(d) Piecewise techniques modeling

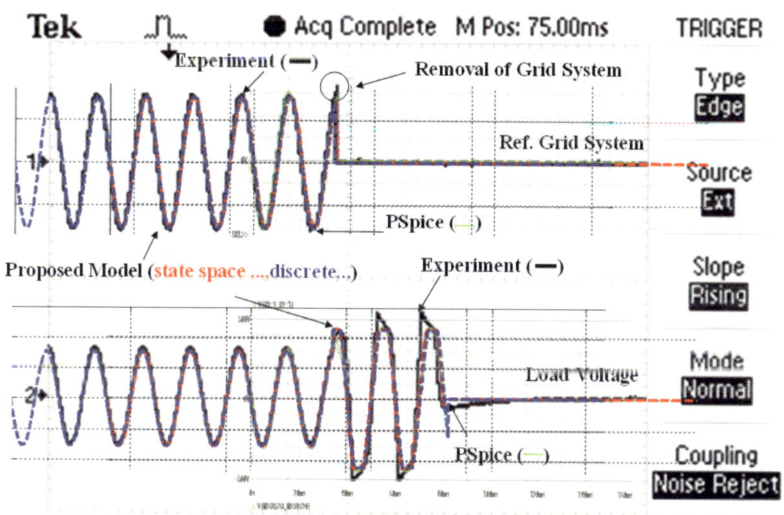

(e) Comparison of grid voltage (top) and load voltage (bottom) from experiment results, the PSpice Program, the state space method and the piecewise method

Fig. 23. Comparison of results for resistive load of 25% inverter output.

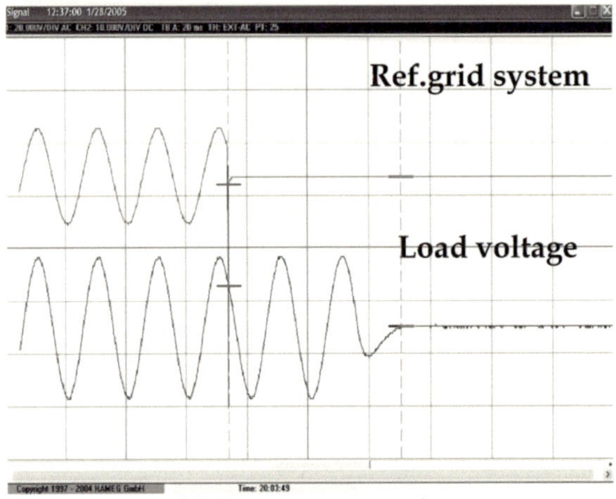

(a) Experiment result

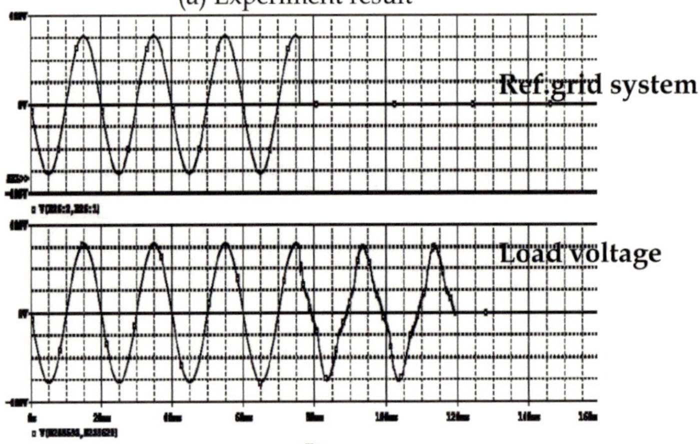

(b) PSpice Program

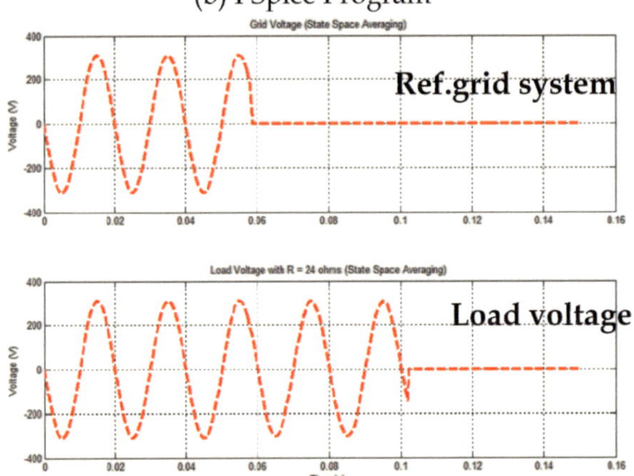

(c) State space averaging modeling

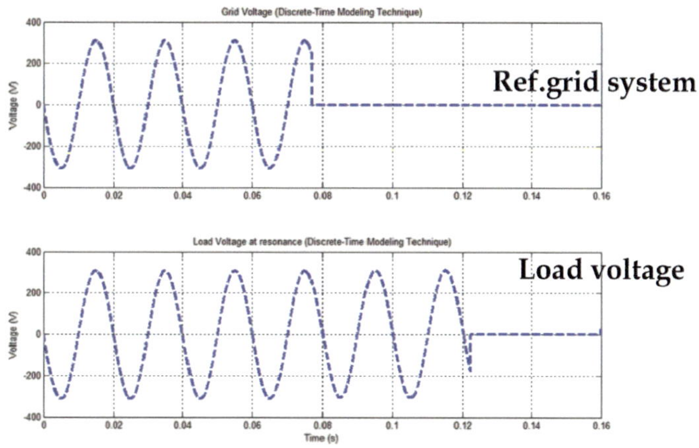

(d) Piecewise technique modeling

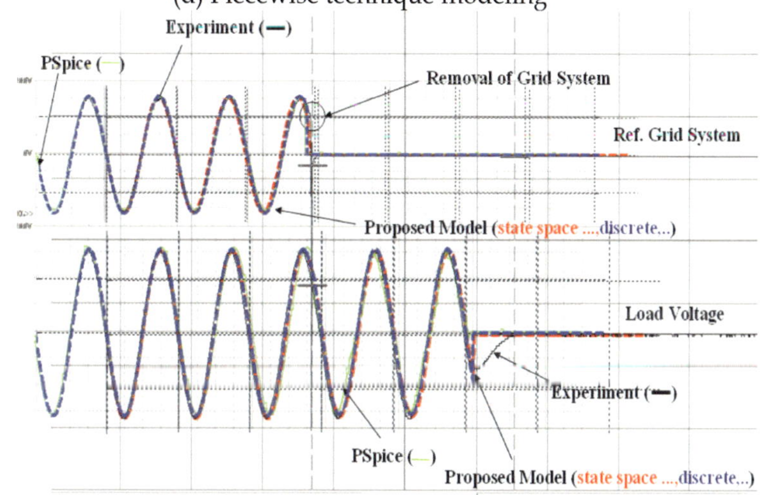

(e) Comparison of grid voltage (top) and load voltage (bottom) from experiment results, the PSpice Program, the state space method and the piecewise method

Fig. 24. Comparison of results for an RLC load resonance case.

Comparisons of the results obtained from the two proposed models are close to the experiment and the PSpice results. When the grid is tripped, the islanding detection from the proposed model can cease to energize within 2-3 cycles or 40-60 ms which pass the criteria of utility's standard as similar to experiment and the PSpice simulation. However, there are some differences due to parasitic element effects which were not considered in the proposed modeling. In case of a resistive load of 125% and 100%, the inverter behaves as current source. The load voltage waveform can give a perfect sinusoidal signal, as shown in Fig.21,22 (a-e). On the other hand, if the inverter operates on voltage source without current control in case of a resistive load of 25%, the load voltage is distorted as shown in Fig.23 (a) and (b) except for the proposed models in Fig.23 (c and d) in which voltage output is set to produce a perfect sinusoidal function. The ratio of load consumption and output inverter can affect the amplitude of load voltage. The higher the ratio, the lower the load voltage amplitude, conforming to the work of Achim Woyte et. al.[8]. The results during islanding phenomena for RLC loads connected in parallel at resonance is shown in Fig.24. It is noticed that the amplitude and frequency of load voltage is kept constant. Therefore, it is difficult to detect islanding by using the passive detection method. However, as compared to the PSpice simulation, the proposed model, which is based on ordinary differential equations, consumes much less for computation time as shown in Table 4. Another advantage of the proposed models is that it does not encounter any convergence problem, which is often the case for PSpice.

Load type	P_{load} (pu.)	PSpice (second)	State space averaged (second)	Piecewise (second)
R	0.25	103	1	6
R	1.0	175	1	5
R	1.25	170	1	5
RLC	1.0	200	1	6

Table 4. Comparison of the time for computation between the PSpice Program and the proposed models

5. Conclusions

The technique to derive a dc-ac full bridge switching converter for a PV grid-connected system are proposed in this paper. An analysis of islanding phenomena due to load variations of R and RLC connections can be easily derived by using the state-space averaging technique and the piecewise technique with feedback current control by setting up the duty cycle with sinusoidal terms around constant value of 0.5. The solution of the two proposed models can be handled via MATLAB/SIMULINK in fast speed and with the absence of the convergence problem, as opposed PSpice simulation. The load voltage behavior simulated by the two proposed models are compared with experiments and the PSpice Program showing good agreement. Small differences, however, were introduced due to parasitic element effects and boundary values of load voltage in the proposed model. Nevertheless, the results demonstrate that a simple and effective model of a grid-connected PV system can be developed. This technique can potentially be further developed for implementation in larger systems consisting of a large array of grid-connected PV modules.

6. Acknowledgements

The authors would like to thank the Joint Graduate School of Energy and Environment (JGSEE), King Mongkut's University of Technology Thonburi (KMUTT) for their kind contribution in providing financial support for this study and the Clean Energy Systems (CES) Group at KMUTT.

7. References

[1] 15 - Year Strategic Plan Promoting New and Renewable Technology Development 15 year (2007-2022), Ministry of Energy, Thailand

[2] Power Development Plan, PDP 2007 Rev 2nd, National Energy Policy Office, Bangkok, 2009
[3] "Renewable Energy Policy: Recent Policies on SPP/VSPP, Seminar on "Renewable Energy: Technology, Markets and Policies in Southeast Asia", BITEC, Bangkok, Thailand, 6 June 2007
[4] PEA/MEA Regulations for the Purchase of Power from Very Small Power Producers (2005)
[5] PEA/MEA Regulations on Uses of Electrical Networks (2008)
[6] PEA/MEA Regulations on Operations of Electrical Networks (2008)
[7] PEA/MEA Regulations on Connections of Electrical Networks (2008)
[8] IEEE929-2000 IEEE Recommended Practice for Utility Interface of Photovoltaic Systems
[9] IEEE 1547.1-2005 Standard Conformance Test Procedures for Equipment Interconnecting Distributed Resources with Electric Power Systems
[10] IEC 62116 Test Procedure of Islanding Prevention Measures for Utility-Interconnected Photovoltaic Inverters
[11] IEC 61727 Photovoltaic Systems- Characteristics of the Utility Interface
[12] Simon S. Ang, Power Switching Converters, Marcel Dekker, 1995
[13] Daniel M. Mitchell, DC-DC Switching Regulator Analysis, McGraw-Hill Book Company, 1988
[14] Muhammad H.Rashid Power Electronics Circuit, Devices and Application, Third Edition, Pearson Education International, 2004
[15] Marian P. Kazmierkowski and Luigi Malesani , Current Control Technique for Three-Phase Voltage-Source PWM Converter: A Survey, IEEE Transactions on Industrial Electronics, Vol. 45,No. 5, pp.691-703(1998)
[16] Chi Kong Tse, Complex Behavior of Switching Power Converter, CRC Press, 2003
[17] Veerapol Monyakul, A New Bidirectional AC-DC Converter with Low Harmonic Input Currents and an Adjustable Power Factor, Doctor of Philosophy Thesis, Oklahoma State University, Stillwater, Oklahoma USA, December,1993
[18] Mei Xu, Roderick V.N Melnik and Uffe Borup, Modeling Anti Islanding Protection Devices for Photovoltaic Systems, Renewable Energy, Vol.26, No.15,pp 2195-2216(2004)
[19] Achim Woyte, Ronnie Belmans and Johan Jijs, Testing the Islanding Protection Function of Photovoltaic Inverters, IEEE Transactions on Energy Conversion Vol.18, No.1, pp. 157-162(2003)

11

Application of a suitable control strategy for grid-connected inverters to the power management of a Microgrid

Daniele Menniti, Ciro Picardi, Anna Pinnarelli and Domenico Sgrò
University of Calabria
ITALY

1. Introduction

The ever increasing energy demand, the necessity of a reduction in costs and higher reliability requirements are driving the present scenario towards Distributed Generation (DG). The DG has been considered as a promising alternative for the coordinated and flexible expansion of the present energy distribution system with reduced cost and improved reliability (Pepermans et al., 2005). In particular, the small DG systems, typically from 1KW to 10 MW and located near to the loads, are gaining popularity due to their higher operating efficiencies and lower emission levels as provider of electrical energy to the consumers. These DG systems are powered by one or more microsources such as: fuel cells, photovoltaic cells, batteries, wind-turbine, micro turbines etc.

A recent evolution resulting by the diffusion of the DG systems is emerged with the concept of Microgrid, which consists in a cluster of loads and paralleled DG systems operating as a single power system that provides power to its local area (Lasseter, 2002). A Microgrid is a systematic organization of DG systems and therefore it has larger capacity and more control flexibility to fulfil system reliability and power quality requirements, in addition to all the inherent advantages of a single DG system.

The above characteristics can be obtained thanks to the use of grid-connected inverters able to quickly manage power generated by the microsources and to generate reactive power near loads, allowing losses reduction. Therefore, high performance control algorithms for power flows control and voltage regulation are required (Li et al., 2006; Katiraei & Iravani, 2006). These algorithms should preferably have no communication links between the paralleled DG systems, which can be located far apart; thus, the control algorithms of each individual DG system should be based on feedback variables that can be measured locally and moreover, they have to ensure a safety operation of the Microgrid avoiding instability problems, which can occur especially when many DG systems are located in a same area.

A good solution for the aforementioned problems can be obtained by the application proposed in this chapter. It is based on the use of a control strategy for grid-connected inverters able to dynamically change the energetic contribution of the microsources, that so adapts oneself to variations of the grid characteristics and contributes to the power management of the Microgrid. The control strategy is developed so as to combine the

advantages of current control and voltage control strategies; as known, the former is to be preferred to ensure stability of the whole conversion system, while the latter one allows a more accurate generation of the reference voltages necessary to apply the PWM voltage technique (Rodriguez et al., 2005). As demonstrated in (Menniti et al., 2007 ; Menniti et al., 2008), the combined use of the two control strategies allows the implementation of a simple and effective single phase control scheme, particularly adapt to deal with critical conditions that can occur in the Microgrids.
The performance of the proposed application is verified in simulation on a Microgrid test, in which several DG systems are contemporary connected to the main bus.
The rest of this chapter is arranged as follows. Section II briefly illustrates the concepts and control issues of Microgrids, while the properties of the grid-connected inverters are given in Section III. Then, in Section IV the Microgrid configuration under examination is shown and in section V the specific used control approach is discussed. Finally, Section VI provides some simulation results, which show the good performances of the proposed power management applied to the Microgrid test.

2. Microgrid Concepts

It is a widespread opinion that small generation should be part of the building energy management systems. In all likelihood, the DG energy output would be ran more cost-effectively with a full range of energy resource optimizing: peak-shaving, power and waste heat management, centralized load management, price sensitive fuel selection, compliance with interface contractual terms, emissions monitoring/control and building system controls. The Microgrid paradigm provides a general platform to approach power management issues.
It has been found that, in terms of energy source security, multiple small generators are more efficient than relying on a single large one for lowering electric bills (Flannery et al., 2004). Small generators are better at automatic load following and help to avoid large standby charges seen by sites using a single generator. Having multiple DG systems on a Microgrid makes the chance of all-out failure much less likely, particularly if extra generation is available.
Moreover, as already explained, the DG systems in Microgrid are generally powered by emerging technologies such as photovoltaic or wind-power and often equipped by inverters to interface with the electrical distribution system. The major issue with these technologies is the nature of the generation; indeed, the availability of their energy source is driven by weather and not by the loads of the systems. These technologies can be labelled as intermittent and ideally they should be operated at maximum output.
An example of a basic Microgrid architecture is shown in Fig. 1, where two paralleled systems DG1 and DG2 are employed. Each DG system is comprised of a dc source, a pulse-width modulation (PWM) voltage source inverter (VSI) and LC filters. In this mode, the two DG systems are controlled to provide local power and voltage support for critical loads A, B and C. This configuration reduces the burden of generation and delivery of power directly from the main grid and enhances the immunity of critical loads to system disturbances in the grid (Flannery et al., 2004 ; Li et al., 2004).
As shown in figure, the Microgrid is connected to the main grid at its point of common coupling (PCC) usually through a static transfer switch (STS).

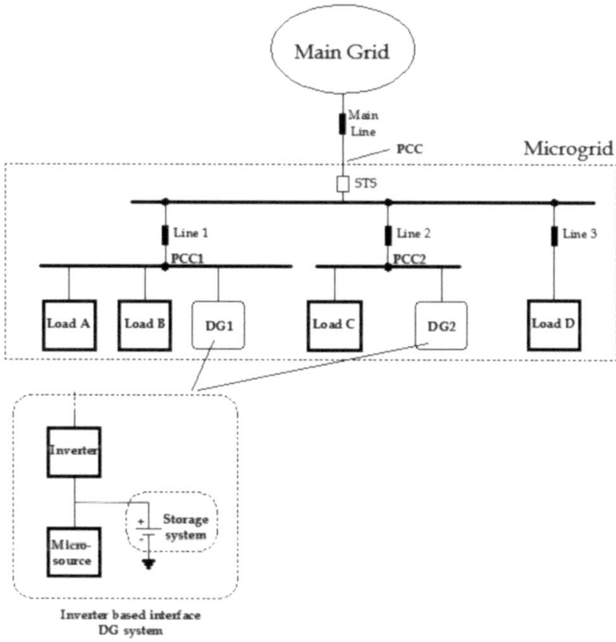

Fig. 1. Microgrid Architecture diagram

An important problem, that can be resolved by the DG systems is the impact of unbalanced grid voltages (usually caused by faults or unbalanced connected) on the overall system performance. If the unbalance in voltages is serious, STS opens to isolate the Microgrid from the main grid; however, also if the voltage unbalance is not so serious, STS may remain closed, resulting in sustained unbalanced voltages at PCC. Such a voltage unbalance can cause an increase in losses in motor loads (and hence motor overheating) and abnormal operation of sensitive equipments in the Microgrid.
Another issue regarding Microgrid is related to the ability of DG systems to increase system reliability and power quality due to the decentralization of supply. Indeed, using a decentralized control for each DG system, expensive communications systems can be avoided and problems connected to transient conditions in the grid (i.e. frequency oscillations) can be reduced.

3. The Grid-connected inverter

With the development of solid-state-based packages, power electronic devices can now convert almost any form of electrical energy to a more desirable and usable form. This is why power electronic interfaces are ideal for DG system applications.
Indeed, many power sources used in DG systems generate electric power in a waveform that the present power distribution grid cannot accept. Sources as photovoltaic cells or fuel cells supply dc electric energy, while the present distribution grid accepts ac electric energy. Often, in wind power generation systems it is necessary to convert the ac voltage from wind

turbine generator to dc voltage, being the frequency and amplitude of the ac voltage from the wind turbine generators variable in time due to the random nature of the wind.
As a consequence, the power electronic interfaces may include both ac-dc conversion systems (rectifiers) and dc-ac conversion systems (inverters).
Another benefit of power electronic devices is their extremely fast response times; in fact, the dc-ac power interface, usually called grid-connected inverter, can respond to power quality events or fault conditions that occur in the grid within the sub-cycle range.
In addition to allowing the transfer of power supplied by the microsource of the DG system, the grid-connected inverter can also permit the control of voltage and reactive power at the PCC of the generation source (Kroposki et al., 2006). In particular, most inverters for DG systems are self commutated and can produce an ac voltage of an arbitrary amplitude and phase. This allows the DG systems to produce any power at any power factor so as a wider operating power factor range than a synchronous generator can be obtained.
The grid-connected inverter can also contain protective functions, for both the distributed energy system and the local electric power system, that allow paralleling and disconnection from the electric power system. Moreover, also some level of metering and control functionality are contained and this shall ensure that the distributed energy system can operate as designed.
Finally, it is worth to highlight that the grid-connected inverter controls must ensure different important requirements: new microsources can be added to the system without modification of existing equipment, set-up can be independently chosen, the Microgrid can connect to or isolate itself from the grid in a rapid and seamless fashion, reactive and active powers can be independently controlled and can meet the dynamic needs of the loads.

4. The Microgrid configuration

In the context described in two previous sections, this paper proposes the application of a suitable control strategy for grid-connected inverters in a simple distributed system in which several DG systems are contemporary connected to the same bus of a Microgrid.
In particular, the Microgrid configuration under examination is shown in Fig. 2 and it consists in four DG systems, based on inverters connected to the main bus by different line impedances.

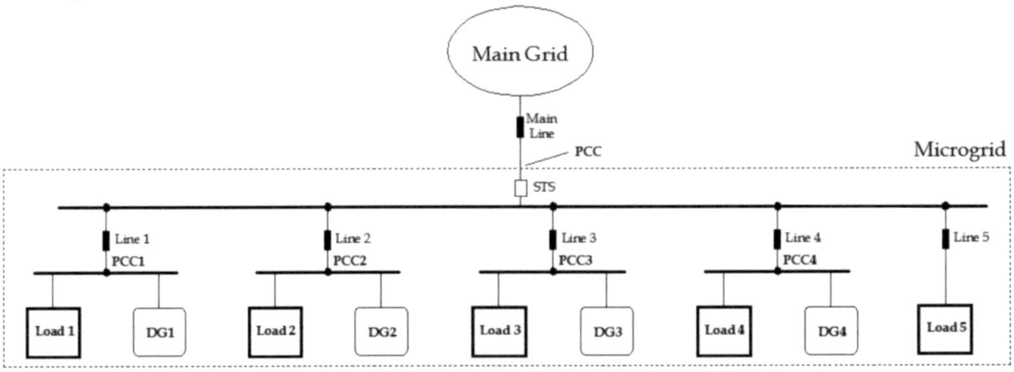

Fig. 2. Microgrid configuration considered.

Each DG is directly connected to a critical load that absorbs a different amount of active and reactive power for each phase. Besides, a linear and balanced load (Load 5) is connected at the main bus.

It is worth to underline that each DG system has merely information about the voltage at its own PCC and the currents absorbed from the critical load connected at the same PCC. Using this information, each DG system should operate in order to correct the power factor and to balance the active power absorbed by each phase of the load, so as to effect a regulation of the voltage and avoid that unbalances scatter in the main system. Obviously, each DG system provides active power to the grid supplying a share of the energy required by the loads in the Microgrid.

Thanks to the potentiality of the grid-connected inverters, the DG systems are able to influence the power flows of the Microgrid very quickly and in significant manner. However, a so fast action can bring out perturbations in the Microgrid and generate instability. Certainly, this can occur when many DG systems are present; for this reason, it is important that the DG systems control manages to avoid negative impacts on the Microgrid stability.

In the following section the control strategy used for the application and proposed by the authors in (Menniti et al., 2007), is explained and illustrated in detail.

5. Strategy control for power management

The configuration of the grid-connected inverter, used in each DG system of Microgrid configuration under examination is shown in Fig. 3.

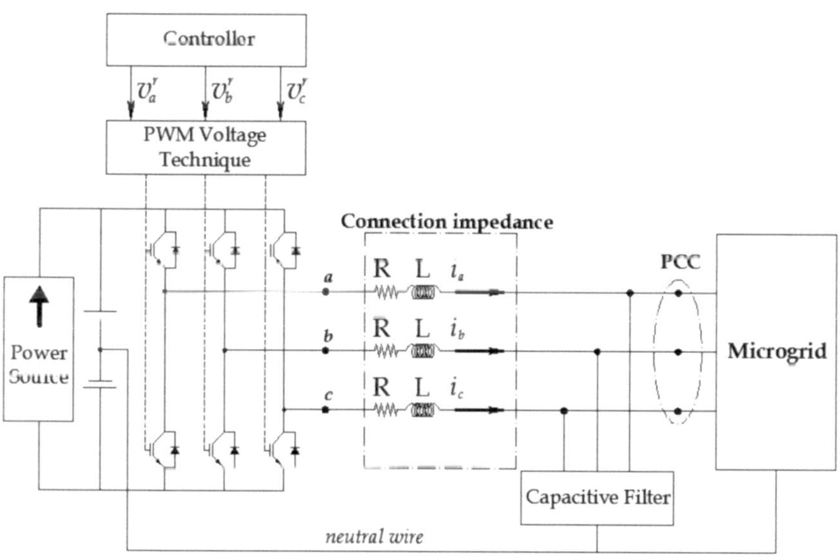

Fig. 3. Grid-connected inverter configuration.

It comprises a three-leg VSI connected on the ac-side to the Microgrid by a suitable connection impedance and on the dc-side to a dc-power source by capacitors. The inverter

works in order to transfer the energy produced by the dc-power source on the Microgrid, controlling the power flows through the connection impedance. This is constituted by three connection impedances each with resistance R and inductance L and a parallel capacitive filter providing a path for some high-order harmonics at the switching frequency.
Considering that the inverter operates in voltage control mode, its controller generates three reference signals v_j^r (where $j = a,b,c$), each of which is referred to the output voltage to be applied on the *j-th* phase, so that the connection impedance current i_j tracks its desired value corresponding to the power flows required between the dc and ac sides. Obviously, it is needed that the output voltages of the VSI track the reference voltages by properly applying the PWM technique; so in the following it will be assumed that $v_j = v_j^r$.
The block diagram relevant to the control strategy applied to the generic phase of the grid-connected inverter, is shown in Fig. 4 (where for sake of simplicity the subscript j is omitted).
As clearly evidenced, it includes two parts: one indicated with *Voltage Control* and the other with *Current Control*. Both the parts need the estimation value for the rms V and the phase φ_v of the correspondent PCC voltage v (obtained by block A operating as Phase Locked Loop) and information about the desired power flows P^* and Q^* to be exchanged with the grid. Moreover, the reference voltage v^r is obtained as sum of two contributions: v_v^r given by the *Voltage Control* and v_c^r given by the *Current Control*.

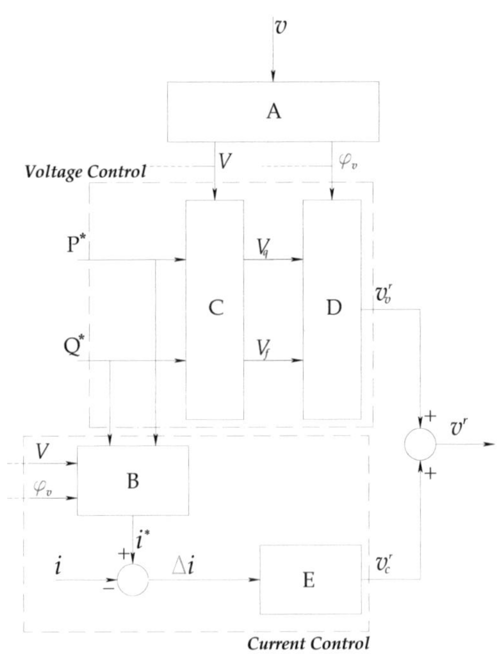

Fig. 4. Block diagram of the control strategy

5.1 Voltage Control

The basic equation of the *Voltage Control* are obtained by considering the simple circuit of Fig. 5. It shows the voltages and the currents relevant to the branch "*a*" of the connection impedance. Obviously, by considering only the components at fundamental frequency, the current flowing through the capacitive filter can be neglected.

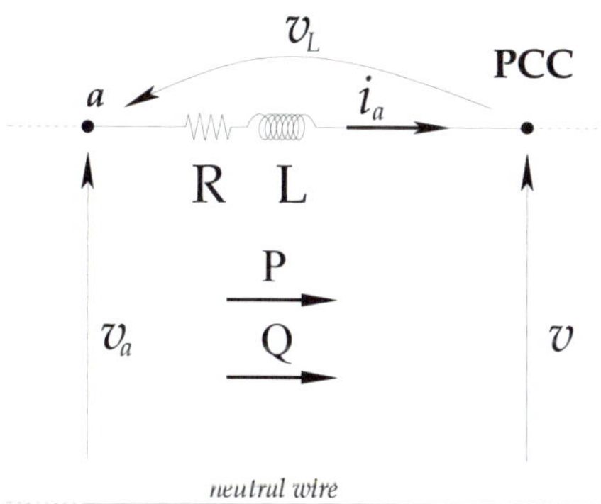

Fig. 5. The equivalent circuit at fundamental frequency.

For sake of simplicity only the powers exchange on the phase "*a*" will be analyzed, because the relationships that rule the power flows through the others phases can be obtained in a similar way.

In particular, in Fig. 5, the fundamental components of the voltages at PCC and at node "*a*" are denoted as v and v_a respectively. Analogously, i_a and v_L denote the current and the voltage on the connection impedance, while P and Q are the fundamental active and reactive power flows from node "*a*" to PCC.

By expressing the quantities v, v_a, i_a and v_L as phasors $\bar{\mathbf{V}}$, $\bar{\mathbf{V}}_a$, $\bar{\mathbf{I}}_a$ and $\bar{\mathbf{V}}_L$ respectively and choosing $\bar{\mathbf{V}}$ as the reference one, the phasor diagram shown in Fig. 6 is obtained.

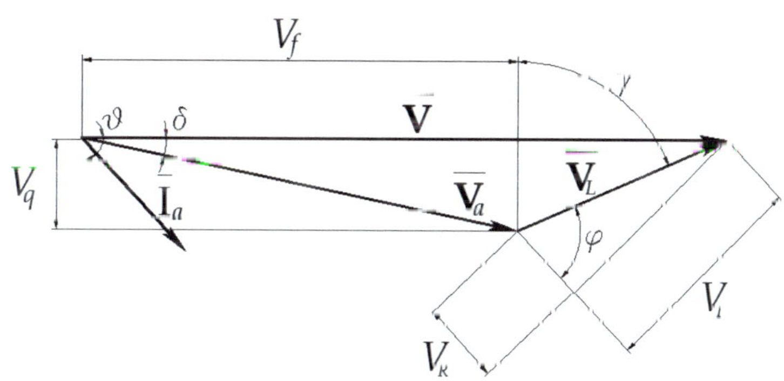

Fig. 6. The phasor diagram relevant to the circuit of Fig. 5.

In detail, in Fig. 6: δ is the phase angle of $\bar{\mathbf{V}}_a$, V_f and V_q are the forward and the quadrature components of $\bar{\mathbf{V}}_a$ with respect to $\bar{\mathbf{V}}$; ϑ is the phase angle of $\bar{\mathbf{I}}_a$; φ is the characteristic angle of the connection impedance and it is equal to $\arctan(X_L/R)$, where

$X_L = \omega L$ represents the reactance of the connection impedance at the particular angular frequency ω ; the angle γ is defined by the following relationship

$$\gamma = \frac{\pi}{2} + \vartheta - \varphi . \tag{1}$$

Moreover, the magnitude of the resistive and the inductive voltage components of $\bar{\mathbf{V}}_L$, V_R and V_I respectively, are obviously defined as

$$V_R = V_L \cos\varphi , \qquad V_I = V_L \sin\varphi , \tag{2}$$

where V_L is the amplitude of $\bar{\mathbf{V}}_L$.
As known, the active and reactive power flows from the node "*a*" to the PCC can be expressed by the following relationships

$$\mathrm{P} = VI_a \cos\vartheta , \qquad \mathrm{Q} = VI_a \sin\vartheta . \tag{3}$$

Considering (2) and the amplitude of the connection impedance current, it is possible to obtain the following law for the active and reactive powers

$$\mathrm{P} = \frac{V}{\sqrt{\mathrm{R}^2 + \mathrm{X}_\mathrm{L}^2}} \frac{V_I}{\sin\varphi} \cos\vartheta , \qquad \mathrm{Q} = \frac{V}{\sqrt{\mathrm{R}^2 + \mathrm{X}_\mathrm{L}^2}} \frac{V_I}{\sin\varphi} \sin\vartheta . \tag{4}$$

From the phasor diagram of Fig. 6, it is easy to observe that

$$\begin{aligned} V_I \cos\vartheta &= V_a \sin\delta + V_R \sin\vartheta , \\ V_L \sin(\varphi - \vartheta) &= V_a \sin\delta = V_q . \end{aligned} \tag{5}$$

Then, considering the first one of (2) and using the Werner formulas, it is possible rewrite

$$\mathrm{P} = \frac{V}{\sqrt{\mathrm{R}^2 + \mathrm{X}_\mathrm{L}^2}} \frac{V_I + V_L \sin(\vartheta + \varphi)}{2 \sin\varphi} . \tag{6}$$

Moreover, from the phasor diagram of Fig. 6, the following angular relationship can be easily obtained

$$\vartheta + \varphi = \gamma - (\pi/2 - 2\varphi) , \tag{7}$$

therefore, by means simple algebraic manipulations, the sinusoidal term of (6) can be written as

$$\sin(\vartheta+\varphi) = 2\sin\gamma\ \sin\varphi\ \cos\varphi - \left(\cos^2\varphi - \sin^2\varphi\right)\ \cos\gamma\ . \tag{8}$$

Consequently, by substituting (8) in (6) and observing, from Fig.6, that

$$\begin{aligned} V_L \sin\gamma &= V - V_f, \\ V_L \cos\gamma &= V_q, \end{aligned} \tag{9}$$

the following relationship for the active power flow can be obtained

$$\mathrm{P} = \frac{V}{\sqrt{\mathrm{R}^2 + \mathrm{X}_\mathrm{L}^2}}\left[\sin\varphi V_q + \cos\varphi\left(V - V_f\right)\right]. \tag{10}$$

In a similar way it is possible to provide the relationship which describe the bond between the voltages on the connection impedance and the reactive power flows supplied by the inverter:

$$\mathrm{Q} = \frac{V}{\sqrt{\mathrm{R}^2 + \mathrm{X}_\mathrm{L}^2}}\left[\sin\varphi\left(V - V_f\right) - \cos\varphi V_q\right]\cdot \tag{11}$$

Neglecting the resistance of the connection impedance, i.e. considering $\varphi = \pi/2$, the expressions for the active and reactive power flows become:

$$\mathrm{P} = \frac{V}{\mathrm{X}_L}V_q, \qquad \mathrm{Q} = \frac{V}{\mathrm{X}_L}\left(V - V_f\right)\cdot \tag{12}$$

These equations show that the active power flow only depends on V_q, while the reactive power flow only depends on V_f; therefore, by controlling the forward and the quadrature component of $\bar{\mathbf{V}}_a$ (i.e. the amplitude and the phase angle of $\bar{\mathbf{V}}_a$ with respect to $\bar{\mathbf{V}}$), it is possible to control the fundamental active and reactive power flows of the circuit of Fig. 5.
It is worth to underline that in the real case the resistance R is not just equal to zero, but surely it is very low, representing a parasitic resistance of an inductive filter; therefore (12) can be considered as acceptable approximations of (10) and (11).
On the basis of the presented analysis, the *Voltage Control* marked in Fig. 4 is simply realized through the two blocks C and D. In fact, on the assumption that the connection filter is purely inductive, the block C calculates V_f and V_q according to the following equations, directly obtained from (12):

$$V_f = V + \frac{\omega \mathrm{L}}{V}\mathrm{Q}^*, \qquad V_q = \frac{\omega \mathrm{L}}{V}\mathrm{P}^* \tag{13}$$

and block D determines the share v_v^r of the reference voltage on the basis of the following relationships:

$$v_v^r(t) = \sqrt{2}\sqrt{V_f^2 + V_q^2}\,\sin(\omega t + \varphi_v + \delta),$$
$$\delta = \arctan(V_q/V_f) \tag{14}$$

5.2 Current Control

As illustrated, the basic equations of the *Voltage Control* have been obtained by using a phasor representation assuming a sinusoidal waveform at the fundamental frequency for the voltages and the currents; since the bond between currents and voltages can be described by phasor diagram only if the system is in steady-state condition, a suitable *Current Control* has been developed in order to provide a corrective action mainly during the transient period.

The block B, included in the *Current Control* (see Fig. 4), uses the values V and φ_v to calculate the instantaneous value of a reference current i^* necessary to obtain the desired power flows P^* and Q^*, by means of the following equation

$$i^*(t) = \sqrt{2}I^*\sin(\omega t + \varphi_i) \tag{15}$$

where

$$\varphi_i = \varphi_v - \arctan\left(\frac{Q^*}{P^*}\right),$$
$$I^* = \frac{\sqrt{[P^*]^2 + [Q^*]^2}}{V}. \tag{16}$$

The block E represents the regulator determining, on the basis of the error $\Delta i = i^* - i$, the other share v_c^r of the reference voltage.

Since the *Current Control* has to act as support of the *Voltage Control* substantially only during the transient period, its contribution can be estimated as proportional to the inductor energy, necessary to carry the actual current value i to its desired value i^*, that can be written as:

$$E = \frac{1}{2}L\Delta i^2. \tag{17}$$

This consideration suggests to use for the block E the following relationship

$$v_c^r = k\Delta i^2\,\mathrm{sgn}(\Delta i), \tag{18}$$

represented by the instantaneous characteristics shown in Fig. 7 and where k is a suitable constant.

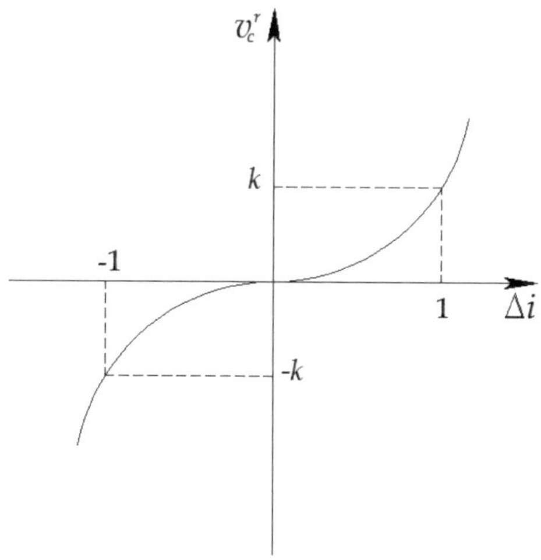

Fig. 7. Instantaneous characteristics of the *Current Control* regulator (block E).

5.3 Management Power Flows

The choice of the desired power flows P^* and Q^* has been effected in order to prove the benefits brought by the DG when many sources are present in a Microgrid. The value Q^* has been chosen equal to the reactive power absorbed by each local load in order to realize the load power factor correction and then to reduce the transmission losses. The desired active power flow P^* has been obtained as sum of two contributions. The former is equal for each phase to one third of the active power provided by the dc-power microsource. The other contribution is calculated in order to balance the power absorbed by an unbalanced load. In particular, the desired active power to be provided by each interfacing inverter on the *j-th* phase has been chosen as:

$$P_j^* = \Delta P_{Lj} + \frac{P_{ms}}{3}, \tag{17}$$

where P_{ms} is the active power given by the microsource and ΔP_{Lj} is the difference between the power absorbed by the local load on the *j-th* phase and the average power absorbed by all the phases, as shown by the following relationships:

$$\Delta P_{Lj} = P_{Lj} - P_{ave}, \qquad P_{ave} = \frac{(P_{La} + P_{Lb} + P_{Lc})}{3}. \tag{18}$$

6. Numerical results

Some numerical results have been carried out to demonstrate the effectiveness of the proposed application. The reactive power compensation obtained by the grid-connected inverters, controlled by the used control strategy for each PCC node and for a generic phase is displayed by Fig. 8. It shows as the reactive power absorbed by every local load is immediately compensated, when the control action for the reactive power management starts at 0.1 sec.; in fact each line current i_{Lh} , with $h=1,..4$ (suitably amplified for necessity of visualization), is carried out in phase with each PCC voltage v and then it is also significantly reduced.

It is important to notice that the current i_{L2} is in opposition of phase with the voltage. This is because DG2 is able to produce more energy than that needed to its load and therefore the active power flow through Line 2 is inverted.

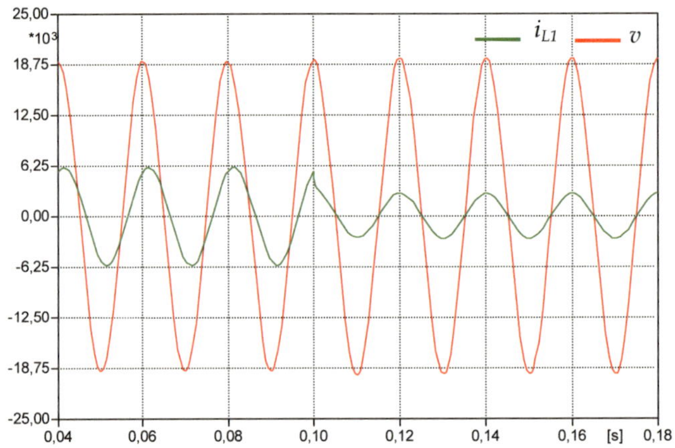

Fig. 8a. Power factor correction for one phase of Load 1.

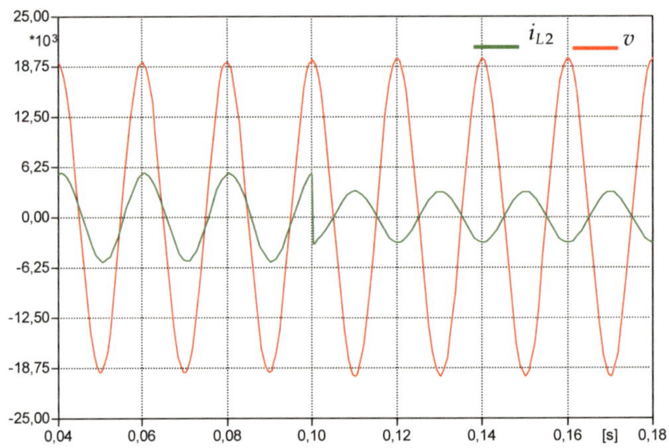

Fig. 8b. Power factor correction for one phase of Load 2.

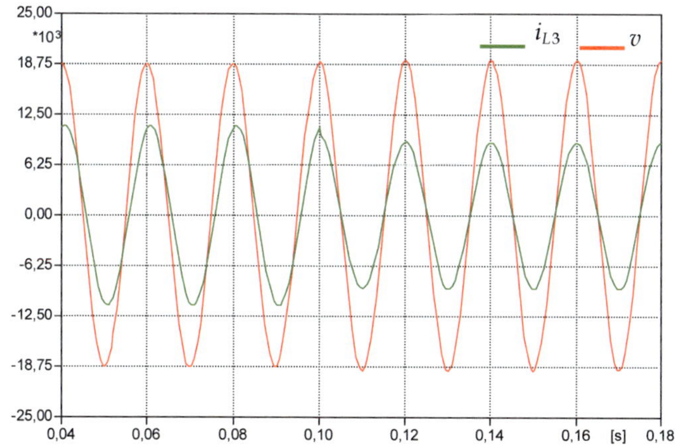

Fig. 8c. Power factor correction for one phase of Load 3.

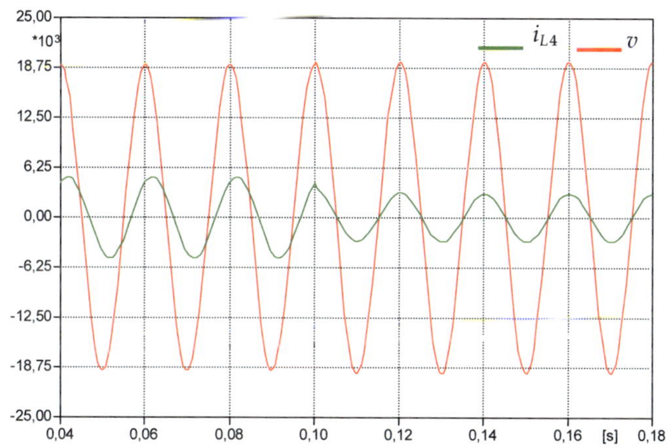

Fig. 8d. Power factor correction for one phase of Load 4.

Another effect of the reactive power compensation, carried out by all the DG systems, is clearly evidenced in Fig. 9. In fact, as a consequence of the reactive compensation the reactive power supplied by the main grid for each phase is significantly reduced. This due to the fact that, in the operating conditions subsequently to the reactive compensation, the only load which needs reactive supply by the grid is the balanced Load 5, being far apart of the DG systems.

A further interesting result is presented in Fig. 10, that shows the effect of the power management used in order to balance the active power absorbed by the critical loads near the DG systems. As a consequence of the power management algorithm, each DG system, together with its critical load, is seen as a balanced Load from the main grid; in this way, the negative and zero current sequences present in the main grid currents are easily cancelled before the compensation action, making so that the power provided by the main grid is balanced (see Fig. 11).

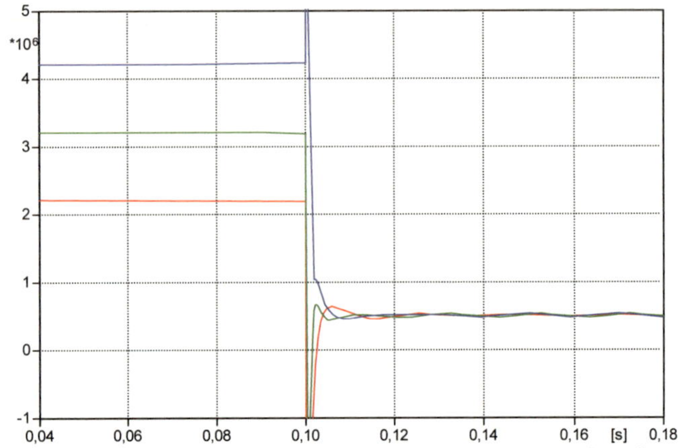

Fig. 9. Reduction of the reactive powers provided by the main grid [Var].

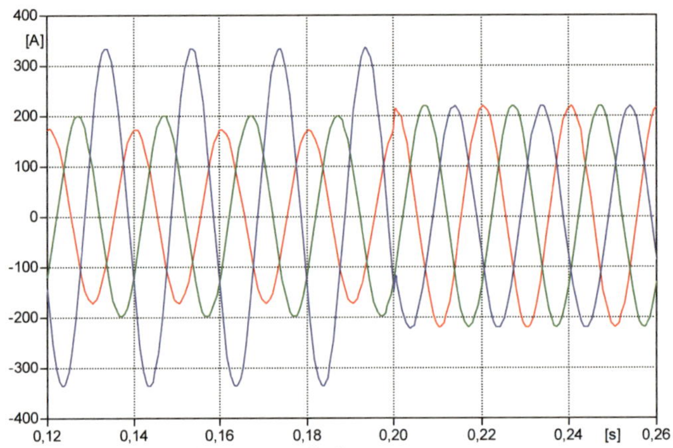

Fig. 10. Balancing of the main grid currents [A].

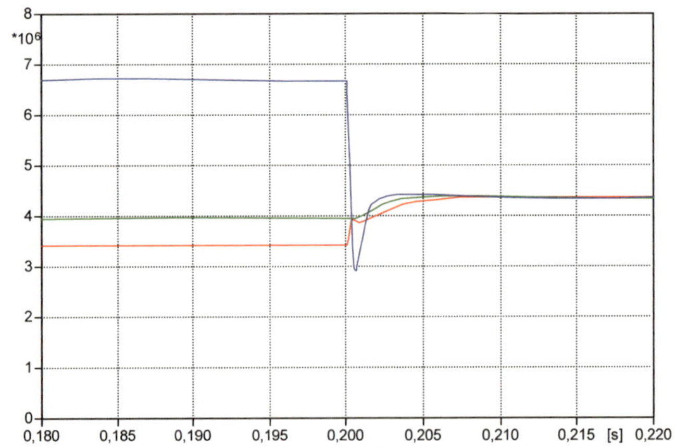

Fig. 11. Balancing of the main grid active powers [W].

Finally, Fig. 12 points out that the power management made by the DG systems avoids that unbalanced currents can unbalance the main bus voltages. Therefore, also the balanced Load 5 absorbs the same value of active power for each phase.

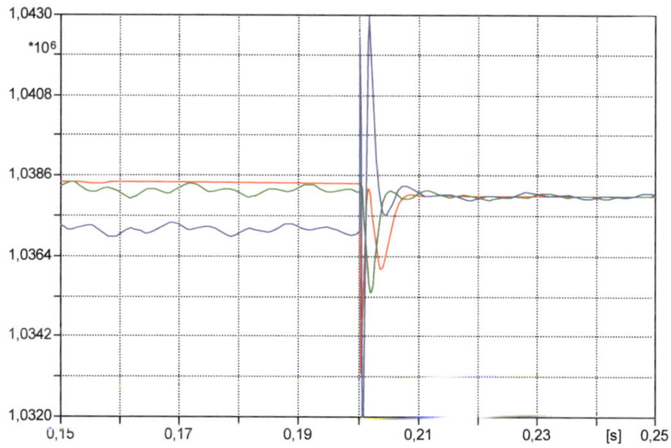

Fig. 12. Active powers absorbed by the Load 5 [W].

7. Conclusions

This chapter has dealt with the application of a suitable control strategy for grid-connected inverters used in DG systems operating in a Microgrid.
First, this chapter has illustrated the concepts and control issues of Microgrids, which consist in clusters of loads and paralleled small DG systems operating as single power systems.
As clearly explained in the chapter, the microsources, used in the DG systems, having different output characteristics, in voltage and current, are connected to the utility grid using voltage source inverters based interfaces, usually known as grid-connected inverters.
Due to the development of the power electronic devices, the grid-connected inverters are able to perform different useful functions such as the control of voltage and reactive power at the PCC of the generation source and a fast response to power quality events or fault conditions.
The control strategy used in the application illustrated in this chapter has been recently presented in literature by the authors and it has been developed so as to combine the advantages of the current control and the voltage control strategies. Moreover, it is important to underline that this control strategy is suitable to deal with three-phase and single-phase systems, assuring at the same time an optimal control of the power flows both in stationary conditions and in transient ones. As a consequence, the control strategy can be applied to large size plants as well as those small size, which are very common in the Microgrids.
In fact, the control strategy has been applied to a Microgrid configuration, consisting in four DG systems, based on interfacing inverters connected to the main bus by different line impedances. Each DG is directly connected to a critical load that absorbs a different

amount of active and reactive power for each phase. Besides, a linear and balanced load is connected at the main bus.

The application has been verified by various numerical simulations. In particular, the simulations have point out as, by using grid-connected inverter, the DG systems have been able to influence the power flows of the Microgrid very quickly and in significant manner. However, thanks to the good characteristics of the control strategy used, such fast performances have been achieved without generating perturbations and instability problems on the Microgrids. In fact, the numerical simulations show that the DG systems effectively provide services as reactive power compensation and load balancing without influence each other.

8. References

Flannery, P.; Venkataramanan, G.; Shi B.(2004). *Integration of Distributed Technologies – Standard Power Electronic Interfaces,* California Energy Commission Consultant Report, April 2004.

Katiraei, F.; Iravani, M. R. (2006). Power Management Strategies for a Microgrid With Multiple Distributed Generation Units, *IEEE Trans. on Power Systems,* Vol. 21, No 4, (November 2006) 1821–1831, 0885-8950.

Kroposki, B.; Pink, C.; DeBlasio R.; Thomas, H.; Simoes, M., Sen, P.K. (2006). Benefits of Power Electronic Interfaces for Distributed Energy System, *Proceedings of Power Engineering Society General Meeting,* pp. 8, June 2006.

Lasseter, R. H. (2002). Microgrids, *PES Winter Meeting* , pp. 305-308, New York (US), January 2002, Institute of Electrical and Elctronics Engineers, New York (US).

Li, Y. W.; Vilathgamuwa, D. M.; Loh, P.C. (2004). Design, Analysis, and Real-Time Testing of a Controller for Multibus Microgrid System, *IEEE Trans. On Power Electronics,* Vol. 19, No 5, (September 2004) 1195-1204, 0885-8993.

Li, Y. W.; Vilathgamuwa, D. M.; Loh, P.C. (2006). A grid-interfacing Power quality Compensator for three-phase three-wire Microgrid applications, *IEEE Trans. on Power Electronics,* Vol. 21, No. 4, (July 2006) 1021-1031, 0885-8993.

Menniti, D.; Picardi, C.; Pinnarelli, A.; Sgrò, D. (2007). Grid-connected inverters for alternative energy sources with a combined voltage and current control strategy, *Proceedings of ICCEP Intern. Conference,* pp. 223-228, Capri (Italy), May 2007.

Menniti, D.; Picardi, C.; Pinnarelli, A.; Sgrò, D. (2008). Power management by grid-connected inverters using a voltage and current control strategy for Microgrid applications, *Proceedings of SPIDAM Intern. Conference,* pp. 1414 - 1419 , Ischia (Italy), June 2008.

Pepermans, G.; Driesen, J.; Haeseldonckx, D., Belmans, R., D'haeseleer, W. (2005). Distributed generation: definition, benefits and issues, *Energy Policy,* Vol. 33, 2005. 787-798.

Rodriguez, J. R.; Dixon, J. W.; Espinoza, J. R.; Pontt, J.; Lezana, P. (2005). PWM Regenerative Rectifiers: State of the Art, *IEEE Trans. on Industrial Electronics,* Vol. 52, No. 1, (February 2005) 5-22, 0278-0046.

Wind Farm Protection Systems: State of the Art and Challenges

Tamer A. Kawady, Naema M. Mansour and Abdel-Maksoud I. Taalab
Electrical Engineering Department, Minoufiya University
Shebin El-Kom, Egypt

1. Introduction

Owing to the rapid increase of the global population and their energy needs, traditional means to satisfy the burgeoning energy demands need careful reevaluation. Coupled with the uneven distribution of resources around the world, economic impacts of large-scale importation and the environmental impacts of continued dependence on nonrenewable fossil fuels, there is an imminent need to transfer, at least partly, the dependence on to renewable energy resources. Among these resources, wind electric conversion systems have emerged as the leader at the present time. According to the Global Wind Energy Council (GWEC) annual report, over 27 GW of new wind power generation capacity came on line worldwide in 2008 representing a 36% growth rate in the annual market, bringing the total global wind power capacity to over 120GW through the end of 2008 as shown in Fig. 1 [1]. This indicates that there is huge and growing global demand for emissions-free wind power, which can be installed quickly, virtually everywhere in the world.

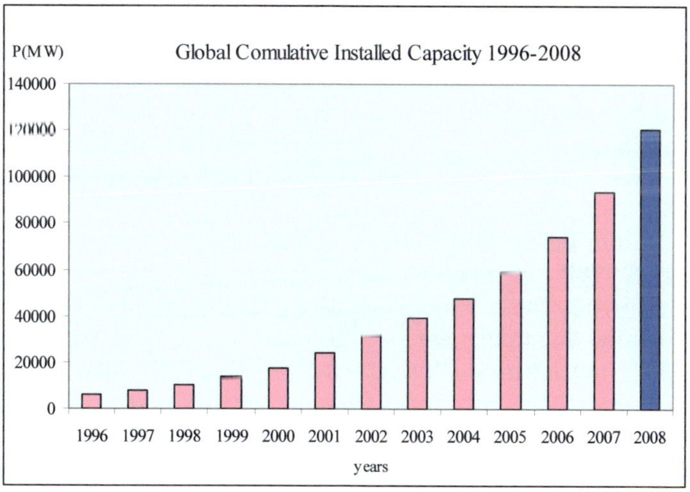

Fig. 1. Global cumulative installed capacity 1996-2008 (source GWEC annual report) [1].

Although, the renewable energy as a percentage of US electricity is only 3%, wind energy now contributes over 42% of all non-hydro renewable generation, up from 33% in 2007 as in Fig. 2. The U.S. Department of Energy's report, 20% Wind Energy by 2030: Increasing Wind Energy's Contribution to U.S. Electricity Supply, found that reaching a level of 20% wind energy by 2030 was feasible under one closely examined scenario [2]. In Europe, according to new European Wind Energy Association (EWEA) statistics released in early February, 2009, the wind penetration of the Europe Union (EU) power is accounted for 36%, or 8,484 MW, of new capacity, beating all other power technologies including gas, coal and nuclear power as shown in Fig. 2 [3]. The EU has agreed on legislation to realise 20% renewable energy by 2020, so to reach the 20% renewable energy target, the EU will need to increase the share of electricity from renewable energy sources from 16% in 2006 to at least 34% in 2020.

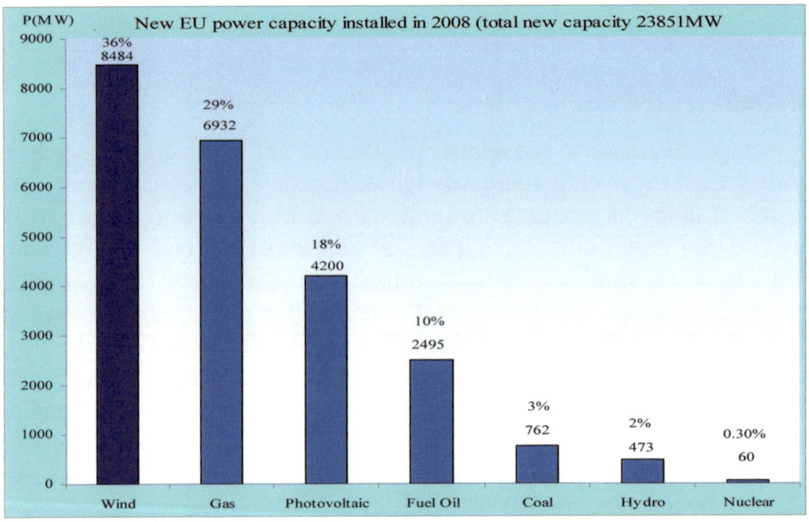

Fig. 2. New EU power capacity installed in 2008 (source EWEA annual report 2008) [2].

The impressive growth in the utilization of wind energy has consequently spawned active research activities in a wide variety of technical fields. Moreover, the increasingly penetration of wind energy into conventional power systems highlights several important issues such as reliability, security, stability, power quality, ... etc. Among these issues, providing wind farms with the proper protection is quite essential. The essential benefits from the dedicated protection functions are to avoid the possible local damage resulting from incident faults and minimize the impact of these abnormal conditions on other sound parts of the network. This consequently enhances the reliability and dependability of the overall grid performance. These terms such as continuity, reliability ...etc. have recently received much attention due to the new de-regulation policies and marketing liberalization. On the other hand, wind farms are characterized with some unique features during their normal and faulty operating conditions. Different factors participate usually into these conditions such as the distributed generation concept, the own behavior of the induction generator, varying wind speed, ... etc. Moreover, the economic perspective plays a major role as well. This consequently highlights different challenges regarding the behavior of their protection and control schemes. The primary objective of this chapter covers two basic

goals. The first goal concerns with describing the outlines of the basic wind farm protection systems that are usually utilized with modern wind farms nowadays. The basic protection zones including the wind farm area, wind farm collection system, wind farm interconnection system and the utility area are described. For each zone, the utilized protective elements are described. Their performance will be fully explored, since these relaying elements for wind farms, in particular, are sometimes characterized with different performance as compared with other ones used conventionally in power systems. The second goal is to emphasize the protection challenges concerning with the utilized protective elements that are usually employed with wind farms.
The outline of this chapter is as follows. The next section presents an overview of the occurred electrical faults in wind farms including their types, natures and statistical analysis. The configuration of the conventional protection systems for wind farms is outlined in section 3 covering the generators, local transformers, wiring circuitry and coupling point. Section 4 highlights the basic challenges of utilized protective elements in wind farms. Some different simulated illustrative examples are presented in section 5 based on a detailed modelling of Al-Zafarana wind farm in Egypt using both SFIG and DFIG generators. Recent trends for improving the protection performance are covered in section 6. Finally the resulted conclusions and the relevant references are provided in sections 7 and 8 respectively.

2. An Overview of Electrical Faults in Wind Farm Systems

The economic perspective plays a major role, in which the enormous cost pressures usually coerce the wind farm designers for economic causes to remarkably reduce the utilized protection schemes. As reported by Bauscke et al. in [4], different levels of damage were recorded resulting occasionally from the drawbacks of the associated protection system. As a result, wind farm providers still utilize simple and none-integrated protection methodologies [5], [6]. Different sources of failures are experienced for wind farms resulting in unwanted disconnections of some wind turbines for relatively large times for locating and maintaining these failures. These causes and their influences were statistically analyzed by Ribrant and Birtling [7] for the Swedish wind power plants as compared with their corresponding analysis in Germany and Finland respectively [8], [9]. As reported from Fig. 3(a), the failures in the electrical systems represent the largest share as compared with other types. The associated downtimes for these failures are ranked as shown in Fig. 3(b). This reveals the expected dangers of such electrical faults and consequently raises the importance of their relevant protective schemes. As concluded from Table 1, 2.38 failures per turbine were recorded yearly in Germany resulting in 62.6 hours of downtime per failure yearly as well. Moreover, those electrical system failures are ranked as the dominant failure causes. With the increasing of the of wind power penetration into power systems, these remarkable downtimes are not acceptable due to their influence on the overall system stability and continuity. This obviously raises the importance of providing wind power plants with the most appropriate protective schemes against such electrical abnormalities.

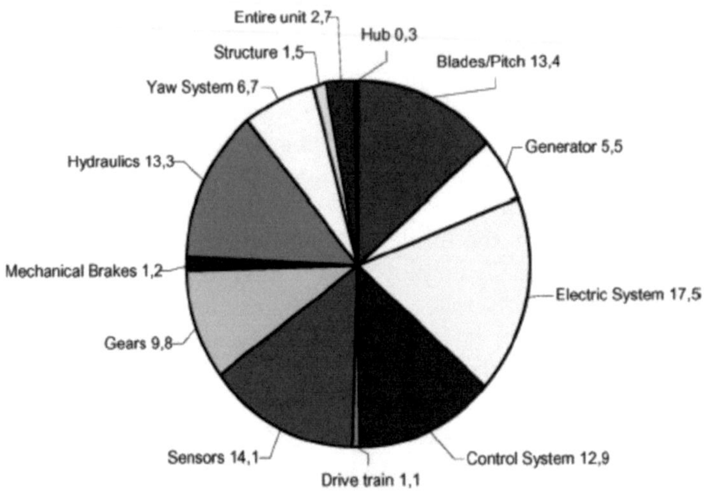

(a)

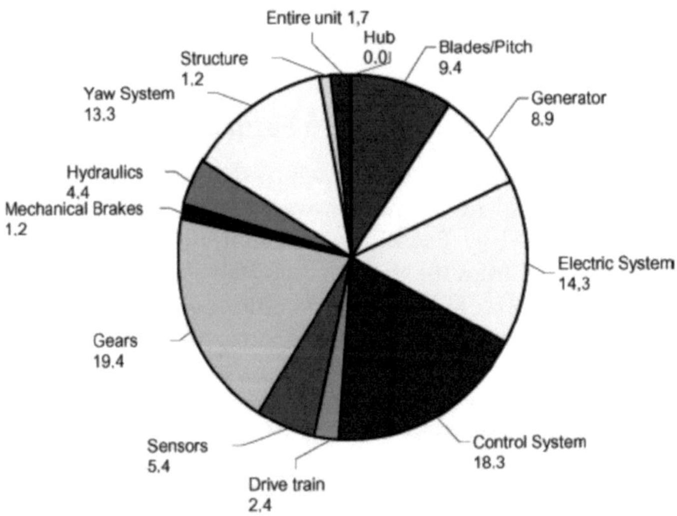

(b)

Fig. 3. Distribution of failures in the Swedish wind power plants (1997-2005) [7].
(a) Distribution according to the failure type
(b) Distribution according to the related downtimes

Country	Sweden	Finland	Germany
Average number of failures per turbine	0.402 times/year	1.38 times/year	2.38 times/year
Average downtime per failure	170 hours/year	172 hours/year	62.2 hours/year
Most failure causes	Electrical system	Hydraulics	Electrical systems
Longest downtime per failure	Drive train	Gears	Generators

Table 1. Statistical summary for the wind plants in Sweden, Finland and Germany.

Since, the increasingly penetration of wind energy into conventional power systems, the availability of the wind turbine, and reliability of the corresponding wind energy conversion systems should be increased. This spawned active research to carry out several investigations of failure statistics in the real field since the study of wind power turbine statistics gives knowledge of reliability performance. The investigation of the statistical studies is very essential for knowledge of the most frequently failures that may aid in the system maintenance planning and reconsidered the used protection schemes.

3. Common protective schemes for wind farms

3.1 Basic Protection Functions

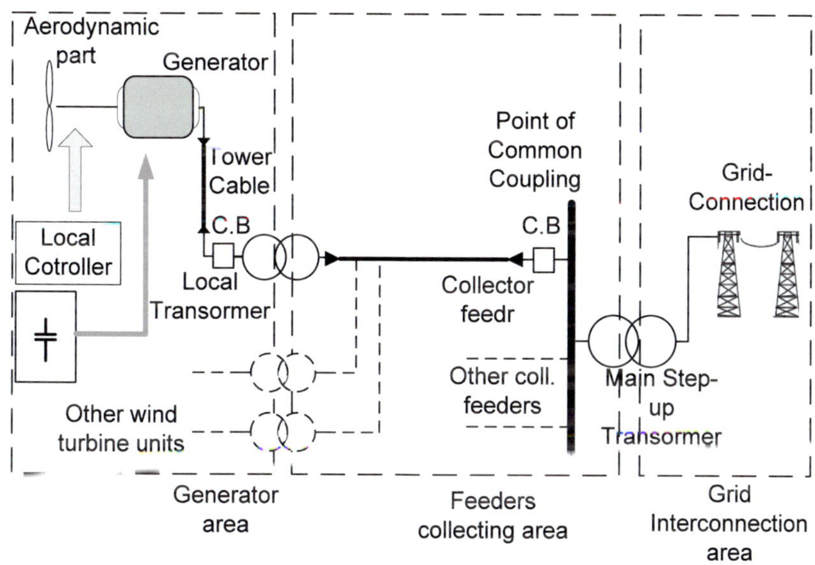

Fig. 4. Typical wind farm construction with its protection zones.

Fig. 4 shows a schematic diagram of a typical wind farm consisting of (n) units of wind turbines. Nowadays, modern wind farms include 20 to 150 units with typical size from 0.5 MW to 1.8 MW wind turbine generators. Larger sizes up to 3 to 5 MW are recently available in the market, in which they were successfully installed in some European countries. The typical generator's terminal voltage may range from 575 to 690 V with frequency of 60 Hz. The generator terminal voltage is stepped up to the Collector Bus system with typical voltage of 34.5 kV. The step up transformer is an oil cooled, pad mounted located at the base of the wind turbine unit. Sometimes, the step up transformer is mounted in the turbine nacelle. Certain considerations should be applied for avoiding the harmonic effects. The typical wind farm collector system consists of a 34.5 distribution substation collecting the output of the distributed wind turbine generators through the incoming feeders. Usually some reactive power compensation units are provided by a collection of switched capacitors. Finally, the collected power is transferred to the utility side via an interconnection step up transformer [6], [10].

The common type of the wind turbine generators that are commercially available nowadays are induction generator (IG), wound rotor synchronous generator (WRSG), and permanent magnet synchronous generator (PMSG). Due the uncontrollable natural characteristic of wind speed, the induction generators IGs are suitable for driving the wind turbines. The two basic types of wind turbines used nowadays are fixed-speed wind turbine (FSWT) that equipped with squirrel cage IGs and variable-speed wind turbines (VSWT) equipped with doubly fed IG (DFIG). Squirrel-cage induction generators work normally within a limited wind speed range, which is one of their main drawbacks in comparison with variable-speed ones. Variable-speed wind turbines are mainly equipped with a doubly-fed induction generators (DFIG) with variable frequency excitation of the rotor circuit. The stator windings are connected directly to the AC grid whereas the rotor windings are coupled through a partial scale back-to-back converter as In Fig. 9. The main advantage of DFIG wind turbines is their ability to supply power at a constant voltage and frequency while variations of the rotor speed. The concept of DFIG for variable-speed wind turbine provides the possibility of controlling the active and reactive power, which is a significant advantage as regards grid integration.

The wind farm protection system is usually divided into different protection zones including the wind farm area, wind farm collection system, wind farm interconnection system and the utility area. First, the induction generator protection is typically accomplished via the generator controlling system covering some certain protection functions such as under/over voltage, under/over frequency, and generator winding temperature (RTDs). Whereas, the generator control system does not contribute for the interconnecting system or the utility zone. The generator step up transformer is usually protected with primary fuses. For those cases when the transformer is mounted in the nacelle, a circuit breaker is integrated with dedicated phase and ground time-overcurrent relays. The collector feeder protection is simplified considering it as a radial distribution feeder using overcurrent protection (50/51). A basic challenge arises due to the distributed generators connected together to the radial feeder in determining the minimum faulty zone. That is in order to keep the remaining sound parts of the farm supplying the power. On the other hand, the protection of the wind farm substation collector bus and main power transformer consists of multi-function numerical relay system including main transformer differential relay, transformer backup overcurrent relay, collector bus differential relay and breaker failure relay. Further details are available in the literatures. Considering the utility area, different protection functions may be used according to the voltage level and the considered protection topology. Direct transfer trip scheme, line differential relay, pilot protection, zones distance relaying, over/under voltage protection, over/under frequency protection, breaker failure protection, synchronous checking and backup overcurrent protection can be used [6], [10]. Taking into consideration that, the wind farm interconnection would be applied to MV distribution network or HV system, the coordination of utility relays and the wind farm will be therefore quite different. Communication system with dedicated SCADA is quite important for wind farm operation. Nowadays, the data from each wind generator control is transmitted via optic cables and spread the substations for general control and monitoring purposes. This provides an ideal situation for providing them with an integrated monitoring and protection system.

3.2 Rotor protection system

Historically grid codes allowed the wind turbines to be disconnected instantaneously with voltage sag below .8 per unit. In 2003, E.ON and VET (Germany) introduced the first FRT code requirements [11]. Later, other international wind energy associations introduced their similar codes as well. Generally speaking, the grid codes required that grid connected wind turbines should withstand with voltage dips on any or all phases in the transmission system as long as the voltage measured at the high-voltage terminals of the grid-connected transformer, or in other words at the common coupling point (CCP), remains above the predetermined level of the grid code [12]-[14]. Different benefits are expected to be gained wit FRT capabilities including enhancing the system stability and fast restoration of system service if the fault is cleared during the allowable time. These capabilities can be achieved by an adapted control strategy.

The crowbar comprises of some certain thyristors that short-circuit the rotor winding and hence thereby limit the rotor voltage and provide an additional path for the fault current. When a disturbance is introduced, high currents are induced into the rotor circuitry from the stator side affecting the dc-link voltage as well. Then, the dc-link over-voltage protection will stop the rotor converter/inverter unit; meanwhile it turns on the crowbar control thyristor. Similarly, the crowbar can be triggered based on the occurring overcurrent through the rotor circuitry. The rotor is now connected to the crowbar and remains connected until the main circuit breaker disconnects the stator from the grid [13], [14]. After clearance of the fault the generator can be line-synchronized again and started in a normal operation mode.

The core of the crowbar operation was described by Akhmatov, Xiang , Holdsworth, Ekanyaki and Niiranen as reported in. Technically, two types of crowbar systems are known including passive and active ones. For passive ones, the crowbar consists of a diode bridge that rectifies the rotor phase currents and a single thyristor in series with a resistor Rcrow. The thyristor is turned on when the DC link voltage Udc reaches its maximum value or the rotor current reaches its limit value. Simultaneously, the rotor of the DFIG is disconnected from the rotor-side frequency converter and connected to the crowbar. The rotor remains connected to the crowbar until the main circuit breaker disconnects the stator from the network. When the grid fault is cleared, the rotor-side converter is restarted, and after synchronization, the stator of the DFIG is connected to the network [15]-[18].

In contrast to a conventional passive crowbar, the active crowbar is fully controllable by means of a semiconductor switch. This type of crowbar is able to cut the short-circuit rotor current whenever needed and thus the DFIG wind turbine is able to ride through a network disturbance. If either the rotor current or dc link voltage levels exceed their limits, the IGBTs of the rotor-side inverter are blocked and the active crowbar is turned on. The crowbar resistor voltage and dc link voltage are monitored during the operation of the crowbar. When both these voltages are low enough, the crowbar is turned off. After a short delay for the decay of the rotor currents, the rotor-side inverter is restarted and the reactive power is ramped up in order to support the grid.

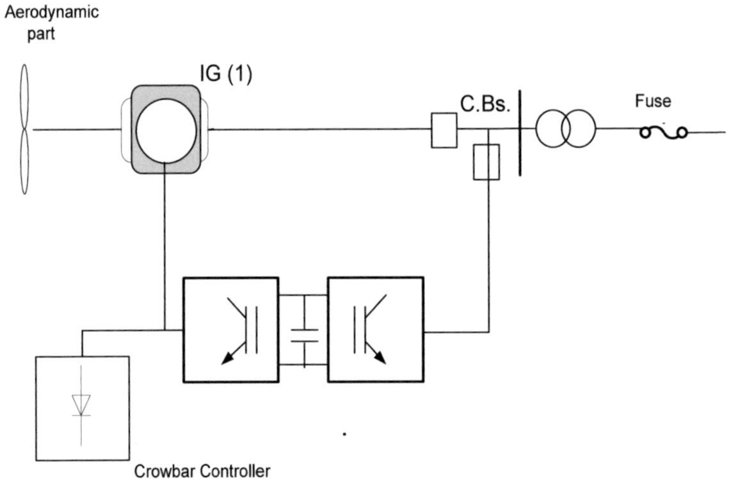

Fig. 5. Crowbar protection system for DFIG units

Practically, crowbar mechanisms raise some problems. The crowbar ignition leads to the loss of the generator controllability through the machine side converter (MSC), since the machine rotor is short-circuited through the crowbar resistors and the MSC is blocked. During this time slot the, generator acts as a common single fed induction generator and consumes reactive power, which is not desirable. Hence, utilizing crowbar mechanisms has recently replaced with employing DC chopper used to limit the DC voltage by short-circuiting the DC circuit through the chopper resistors. This was demonstrated in Fig. 6. Further information is available in the literatures [11].

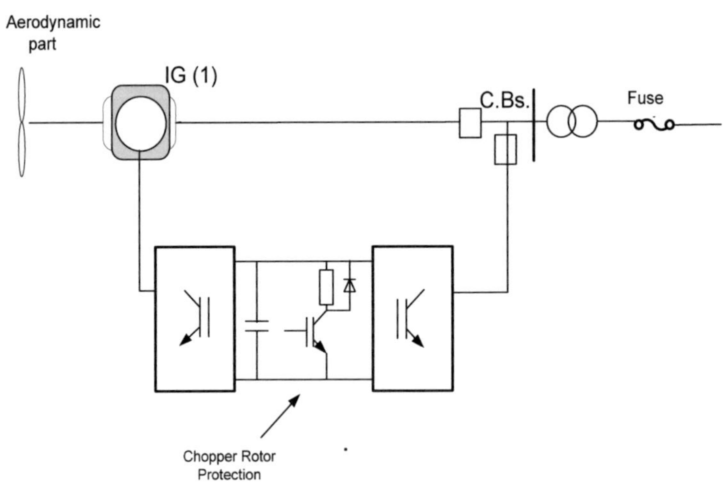

Fig. 6. Chopper rotor protection system for DFIG units

3.3 Wind turbine controller role

Typically, each turbine set is equipped with a multi-function numerical integrated controller unit. This controller provides different control actions regarding concerning with the generator system, power factor correction, yaw operation, hydraulic system, pitch mechanism, ... etc. In addition to its control functions, it provides different protection functions such as,

- Over/Under voltage protective relays (27/59).
- Over/Under frequency protective relays (81O/U).
- Monitoring the electrical parameters (currents, voltages, power ... etc.) and disturbance recorders.
- Communication with the main control system
- Power flow control

The aforementioned controller unit interacts with the utilized protective elements with the generator set for providing the generator with the appropriate protection as possible. According to the manufacturer designs (or operators experience), the allowable setting for these protective elements are set. As example, some manufactures recommend to instantaneously trip the generator set for a 25% overvoltage conditions and with a 0.1 second for a 20% conditions. Also, a 0.2 second for tripping is recommended for over/under frequency conditions with ±2 Hz frequency deviation and instantaneous tripping for ±3 Hz frequency deviation or above.

3.4 Monitoring, Command, and Control Systems

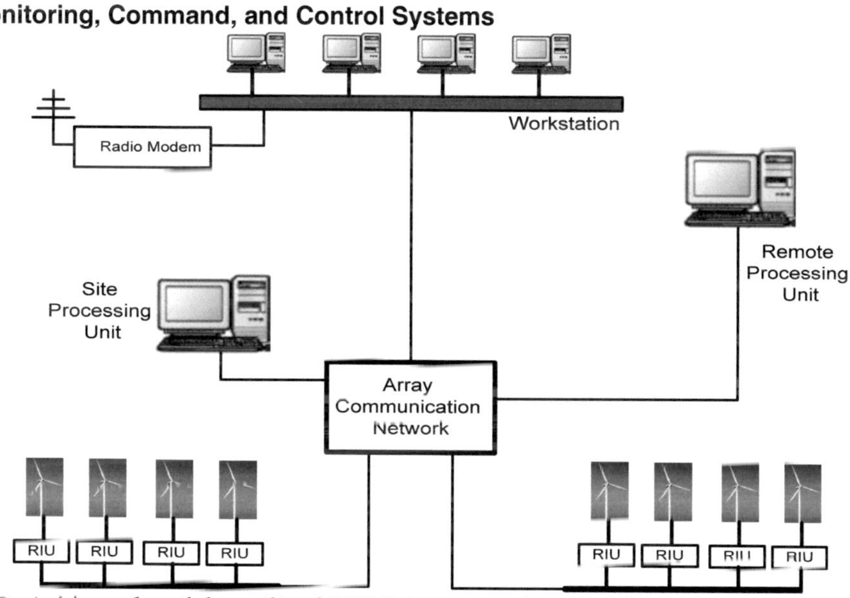

Fig. 7. Basic hierarchy of the utilized SCADA systems for wind farms.

Fig. 7 illustrates the basic outline of the utilized SCADA system that usually utilized with wind farms. Actually, the configuration of the communications system, along with the SCADA interfaces for large wind farm projects is complex depending on different factors including the utilized numerical protection and control devices hierarchies. The data from each of the wind turbine control systems are collected with its local Remote Interface Unit (RIU) and send over a fiber optic cable installed in the same duct bank of the medium voltage collector feeder. The data is then transmitted to the main workstation for monitoring and control purposes [6]. The main workstation facilitates to monitor all environmental and electrical information at each turbine set and to control their operation using either automatic or manual process. Further spreading of the wind farm data is possible using radio or other communication channels. More sophisticated situations arise for off-shore farms in particular.

3.5 Protection of the local step-up transformer

The local step-up transformer at each generator set is usually connected with Delta-Grounded Wye (or Grounded Wye-Grounded Wye) according to generator manufacturer's requirement. Current Limiting Fuses (CLFs) are commonly employed for protecting such transformer and its corresponding sub-connecting feeder for each induction generator set. This is mainly due to its reliability, simplicity and low cost. This economic perspective, on the other hand, is quite essential for wind farms, in particular, due to the massive numbers of such power system elements in large wind farms.

3.6 Collector feeder protection

The collector feeder is simply considered as a radial distribution. It is usually protected using overcurrent protection (50/51) element. A basic challenge arises due to the distributed generators connected together to the radial feeder in determining the minimum faulty zone. That is in order to keep the remaining sound parts of the farm supplying the power. The interconnection transformer has usually a 3-winding configuration with a grounded Wye - grounded Wye connection. The tertiary winding is connected Delta. The Delta Tertiary is used to stabilize the neutrals of the transformer and provide zero sequence current to ground fault on both sides of the main transformer bank. If a Delta-Wye transformer is used for the main interconnecting transformer, a grounding transformer may be installed on the Delta side of the transformer to provide stabilization of the transformer primary neutral. The protection for the wind farm distribution substation consists of multifunction numerical systems including a main transformer differential relay, transformer Time-Overcurrent relay for the main transformer back-up protection, collector bus differential relay, distribution Time-Overcurrent relay, and breaker failure protection. It should be considered that, the wind farm interconnection would be applied to MV distribution network, HV system ... etc. Therefore, the coordination of utility relays and the wind farm will be quite different [19].

4. Basic protection challenges

The essential benefits from the dedicated protection functions are to avoid the possible local damage resulting from incident faults and minimize the impact of these abnormal conditions on the other sound parts of the network. It enhances consequently the reliability and dependability of the overall grid performance. In spite of the obvious importance of the electrical protection of wind power plants, it surprisingly has not garnered a sufficient attention tell present. The economic perspective plays a major role, in which the enormous cost pressures usually coerce the wind farm designers for economic causes to remarkably reduce the utilized protection schemes. Different viewpoints arise as the causes for these problems as summarized below.

4.1 Distribution system topology

The distribution connectivity nature, where infeed points are tapped sequentially from the same main feeder, may strongly influence the profile of the occurring fault. This is mainly due to the simultaneous feeding of the fault even with other unfaulted units.

4.2 Protection system configuration

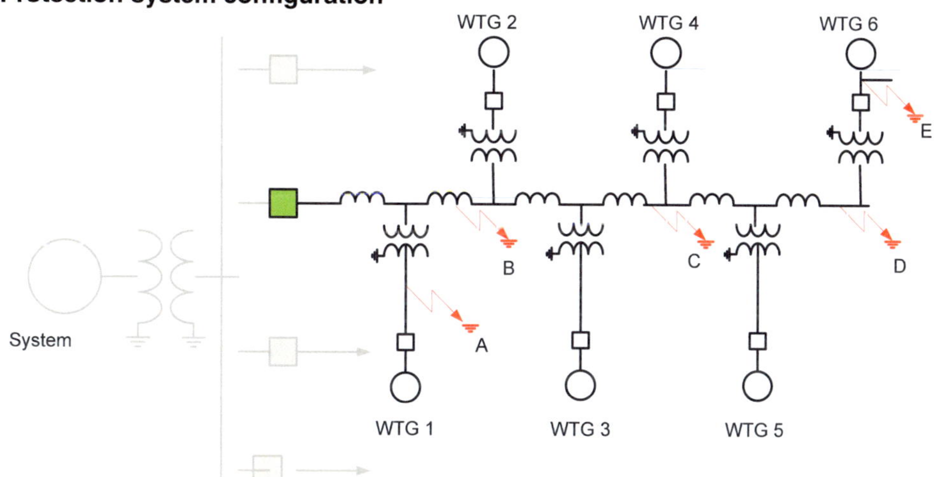

Fig. 8. Fault location effects on the protection of the collecting feeder

The economic factor and the distribution topology of the associated electrical networks result in utilizing simple protection methodologies with the local breaker/fuse combination for each generator-transformer unit. Then discriminating and locating those faults occurring at positions A, B, C, D and E (shown in Fig. 8) are usually lost. This results in disconnecting the whole collecting feeder for such faults.

4.3 Distributed generation effects

Since the collector feeder can be considered as a radial distribution feeder connecting several wind generation units, it is expected to raise those similar protection problems of common distributed generations. Relaying mis-coordination and mal-operation are the most common ones.

4.4 Control system requirements

Due to the impacts of the interconnection with the main power grid, some certain requirements such as "Fault Ride Through" are utilized to keep the system stability. The interaction among these operation modes and the employed relaying schemes should be considered to realize the proper and well coordinated protection performance.

4.5 Dynamic behavior of the induction generator.

As known, induction machines have their own dynamic performance as compared with conventional synchronous ones. Moreover, the continuous wind speed variations and the interaction of the associated power electronics (for DFIG ones) collaborate together for providing the behavior of these machines during fault periods. More sophisticated and well coordinated relaying schemes should be provided to realize the most appropriate protection methodology for wind farm elements.

Insufficient protective elements, non-integrated control scenarios and improper coordination among protective and control strategies may lead to serious problems for large grid-connected wind farms. As example for these problems is the accident happening in North Germany on November 4, 2006. The UCTA interconnected grid was affected by a serious incident originating from the North German transmission grid that led to power supply disruption for more than 15 million households, splitting a synchronously connected network into three islands (two under-frequency and one over-frequency). After cascading overloads and tripping, two of three large separated systems (Western Island and North-Eastern Island) ended up with a significant amount of wind generation resources. Western Island (under-frequency state): During the incident, about 40% of the wind power units tripped. Moreover, 60% of the wind power units connected to the grid tripped just after the frequency drop (4,142 MW). Wind power units were automatically reconnected to the grid when the conditions of voltage and frequency were in the accepted range. North-Eastern Island (over-frequency state): Significant imbalance in this subsystem caused rapid frequency increase and triggered the necessary primary, standard and emergency control actions of tripping wind generation units sensitive to high frequency values. Tripping these units (estimated value of 6,200 MW) helped to decrease the frequency value during the first few seconds of disturbance. Further information for this event was available in [20]. The following section demonstrates different simulation examples for a typical 305 MW in Al-Zafarana, Egypt aiming to visualize the basic challenges of protective relays utilized for wind farms.

5. Simulated illustrative examples

5.1 Modeling of Al-Zaafarana wind farm

A 305 MW wind farm was recently established in Al- Zaafarana (220 south east of Cairo, Egypt) and connected to the 220 kV Egyptian grid. This promising area (shown in Fig. 9) is distinctive with different superior features such as an average annual wind speed of 9.5 m/s, and its excellent geographical and environmental features. The farm was structured through six stages of 30, 33, 30, 47, 85 and 80 MW respectively. Except the latter two stages, other ones are with fixed speed and variable pitch operation. Currently, two further stages are being constructed adding extra 240 MW to the farm. The fourth and fifth stages were selected for simulation purposes representing typical examples for fixed and variable speed operations respectively. The fourth stage consists of 71 wind turbines (with a 660KW squirrel cage induction generator for each turbine) providing a total power of 47MW. The fifth stage consists of 100 wind turbines (with a 850 KW DFIG units for each turbine) providing a total power of 85MW. Fig. 10 (a) and (b) demonstrate the distribution of turbine units with 4 and seven collecting feeders for both stages respectively. Each wind turbine is connected to a 690V/22 KV local step-up transformer and then integrated with the grid through 22/220 kV step-up transformers [21].

Fig. 9. Al-Zafarana wind farm location

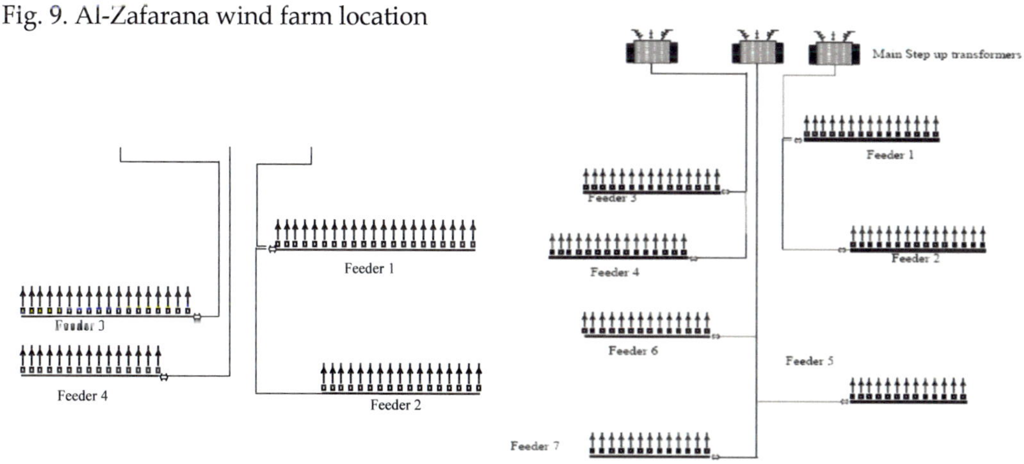

(a) (b)

Fig. 10. Schematic of the fourth and fifth stages of Al-Zaafarana wind farm.

(a) Fourth stage with 47 MW
(b) Fifth stage with 85 MW

Fig. 11 (a) and (b) show the detailed schematic of each wind turbine unit constructed with the wind turbine and asynchronous machine blocks in MATLAB for both SFIG and DFIG respectively. Dedicated pitch control and converter/inverter blocks are implemented as shown in the aforementioned figures. The nominal wind speed was assigned to 9.5 m/sec "the annual average wind speed in its corresponding location", whereas the "cut-in" wind speed was assigned to be 4.5 m/sec.

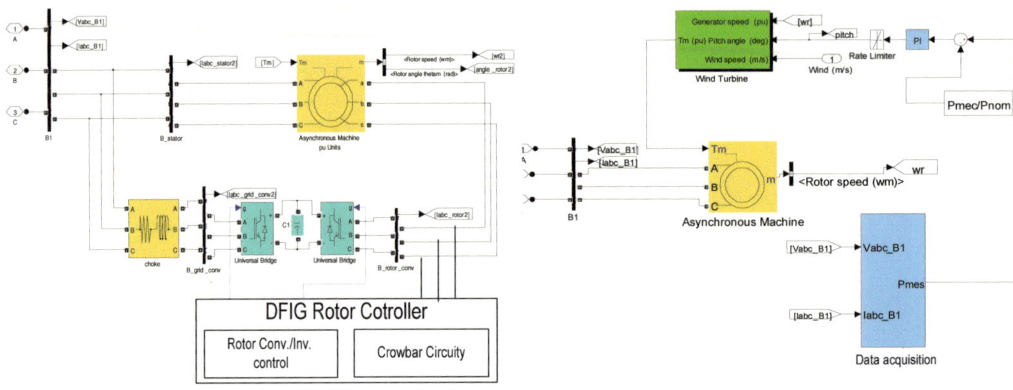

(a) (b)

Fig. 11. Simulink-based diagram of Al-Zaafarana wind farm Generator modeling
(a) DFIG machines
(b) SFIG machines

The relatively large number of wind turbine units, in which each of them was constructed with different individual items "Turbine, generator, local transformer, feeding cable, …" raises the obvious need for employing a reduced modeling for the selected stages. On the other hand, reduced model should be conditioned with the following restrictions:

- Model Accuracy for each individual power system element should be kept in its higher level.
- The essential concepts for distributed generation must be satisfied.
- Equivalence of currents for each individual unit as well as overall farm currents for both detailed and reduced model should be realized.
- Equivalence of the generated power for each individual unit as well as for the overall farm for both detailed and reduced model should be realized.
- Total power losses (due to connecting cables) should be considered.

For each stage, only one collecting feeder was constructed using detailed modeling for three units (the first, second and last units), whereas other units in the same feeder were lumped by a single equivalent unit. Other collecting feeders were also represented by their total equivalent power as well. For those lumped units, cable lengths were considered for keeping the total power losses equal to those resulted with the corresponding detailed model. The response of the reduced model was validated compared with the corresponding detailed one via different simulation examples for both faulty and non-faulty operating conditions. Details for the proposed modeling methodology were fully addressed in [22].
For either simulated wind farms stages, the behavior of both SFIGs and DFIGs were thoroughly investigated under various faulty and non-faulty operating conditions. The prepared simulation cases covered a wide variety of operating conditions including fault type, fault location, fault resistance and wind speed variations. These fault cases were prepared using the developed reduced model for both stages at different positions. For each

test case, three phase voltages and currents were recorded at various locations. This facilitated to explore the overall performance of the wind farm properly.

5.1 Performance of SFIG units

5.1.1 Ground faults

Ground fault is generally the most common fault type in electrical networks, whereas its behavior depends mainly on the fault position, soil resistivity, fault resistance and the applied grounding methodology. For a solid A-G fault at the generator terminals, the currents and the voltages at the generator terminals are illustrated in Fig. 12. No sensible fault current was remarked as a result of the ungrounded stator winding. The resulting overvoltage permitted the local controller to open the local C.B. within 100 ms.

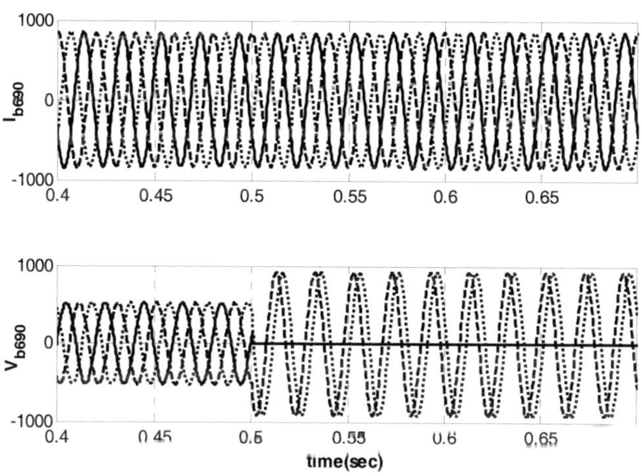

Fig. 12. Response to A-G fault at the generator terminals.

On the other hand, repeating the solid A-G fault before the fuse yielded the shown fault currents in Fig. 13 (a) and (b) fed from the associated local generator and other generating units (in addition to the main grid) respectively. Surprisingly, the fault current fed from the local generator was not sufficient to permit tripping of its local breaker (CB1) as remarked from Fig. 13 (a). On the other hand, the accumulated fault current from both the other generating units and the grid network is sufficient to melt the local fusing element as shown remarked from Fig. 13 (b). More complex situations were visualized with non-solid ground faults resulting from the occurred lower fault currents even with small fault resistance values. Also, repeating the fault before the local generator breaker (along the tower cable) is a challenge as well. Then, the need for more advanced protecting schemes for detecting such faults as well as for minimizing the tripped generation units is obvious.

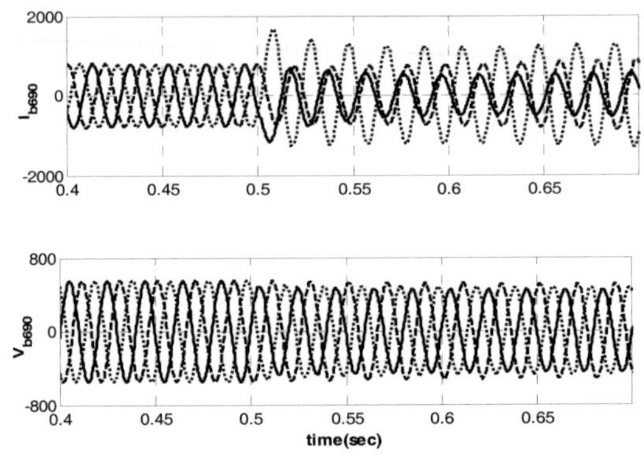

(a)

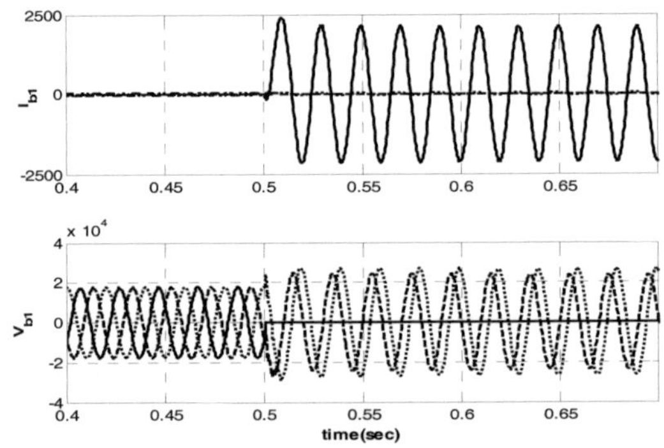

(b)

Fig. 13. Response to A-G fault before the fusing element.
(a) Voltages and currents at generator terminals.
(b) Voltages and currents at the medium voltage side

5.1.2 Phase faults

For A-B fault at the generator terminals, the resulting fault current exceeded the pre-determined current setting for the associated generator breaker as well as the fuse of the faulted unit as described in Fig. 14 (a) and (b) for both fault feeding currents. The fault was accordingly tripped from both sides. Similarly to ground fault conditions, other connected turbines to the same collecting feeder participated also in feeding the fault. These units fed the fault individually with almost the same fault current level. However, their local fuses were not permitted to trip their branches. This was resulted from the obviously larger fault current passing through the own fusing element of the faulty unit (summation of other fault feeding currents), which accelerated its tripping action. Similar behavior was obtained with repeating the same fault condition (A-B fault) before the fuse element. It resulted from exceeding both counterparts of the fault current the setting boundaries of the associated breaker and the fuse.

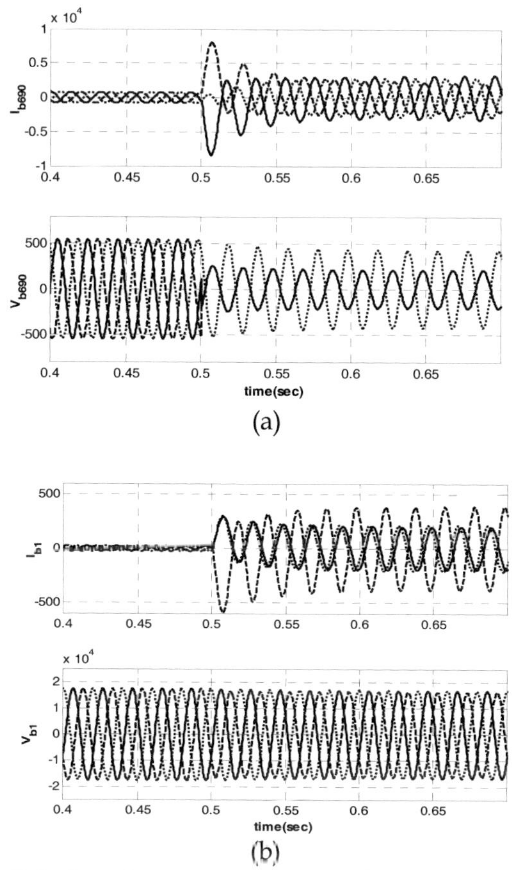

Fig. 14. Response to A-B fault at the generator terminals.
(a) Voltages and currents at the generator terminal.
(b) Voltages and currents at the medium voltage side.

With repeating the same fault condition after the fusing element, each participated wind generators fed almost the same fault current through its corresponding fuse. This resulted in completely losing the overall collecting feeder rather than tripping the faulty branch only. This typical distributed generation figure still represents a challenge for the utilized conventional protection elements. For a three phase fault on the generator terminals, Fig. 15 illustrated the response of the faulty unit demonstrating the associated voltages, currents and generated power. Fortunately, the three phase voltage and current quantities at the generator terminals were rapidly decreased to zero. The local controller of the associated generator disconnected its local breaker successfully due to the occurred undervoltage condition. Repeating the same fault before the fusing element yielded similar voltage and current profiles for the corresponding generator. Fortunately, the large fault current feeding from other generation units in addition to the grid network exceeded the fuse setting. On the other hand, other generating units sharing the same step up transformer had similar voltage and current profiles. This resulted in disconnecting these units by their undervoltage control, if the fuse associated with the faulty unit failed to operate.

Unfortunately, repeating the same fault condition after the fuse element resulted in disconnecting the whole collecting feeder. Also, other collecting feeders sharing the same step-up transformer had similar situation.

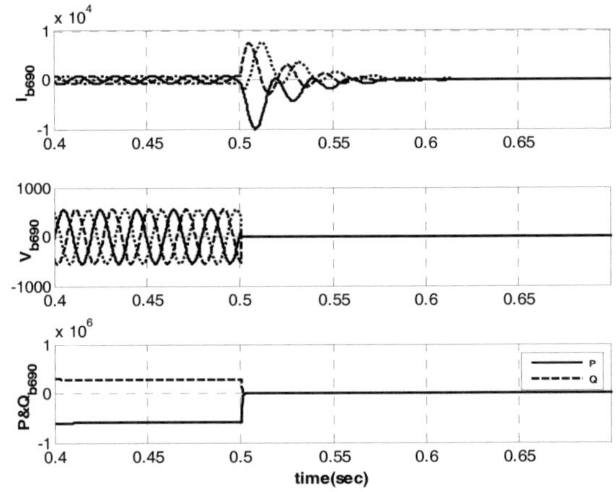

Fig. 15. Three phase fault at the generator terminals

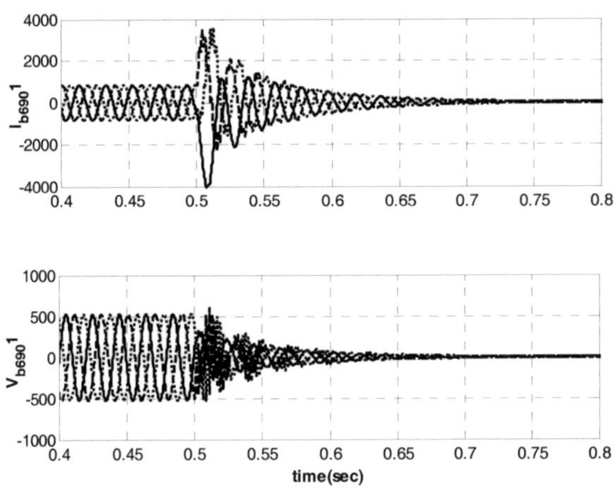

Fig. 16. Unit response to a three phase fault 5 km apart from the wind farm

5.1.3 Grid faults

In order to investigate the impact of network faults, a three phase fault was applied 5 km apart from the grid connection bus. The voltages and currents of all wind generation units were rapidly decreased to zero as shown in Fig. 16. Similarly, grid phase-phase faults resulted into under-voltage situations which may result in disconnecting all units as well. This of the generating units may strongly affect the system stability, particularly with large wind farms. Further details are available in [23], [24].

5.2 Performance evaluation of DFIG units

Depending on the developed model of the selected DFIG stage, the behavior of the modeled DFIG stage in conjunction with the related FRT mechanism was thoroughly investigated under various faulty and non-faulty operating conditions. For each case, voltage and current quantities for both stator and rotor circuitries were recorded as described in the following sub-sections.

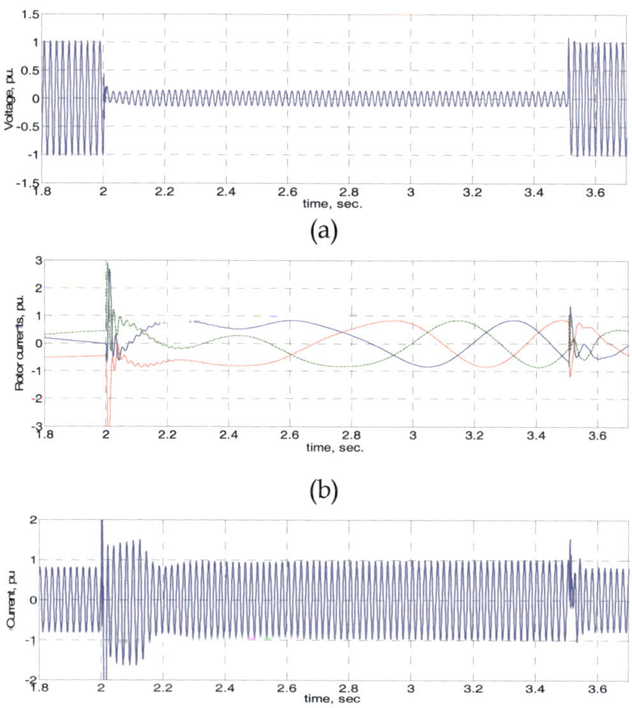

(c)

Fig. 17. Simulation response due to a solid 3-phase grid-fault without crowbar initialization.
(a) Stator phase voltage, pu.
(b) Rotor phase currents, pu.
(c) Stator phase current, pu.

5.2.1 Grid faults

During grid faults, the occurred faults resulted in a sufficient drop of phase voltage so that the associated crowbar mechanism was initiated to protect the rotor windings from the excessive fault current. As illustrated in Fig. 17, both rotor and stator windings suffered from the increased currents resulted from a solid three phase grid fault occurring beyond the main collecting step up transformer. The corresponding crowbar scheme was inhibited during this test case. However, the occurred current levels were not sufficient for initiating the associated fuses or local breakers at each generating unit. Utilizing the crowbar scheme resulted in rapidly decreasing the rotor currents to zero as described in Fig. 18. As soon as the crowbar scheme was initiated as the machine reacted exactly as a SFIG one. Hence, the stator currents we decreased to zero as remarked in Fig. 18 (b). Consequently the local protection at each generator set (fuses and local breakers) was blocked.

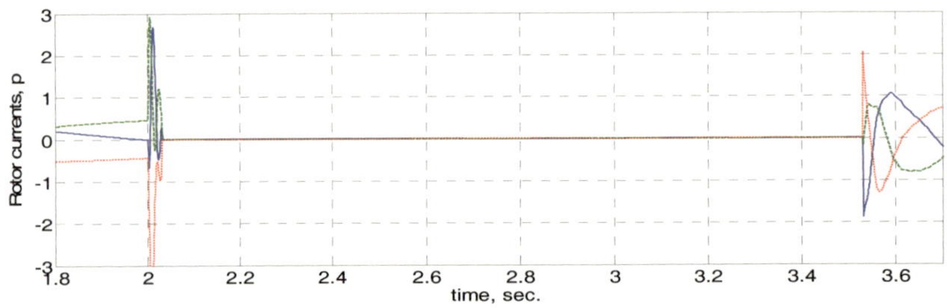

(a)

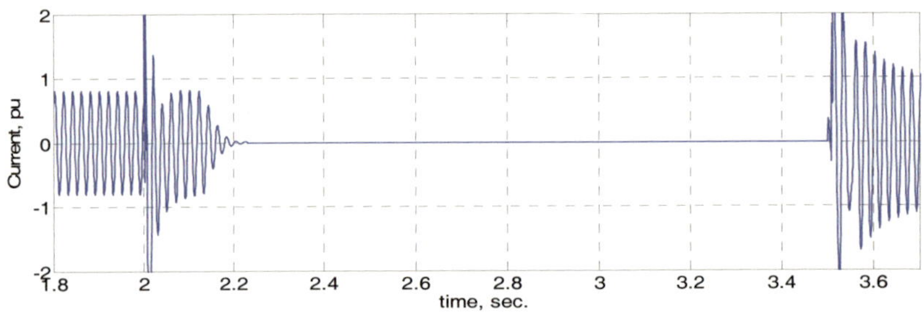

(b)

Fig. 18. Simulation response due to a solid 3-phase grid-fault with crowbar initialization.
(a) Rotor phase current, pu.
(b) Stator phase current, pu.

5.2.2 Wind farm faults

The DFIG response for a 2-phase solid fault beyond the local step-up transformer and the fusing element was investigated in Fig. 19. As remarked from the results, the occurred voltage drop initiated the crowbar mechanism. As noted from Fig. 19(c), the resulted stator fault current was not enough to initialize the utilized fuse element. At non-solid faults usually the fault current decreases due to an increased fault resistance. These faults should be considered for evaluating the behavior of the DFIG machines equipped with FRT mechanisms. When a fault resistance is inserted into the fault current path, the decrease of the fault current is accomplished with a decrease of the occurring voltage drop at the generator terminals. Consequently, the FRT mechanism may incorrectly be initiated for faults occurring inside the wind farm. This results in inhibiting the operation of the related overcurrent protection due to the reduced fault current. This resulted in inhibiting the operation of the associated fusing element. These situations of network faults were demonstrated well in [22].

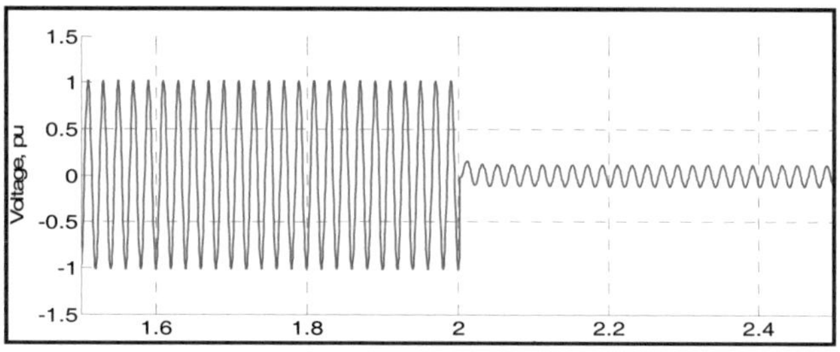

time, sec

(a)

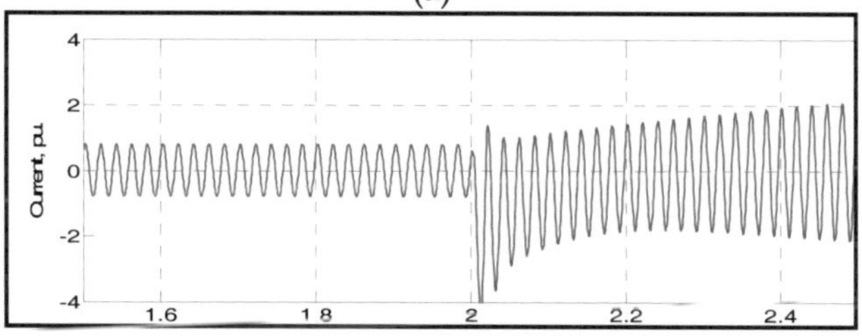

time, sec

(b)

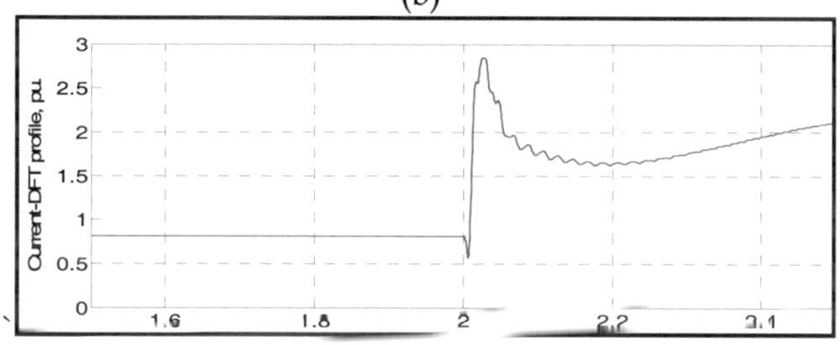

time, sec

(c)

Fig. 19. Simulation response due to a solid 2-phase fault beyond the local transformer.
(a) Stator phase voltage, pu.
(b) Stator phase current, pu.
(c) Stator phase current peak profile with DFT

6 Trends for improving wind farm protection systems

Due to the own behavior of induction generators as well as the specific topology of distributed generation concept for wind farms, employing CLFs for protecting the local transformer for each generating set was characterized with some certain shortcomings. Three different suggestions are proposed to eliminate the aforementioned protection problems described as follows.

6.1 Redesigning the distribution network

Different configurations of connecting wind turbine units were typically employed including Radial design, Single sided-ring design and Double sided-ring design, ... etc as described in the literatures. All of them are characterized with conventional distributed generation profile. This consequently leads those un-faulty wind generators sharing the same collecting feeder to participate with the faulty unit for supplying the fault current. Larger portions of wind turbine units may be redundantly disconnected. Redesigning this wiring connectivity may eliminate the associated problems with the conventional distribution network topology.

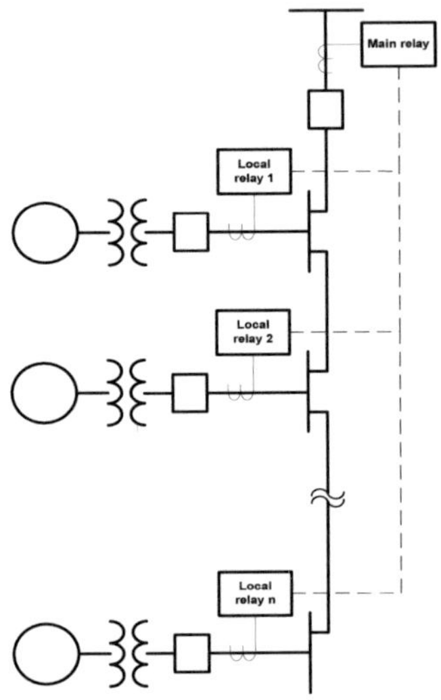

Fig. 20. Proposed communication-based relaying employment

6.2 Employments more protective elements

Utilizing more protective elements may represent solutions for all highlighted problems. As example, directional elements may eliminate the effects of distributed generator effects remarkably. Compromising between the economic prospective and the aimed performance is essential for realizing a practical and acceptable behavior.

6.3 Integrated protection with Enhanced communication employment

The proper employment of communication facilities in conjunction with the more sophisticated and integrated protective schemes may impressively eliminate the aforementioned problems. As described in Fig. 20, each of the local relays at each generator unit coordinates its response simultaneously with other generating units as well as wit the main protective element at the collecting bus. Fault location computation as well as the final tripping decision are then decided upon the status of these received signals. Depending on the existed communication links utilized with the used SCADA system of existed wind farms facilitates these steps.

7. Conclusions

This chapter emphasized the basic outline of the common configuration of protective relays that are usually utilized with modern wind energy conversion systems. Electrical faults occurring into the different zones of wind farms were described. Accordingly, different problems arise with the simple and non-integrated protection schemes that are usually utilized with wind farms. The associated challenges of those protective elements were discussed and their relevant problems were visualized. Among these problems, unwanted disconnection of wind generation units, rather than disconnecting the faulty unit only, is not acceptable. This negatively impacts the continuity and the stability of the overall system. Some simulation examples were presented for demonstration purposes. These simulations were developed based on a real 305MW wind farm in Alzafarana-Egypt including both SFIG and DFIG configurations. For DFIGs, in particular, utilizing crowbar mechanisms may affect the behavior of conventional overcurrent protection elements against network faults occurring into the local connecting circuitry of the wind farm. The study emphasized the need for enhancing the existed protection schemes for wind farms to realize better power system performance as well as minimize the possible damages resulting from the fault occurrence. Intelligent techniques, enhancing the existed protection schemes for wind farms, redesigning the wind farm wiring topology and integrated protection schemes may play definite roles towards eliminating these problems.

8. References

[1] Global Wind Energy Council (GWEC), "Global Wind Energy 2008 Report", http://www.gwec.net.

[2] American Wind Energy Association, "Annual Wind Industry Report, 2008", http://www.awea.org.

[3] European Wind Energy Association (EWEA), "EWEA ANNUAL REPORT 2008", http://www.ewea.org.

[4] Stefan Bauschke1, Clemens Obkircher, Georg Achleitner, Lothar Fickert and Manfred Sakulin, "Improved Protection system for electrical components in wind energy plants", 15th International Conference on Power System Protection, PSP '2006, Bled-Slovenia, 6-8 Sept. 2006

[5] R. Ramakumar, N. Butler, A. Rodriguez and S. Venkata, "Economic aspects of advanced energy technologies", Proceedings of the IEEE, Volume 81, Issue 3, March 1993, pp. 318 - 332.

[6] D. Hornak, N. Chau, "Green power - wind generated protection and control considerations", Protective Relay Engineers, 2004 57th Annual Conference for 30 Mar-1 Apr 2004, pp. 110 - 131.

[7] Johan Ribrant and Lina Margareta Bertling, "Survey of Failures in Wind Power Systems With Focus on Swedish Wind Power Plants During 1997–2005", IEEE TRANSACTIONS ON ENERGY CONVERSION, VOL. 22, NO. 1, MARCH 2007.

[8] H. Holttinen, T. Lakso, and M. Marjaniemi. (2005, Dec. 18) Tuulivoiman Tuotantotilastot Vuosiraportti, "Performance of wind power plants", annual report). 2000–2005, VTT, Espoo, Finland, [Online]. Available: http://www.vtt.fi (2005–12-18).

[9] M. Durstewitz (Ed.), "Windenergie report Deutschland 2004–2005", Institut foer Solare Energieversorgungstechnik (ISET), Kassel, Germany, N. E. Carlstedt, private communication, Nov. 2005.

[10] S. Musunuri, "Protection requirements for a large scale wind park", Protective Relay Engineers Confeence, 2009-30-03 to 02-04, 2009, pp. 478 - 491.

[11] Erlich, W. Winter, A. Dittric, "Advanced Grid Requirements for the Integration of Wind Turbines into the German Transmission System", IEEE General Meeting, GM2006.

[12] Andreas Dittrich Alexander Stoev, "Comparison of fault ride-through strategies for wind turbines with DFIM generators", Power Electronics and Applications, 2005 European Conference on, Dresden, Germany.

[13] V. Akhmatov, "Analysis of Dynamic Behaiour of Electric Power System with Large Amount of Wind Power", PhD Thesis, Electric Power Engeneering, Orsted DTU Technical University of Denmark, April 2003, Denmark.

[14] Niiranen, J. 2004. "Voltage Dip Ride Through of Doubly-Fed Generator Equipped with Active Crowbar", Nordic Wind Power Conference, 1-2 March 2004, Chalmers University of Technology, Göteborg, Sweden.

[15] Niiranen, J. 2005. "Experiences on Voltage Dip Ride through Factory Testing of Synchronous and Doubly Fed Generator Drives", Proceedings of 11th European Conference on Power Electronics and Applications. Dresden, Germany, 11-14 September 2005.

[16] Anca D. Hansen and Gabriele Michalke, "Fault ride-through capability of DFIG wind turbines", Renewable Energy, Elsevier, Vol. 32 (2007), pp. 1594–1610

[17] Niiranen, J. 2006. "About the Active and Reactive Power Measurements in Unsymmetrical Voltage Dip Ride Through Testing", Nordic Wind Power Conference, 22-23 May 2006, Espoo, Finland.

[18] Xiang, D., Ran, L., Tavner, P.J., Bumby, J.R. 2004. "Control of a Doubly-fed Induction Generator to Ride-through a Grid Fault", Proceedings of ICEM 2004, Cracow, Poland, 5-8 September 2004.

[19] S. Haslam, P. Crossley and N. Jenkins, "Design and evaluation of a wind farm protection relay", Generation, Transmission and Distribution, IEE Proceedings, Volume 146, Issue 1, Jan. 1999, pp. 37 – 44.

[20] Union for the Co-ordination of Transmission of Electricity (UCTE), "FINAL REPORT - SYSTEM DISTURBANCE ON 4 NOVEMBER 2006", UCTE, Brussels – Belgium, www.ucte.org.

[21] Tamer A. Kawady, Naema Mansour, Aly Osheiba, Abdel-Maksoud Taalab and Rama Ramakumar, "Modeling and Simulation Aspects of Wind Farms for Protection Applications", 40th ANNIVERSARY FRONTIERS OF POWER CONFERENCE October 29-30, 2007, Stillwater, Oklahoma, USA

[22] Tamer Kawady, "An Interactive Simulation of Grid-Connected DFIG Units for Protective Relaying Studies", IEEE PES/IAS Sustainable Alternative Energy Conference-2009, Valencia, Spain, 28-30 Sept., 2009.

[23] Tamer A. Kawady, Naema Mansour, Abdel-Maksoud Taalab, "Performance Evaluation of Conventional Protection Systems for Wind Farms" IEEE/PES General Meeting, GM-2008, Pittsburg, 20-24 July, 2008.

[24] Tamer A. kawady, Naema Mansour and Abd El-Maksoud I. Taalab, "Evaluating the Role of Current Limiting Fuses for Wind Farm Protection Applications", IEEE-PES General Meeting-2009, Calgary, Canada, 26-30 July, 2009.

13

Protection in distributed generation

M.Paz Comech, Miguel García-Gracia,
Samuel Borroy and M.Teresa Villén
CIRCE (Centre of Research for Energy Resources and Consumption)
University of Zaragoza
Spain

1. Introduction

Distributed generation (DG) is related to the use of small generating units installed close to consumption centres. Nowadays, there is an increasing interest in connecting generation to the distribution networks due to the environmental concern increase.

Most protection systems for distribution networks assume power flows from the grid supply point to the downstream low voltage network (Jenkins et al, 2000). Protection is normally based on overcurrent relays with settings selected to ensure discrimination between upstream and downstream relays. A fault on a downstream feeder must be cleared by the relay at the source end of the main feeder. It must not result in the operation of any of the relays on an upstream feeder unless the downstream relay fails to clear the fault. This will result in a blackout in a part of the network that should not have been affected by the fault.

If generation is embedded into the distribution system, the fault current seen by the relay may increase or decrease depending on the location of the relay, the fault and the distributed generators. To ensure an appropriate coordination between relays, grading studies must take into account the maximum and the minimum infeed from all the embedded generators and from the grid supply point.

This chapter describes the protection system problems in distribution systems when connecting distributed generation. First section describes the protection system characteristics in distribution systems without distributed generation, followed by the IEEE requirements for distributed generators under fault conditions (IEEE, 2003). After that the influence of distributed generation on system protection is analysed, taking into account grading, recloser operation and fault detection problems. Fault detection problems are specially relevant for wind turbines. Finally, some conclusions and future solutions that are being developed are described.

2. Protection in distribution systems

The objective of the utilities is to deliver electric energy in a safe, reliable, and economical manner (IEEE 2007). Protective relaying is applied to distribution lines to achieve this objective. The main objectives of a protection system are to detect and respond to all possible types of fault conditions that could occur, while affecting the minimum number of customers, and not limiting the capability of the system to carry load current. Since attempting to accomplish some of these objectives makes it impossible to accomplish others, compromises need to be made. The limits of these compromises are the criteria used to determine locations for the fault-interrupting devices, and the sensitivity and operating speed of the fault detecting devices.

There are different protective schemes applied to distribution systems, the overcurrent protection and the fuse saving/blowing scheme are the most used. These schemes are detailed below.

2.1 Overcurrent scheme

Overcurrent protection is the simplest scheme used to protect distribution lines. There are two types of overcurrent relays applied to distribution systems: phase overcurrent and ground overcurrent relays.
These relays can be directional or non-directional depending on system configuration and protection requirements. For radial distribution, non-directional overcurrent relays are applied; while for looped system networks, directional overcurrent relays are more appropriate.

In almost all cases, the distribution feeder protection begins at the substation with feeder breaker or recloser. The recloser is normally equipped with an inverse-time overcurrent device which detects faults and orders the trip of the breaker. Then, after a preset delay, it recloses the breaker. This device should be coordinated upstream with the protection of high voltage transformer and with downstream devices. Downstream devices are generally reclosers or fuses. All of the protective devices placed in series should coordinate on a time - current basis with each other in order to minimize the impact of an outage due to a fault. Fig. 1 shows the overcurrent scheme applied in a radial network.

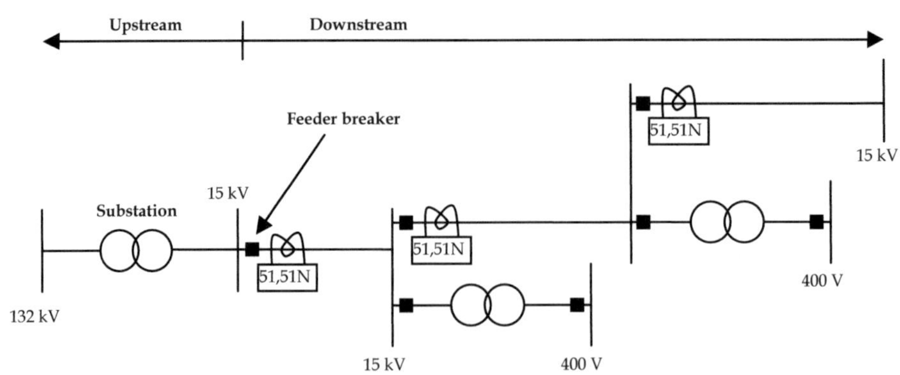

Fig. 1. Overcurrent relays applied to radial distribution network

2.2 Fuse saving/blowing schemes

Protection schemes for distribution feeders based on fuses are generally divided in two types:

a) Fuse saving schemes
b) Fuse blowing schemes

An example system for fuse blowing/fuse saving schemes is given in fig. 2.

Fuse saving schemes are effective on long multi-tapped rural distribution feeders with primarily residential loads that are not as sensitive to momentary outages as industrial loads. In a fuse-saving scheme, breakers or reclosers are set such that they trip before the fuse operates and then automatically reclose (Girgis & Brahma, 2001). In many cases, faults are only temporary and the line will successfully reclose, causing only a momentary disruption.

Fuse blowing schemes are used to limit the number of main feeder trip reclose cycles in distribution feeders where loads are sensitive to momentary outages and significant disruption occurs if the line is momentarily deenergized. This scheme is often used on feeders serving industrial plants and urban load centres where a number of trip-reclose cycles could result in equipment damage or added risk of personal injury. In this type of schemes, the overcurrent relay or recloser control curves are set above fuse curves such that fuses operate first to clear the faulted line section and the substation device serves as a back-up in case the fuse fails, in addition to its function of operating for faults on the main feeder trunk.

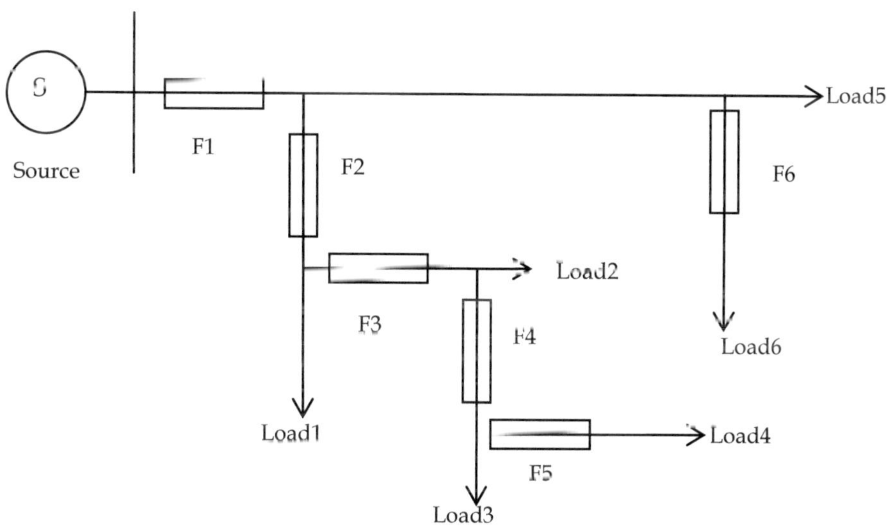

Fig. 2. Example system for fuse scheme

3. Requirements for Distributed Resources in Electric Power Systems

IEEE standard for interconnecting distributed generators with electric power systems (IEEE, 2008) provides interconnection technical specifications and requirements. As it is pointed out in the standard, its existence does not imply that there are no other ways to operate in subjects related to its scope. However it can be assumed as a guide, so standard technical criteria, specially those which involve protection system, are summed up in this section.

3.1 General requirements

The following are some of the most relevant general requirements for distributed generators (DGs) specified in the IEEE standard. The requirements involved in fault conditions are described in sections below.

- *Voltage requirements*: The DGs shall not cause variations in the area electric power system (area EPS) service voltage at other areas.

Grounding: The grounding scheme of the DGs interconnection shall not cause overvoltages that exceed the rating of the equipment connected to the area EPS and shall not affect the coordination of the ground fault protection on that area.
Synchronization: The connection of the DGs shall not cause a voltage fluctuation at the point of common coupling greater than ±5% of the prevailing voltage level of the area power system at that point.
Area energization: The DGs shall not energize the area power system if it is de-energized.

3.2 Requirements under fault conditions

The standard specifies the response of the distribution resources when abnormal conditions, such as faults, arise on the area EPS. According to the standard, the DG units shall cease to energize the area power system for faults on the area circuit to which they are connected. If a fault is detected in a circuit, and the circuit has been de-energized by its protection system in order to clear the fault, no source embebed on the circuit must energize it.

To achieve the interruption of the energization of a faulted circuit and fulfil the standard requirement, the distribution generation unit can be isolated by the following means:
Remote fault detection by the power system circuit protections and subsequent direct transfer trip of the DG unit, using communications.
Local fault detection by the DG unit and subsequent disconnection from the system.

The ability of the DG units to detect external faults depends strongly on the technology of the generator and the type of fault. The behaviour of different types of generators against faults are described next:
Synchronous generator: These units produce fault currents during extended periods of time. The initial fault contribution can reach more than six times the generator full-load current and can decay over several seconds below generator full-load current as the generator field collapses. Furthermore, during the fault, the voltage on the generator is significatively depressed, which can be used to aid in fault detection.

Based on the described current contribution and voltage depression, different methods are commonly used by the DG units based on synchronous generator to detect faults, such as voltage-controlled overcurrent relays and voltage-restrained overcurrent relays.

Voltage-controlled overcurrent relays are adjusted for a fixed sensitivity. The overcurrent function is enabled only when the voltage is depressed, which is controlled by an undervoltage element. As the sensitivity of the overcurrent relay is a fixed value, the adjusted current must be the minimum magnitude of faults that persist for a long time, which can cause coordination difficulties.

Voltage restrained overcurrent relays can be adjusted at different sensitivity levels depending on the voltage depression. Typically, at 25% voltage, the sensitivity is four times the one at the rated voltage.

Induction generator and double-fed asynchronous generator: The ability of protective relays to detect faults on the basis of fault-current contribution from this kind of generators depends on the persistence of the fault-current contribution and the required time for relay operation. The persistence of fault current depends on the type of fault (single- or multiphase), the severity of fault, and the means by which the reactive power requirements of the induction generator are supplied.

Due to the importance of these generators, mainly used in wind farms, section 4.3 details their behaviour against faults.

Inverters: Most inverters can not supply important currents under external fault conditions; usually no more than 1.2 to 1.5 times their rated load current. Fault detection schemes using overcurrent principles, which are universally applied, are not usually effective in this case. DG units that use this technology must rely on other methods such as abnormal voltage or frequency sensing to detect faults on the area power system.

In case of external faults, it is clear that the capacity of the distributed generators protections to detect them typically increases with the level of fault current supplied from these energy sources.

On the other hand, the contribution of the DGs to external faults can affect the behaviour of the protections of the area power system. It is a fact that once a DG unit is connected to a power system circuit, it is an energy source that, depending on its type of generator technology, can also supply current to a fault in their evacuation circuit or even in another point of the local power system. The fault-current contribution of the DG unit is the amount of current that the DG will supply to a fault in the circuit. If the DG fault-current contribution is substantial, it will contribute significantly to the total current supplied to the fault. Furthermore, the area power system contribution to the fault will be reduced by the presence of the DG. This may affect the time taken by the protections of the area power system to detect the fault or, in an extreme case, to prevent them from detecting the fault.

3.3 Requirements on reclosing coordination

The DG units are required to cease energizing the circuit to which they are connected before any reclosing attempt of that circuit.

Automatic reclosing is widely used in distribution networks in order to clear transient faults and restore the supply with the lower impact. Due to the common radial design in traditional feeders, where the power system is the only source of power, no voltage or sync-check supervisions are needed to perform the reclose attempts. The installation of DGs involves changes on these basic design principles. The assumption that the power system is the only power source for the feeder is invalidated by the presence of disperse generators which can also energize the circuit. In order to avoid an important damage, all DG protective devices are required to be carefully coordinated with the recloser of the circuit to which they are connected. If a temporary fault occurs in the feeder and the DG does not trip off and extinguish the fault arc prior to the circuit reclosing attempt, the reclosing attempt will be unsuccessful and the automatic restoration of that circuit may be jeopardized.

3.4 Requirements under voltage and frequency abnormal conditions

DG units are required to cease to energize the area power system under some voltage conditions. Table 1 specifies the clearing times required depending on the voltage value.

Voltage range (% of base voltage)[a]	Clearing time (s)[b]
V < 50	0.16
50 ≤ V < 88	2.00
110 < V < 120	1.00
V ≥ 120	0.16

[a] Base voltages are the nominal system voltages stated in ANSI C84.1 1995,Table 1
[b]DG ≤ 30 kW, maximum clearing times; DG > 30kW, default clearing times.

Table 1. Interconnection system response to abnormal voltages

When the system frequency is in the range given in Table 2, the DG shall cease to energize the area power system within the clearing time indicated (It must be pointed out that the nominal values considered in the standard are based on American system, where nominal frequency is 60 Hz).

DG size	Frequency range (Hz)	Clearing time (s)[a]
≤ 30 kW	> 60.5	0.16
	< 59.3	0.16
>30 kW	> 60.5	0.16
	< {59.8-57} (adjustable set point)	Adjustable 0.16 to 300
	< 57.0	0.16

[a]DG ≤ 30 kW, maximum clearing times; DG > 30kW, default clearing times.

Table 2. Interconnection system response to abnormal frequencies

4. Influence of Distributed Generation in the protection system

The interconnection of Distributed Generation (DG) brings a great change to the configuration of the utility distribution network (Yuping Lu, et al. 2007).

As it has been mentioned before, nowadays, the most common configuration in distribution systems is radial. In this type of configurations only one source feeds a downstream network (Girgis & Brahma, 2001), (Dugan & Mcdermott, 2002). With the connection of DG, in case of fault, the system can lose the radial configuration, since the DG sources contribute to the fault and therefore, the system coordination could be lost (Girgis & Brahma, 2001).

The connection of DG in distribution networks must take into account the following subjects:

- Protection behaviour (coordination problems)
- Adequate ratings of power equipments
- Islanding
- Detection problems
- Operation procedures

The connection of new generation sources in the distribution system modifies the power flow, customer's voltage conditions and the requirements of the utility equipment (Barker& de Mello, 2000). In a fault situation, distributed generators modify the current contribution to fault, and therefore it influences in the behaviour of network protection (Yuping Lu, et al. 2007), (Girgis & Brahma, 2001). The influence will depend on the number, type, location and size of DG (Yuping Lu, et al. 2007), (Girgis & Brahma, 2001). Thus, the characteristics of power equipments and the coordination system, which were established without considering the contribution of distributed generation, must be checked when DG is going to be connected (Yuping Lu, et al. 2007).

The requirements of network power equipments were established in a situation where DG did not exist. In a new situation where DG is present into the system, the initial design may not be adequate because the presence of DG modifies the power flow either in permanent or in fault situation.

On a radial system without DG, the relay protection will see downstream faults because only in that direction will appear power flow. In this situation the fault is cleared opening the breaker of the main feeder because only the main feeder can contribute to the fault. When DG is present in the system, an additional power flow appears from load side to source side and vice versa, therefore the opening of the main feeder breaker does not assure that the fault is cleared (Girgis & Brahma, 2001), (Dugan & Mcdermott, 2002).

Another possible situation when DG is connected is the island operation. It is produced when, once the breaker of the main feeder has opened after the fault, the DG continues feeding an isolated part of the network. The islanding is not desirable as it can produce safety and power quality problems in the isolated area. When a DG is working in island, frequency and voltage of this new part of the system can change and damage the system and loads connected in this area. Furthermore, if the distributed generator continues connected during a reclosing, serious damages can appear because of the difference between the grid and the island conditions (voltage, frequency, phase,...) (Barker & de Mello, 2000).

Another problem related to DG may be the different behaviour, in fault situation, between the different types of generators, in particular between synchronous and asynchronous generators. These differences are explained in section 4.3.

Finally, in areas where the supply depends on distributed generation, if a fault occurs and DG is disconnected following the requirements of IEEE standard (IEEE, 2003), the network could not be able to feed the total load when the utility breaker closes after the fault is cleared. Therefore, changes in operating procedure will be necessary to restore power in these situations (Dugan &. Mcdermott, 2002).

In this section coordination problems and automatic reclosing aspects related to the connection of DG to distribution networks are analyzed. Moreover, the behaviour of different types of generators before faults is described.

4.1 Coordination problems

In this section, an example illustrates the loss of coordination when DG is connected in a distribution network. First of all, the behaviour of distributed network without considering DG is analyzed, and after that, the loss of coordination due to the connection of DG is shown.

As it has been commented in section 2, the most used protective schemes in distribution networks are fuse blowing and fuse saving schemes. Therefore, an example network is analyzed using both protective schemes.

Case 1: Distribution network protected by fuse blowing schemes (Fuse-Fuse Coordination).

A fuse is characterised by Minimum Melting (MM) and Total Clearing (TC) characteristics. The Minimum Melting characteristic gives the time in which fuse is melted for a given value of fault current and Total Clearing characteristic gives the fault clearing time of fuse considering fault arc extinction for a given value of fault current (Girgis & Brahma, 2001). Fig. 3 shows the example network protected by fuses.

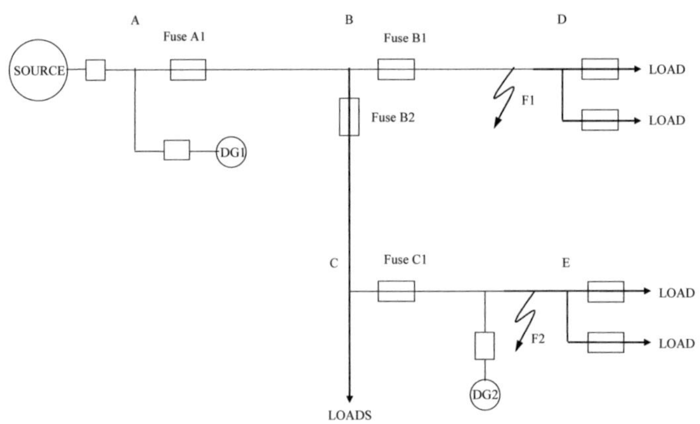

Fig. 3. Fuse blowing scheme

Initially, it is considered that DG is not connected. In this scenario, the network protection is coordinated for a fault on feeder 1 (F1) if both, fuse B1 acts before fuses A1, B2 and C1, and the TC characteristic of fuse B1 is lower than the MM characteristic of fuses A1, B2 and C1 for all range of coordination (I_{fmin}, I_{fmax}).
Fig. 4 shows the coordination graphic for fuses A1 and B1.

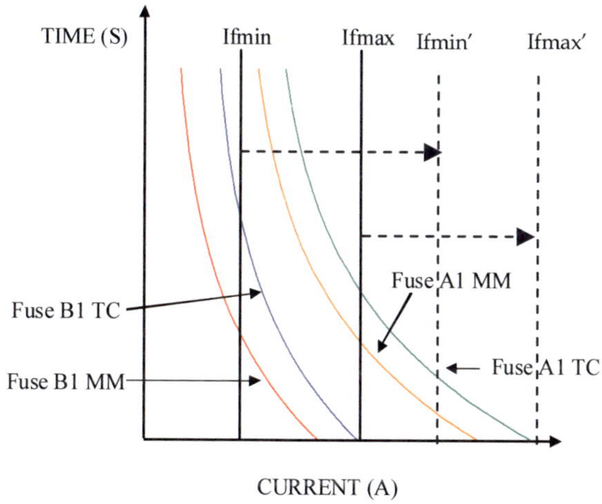

Fig. 4. Coordination graphic of fuses A1 and B1

As it has been commented, when DG is connected in a distributed network, the range of fault current (I_{fmin}, I_{fmax}) is modified and current flows in two directions (from source side to load side and vice versa). In this new situation it could happen that the system, initially coordinated, losses the coordination.

To illustrate how DG affects system fault current and system protection behaviour, the distributed network showed in fig. 3 is analyzed for faults in different locations.

- Fault in section BD (Fault F1)

  For faults in section BD, the current fault levels, I_{fmin} and I_{fmax}, increase because of the presence of generators in source side ($I_{B1}=I_{A1}+I_{B2}$).
  In this situation, the coordination between fuses could be lost if the increment of current fault makes the range of coordination (I_{fmin}',I_{fmax}') exceed the extent of the fuses curves, as it is shown in fig. 4.

- Fault in section CE (Fault F2)

  Fuses C1 and B2 detect downstream and upstream faults. Moreover, the current seen by both of them is always the same. ($I_{B2}=I_{C1}$).
  In a coordinated system, it is desirable that for faults on feeder 2 (F2), the fuse C1 acts before the fuse B2, and for faults on main line (F1), the fuse B2 acts before the fuse C1. As the current seen by both of them is the same, this requirement can not be fulfilled.

Case 2: Distribution line protected by fuse saving schemes (Recloser-Fuse Coordination)

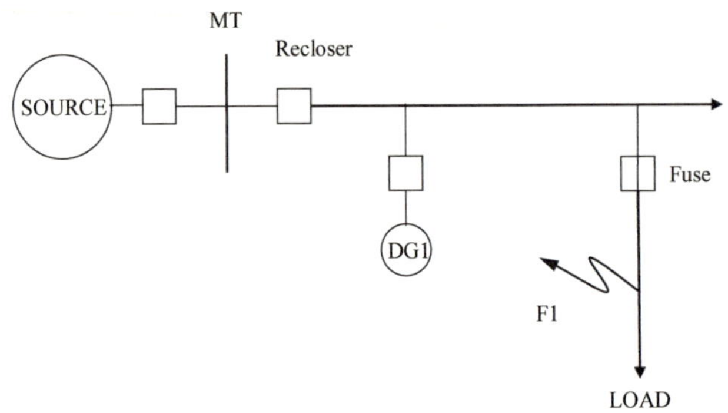

Fig. 5. Fuse saving scheme

Fig. 5 shows a distribution line where the main feeder is protected by a recloser and the load feeder is protected by a fuse. In this configuration the recloser has to act against temporary faults and the fuse against permanent ones.

For the analysis of this configuration, a recloser actuation sequence Fast-Fast-Slow-Slow is supposed. According to this sequence, if faults occur in the load feeder, the recloser acts opening the feeder breaker according to its fast overcurrent curve. The feeder breaker stays in this state for a defined time until the recloser orders to close it, allowing temporary faults to be cleared. If the fault persists, the recloser acts again. If after the second fast actuation of the recloser the fault is not cleared, the fault is assumed to be permanent. Therefore, after the second fast actuation of the recloser, it changes its overcurrent curve to the slow one, so that the fuse acts faster than the recloser. For the correct performance of the described scheme, the recloser and the fuse must be coordinated, as it is shown in Fig. 6.

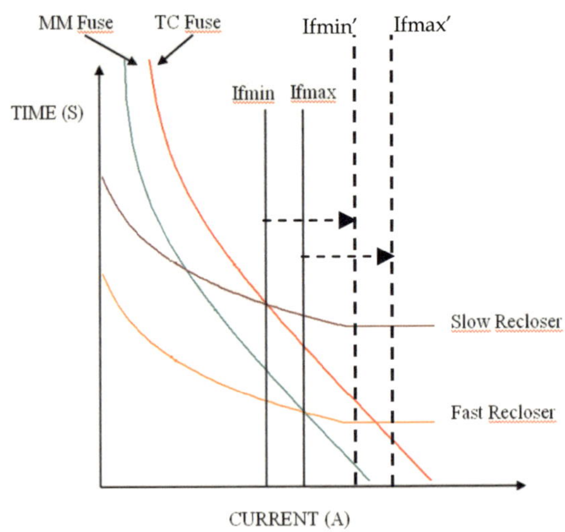

Fig. 6. Characteristics of the fuse (TC and MM) and the recloser (Fast and Slow) for a coordinated system

As it is showed in Fig. 6, for a fault in the load feeder, if fault current is inside the limits (I_{fmin} and I_{fmax}), the system will be coordinated. From the curves is clear that, in that range of current, fast recloser curve is faster than the fuse one and that the slow recloser is slower than the fuse. If the fuse fails in its actuation, the slow recloser should act as backup protection.

When DG is not connected, the currents seen by the recloser and the fuse for faults in the load feeder are the same so, while current fault is inside coordination range (I_{fmin} and I_{fmax}), the system is coordinated, as it is showed in Fig. 6.

Considering that DG is connected, the new situation results into the following changes:

- The minimum and the maximum fault currents for a fault in load feeder are modified because of the presence of DG.
- The fault current seen by the recloser is different to the current seen by the fuse.

If fault level increases due to DG, it could happen that fault currents are outside the coordination range. As it is showed in Fig. 6, in this situation the system loses the coordination because the MM characteristic of the fuse is lower than the fast recloser curve.

The different currents seen by the recloser and by the fuse depend on the size, location and type of DG. This difference of current can cause coordination problems to appear, if the current seen by both protections makes the fuse to act before the fast recloser actuation.

The described problems get worse if intermittency of renewable energy sources, which are mostly used in distributed generators, is considered.

4.2 Automatic reclosing problems

For a normal system operation, in a scenario where islanding is not contemplated, as it is recommended in the IEEE Std. 1547 (IEEE, 2003), the correct sequence of events in case of fault is the following (Pregelj et al, 2006):

1. The feeder circuit breaker is tripped, ordered by its protection devices
2. DG connected to the feeder is tripped
3. The recloser reconnects the feeder automatically
4. DG reconnects after normal operating voltage and frequency are established.

As it has been advanced in section 3.3, automatic reclosing is a widely extended practice in distribution networks and, in fact, it is necessary to clear transient faults and restore the supply with the lower impact. Under the basic assumption that upstream substation is the only power source for the distribution feeder, reclosing practices usually do not consider voltage or synchronism supervision, and most of the operators use instantaneous reclosing in distribution networks.

The presence of DG connected to distribution circuits can involve a serious problem for its reclosers performance, as it invalidates the aforementioned assumption. DG could maintain energization of the feeder after the feeder breaker trip. Avoiding potential reclosing problems is one of the main reasons for the IEEE standard requirement (IEEE, 2003) according to which, the DG units shall cease to energize the area power system for faults on the circuit to which they are connected.

Effective fault arc extinction may be prevented if the DG units keep energizing the circuit after the feeder breaker trip due to a fault. If the fault is temporary but the DG does not trip, the arc may be maintained even being the feeder breaker opened. In that case, when the reclosing attempt is made, the fault arc is not completely extinguished, so the fault becomes energized again. This involves the failure of the reclosing attempt and the corresponding damage for the system (Conti, 2008).

On the other hand, even if the fault is totally cleared at the moment of reconnection of the feeder circuit breaker, out-of-phase recloses can take place if DG units energize the circuit, since usually no sync check is made in distribution feeders. The problems derived from an out-of-phase reclose may affect both the DG units and the power system:

- In case the DG units connected to the feeder reclosed are rotator generators, they could suffer high electromechanical torques, which could damage them.
- The feeder may be subjected to important transient overvoltages, due to the different voltage phase angles. These overvoltages could not only damage the feeder devices, but also affect the customer systems. Also severe magnetic inrush currents could flow through transformers and motors connected to the feeder.

Two immediate possibilities can be raised to avoid the reclosing problems, regarding the DG units and the reclosers (IEEE, 2008). The first one, according with the IEEE standard and described in section 3.2, is that DG units must trip when a fault occurs in the circuit to which they are connected. At the same time, they should not trip if a fault occurs in other distribution feeders, even although they are next to the feeder. As it is explained in section 3.2, most generation technologies present difficulties to detect faults. In these cases, the DG units may have severe problems to disconnect themselves under fault conditions. The use of communication aided schemes can be a solution for these type of generators since it permits, at the same time that the feeder circuit breaker is tripped, a direct transfer trip to be sent to the DG units connected to the circuit, regardless of whether they have detected the fault or not. The implementation of the standard IEC61850 (IEC, 2003) is going to provide the exchange of information and signals between all devices interconnected to the Ethernet network, so communication schemes shall be favoured and improved. Fig. 7 shows a simplified example of direct transfer trip scheme.

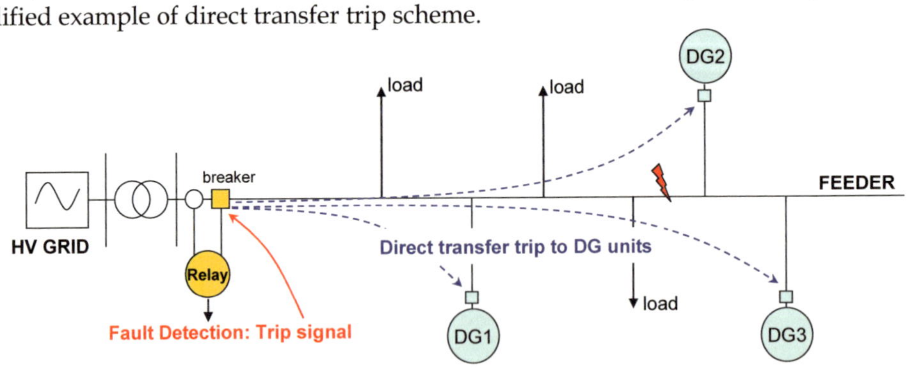

Fig. 7. Example of direct transfer trip scheme

On the other hand, reclosing schemes may be slightly modified in order to integrate the presence of DG connected to the circuit. A new configuration may include voltage supervision prior to sending the closing order to the breaker, so that the breaker is not closed until the voltage downstream is under a prefixed value. Using the voltage supervision assures that at the reclosing attempt no DG unit is energizing the circuit.

For higher safety, a direct transfer trip scheme like the above explained and a voltage supervision reclosing scheme can be combined.

4.3 Asynchronous and synchronous generators. Fault detection problems

One of the main problems in system protection when installing distributed generation is the ability of protective relays to detect external faults, especially in asynchronous generators and in generators connected through converters. This is the case of wind turbines. Fig. 8 shows the most installed configurations of wind turbines.

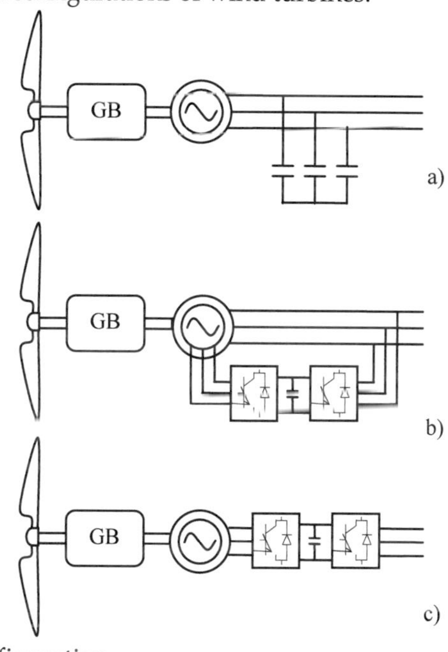

Fig. 8. Wind turbine configuration

Fig. 8a shows the fixed-speed wind turbine with asynchronous squirrel cage induction generator (SCIG) directly connected to the grid via transformer. Fig. 8b represents the limited variable speed wind turbine with a wound rotor induction generator and partial scale frequency converter on the rotor circuit known as doubly fed induction generator (DFIG). Fig. 8c shows the full variable speed wind turbine, with the generator connected to the grid through a full-scale frequency converter.

Protection engineering software tools used to set and coordinate the protection system usually do not include wind generator models. Only the synchronous generator model is available in some of them (Electrocon, 2000). This model supposes the short circuit contribution constant and equal to the subtransient short circuit current. Therefore, at the present time, there are two options when the protection coordination must be performed in areas with high wind power penetration: to neglect the effects of the wind turbines contribution or to model the asynchronous wind turbine as synchronous generator.

4.3.1 Wind turbines before short circuit contribution

Wind turbines contribution to short circuit is analyzed in this section. Fig. 9 shows the current contribution of an SCIG generator.

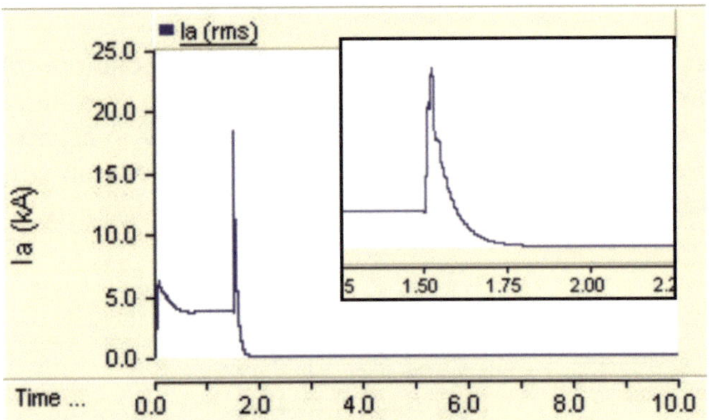

Fig. 9. Asynchronous generator rms current contribution

Unlike synchronous generators, induction generators do not have field windings to develop the required electro-magnetic field in the air gap of the machine, so induction generators can not work without external power supply. Therefore, under fault conditions their air gap flux drops quickly and their contribution to the fault is usually negligible after two or three cycles (Anderson, 1998). During the first cycles the contribution of the asynchronous machine is not despicable (Anderson, 1973), (Das, 2002), therefore, neglecting the induction initial short circuit current could lead to errors in the choice of the protective relaying, switching equipment and phase settings of the protective relaying.

In order to compare, Fig. 10 shows a synchronous generator short circuit current contribution.

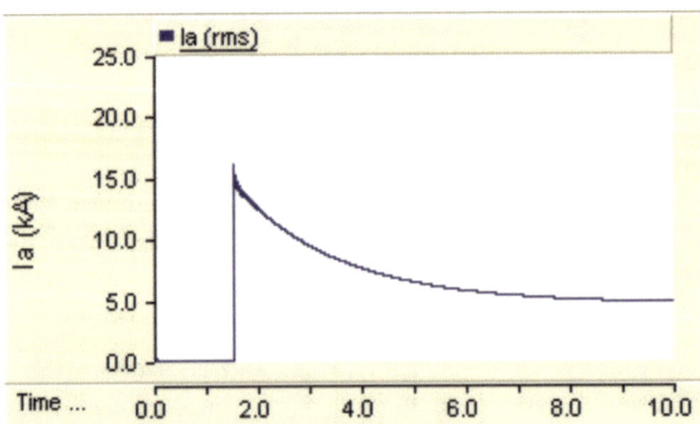

Fig. 10. Synchronous generator rms current contribution

Comparing both figures, it is observed than the main difference between the short circuit current contributions of a synchronous generator (Fig. 10) and an asynchronous one (Fig. 9) is the speed with which it drops. Then, the detection of the fault by the protective device must be done before the current contribution drops. The response time of the overcurrent relays is around 40 to 60 ms, therefore, if the relay settings for the instantaneous pick up current do not take into account the response time of the relay and the time in which the contribution drops, the protection could not detect the fault.

DFIG stator is connected directly to the network while its rotor is connected to the network by means of a power converter which performs the active and reactive power control. A voltage dip causes large currents in the rotor of the DFIG to which the power electronic converter is connected, so a high rotor voltage will be needed to control the rotor current. When this required voltage exceeds the maximum voltage of the converter, it is not possible any longer to control the desired current (Morren & de Haan, 2007). This implies that a large current can flow, which could destroy the converter.

In order to avoid breakdown of the converter switches, a crowbar is connected to the rotor circuit. When the rotor currents become too high, the converter is disconnected and the high currents do not flow through the converter but rather into the crowbar resistances. Then the generator operates as an induction machine with a high rotor resistance. When the dip lasts longer than a few hundreds of milliseconds (Tmax_crowbar), the wind turbine can even support the grid during the dip (Morren & de Haan, 2005).

This behaviour is shown in Fig. 11, a 200 ms short circuit has been simulated. The bold line shows the crowbar state, at t = 0.5 s, the fault is produced and the crowbar is activated. After Tmax_crowbar ms, the crowbar is deactivated; the rotor is connected to the converter and the control of the converter limits the current contribution to its nominal values.

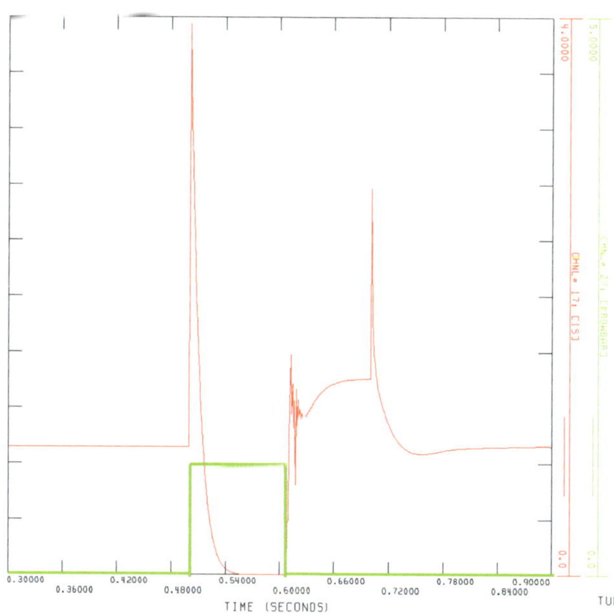

Fig. 11. DFIG rms short circuit contribution

Full converter contribution depends on the converter behaviour before short circuit. Most converters can not supply important currents under external fault conditions; usually no more than 1.2 to 1.5 times their rated current of the converter. For that reason, it can be modeled as a synchronous generator with the option "Limit Maximum Phase Current" activated. In this case, fault detection schemes using overcurrent principles that are universally applied, are not usually effective. Wind turbines units that use this technology must rely on other methods such as abnormal voltage or frequency sensing to detect faults on the area power system.

4.3.2 Wind turbine overcurrent protection analysis

In order to analyse the protection behaviour before wind turbine contribution, and the ability of protection software to model this response, the same network has been modelled in PSCAD/EMTDC and in CAPE to compare the "real" behaviour of the system with the one expected when the protection system is analysed using the models available in protection software tools, with the aim of checking, if the representation of the wind generator as a synchronous generator is valid in this case.

The wind turbine model in PSCAD/EMTDC consists of the asynchronous generator model, the drive train model and the rotor model. The generator and the drive train models have been developed by using the models available in the PSCAD/EMTDC library.

Shaft system influence must not be neglected since the shaft oscillations result in fluctuations of the voltage, machine current, rotor speed and other electrical and mechanical parameters. The shaft system has been modelled by using the PSCAD/EMTDC library multi-mass model. The literature about modelling indicates that a two mass model is adequate to model the drive train in wind turbines (Akhmatov, 2003).

The rotor is a complex aerodynamic system that can be modelled with different detail levels (Petru & Thiringer, 2002). When electrical behaviour is the main point of interest, an algebraic relation between wind speed and mechanical power extracted is assumed to model the rotor behaviour, which is described by the following equation (Heier, 1998), (Slootweg, 2003):

$$P = 1/2 \cdot \rho \cdot A \cdot Cp(\lambda) \cdot V_w^{\ 3} \tag{1}$$

Where P is the power extracted from the wind [W], ρ the air density [kg/m^3]; A the swept area [m^2] and Vw the wind speed [m/s]. Cp is the performance coefficient that depends on λ, the tip speed ratio:

$$\lambda = \frac{\omega_{tur} R}{V_w} \tag{2}$$

Fig. 12 shows the modelled network and the faults (A, B, C, D and E) simulated in PSCAD/EMTDC; the same network has been modelled in CAPE. The red squares represent the points in which the overcurrent protections are connected.

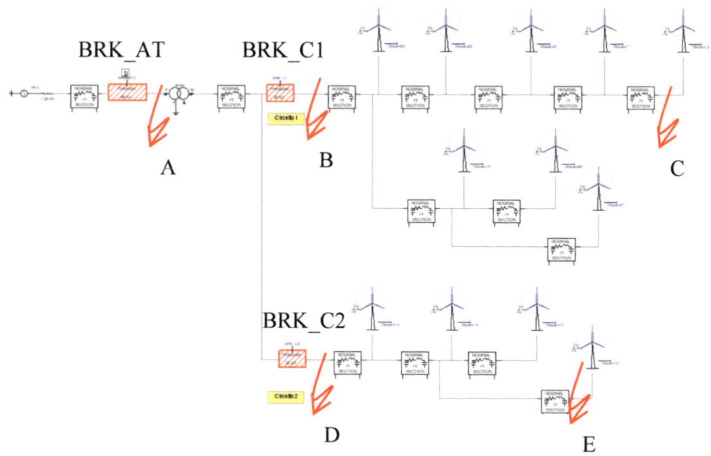

Fig. 12. Modelled network

Fig. 13 shows test A results obtained with CAPE and PSCAD/EMTDC. The value of the current contribution from the wind farm is slightly minor in the PSCAD/EMTDC simulation (109.4 A).

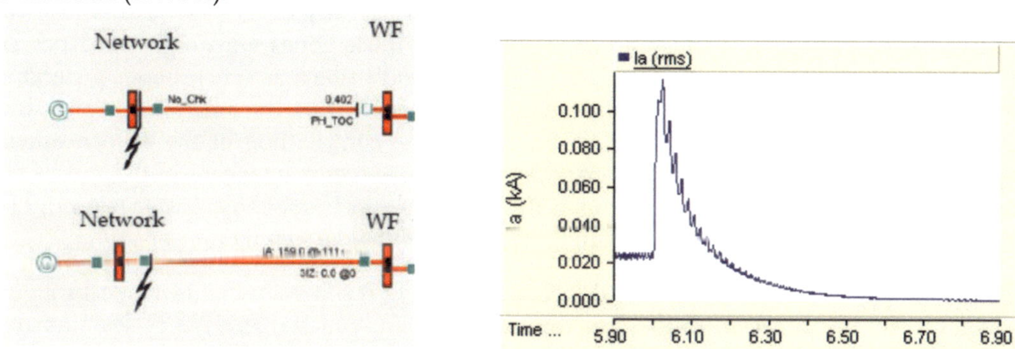

Fig. 13. CAPE and PSCAD results for test A

Table 3 shows the comparison between the tripping time obtained by PSCAD/EMTDC and CAPE. Breaker opening time (60 ms) has been taken into account in these results. When the instantaneous protection operates (tests B, C, D, E) the results obtained by both softwares are similar, with a 15 ms difference in test E. In test A, the inverse time overcurrent protection (51) operates in CAPE, but, as it has been seen in the previous sections and in Fig. 8, when the trip signal is active in CAPE (0.402-0.06), the short circuit current contribution is almost zero and the real protection would not trip. Therefore, the result obtained by CAPE in test A is incorrect.

			CAPE	PSCAD
A	BRK_AT	Inverse time	0,402	No trip
B	BRK_C1	Instantaneous	0,090	0,0988
C	BRK_C1	Instantaneous	0,090	0,1013
D	BRK_C2	Instantaneous	0,090	0,0994
E	BRK_C2	Instantaneous	0,090	0,1050

Table 3. Overcurrent times by CAPE and PSCAD/EMTDC

For protection system analysis, CAPE recommends the synchronous generator model with the option "Limit Maximum Phase Current" activated to model the wind turbine behaviour. This model can be useful to model the full converted wind turbines (Fig. 8c), since the converter limits the short circuit current to 1.2 or 1.3 times its rated current value. Nevertheless, the DFIG and SCIG short circuit behaviour do not show a good agreement with the shown by this model.

Induction machine short circuit contribution drops after a few cycles, but it is not negligible in this few first cycles. Due to it, wind farm contribution must be taken into account to set the instantaneous overcurrent protection, but not for the inverse time overcurrent protection.

If the asynchronous generator current contribution is neglected, the errors in the instantaneous protection setting can be significant in those zones with high wind power penetration. In case of modelling the asynchronous wind turbine as synchronous generator with a correct value for transitory impedance, like in this analysis, the initial behaviour (first cycles after the fault) is correct, but in few cycles the contribution of the asynchronous generators drops to zero and after those cycles the representation by means of this model no longer is adapted. In that order of time, if the model of synchronous generator is used, the inverse time overcurrent protection would not be coordinated precisely.

One possibility for asynchronous wind turbine modelling is shown in Fig. 14: a synchronous generator model connected to the grid through a switch, which opens 2-3 cycles after the fault. Table 4 shows the results obtained by the different models for tests A and B of the previous example. The behaviour of the proposed model shows a good agreement with the results obtained by PSAD/EMTDC. If the wind turbine contribution is neglected, the instantaneous overcurrent protection operation is not correct. If a synchronous model is used, the inverse curve overcurrent calculation would be wrong.

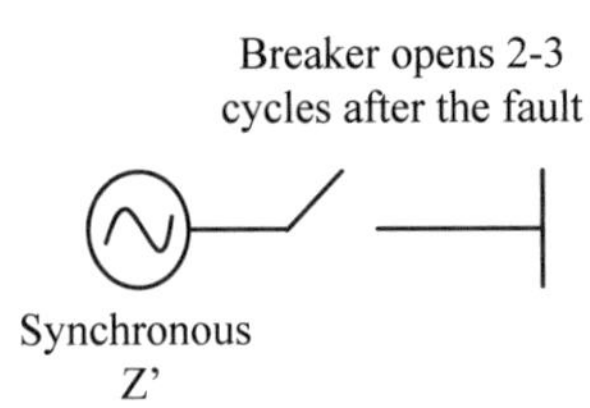

Fig. 14. Proposed model for the protection analysis of asynchronous wind turbines

		Trip time in CAPE with the proposed model	Trip time in CAPE neglecting wind turbine contribution	Trip time in CAPE with the synchronous model	Trip time in PSCAD
A	Inverse time	Infinite	Infinite	Infinite	0,400
B	Instantaneous	0,090	Infinite	0,0915	0,090

Table 4. Tripping time with the different models

5. Conclusions and future

The expectable great penetration of dispersed generators in distribution systems can lead to conflicts with the current protection schemes, since they were designed to work in a different scenario and under different conditions.

Most of the problems described in this chapter would be solved with the implementation of communication schemes and providing the power system with some "intelligence". In this line, the standard *IEC61850* and new developments like *Smart Grids* are expected to bring new tools to integrate DG in distribution systems and overcome the operation problems that could arise otherwise.

Distributed generation integration requires a smart grid to be feasible. This smart grid solves the possible problems that can affect the optimum behaviour of the system by operating the continuous information of the state of the different installations. The smart grid must contribute to the quality and the reliability of the service in the distribution grids as well as to improve the energetic efficiency; nevertheless, the integration of these systems requires a technological effort in automation and communication due to the fact that the exigencies of costs on the distribution systems cannot be comparable to high voltage costs.

Furthermore, the new international standard for communication networks and systems in substation, IEC61850 protocol, is expected to bring a profound evolution in electrical power systems. The new functionalities supported by an intelligent application of IEC61850 must result in a significant improvement of stability and reliability of power system. Smart grids shall take advantage of these features to achieve the commented objectives. Fig. 15 shows an example of communications scheme for a smart grid using IEC61850.

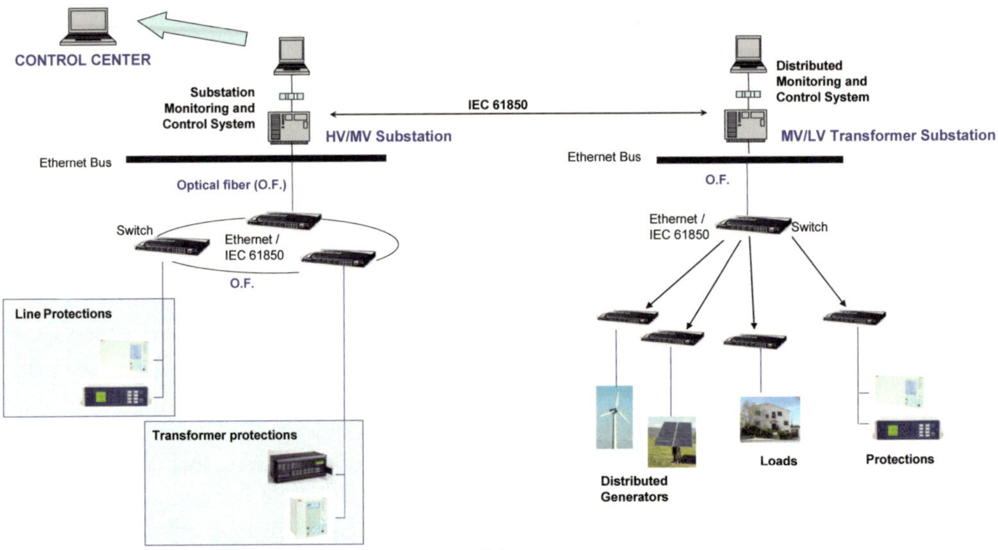

Fig. 15. Communication scheme of a smart grid

6. References

Akhmatov, V. (2003), Analysis of Dynamic Behavior of Electric Power Systems with Large Amount of Wind Power, PhD Thesis, 2003

Anderson, P. (1973), Analysis of faulted power systems, Iowa State University Press, 1973

Anderson, P.M. (1998), Power system Protection, IEEE Press, 1998

Barker, P.P. & De Mello,R.W. (2000). Determining the impact of distributed generation on power systems. I. Radial distribution systems, Proceedings of IEEE Power Engineering Society Summer Meeting, pp.1645-1656 vol.3, 0-7803-6420-1, Seattle,WA, July 2000, Seattle,WA.

Chaitusaney,S. & Yokoyama,A. (2008). Prevention of Reability Degradation from Recloser-Fuse Miscoordination Due To Distributed Generation. IEEE Transactions on Power Delivery, Vol. 23, No 4 (Oct. 2008), pp. 2545-2554, 0885-8977.

Conti, S. (2008), Analysis of distribution network protection issues in presence of dispersed generation, Electric Power Systems Research 79 (2009) 49–56.

Das, J.C. (2002), Power system analysis: short circuit, load flow and Harmonics, Marcell Dekker, 2002

Dugan, R.C. & McDermott, T.E. (2002). Distributed Generation. IEEE Industry Applications Magazine, Vol. 8, No. 2, (March-April 2002), pp. 19-25, 1077-2618.

Eirgrid (2007), Grid Code. Version 3.0, September 2007, http://www.eirgrid.com

Electrocon, (2005), CAPE online help, Ann Arbor, Michigan 2005

Elkraft System, Eltra, (2004), Wind Turbines connected to grids with voltages above 100 kV. Technical regulation for the properties and the regulation of wind turbines, Nov. 2004, Regulation T.F.3.2.5. www.enrginet.dk

E.ON Netz (2006), Grid Code. High and extra high voltage, Bayreuth 1 April 2006, http://eon-netz.com

Girgis, A. Brahma, S. (2001), Effect of distributed generation on protective device coordination in distribution system, Proccedings of Large Engineering Systems Conference on Power Engineering, LESCOPE '01, pp. 115-119, July 2001

Heier, S. (1998), Grid Integration of Wind Energy conversion systems, Chicester, U.K. Wiley 1998

IEC, IEC61850 Standard. Communication Networks and Systems in Substations. Available: http://www.iec.ch.

IEEE Power Engineering Society (2007), IEEE Guide for Protective Relay Applications to Distribution Lines, IEEE Std C37.230TM-2007, ISBN 978-0-7381-5711-5

IEEE (2003), IEEE Standard for Interconnecting Distributed Resources with Electric Power Systems, IEEE Std 1547TM-2003 (R2008), ISBN: 0-7381-3721-9

IEEE (2008), IEEE Application Guide for IEEE Std 1547TM, IEEE Standard for Interconnecting Distributed Resources with Electric Power Systems, IEEE Std 1547.2TM-2008, ISBN: 978-0-7381-5866-2

Jenkins, N. Allan, R. Crossley, P. Kirschen, D. Strbac, G. (2000) Embedded Generation, The Institution of electrical engineers, ISBN 0 85296 774 8, Herts, United Kingdom

Morren, J., de Haan, S.W.H. (2007) Short-Circuit current of wind turbines with doubly fed induction generator, IEEE Trans. On Energy convers, vol. 22, no. 1, march 2007

Morren, J., de Haan, S.W.H (2005) Ridethrough of wind turbines with doubly fed induction generators during a voltage dip, IEEE Trans. Energy Convers., vol. 20, no. 2, pp. 435-441, Jun. 2005

National Grid Electricity Transmission plc. (2007), The Grid Code, Issue 3, Revision 22, 17 Sept. 2007, www.nationalgrid.com/uk/

Petru, T. Thiringer, T. (2002), Modeling of Wind Turbines for Power Systems Studies, IEEE Transactions on Power Systems, vol 17, NO 4, Nov 2002

Pregelj, P. Begovic, M. Rohatgi (2006), Recloser Allocation for Improved Reliability of DG-Enhanced Distribution Networks, IEEE Transactions On Power Systems, Vol. 21, No. 3, August 2006.

Shaw Power Technologies, Inc. (Shaw PTI), Online Documentation PSS/E-29

Slootweg, J.G. Polinder, H. Kling, W.L. (2003), Representing Wind Turbine Electrical Generating Systems in Fundamental Frequency Simulations, IEEE Transactions on Energy Conversion, Vol 18, NO.4, Dec 2003

Verband der Netzbetreiber - VDN (2004), REA generating plants connected to the high and extra high voltage network, August 2004, Berlin, http://www.vdn-berlin.de/eeg_hh_en.asp

Yuping Lu; Lidan Hua; Jian Wu; Gang Wu & Guangting Xu (2007). A Study on Effect of Dispersed Generator Capacity on Power System Protection. Proceedings of IEEE Power Engineering Society General Meeting,2007, pp. 1-6, 1-4244-1296-X, Tampa,FL, June 2007.

Capacity Estimation Methods Applied to Mini Hydro Plants

Rafael Peña and Aurelio Medina
Universidad Michoacana de San Nicolás de Hidalgo
México

1. Introduction

Mini hydro generation is one of the most cost-effective and reliable energy technologies in distributed generation systems to be considered for providing clean electricity generation. These systems constitute a viable alternative to address generation and electric power supply problems to isolated regions or small loads.

In particular, the key advantages that small hydro has over other technologies considered in distributed generation systems, such as, wind, wave and solar power are (BHA, 2005):

- A high efficiency (70 - 90%), by far the best of all energy technologies.
- A high capacity factor (typically >50%), compared with 10% for solar and 30% for wind.
- A high level of predictability, varying with annual rainfall patterns.
- Slow rate of change; the output power varies only gradually from day to day (not from minute to minute).
- It is a long-lasting and robust technology; systems can readily be engineered to last for 30 years or more.

It is also environmentally benign. Small hydro is in most cases "run-of-river"; in other words any dam or barrage is quite small, usually just a weir, and little or no water is stored. Therefore run-of-river installations do not have the same kind of adverse effects on the local environment as large-scale hydro plants.

There is no consensus on the definition of mini hydro plants. Some countries like Portugal, Spain, Ireland, and now, Greece and Belgium, accept 10 MW as the upper limit for installed capacity. In Italy the limit is fixed at 3 MW (plants with larger installed power should sell their electricity at lower prices); in France the limit was established at 8 MW and UK favours 5 MW. Hereunder will be considered as Mini Hydro any scheme with an installed capacity of 10 MW or less.

Today, among all the renewable energies, hydropower occupies the first place in the world and it will keep this place for many years to come. Figure 1 shows the electricity generation from renewable energies; it can be seen that the hydro generation represents the 94.3% of the total of the generation using renewable resources; from this percentage, the 9.3% is generated by mini hydro plants. Also, the market for small power plants is more attractive than ever, due to the power market liberalization in the world, which opens the opportunity for the industry to generate electricity to full fill its basic needs.

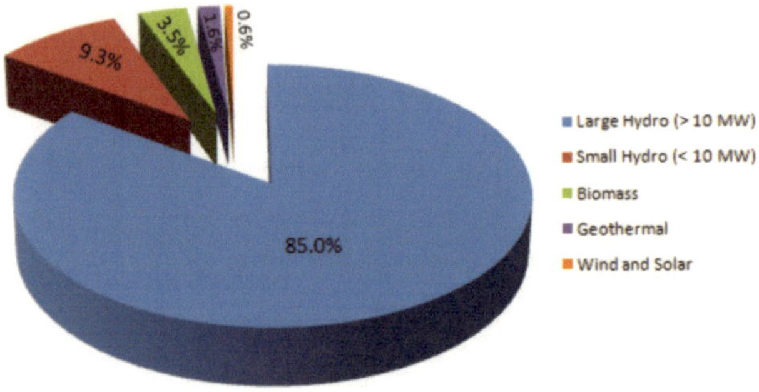

Fig. 1. Electricity generation from renewable energy (BHA, 2005)

An important aspect on this type of systems is the planning, design and evaluation of the potential energy available in a selected place (Puttgen et al., 2003). The best geographical areas for mini hydro plants are those where there are steep rivers, streams, creeks or springs flowing year-round, such as in hilly areas with high year-round rainfall. To assess the suitability of a site for a mini hydro power system, a feasibility study should be made.
Feasibility studies show how the water flow varies along the years and where the water should be taken to obtain its maximum profit; they also show the power amount that can be obtained from the water flow, as well as the minimum and maximum limits of the profitable power.
There are other important factors that should be addressed when deciding if a mini hydro plant would work at a specific site:

- The potential for hydropower at the site.
- The requirements for energy and power.
- Environmental impact and approvals.
- Equipment options.
- Costs and economics.

Keep in mind that each micro-hydropower system cost, approvals, layout and other factors are site-specific and unique for each case.
In the literature concerning the design and selection of the main components of a mini hydro plant (Penche, 1998; Khennas et al., 2000; ITDG, 1996), the size of the generator is chosen based on the water flow time series organized into a relative frequency histogram, and with a pre-selected plant factor, whose value usually varies from 0.70 to 0.85 for this type of systems (ITDG, 1996). From this information, it is also possible to calculate the theoretical average power and the average annual generation that can be obtained from the site (e.g. river).
In this chapter, forecast methodologies based on data measurements from a monthly water flow time series are applied to predict the behaviour of the water for a particular river where it is desired to install the mini hydro plant. First, forecast techniques are discussed and then a example of the application of the techniques to the water flow time series is presented. The proposed procedure aims enhancing the estimation of the generator capacity as the historical data of water flow is now complemented with the results obtained via

forecast techniques (Peña et al., 2009). For completeness, a selection of the most important electro-mechanical elements of a proposed mini hydro plant is also provided.

2. Water Flow Time Series

In order to determine the hydro potential of water flowing from the river or stream, it is necessary to know the flow rate of the water and the head through which the water can fall. The flow rate is the quantity of water flowing past a point at a given time. Typical units used for flow rate are cubic metres per second (m^3/s), litres per second (lps), gallons per minute (gpm) and cubic feet per minute (cfm). The head is the vertical height in metres (m) or feet (ft.) from the level where the water enters the intake pipe (penstock) to the level where the water leaves the turbine housing.

The measurements or historical data of water flow can be organized into a water flow time series. A time series is a series of measurements, observations, and recordings of a set of variables at successive points in uniform time intervals (Hamilton, 1994).

Fig. 2 shows the water flow time series used in this chapter. This historical data corresponds to measurements taken from the Cardel Hydrometric Station, in La Antigua River, Veracruz, México. The time series has 420 monthly observations, from the period of January, 1951 to December, 1985 (CONAE, 2005).

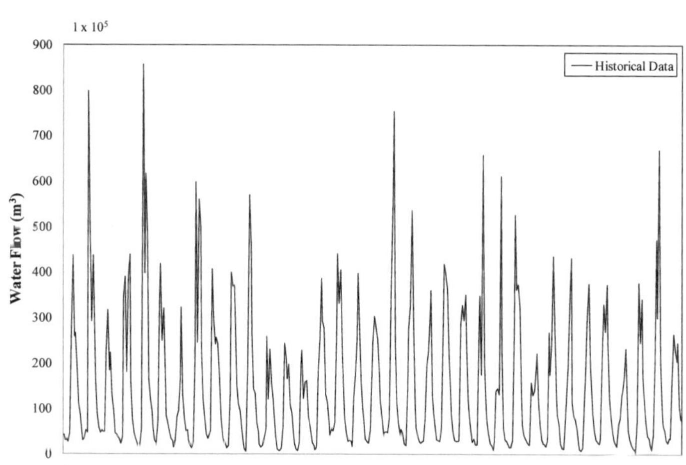

Fig. 2. Water flow time series

3. Capacity Estimation: Classic Method

Records of water flow variations along the years are taken in hydrometric stations located in the main rivers. These stations take data about the hydrologic situation of the area including the water flow variations of the river; this is periodically measured, in some cases on a day-to-day basis. The water flow records are very useful to allow forecasting the future behaviour of the river. This data is also taken into account to decide if a mini hydro plant can be installed in a specific place.

From historical records of water flow, a Flow Duration Curve (FDC) can be built. The flow duration curve is a plot that shows the percentage of time that the water flow in a river is likely to equal or exceed a specified value of interest (the area below the curve is a measure of the potential energy of the river or stream). For instance, the FDC can be used to assess the expected availability of water flow over time and the power and energy at a site and to decide on the "design flow" in order to select the turbine. Decisions can also be made on how large a generating unit should be. If a system is to be independent of any other energy or utility backup, the design flow should be the flow that is available 70% of the time or more. Therefore, a stand-alone system such as a mini hydro plant should be designed according to the flow available throughout the year; this is usually the flow during the dry season. It is possible that some streams could dry up completely at that time.

Figure 3 shows the FDC for the time series under study. From this figure it is possible to observe that for nearly 80% of the time the water flow is equal or below to 28.4×10^5 m^3. Also, it shows that if a mini hydro plant is to be installed in this river and it is desirable to have it working 70% of the time, then a value of 42.28×10^5 m^3 should be chosen as the design water flow.

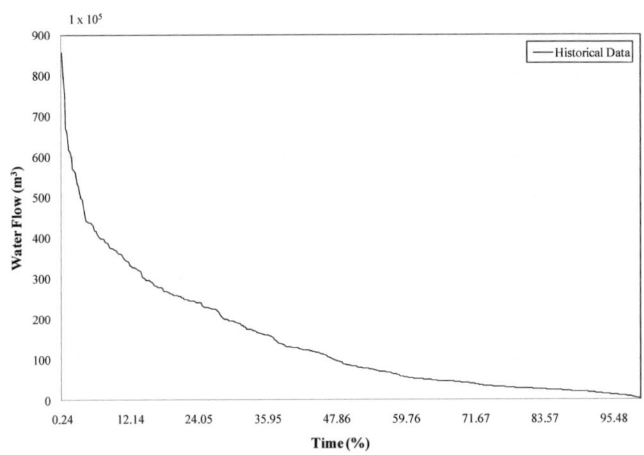

Fig. 3. Flow duration curve

Another important graph derived from the historical data is the relative frequency histogram. The relative frequency is defined as the number of times that a value occurs in a data set. Figure 4 shows the relative frequency histogram calculated from the historical data under study. From this bar chart it is possible to *a priori* visualize the data concentration, and the minimum and maximum values of the time series, e.g. from Figure 4, it can be seen that the value with more repetitions in the time series is 24.2×10^5 m^3.

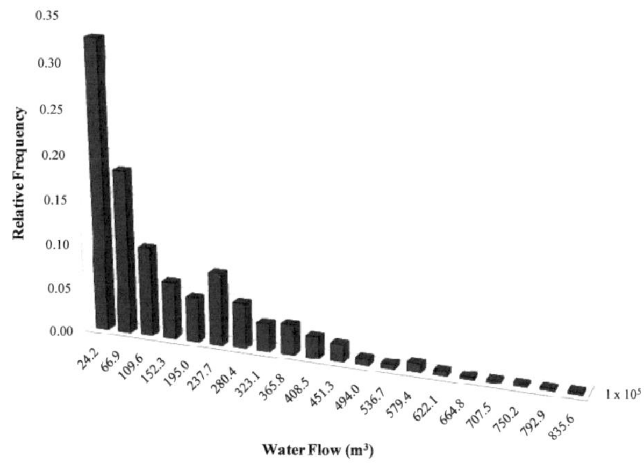

Fig. 4. Relative frequency histogram

4. Forecast Techniques

As mentioned before the water flow time series is frequently used for the design and capacity estimation of mini hydro plants. However, records of these variables from a previously selected place are not always available or there is just a few historical data. Besides, climate changes around the world have provoked, for instance, that drought and rainy seasons are not as periodic as they used to be; this makes the water flow levels to change drastically from one season to another, and in some cases rivers even tend to dry-out.

The reasons given before and other unexpected events justify the application of forecast techniques to appropriately process the available hydro resources at a specific geographic site, so that an adequate capacity estimation of a mini hydro plant can be achieved (and also to know the future behaviour of a selected river).

In order to deal with the forecast problem, various forecast techniques have been used: Kalman Filters (Sorensen & Madsen, 2003), Box-Jenkins methodologies (Montañés et al., 2002), Neural Networks (Xie et al., 2006), etc. most of them providing satisfactory results. In this chapter, ARIMA (Zhou et al., 2004), Neural Networks (Azadeh et al., 2007), and Genetic Programming (Flores et al., 2005) methodologies are presented and then applied to the water flow time series to forecast the behaviour of the water flow in the years to come. The best forecast obtained with the application of these methods is then used to estimate the capacity of a proposed mini hydro plant.

4.1 ARIMA Model

The acronym ARIMA stands for "Auto-Regressive Integrated Moving Average", whose model is a generalization of an auto-regressive moving average or ARMA model. These models are widely used in the time series forecast problem and they are usually part of the Box-Jenkins methodology (Montañés et al., 2002). The ARIMA model is generally referred as an ARIMA(p, d, q) model, where p, d, and q are values used to define the number of auto-regressive, integrated, and moving average terms of the model, respectively.

In order to describe the mathematics involved in an ARIMA model, it is important to define an ARMA(p, q) model for a data time series $X(n)$, where n is an integer index to indicate a specific data within the time series, then an ARMA(p, q) model is given by (Ramachandran & Bhethanabotla, 2000)

$$(1-\sum_{i=1}^{p} \phi_i L^i)\, X(n) = (1+\sum_{i=1}^{q} \theta_i L^i)\, \varepsilon_n \tag{1}$$

where L is the lag operator, ϕ_i are the auto-regressive parameters of the model, θ_i are the moving average parameters of the model, and ε_n are error terms. The error terms ε_n are usually known as white noise and they are assumed to be independent with zero co-variances, and identically distributed sampled data from a normal distribution with zero average.

The ARIMA(p, d, q) model can be obtained integrating Equation (1). That is,

$$(1-\sum_{i=1}^{p} \phi_i L^i)\, (1-L)^d\, X(n) = (1+\sum_{i=1}^{q} \theta_i L^i)\, \varepsilon_n \tag{2}$$

where d is a positive integer that controls the level of differentiation. Note that if d = 0, this model is equivalent to an ARMA model.

There are three basic steps to the development of an ARIMA model (Brockwell & Davis, 2002):

1) *Identification/model selection*: the values of p, d, and q must be determined. The principle of parsimony, also known as principle of simplicity, is adopted; most stationary time series can be modeled using very low values of p and q.
2) *Estimation*: the θ and the ϕ parameters must be estimated, usually by employing a least squares approximation to the maximum likelihood estimator.
3) *Diagnostic checking*: the estimated model must be checked for its adequacy and revised if necessary, implying that this entire process may have to be repeated until a satisfactory model is found.

The most crucial of these steps is identification, or model selection. This step requires the researcher to use his or her personal judgment to interpret some selected statistics, in conjunction with a graph from a set of autocorrelation coefficients, to determine which model the data suggest is the appropriate one to be employed. In this respect the ARIMA model is an art form, requiring considerable experience for a researcher to able to select the correct model.

Using the R-Project software (R-Project, 2009), a model that allows the determination of each ARIMA parameters was implemented; 370 data points were used to obtain the model and the last 50 values of the time series were used to compare the forecast obtained by the model with the results from the historical time series.

For an ARIMA (3, 1, 2) model, e.g. with three auto-regressive parameters (AR_X), one integrator parameter ($INTG_X$) and two moving average parameters (MA_X), the coefficients shown in Table 1 were obtained.

AR_1	AR_2	AR_3	MA_1	MA_2	$INTG_1$
-0.1559	0.2239	0.0753	-0.5109	-0.4891	-0.4652

Table 1. ARIMA technique coefficients

The calculated forecast data obtained with the ARIMA model and the historic data are compared in Fig. 5. The forecasted data satisfactorily matches the original time series, however, the form of the curve for the obtained forecast tends to be periodic, that is, it fails to adequately reproduce the peaks taking place in the original time series.

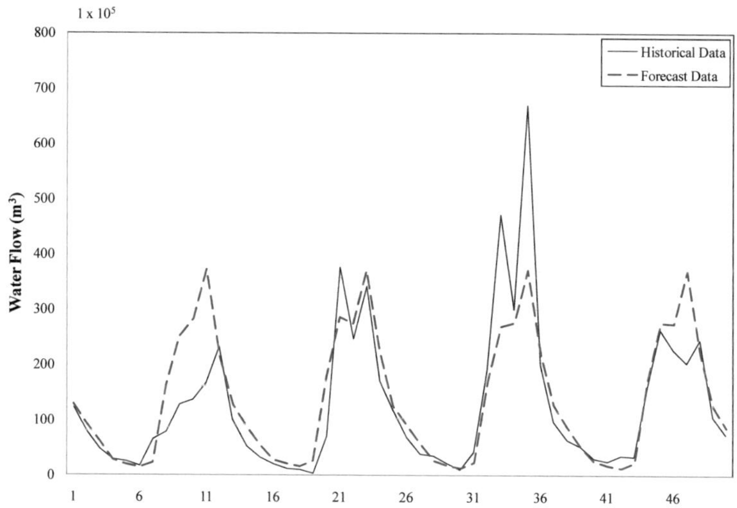

Fig. 5. ARIMA forecast data

4.2 Neural Networks

A neural network is a computational model that closely resembles the neuron cell structure of the biological nervous system. Neurons in a neural network are linked between them, so information can be transmitted from one neuron to others. An artificial neuron is a device with many inputs and one output. The neuron has two modes of operation; the training mode and the using mode. In the training mode, the neuron can be trained to fire (or not), for particular input patterns. In the using mode, when a taught input pattern is detected at the input, its associated output becomes the current output. If the input pattern does not belong to the taught list of input patterns, the firing rule is used to determine whether to fire or not.

The firing rule is an important concept in neural networks and accounts for their high flexibility. A firing rule determines how one calculates whether a neuron should fire for any input pattern. It relates to all the input patterns, not only the ones on which the node was trained.

A simple firing rule can be implemented by using Hamming distance technique (Beale & Jackson, 1990). The rule goes as follows:

- Take a collection of training patterns for a node, some of which cause it to fire (the 1-taught set of patterns) and others which prevent it from doing so (the 0-taught set). The patterns not in the collection cause the node to fire if, on comparison, they have more input elements in common with the 'nearest' pattern in the 1-taught set than with the 'nearest' pattern in the 0-taught set. If there is a tie, then the pattern remains in the undefined state.

Neurons in a neural network are linked between them, so information can be transmitted from one neuron to others. Given a data training set the neural network can learn the historic data through a training process and, with a learning algorithm the weights of the neurons are adjusted; through this procedure, the neural network acquires the capacity to predict answers of the same type of phenomenon.

Activation functions are also needed in order to train the neural network; different activation functions are frequently used, including step, linear (or ramp), threshold, and sigmoid functions.

In this research, the back-propagation learning algorithm (the most commonly used), was employed where the neural network forms a mapping between the inputs and the desired outputs from the training set by altering the weights of the connections within the network (Hilera & Martínez, 2000).

Different types of neural networks have been developed depending on the characteristics and the type of application. Common applications of this computational model are: speech recognition, control systems, classification of patterns, identification of systems, time series forecast, etc. (Hilera & Martínez, 2000).

A general structure of the connections of a neural network can be observed in Figure 6. During the phase of training, the connections weights W_{ij} are iteratively calculated; these connections link the neurons of the inputs E_i with the neurons of the layer O_j. If an input is received, this is sent across the neurons of the layer, and the weights of the connections are then adjusted, generating an output S. This output is compared with the real input, generating an error, which is back-propagated from the output to the input; this iterative process is carried-out until the neural network is able to reproduce the input.

The mathematical approach of the neural network working is as follows: A neuron in the output layer determines its activity by following a two step procedure.

- First, it computes the total weighted input E_i, using the Equation:

$$E_i = \sum_j y_j W_{ij} \tag{3}$$

  where y_j is the activity level of the jth neuron in the previous layer.

- Next, the neuron calculates the activity y_i, using a function of the total weighted input. Typically the sigmoid function is used:

$$y_i = 1/(1 + e^{-E_i}) \tag{4}$$

  Once the activities of all output neurons have been determined, the network computes the error *err*, which is defined by the expression:

$$err = (1/2)\sum_j (y_j - d_j)^2 \tag{5}$$

  where d_j is the desired output of the ith neuron.

Now, the back-propagation algorithm can be applied; it consists of four steps:

1) Compute how fast the error changes as the activity of an output neuron is changed. This error derivative (*EA*) is the difference between the actual and the desired activity.

$$EA_i = (\partial err \,/\, \partial y_i) = y_i - d_i \tag{6}$$

2) Compute how fast the error changes as the total input received by an output neuron is changed. This quantity (*EI*) is the answer from step 1 multiplied by the rate at which the output of a neuron changes while its total input is changed.

$$EI_i = (\partial err \,/\, \partial E_i) = (\partial err \,/\, \partial y_i) \times (\partial y_i \,/\, \partial E_i) = EA_i \, y_i \, (1 - y_i) \tag{7}$$

3) Compute how fast the error changes as a weight of the connection into an output neuron is changed. This quantity (*EW*) is the answer from step 2 multiplied by the neuron activity level from which the connection emanates.

$$EW_{ij} = (\partial err \,/\, \partial W_{ij}) = (\partial err \,/\, \partial E_i) \times (\partial E_i \,/\, \partial W_{ij}) = EI_i \, y_j \tag{8}$$

4) Compute how fast the error changes as the activity of a neuron in the previous layer is changed. This crucial step allows back-propagation to be applied to multilayer networks. When the activity of a neuron in the previous layer changes, it affects the activities of all the output neurons to which it is connected. So, to compute the overall effect on the error, we add together all these separate effects on output neurons. But each effect is simple to calculate; it is the answer in step 2 multiplied by the weight on the connection to that output neuron.

$$EA_j = (\partial err \,/\, \partial y_j) = \sum_i (\partial err \,/\, \partial E_i) \times (\partial E_i \,/\, \partial y_j) = \sum_i EI_i \, W_{ij} \tag{9}$$

By using steps 2 and 4, we can convert the *EA*s of one layer of neurons into *EA*s for the previous layer. This procedure can be repeated to get the *EA*s for as many previous layers as desired. Once we know the *EA* of a neuron, we can use steps 2 and 3 to compute the *EW*s on its incoming connections.

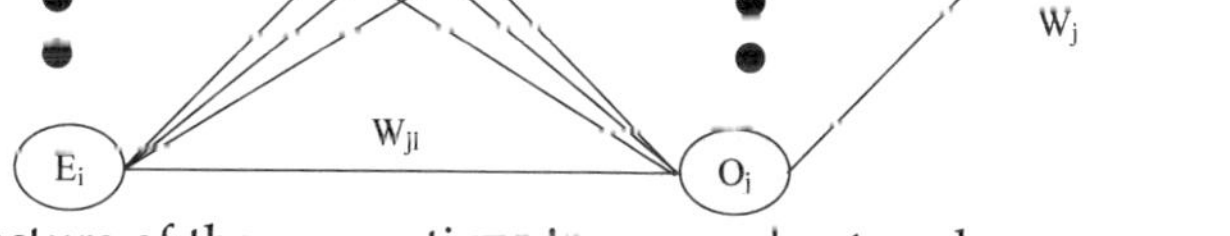

Fig. 6. General structure of the connections in a neural network

For the case under consideration a neural network with 3 layers and 30 neurons for each layer was implemented in C language; 370 data from the Cardel Hydrometric station time series were used to train the neural network and 50 values were used for the validation of the obtained forecast.

Figure 7 shows the results obtained with this computational technique and the corresponding comparison against the historical time series. From this figure it is possible to observe that the neural network is able to forecast future data more accurately than the ARIMA method.

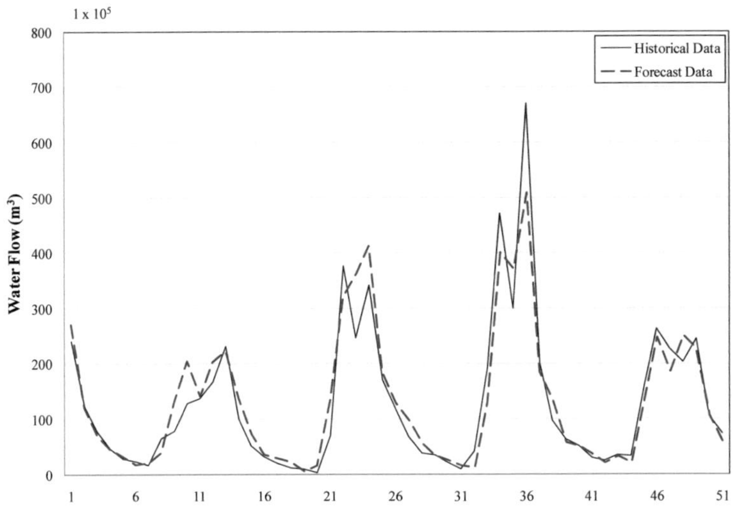

Fig. 7. Neural network forecast data

4.3 Genetic Programming by ECSID Software

Genetic Algorithms (GA) (Holland, 1992) and Gene Expression Programming (GEP) (Ferreira, 2006) are evolutionary tools inspired in the Darwinian principle of natural selection and survival of the fittest individuals. These methods use an initial random population and apply genetic operations to this population until the algorithm finds an individual that satisfies some termination criteria.

In order to simulate the evolutionary process both GA and GEP follow the next steps:

1) *Initialization*: Creates an initial random population.
2) *Evaluation*: Evaluates all the individuals and test whether or not the best one satisfies the termination criteria.
3) *Selection*: Use fitness proportional selection and apply the genetic operations to this population.

GA uses a fixed chromosome structure, which can be an array of bits, numbers, characters, etc. To use GA the problem is codified as a fixed chromosome and then the problem is solved using an evolutionary process. The genetic operations more widely used are crossover, selection and mutation.

GEP is similar to Genetic Programming (GP) (Koza, 1992); it is an evolutionary algorithm that evolves computer programs. The basic idea behind GEP is a clever representation for the chromosomes (a string instead of a tree), which leads to an easier implementation.

ECSID stands for "Evolutionary Computation based System IDentification"; it is a program that makes mathematical models from an observation data set (Flores et al., 2005).

ECSID obtains a formula that models the training data set, using a slide window prediction method (Jie et al., 2004). The slide window prediction method uses a window of size 16; the

window contains the actual data, and the model computes the synthetic time-series and the prediction errors for that time window.
Equation (10) shows the general models generated by the software, where $f(i)$ represents the time series at instant i, $e(i)$ is the vector of prediction errors, h is the window size, and a, b are unknown coefficients to be find out by the software.

$$f(i+1) = \sum_{i=t-h}^{t-1} a_i f(i) + \sum_{i=t-h}^{t-1} b_i e(i) \tag{10}$$

In order to make a forecast with this program, 337 data were used to obtain a model that represents the time series, and 83 data for the validation of the forecast obtained with the model.
From the ECSID model, a forecast study was conducted, with the results compared against real data. The comparison is shown in Figure 8, where the predicted data tends to an irregular triangular waveform, which is repeated in constant oscillation periods of 12 data (one year). For the first 50 points, the model achieves an acceptable reproduction of the original time series under study.

5. Application to the Capacity Estimation of Mini Hydro Plants

With the results obtained from the previous section (forecast data) and the available historical data, it is possible to determine the design water flow Q_i for the turbines of the mini hydro plant. It is also necessary to know the plant factor *pf*, which is the percentage of time that the plant is expected to be generating electric power at full capacity. The typical plant factor for this type of hydro plants varies from 0.70 to 0.85 (ITDG, 1996).
The following procedure was used to determine the value of Q_i:

- A ten-year forecast is conducted using the time series; this period usually corresponds to the useful life of hydraulic turbines.
- The obtained forecast data are added to the historical data, in this way, a more complete set of data of the river water flow is available.
- A new flow duration curve can be built using the new data set.
- Different values of *pf* are now selected and the respective value of Q_i from the water flow duration curve is chosen.

A summary of the design water flow obtained for the new time series calculated from the different forecast techniques (at different typical plant factors), and the average monthly water flow Q_m is shown in Table 2; the water flow is in 1×10^5 m^3. This Table illustrates that the neural network technique provided the most accurate results. The ARIMA and ECSID results suggest that the water flow in the river tends to be lower each year, that is, a smaller design water flow should be selected.

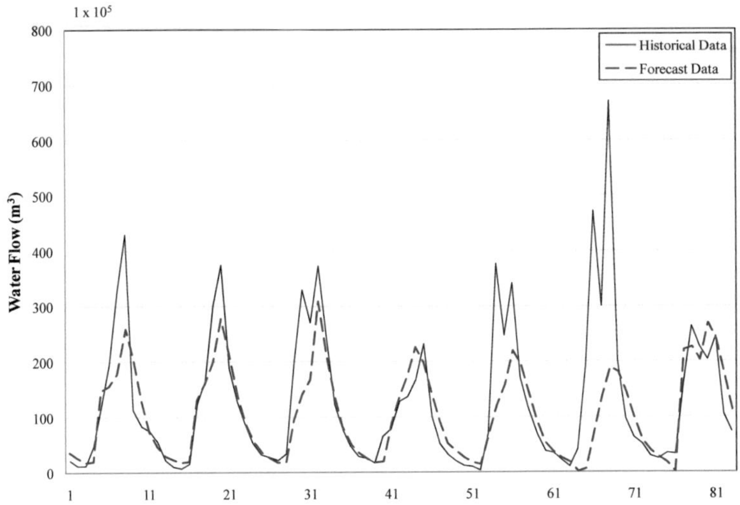

Fig. 8. ECSID forecast data

The theoretical average power P_m obtained from the river can be estimated using Equation (11) (IDAE, 1996):

$$P_m = 9.81 Q_m H \eta \gamma \quad (11)$$

where H is the height (m); η total efficiency of the system (p.u.); and γ is the specific weight of water (1 000 kg/m³). The Average Annual Generation of the system AAG is determined by (Penche, 1998),

$$AAG = (9.81 Q_i H \eta \, 8760 \, pf)/1{\times}10^6 \quad (12)$$

Forecast Technique	Q_m	Q_i			
		Plant factor (pf)			
		0.70	0.75	0.80	0.85
Historic data	147.894	42.283	32.874	28.398	24.414
ARIMA	115.733	31.326	27.356	23.161	19.488
Neural Network	144.854	40.501	32.793	28.228	23.948
ECSID	135.733	38.442	30.907	24.918	22.872

Table 2. Design water flow

For the case study, there are different places close to the Cardel hydrometric station with heights varying from 5 m to 15 m. These are locations where the construction of the civil work of the mini hydro plant can be built (CSVA, 2008).

Equations (11) and (12) are applied to obtain the theoretical total generation of the system. The parameter η includes the effects of the losses in the whole system, e.g. for our case η is assumed to be 1.0 p.u.

Table 3 summarizes the power generation that can be obtained from this river based on the results achieved with the neural network method, considering a plant factor of 0.8, and a design water flow of 28.228 × 105 m^3 at different heights.

	5 m	10 m	15 m
P_m	355.254 kW	710.509 kW	1.065 MW
AAG	5.283 GWh	10.567 GWh	15.851 GWh

Table 3. Average power and average annual generation of the system at different heights

It can be observed from Table 3 that the ideal height to be considered for the design and the selection of the mini hydro plant components is between 10 to 15 meters (CSVA, 2008) where the maximum power from the river can be obtained. For the case of study a capacity of 1 MW was selected which corresponds to the maximum height (15 m).

6. Selections of Electromechanical Equipment

Fig. 9(a) illustrates the high head scheme of a mini hydro plant; this scheme uses weirs to divert water to the intake, from where it is conveyed to the turbines, via a pressure pipe or penstock. Fig 9(b) shows the electrical diagram of a mini hydro plant. The basic electro-mechanical equipment in these plants comprises the turbines, generators, transformers and the interconnecting power line. An appropriate design and selection of the mini hydro plant components based on the forecast techniques and the results reported in Section 4 and 5, are described next.

6.1 Hydraulic Turbine

For the proposed case of study, the water flow design of 10.890 m^3/s and a height of 15 m were selected. To know the specific characteristics of the selected turbine, it is necessary to have the charts, graphs and characteristics from the turbine manufacturer. In the Alstom's web page (Alstom, 2009), a program that selects and presents the characteristics for the turbine-generator group used in a mini hydro plant can be accessed.

Providing the water flow and height data, the program Mini Aqua Configurator (Alstom, 2009) shows a graph with the type of available turbines, as illustrated by Fig. 10, showing the exact position and the type of the selected turbine (square in the central part of the graph).

The type of turbine to be selected for this case study can be Kaplan or Francis. The use of the Kaplan turbine is recommended since it is generally cheaper. The power generation estimated by the program for the turbine-generator group is 1.48 MW. Other relevant data are: 720 rpm, 88% of efficiency and 1320 mm for the diameter of the head turbine.

6.2 Generator

The selection of the generator for the mini hydro plant is mainly based on the turbine characteristics. Thus, a synchronous generator was selected with the following specifications: 1.5 MVA, 380 V, 60 Hz, 720 rpm and 0.9 power factor.

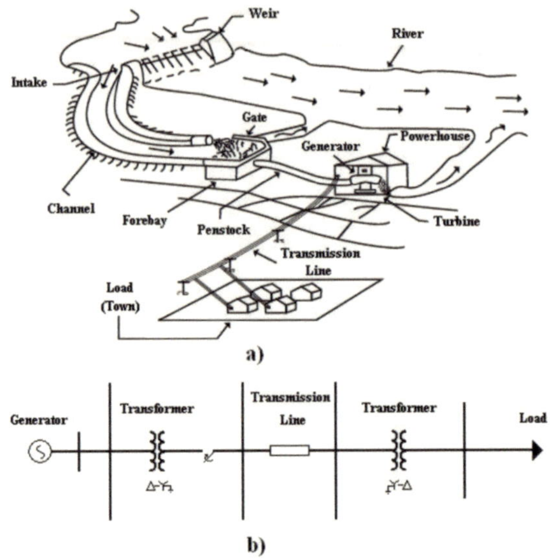

Fig. 9. Mini hydro plant: a) High head scheme, b) Electrical diagram

6.3 Transformers

For the proposed mini hydro plant, the use of three transformers is assumed; e.g. step-up, step-down transformers with a nominal capacity of 1.5 MVA and a transformer of 45 kVA for the self services in the plant. The proposed transformers have the following characteristics: 380 V / 13.2 kV to 60 Hz, Both the transformers will be connected in delta-wye grounded configuration.

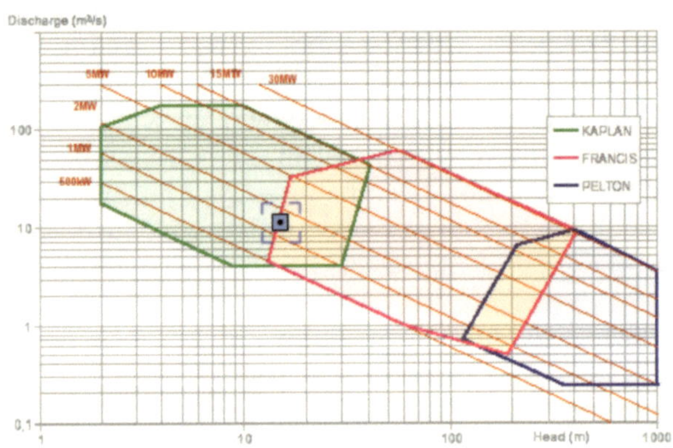

Fig. 10. Diagram of the available turbines in Alstom for mini hydro plants (Alstom, 2009)

6.4 Interconnection Power Line

According to CFE (Comisión Federal de Eléctricidad, México) construction norms (CFE, 1995), for this type of system the following primary distribution arrangement is recommended with the following characteristics: 3 phases, star connected, with solid connection to ground at the substation site. A simple cross-arm post type (TS) will be used for the distribution line.

7. Conclusion

In this chapter, a procedure for estimating the capacity of distributed generation based on mini hydro plants has been presented. This procedure has been successfully applied for a practical case where this type of distributed generation can be installed.
The classic method used to estimate the capacity of a mini hydro plant was also introduced and then time series forecasting techniques were applied in order to enhance the historic data available. Based on the results of these forecast techniques and the available historical data, an estimation of the design water flow was performed, and a possible capacity for a mini hydro plant calculated.
The time series forecast analysis has been conducted using an ARIMA model, a neural network and Genetic Programming by means of the software ECSID. For each case study, the obtained forecast results have been validated against actual measured data. The forecast technique based on neural networks has proven to give the best results for the case under consideration. Also, the selection of the most important electro-mechanic components of a proposed mini hydro plant based on the selected design water flow has been described.

8. References

Alstom. (2009). *Mini Aqua Configurator.* [Online]. Available: http://www.hydro.power.alstom.com/configurator/project_selection.html

Azadeh, A.; Ghaderia, S.F. & Sohrabkhania, S. (2007). Forecasting electrical consumption by integration of Neural Network, time series and ANOVA. *Applied Mathematics and Computation,* Vol. 182, No. 2, (March 2007), pp. 1753-1761, ISSN 0096-3003

Beale, R.; Jackson, T. (1990). *Neural computing: an introduction,* IOP Publishing Ltd, ISBN 0-85274-262-2, Bristol, UK

BHA. (2005). *A guide to UK mini-hydro developments,* The British Hydropower Association. London, UK

Brockwell, P. & Davis, R. (2002), *Introduction to Time Series and Forecasting,* Springer-Verlag, ISBN 0-387-95351-5, New York, USA

CFE. (1995). *Normas de Distribución, construcción de líneas aéreas (Distribution norms, construction of electric lines),* Comisión Federal de Electricidad, Mexico

CONAE. (2005). *Situación Actual de la Minihidráulica Nacional y Potencial en una región de los Estados de Veracruz y Puebla (Current Situation of the National Mini-hydro and the Potential Available for the States of Veracruz and Puebla, Mexico),* Comisión Nacional de Ahorro de Energía, Mexico

CSVA. (2008). *Consejo del Sistema Veracruzano del Agua.* [Online]. Available: http://www.csva.gob.mx

Ferreira, C. (2006). *Gene Expression Programming: Mathematical Modeling by an Artificial Intelligence,* Springer-Verlag, ISBN 3540327967, Berlin, Germany

Flores, J.J.; Graff, M. & Cardenas, E. (2005). Wind Prediction Using Genetic Algorithms and Gene Expression Programming. *Proceedings of AMSE 2005,* pp. 17-26, Morelia, Mexico, April 2005

Hamilton, J.D. (1994). *Time series analysis,* Princeton University Press, ISBN 0-691-04289-6, New Jersey, USA

Hilera, J.R. & Martínez, V.J. (2000). *Redes neuronales artificiales. Fundamentos, modelos y aplicaciones (Artificial Neural Networks, fundaments, models and applications),* Alfaomega Ra-Ma, ISBN 8478971556, Madrid, Spain

Holland, J. H. (1992). *Adaptation in Natural and Articial Systems,* University of Michigan Press, ISBN 0-262-58111-6, Massachusetts, USA

IDAE. (1996). *Manual of renewable energies: Mini hydroelectric power stations,* Ministry of Economy, Madrid, Spain

ITDG. (1996). *Manual de Mini y Microcentrales Hidráulicas; una guía para el desarrollo de proyectos (Mini and Micro-Hydraulic Plants; a project development guide),* Intermediate Technology Development Group, Peru

Jie, Z.; Changjie, T. Chuan, L. Anlong C. & Chang'an, Y. (2004). Time series prediction based on gene expression programming. *Proceedings of WAIM 2004,* pp. 55-64, ISBN 978-3-540-22418-1, Dalian, China, September 2004

Khennas; Smail & Barnett, A. (2000). *Best practices for sustainable development of micro hydro power in developing countries,* World Bank / ESMAP, Madrid, Spain

Koza, J.R. (1992), *Genetic Programming: On the Programming of Computers by Means of Natural Selection (Complex Adaptive Systems),* The MIT Press, ISBN 0-262-11170-5, Massachusetts, USA

Montañés, E.; Quevedo, J.R. Prieto, M.M. & Menéndez, C.O. (2002). Forecasting Time Series Combining Machine Learning and Box-Jenkins Time Series. *Proceedings of IBERAMIA 2002,* pp. 491-499, ISBN 978-3-540-00131-7, Seville, Spain, January 2002

Penche, C. (1998). *Layman's book: on how to develop a small hydro site,* European Small Hydropower Association (ESHA), Brussels, Belgium

Peña, R.; Medina, A., Anaya-Lara, O. & McDonald, J.R. (2009). Capacity estimation of a minihydro plant based on time series forecasting, *Elsevier Renewable Energy,* Vol. 34, No. 5, (May 2009), pp. 1204-1209, ISSN 0960-1481

Puttgen, H.B.; MacGregor, P.R. & Lambert, F.C. (2003). Distributed generation: Semantic hype or the dawn of a new era? *IEEE Power and Energy Magazine,* Vol. 1, No. 1, (January-February 2003), pp. 22-29, ISSN 1540-7977

Ramachandran, R. & Bhethanabotla, V.N. (2000). Generalized autoregressive moving average modeling of the Bellcore data. *Proceedings of LCN 2000,* pp. 654-661, ISBN 0-7695-0912-6, Tampa, USA, November 2000

R-Project. (2009). *The R Project for Statistical Computing.* [Online]. Available: http://www.r-project.org/

Sorensen, J.V.T. & Madsen, H. (2003). Water level prediction skill of an operational marine forecast using a hybrid Kalman filter and time series modeling approach. *Proceedings of OCEANS 2003,* pp. 790-796, ISBN 0-933957-30-0, San Diego, USA, September 2003

Xie, J.; Tang, X. & Li, T. (2006). Low Power Design based on Neural Network Forecasting for Interconnection Networks. *Proceedings of WCICA 2006,* pp. 2979-2983, ISBN 1-4244-0332-4, Dalian, China, June 2006

Zhou, M.; Yan, Z. Ni, Y. & Li, G. (2004). An ARIMA approach to forecasting electricity price with accuracy improvement by predicted errors. *Proceedings of IEEE Power Engineering Society General Meeting 2004,* pp. 233-238, ISBN 0-7803-8465-2, Denver, USA, June 2004

15

Optimal Coordination and Penetration of Distributed Generation with Multi Shunt FACTS Compensators Using GA/Fuzzy Rules

Belkacem Mahdad *, Tarek Bouktir ** and Kamel Srairi *
*Department of Electrical Engineering, Biskra University
**Department of Electrical Engineering, Oum Elbouaghi University
ALGERIA

1. Introduction

In an open electricity market, every consumer will be able to buy his own electricity from any source desired with the result that the unplanned power exchanges are increasing. In order to cope with these kind of problems and increase usable power distribution capacity, distribution generation technology (DG) and Flexible AC transmission systems (FACTS) where developed and introduced to the market.

Optimal placement and sizing of distribution generation is a well-researched subject which in recent years interests many expert engineers. Efficient placement and sizing of distribution generation (DG) in practical networks can result in minimizing operational costs, environmental protection, improved voltage regulation, power factor correction, and power loss reduction (Méndez et al., 2006). DG is defined as any source of electrical energy of limited size interconnected to the distribution system. DG technologies include photovoltaic systems, wind turbines, fuel cells, small micro-sized turbines, sterling-engine based generators and internal combustion engine-generators (Vovos et al., 2007). In practical installation and integration of DG in power system with consideration of FACTS devices, there are five common requirements as follows (Mahdad et al., 2007):

- What Kinds of DG and FACTS devices should be installed?
- Where in the system it should be placed?
- How to estimate economically the number, optimal size of DG and FACTS to be installed in a practical network?
- How to coordinate dynamically the interaction between multiple DG, FACTS devices and the network to better exploit the DG and FACTS devices to improve the index power quality?
- How to review and adjust the system protection devices to assure service continuity and keep the index power quality at the margin security limits?

The global optimization techniques known as genetic algorithms (GA), simulated annealing (SA), tabu search (TS), and evolutionary programming (EP), which are the forms of probabilistic heuristic algorithm, have been successfully used to overcome the non-convexity problems of the constrained ED (Bansal, 2005), (Huneault & Galiana, 1991). The GA method has usually better efficiency because the GA has parallel search techniques. Due to its high potential for global optimization, GA has received great attention in solving optimal power flow (OPF) problems. Fig. 1 shows the global strategy of the proposed approach to enhance the optimal power flow (OPF) in the presence of multi shunt FACTS devices and a multi distribution generation. A number of approaches for placement of DG to minimize losses have been proposed (Keane & O'Malley, 2006). Wang & Nehrir (2004) proposed a method which places DG at the optimal place along feeder and within networked systems with consideration of losses. Choudhry & Hanif (2006) proposed a strategy for voltage control for distribution networks with dispersed generation. Keane & O'Malley (2005) developed a methodology to optimally allocate DG capacity on the distribution network. The constraints taken in consideration were the voltage deviation, thermal limit, short circuit capacity. The methodology guarantees that the network capacity is maximized. Kuri & Redfern (2004) proposed a methodology based GA to place generators of discrete capacities in order to minimize losses and costs. Harrison et al. (2007) suggested a heuristic approach where an investment based objective function determines optimal DG site and size.

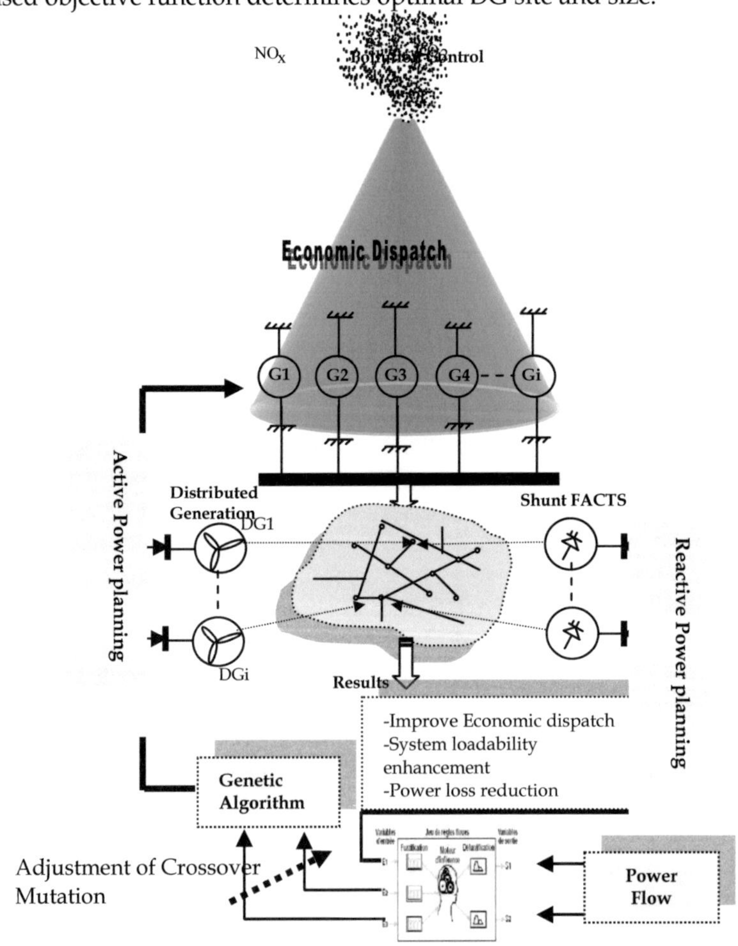

Fig. 1. Global strategy of OPF proposed with consideration of shunt FACTS and DG.

It is clear from the approaches cited in the literature that they offer optimal solution to the penetration of DG in a practical network, but not many approaches treat the problem of optimal coordination of multi DG with multi shunt FACTS devices to minimize fuel cost and improve the system security.
This Chapter presents a dynamic methodology for optimal allocation and sizing of multi DG units coordinated with multi shunt FACTS devices for a given practical distribution network, so that the cost of active power can be minimized. The proposed approach is based on a combined Genetic/Fuzzy Rules. The genetic algorithm generates and optimizes combinations of distributed power generation to be integrated to the network to minimize power losses, and in second step simple fuzzy rules based in practical expertise rules to control the reactive power of a multi dynamic shunt FACTS Compensators (SVC, STATCOM) designed to improve the system security. This proposed approach is implemented with Matlab program and applied to small case studies, IEEE 25-Bus and IEEE 30-Bus.The results obtained confirm the effectiveness of the proposed approach in sizing and integration of an assigned number of DG units in a practical network.

2. Active Power Planning

The active power planning problem is considered as a general minimization problem with constraints, and can be written in the following form:

$$\text{Minimize } f(x) \tag{1}$$

$$\text{Subject to: } g(x) = 0 \tag{2}$$

$$h(x) \leq 0 \tag{3}$$

$f(x)$ is the objective function, $g(x)$ and $h(x)$ are respectively the set of equality and inequality constraints. x is the vector of control and state variables. The control variables are generator active and reactive power outputs, bus voltages, shunt capacitors/reactors and transformers tap-setting. The state variables are voltage and angle of load buses. For optimal active power dispatch, the objective function f is the total generation cost expressed as follows:

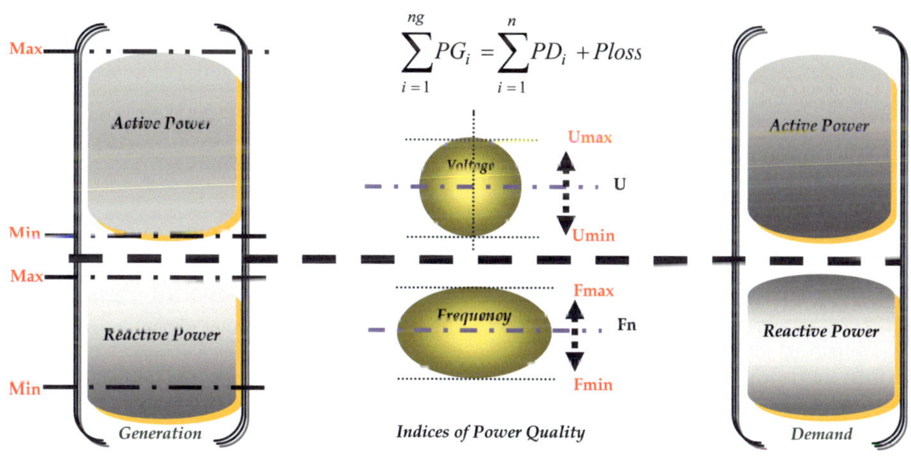

Fig. 2. Role of optimal power flow (OPF)

$$\text{Min } f = \sum_{i=1}^{N_g} \left(a_i + b_i P_{gi} + c_i P_{gi}^2 \right) \tag{4}$$

where N_g is the number of thermal units, P_{gi} is the active power generation at unit i and a_i, b_i and c_i are the cost coefficients of the ith generator.

The equality constraints $g(x)$ are the power flow equations, expressed as follows:

$$P_{gi} - P_{di} - \sum_{j=1}^{N} |V_i||V_j||Y_{ij}| \cos(\delta_i - \delta_j - \delta_{ij}) = 0 \tag{5}$$

$$\text{and } Q_{gi} - Q_{di} - \sum_{j=1}^{N} |V_i||V_j||Y_{ij}| \sin(\delta_i - \delta_j - \delta_{ij}) = 0 \tag{6}$$

The inequality constraints $h(x)$ reflect the limits on physical devices in the power system as well as the limits created to ensure system security:

- Upper and lower limits on the active power generations:

$$P_{gi}^{\min} \le P_{gi} \le P_{gi}^{\max} \tag{7}$$

- Upper and lower limits on the reactive power generations:

$$Q_{gi}^{\min} \le Q_{gi} \le Q_{gi}^{\max} \tag{8}$$

- Upper and lower bounds on the tap ratio (t).

$$t_{ij}^{\min} \le t_{ij} \le t_{ij}^{\max} \tag{9}$$

- Upper and lower bounds on the shifting (α) of variable transformers:

$$\alpha_{ij}^{\min} \le \alpha_{ij} \le \alpha_{ij}^{\max} \tag{10}$$

- Upper limit on the active power flow (P_{ij}) of line i-j.

$$|P_{ij}| \le P_{ij}^{\max} \tag{11}$$

- Upper and lower bounds in the bus voltage magnitude:

$$V_i^{\min} \le V_i \le V_i^{\max} \tag{12}$$

- Upper and lower bounds in the Shunt FACTS parameters

$$X^{\min} < X_{FACTS} < X^{\max} \tag{13}$$

3. Strategy of the GA-Coordination Fuzzy Rules for DG-Shunt Facts

3.1 Principle of the Proposed Approach

A flexible methodology based in two-subproblem algorithm to solving the new formulation of optimal power flow (OPF) problem incorporates DG and shunt flexible AC transmission system (FACTS) is presented in Fig 3. The controllable FACTS devices considered include shunt compensators (SVC) and Static Compensator (STATCOM). The proposed algorithm decomposes the solution of such a modified OPF problem into three linked sub problems. The first subproblem is an active power generation by efficient Genetic Algorithm, and the second subproblem is an active power planning of multi distributed generation to be integrated to the network to minimize power losses and the third subproblem is a reactive power planning coordinated with an efficient power flow problem to make fine adjustments on the optimum values obtained from the Genetic Algorithm. This will provide updated voltages, angles and point out generators having exceeded reactive limits.

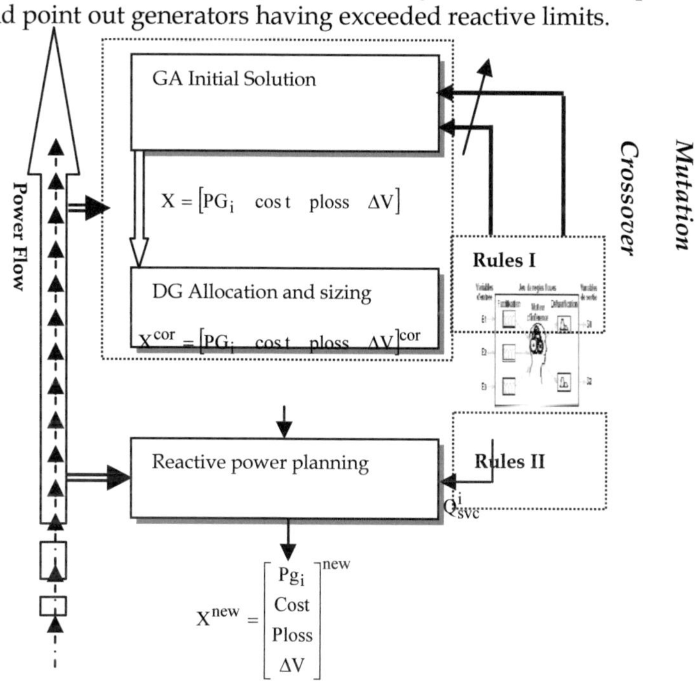

Fig. 3. Formulation mechanism of the proposed approach

Where:

X : Initial vector solution generated by GA.

X^{cor} : The corrected vector solution introduced by DG.

X^{new} : The final vector solution after reactive power planning.

ΔV : Voltage deviation.

3.2 Objective Functions

1) Fuel Cost Minimization

The objective function of the first subproblem is to minimize the total fuel cost without consideration of DG units.

$$\text{Min } J_1 = \sum_{i=1}^{NG} f_i \quad (14)$$

While f_i is the fuel cost ($/h).

NG is the number of generators.

2) Power Loss Minimization

The objective function is modified to minimize the power losses while at the same time fuel cost minimized, the active power of the slack bus is adjusted in coordination with the active power delivered by the DG units. The objective function can be formulated as:

$$\text{Min}\, J_2 = Ploss \tag{15}$$

while $Ploss$ is the power loss (MW)

4. Reactive Power Dispatch for Voltage Support

The goal here is to assure the minimum reactive exchanged between the dynamic shunt compensator and the network. Based in experience (Mahdad et al. (2007)) there is a maximum load increase on load margin with respect to the compensation level, the minimum reactive power is defined as the least amount needed from network system to maintain the same degree of system security. The problem can be formulated as a reactive power dispatch problem as follows.

$$\text{Max}\ RIS = \frac{LoadFactor(KLd)}{\sum_{i=1}^{NSVC} Q_{svc}^{i}} \tag{16}$$

Subject to:

$$V_i^{\min} \le V_i \le V_i^{\max} \tag{17}$$

$$Q_{SVC}^{\min} \le Q_{SVC}^{j} \le Q_{SVC}^{\max} \tag{18}$$

Where;
RIS: reactive index sensitivity.
$NSVC$: the number of shunt compensator.
KLd : loading factor.
Q_{SVC} : Reactive power exchanged.

To solve the above optimization problem, we adopt a coordination rules based on heuristic strategy. A global data base is generated during the successive action.

Fig. 4 shows the principle of the proposed reactive index sensitivity to improve the economical size of shunt compensators installed in practical network.

In this Figure, the curve represents the evolution of minimum reactive power exchange based in system loadability; the curve has two regions, the feasible region which contains the feasible solution of reactive power. At point 'A', if the SVC outputs less reactive power than the optimal value such as at point 'B', it has a negative impact on system security since the voltage margin is less than the desired margin, but the performances of SVC

Compensator not violated. On the other hand, if the SVC produces more reactive power than the minimum value ($Q_{\min}$), such as point 'C', it contributes to improving the security system with a reduced margin of system loadability, this reactive power delivered accelerates the saturation of the SVC Compensator.

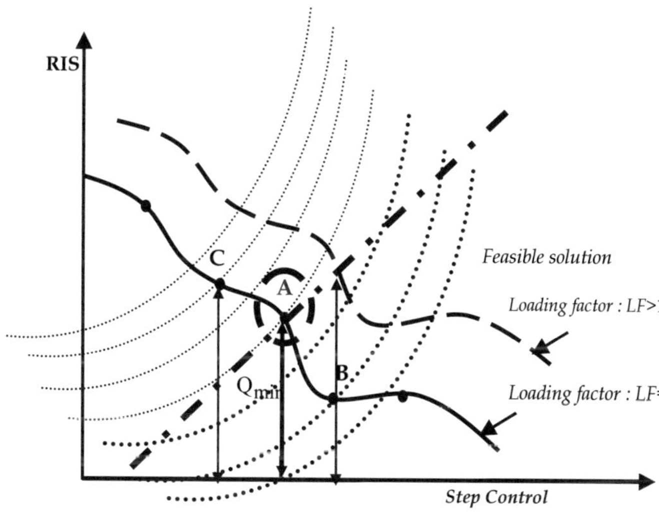

Fig. 4. Schematic diagram of minimum reactive power sensitivity

5. GA Solution for the Economic Dispatch

GAs are general purpose optimization algorithms based on the mechanics of natural selection and genetics [14]. They operate on string structure (chromosomes), typically a concatenated list of binary digits representing a coding of the control parameters (phenotype) of a given problem. Chromosomes themselves are composed of genes. The real value of a control parameter encoded in a gene, is called an allele.
A genetic algorithm is governed by three factors: the mutation rate, the crossover rate and the population size. GAs are search processes, which can be applied to unconstraint problems. Constraints may be included into the fitness function as added penalty terms [15].

a) Chromosome Type

Implementation of a problem in a GA starts from the parameter encoding. The encoding must be carefully designed to utilize the GA's ability to efficiently transfer information between chromosome strings and objective function of the problem. The encoded parameter is the power generation. Fig. 5 shows the structure of the proposed chromosome.

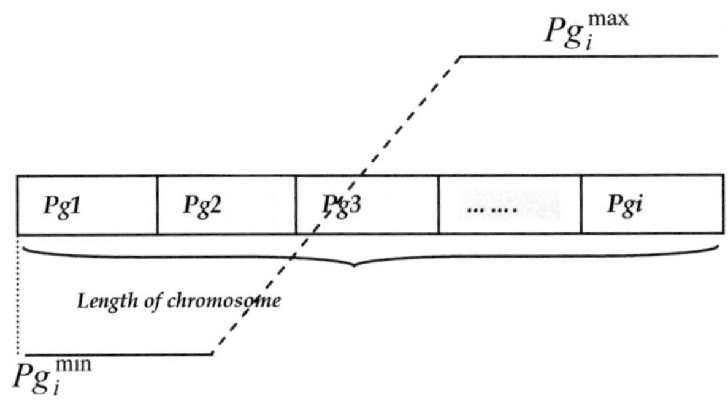

Fig. 5. Chromosome structure.

b) Fitness of Candidate Solution

Evaluation of a chromosome is accomplished by decoding the encoded chromosome string and computing the chromosome's fitness value using the decoded parameters. The fitness function adopted is given as:

$$Fitness = \frac{M}{f_i + Penality_Q^V} \tag{19}$$

where M is the maximum possible cost of generation, Objective function is the generation cost and $Penality\ _Q^V$ denotes a penalty for violating voltage limits V_j^{min}, V_j^{max}, and the penalty on the slack node for violating reactive power limit.

6. Fuzzy Logic Method

The use of fuzzy logic has received increased attention in recent years because of it's usefulness in reducing the need for complex mathematical models in problem solving (Ng et al. (200)). Fuzzy logic employs linguistic terms, which deal with the causal relationship between input and output variables. For this reason, the approach makes it easier to manipulate and solve problems.

6.1 Fuzzy Rules for Crossover and Mutation Adjustment

For better results and to get faster convergence, conventional GA modes have been modified. In recent years various techniques have been studied to achieve this objective, these include (Bakistzis et al. (2002)):

- Using advanced string coding.
- Generating initial population with some prior knowledge (Todorovski & Rajičič, (2006)).
- Establishing some better evaluation function.
- Including new operators such as elitism, multi point or uniform crossover and creep mutation (Yalcinoz et al., (2001)).

This approach proposes a flexible Genetic Algorithm based on fuzzy logic rules with the ability to adjust continuously the crossover and mutation parameters. Fig. 6 presents the proposed block diagram of a fuzzy controlled genetic algorithm.

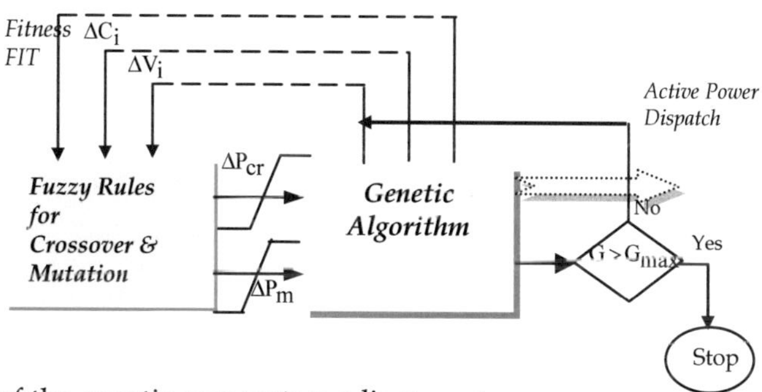

Fig. 6. Block diagram of the genetic parameters adjustment.

Crossover and mutation are considered critical for GA convergence. A suitable value for mutation provides balance between global and local exploration abilities and consequently results in a reduction of the number of iterations required to locate the good near solution. Experimental results based in application of GA to many practical networks at normal and abnormal conditions with load incrementation indicated, that it is better to adjust dynamically the value of the two parameters crossover and mutation.

It is intuitive that for a small variation in the chromosomes in a particular population, the effect of crossover during this critical stage becomes insignificant therefore, creating diversity in the population is required by increasing mutation (High value) probability of the chromosome and reducing (Low value) the value of crossover, note that the terms, small and high are linguistic.

The proposed approach employs practical rules interpreted in fuzzy logic rules to adjust dynamically the two parameters (crossover and mutation) during execution of the GA standard algorithm.

6.2 Membership Function Design

The membership function adopted by engineer differences from person to person and depends in problem difficulty therefore they are rarely optimal in terms of reproduced desired output.

1) Inputs and Outputs of Crossover and Mutation Fuzzy Controller

The inputs of the crossover fuzzy controller are changes of chromosomes fitness, the diversity in the cost generation, and voltage deviation, the output is the rate variation in crossover. The inputs and outputs of mutation are the same of crossover fuzzy controller. A sample rules for crossover and mutation changes is presented in Fig. 7.

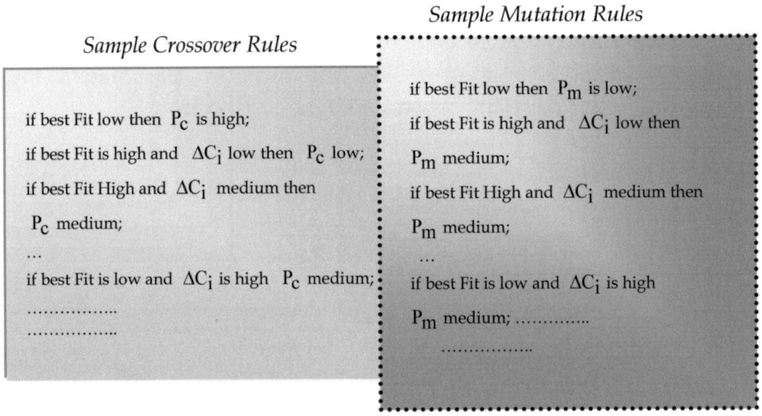

Fig. 7. Sample rules for crossover and mutation tuning

In this study, the upper and lower value for crossover probability and mutation probability, respectively are changed based in membership function for each variable from 0.9 to 0.4 and from 0.01 to 0.1. The consequent effects on the final value of crossover probability and mutation probability are calculated using the following equations:

$$P_c^{(t)} = P_c^{(t-1)} + \Delta P_c \tag{20}$$

$$P_m^{(t)} = P_m^{(t-1)} + \Delta P_m \tag{21}$$

where, $P_c^{(t)}$, $P_m^{(t)}$ are respectively the crossover and mutation probability at the iteration 't'.

6.3 Fuzzy Rules for Reactive Power Planning

The fuzzy variables associated with Reactive Power Planning Problem 'RPP' of a multiple dynamic shunt compensator are stated below.

1) Fuzzy Input Variables

- Bus voltage
- Active Power loss
- Reactive Power loss

2) Fuzzy Output Variables

- Voltage regulation for the shunt compensator.

3) Membership Function

A membership function uses a continuous function in the range [0-1]. It is usually decided from human expertise and observations made and it can be either linear or non-linear. This choice is critical for the performance of the fuzzy logic system since it determines all the information contained in a fuzzy set. Engineers experience is an efficient tool to achieve a design of an optimal membership function, if the expert operator is not satisfied with the conception of fuzzy logic model, he can adjust the parameters used to the design of the memberships functions to adapt them with new database introduced to the practical power system. Fig. 8 shows the general block diagram of the proposed coordinated fuzzy approach applied to enhance the system loadability with minimum reactive power exchanged. Fig. 9 shows the combined of the voltage, active power and reactive power as input to the shunt compensator controller.

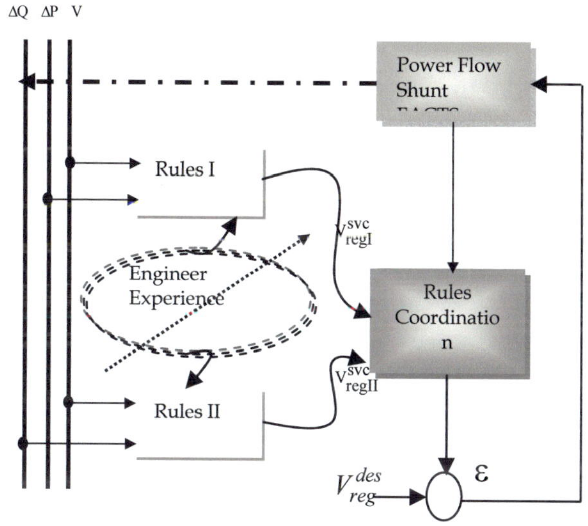

Fig. 8 Diagram of the proposed coordinated fuzzy approach

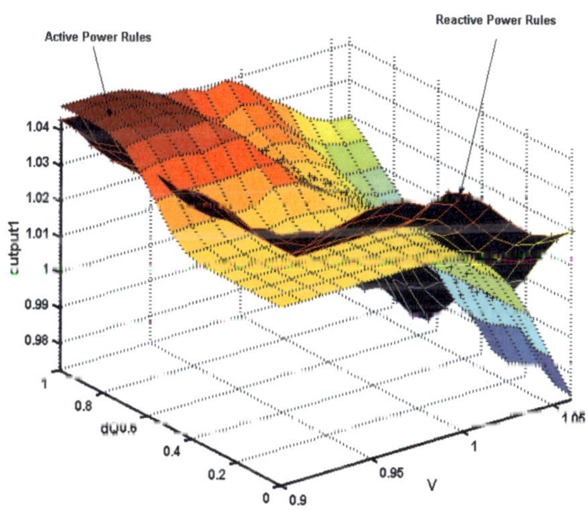

Fig. 9. Combination voltage, active and reactive power rules.

The solution algorithm steps for the fuzzy control methodology are as follows:

1) Initial database:
Introduce the initial vector solution for power generation $X^{cor} = [PG_i \quad \cos t \quad ploss \quad \Delta V]^{cor}$

Introduce the initial vector solution for distribution generation units DG_i.

2) Perform the initial operational power flow to generate the initial database $(V_i, \Delta P, \Delta Q)$.

3) Identify the candidate bus using continuation load flow.

4) Install the specified dynamic shunt compensator to the best bus chosen, and generate the reactive power using power flow based in fuzzy expert approach.

7. Shunt Facts and DG Modeling

7.1 Steady State Model of DG

The proposed approach requires the user to define the number of DG units to be installed based in voltage stability index (loading factor). The genetic algorithm generates and optimizes combination of DG sizes. For each combination of solution. Power losses and minimal cost used as a fitness function for the GA.
DG units modelling depend on the constructive technology and their combined active and reactive power control scheme (Golshan & Arefifar, (2006)).
In this study DG has been considered as not having the capability to control voltages, and therefore, it has been modelled in power flow study as a negative load, as a PQ node. Dynamic shunt compensators (SVC and STATCOM) modelled as a PV node used in coordination with DG to control the voltage by a flexible adjustment of reactive power exchanged with the network. Fig. 10 shows the proposed combined model of DG and SVC.

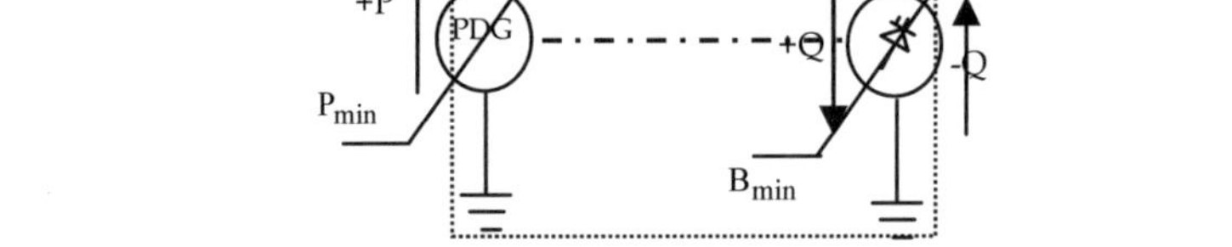

Fig. 10. The proposed combined model of DG/SVC Compensators integrated in power flow algorithm

7.2 Static VAR Compensator (SVC)

The steady-state model proposed in (Feurt-Esquivel et al., (1998)) is used here to incorporate the SVC on power flow problems. This model is based on representing the controller as a variable impedance, assuming an SVC configuration with a fixed capacitor (FC) and

Thyristor Controlled Reactor (TCR) as depicted in Fig. 11. Applying simultaneously a gate pulse to all thyristors of a thyristor valve brings the valve into conduction. The valve will block approximately at the zero crossing of the ac current, in the absence of firing signals. Thus, the controlling element is the Thyristor valve. The thyristors are fired symmetrically, in an angle control range of 90 to 180 with respect to the capacitor (inductor) voltage.

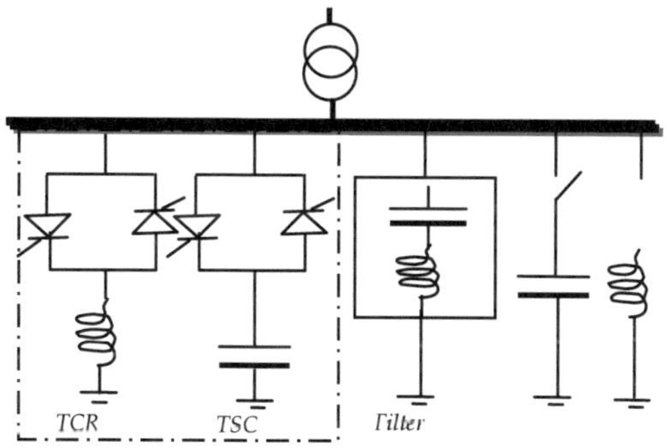

Fig. 11. Basic circuit representation of SVC Compensator

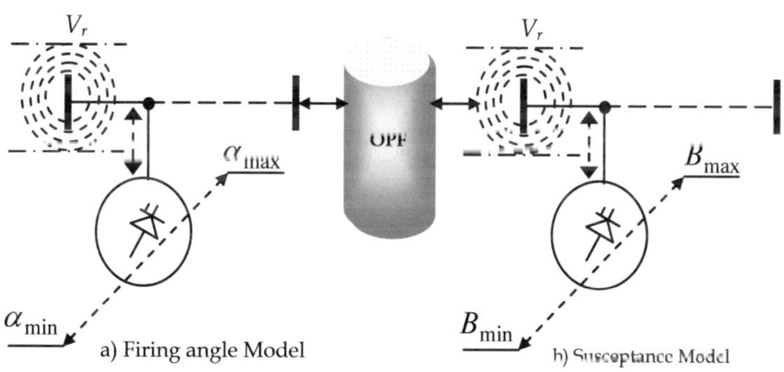

Fig. 12. Steady state model representation of SVC Compensator

$$V = V_{ref} + X_{sl} I \tag{22}$$

X_{sl} is in the range of 0.02 to 0.05 p.u. with respect to the SVC base. The slope is needed to avoid hitting limits. At the voltage limits the SVC is transformed into a fixed reactance. The total equivalent impedance Xe of SVC may be represented by

$$X_e = X_C \frac{\pi / k_X}{\sin 2\alpha - 2\alpha + \pi(2 - 1/k_X)} \tag{23}$$

where $k_X = x_C / x_L$ and $B_e = 1/X_e$

The SVC is usually connected to the transmission system through a step-down transformer, which is treated in a similar manner as other transformers in the system.

Steady-state limits of the firing angle are $90^0 < \alpha < 180^0$

Where partial conduction is obtained. Firing angles less than 90^0 are not allowed, as they produce unsymmetrical current with a high dc component.

7.3 Multi Function Control

The objective function of the multi control functional operation of a coordinated multi DG with shunt FACTS devices is the combination from the prescribed control targets:

$$F_{DG/SVC} = \alpha_1 \left| P - P^{des} \right| + \alpha_2 \left| Q - Q^{des} \right| + \alpha_3 \left| V - V^{des} \right| \quad (24)$$

where P^{des}, Q^{des}, and V^{des} are the control targets of active and reactive power flow along line, and voltage of bus K, respectively. Fig. 13 illustrates the three combined voltage active and reactive power control.

Coefficients α_1, α_2, and α_3 can take 1 or 0 based in the control strategy adopted.

For a power system with N_{DG} and N_{SVC} devices integrated in practical network to enhance the power flow control, the optimization objective is:

$$Min \quad F \quad (25)$$

The mathematical descriptions of the three control modes of coordinated multi DG with shunt Compensators are presented as follows.

Target 1: Bus Voltage Control

The bus Voltage control constraint is given by

$$V_m - V_m^{des} = 0 \quad (26)$$

where V_m^{des} is the desired bus voltage control

Target 2: The active Power Flow Control

$$P_{mk} - P_{mk}^{des} = 0 \quad (27)$$

where P_{mk}^{des} is the desired active power control

Target 3: The Reactive Power Flow Control

$$Q_{mk} - Q_{mk}^{des} = 0 \quad (28)$$

where Q_{mk}^{des} is the desired reactive power control.

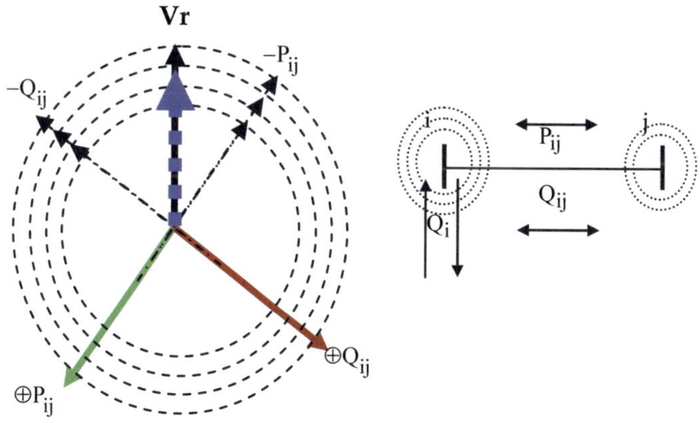

Fig. 13. Three control mode: voltage, active and reactive power control

8. Case Studies

The combined GA/Fuzzy rules is coded in Matlab program, two test cases were used to demonstrate the performance of the proposed algorithm. Consistently acceptable results were observed.

8.1 Case Studies on the IEEE 30-Bus System

The first test is the IEEE 30-bus, 41-branch system, for the voltage constraint the lower and upper limits are 0.9 p.u and 1.1 p.u., respectively, (expect for PV buses where (V_{max} =1.1 p.u.). For the purpose of verifying the efficiency of the proposed approach, we made a comparison of our algorithm with others competing OPF algorithm. Bakistzis et al. (2002) presented an enhanced GA (EGA), Bouktir et al. (2008) presented a standard GA, and an Ant Colony Optimization (ACO) proposed by Bouktir et al. (2007) to solve the economic dispatch problem. Saini et al. (2006) developed a Fuzzy GA (FGA) for OPF. The operating cost in our proposed approach is 801.3445 $/h and the power loss is 9.12 MW which are better than the others methods reported in the literature. Results depicted in Table 1 shows clearly that the proposed approach gives better results. Table 2 shows the results of the reactive power generation and phase angle for the PV bus. Table 3 shows the best solution of dynamic shunt compensation obtained at the standard load demand (Pd=283.4 MW) using reactive power planning.

Variables	GA/FRules	SGA	EGA	ACO	FGA
P1(MW)	180.12	179.367	176.20	181.945	175.137
P2(MW)	44.18	44.24	48.75	47.0010	50.353
P5(MW)	19.64	24.61	21.44	20.5530	21.451
P8(MW)	20.96	19.90	21.95	21.1460	21.176
P11(MW)	14.90	10.71	12.42	10.4330	12.667
P13(MW)	12.72	14.09	12.02	12.1730	12.11
Cost ($/hr)	801.3445	803.699	802.06	802.578	802.0003
Ploss (MW)	9.120	9.5177	9.3900	9.8520	9.494

Table 1. Results of the Minimum Cost and Power Generation Compared to SGA, EGA, ACO and FGA for IEEE 30-Bus

Variables	**GA/Fuzzy Rules**	SGA	FGA
Q1(Mvar)	-4.50	-3.156	-6.562
Q2(Mvar)	30.71	42.543	22.356
Q5(Mvar)	22.59	26.292	30.372
Q8(Mvar)	37.85	22.768	18.89
Q11(Mvar)	-2.52	29.923	21.737
Q13(Mvar)	-13.08	32.346	22.635
θ1(deg)	0.00	0.000	0.00
θ2(deg)	-3.448	-3.674	-3.608
θ5(deg)	-9.858	-10.14	-10.509
θ8(deg)	-7.638	-10.00	-8.154
θ11(deg)	-7.507	-8.851	-8.783
θ13(deg)	-9.102	-10.13	-10.228

Table 2. Results of the Reactive Power Generation and Phase Angle of GA/Fuzzy Rules Compared to SGA, FGA for IEEE 30-Bus.

Shunt N°	1	2	3	4	6	7	8	9
Bus N°	10	12	15	17	21	23	24	29
Best Qsvc [pu]	0.155	0.0798	0.03012	0.0495	0.0615	0.0384	0.0458	0.025
bsh [pu] [14]	0.05	0.05	0.03	0.05	0.05	0.04	0.05	0.03

Table 3. Comparative Results of the Shunt Reactive Power Compensation for IEEE 30-Bus

8.2 Case Studies on the IEEE 25-Bus test System without SVC and DG Installation

The proposed approach has been tested on IEEE 25-bus electrical network. It consists of 25 buses, 35 branches (lines and transformers), 5 generators and 24 loads.

1) Initial OPF with Genetic Algorithm

In this first step, DG and dynamic shunt compensation are not taken in consideration, for the voltage constraint the lower and upper limits are 0.94 p.u and 1.06 p.u., respectively. The GA population size is taken equal 30, the maximum number of generation is 100, and crossover and mutation are applied with initial probability 0.95 and 0.01 respectively. 10 test runs were performed; the convergence of this initial OPF is shown in Fig. 14.

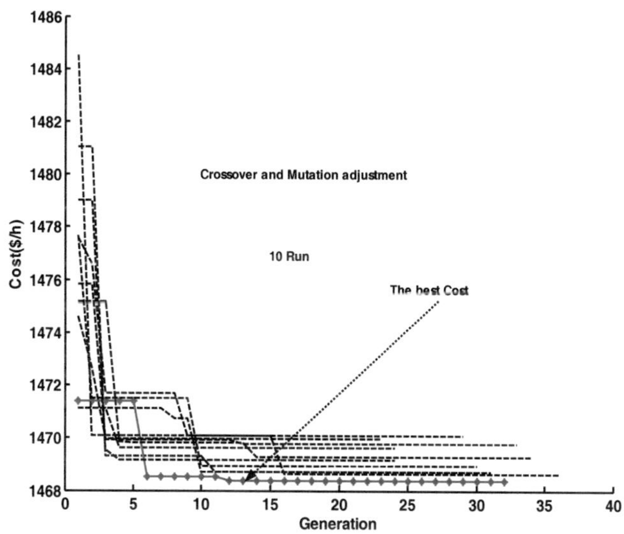

Fig. 14. Convergence for GA for the initial power generation subproblem with crossover and mutation adjustment (IEEE 25-bus).

Bus	|V| (p.u)	Pg (MW)	Qg (Mvar)
1	1.051	164.95	23.357
2	1.052	89.00	7.3780
3	1.043	81.82	31.718
4	1.012	20.98	24.699
5	1.042	184.58	15.600
PD (MW)	530.00		
Ploss (MW)	11.332		
Cost ($/hr)	1467.9		

Table 4. shows the results of the generators voltage, active and reactive power generation without DG and shunt FACTS devices.

2) Optimal Placement of Shunt FACTS and DG

Before the insertion of SVC and DG devices, the system was pushed to its collapsing point by increasing both active and reactive load discretely using continuation load flow. In this test system according to results obtained from the continuation load flow, we can find that based in Fig. 13 that buses 15, 22, 24, 25 are the best location points for initial installation of DG and reactive power planning for multi SVC Compensators.

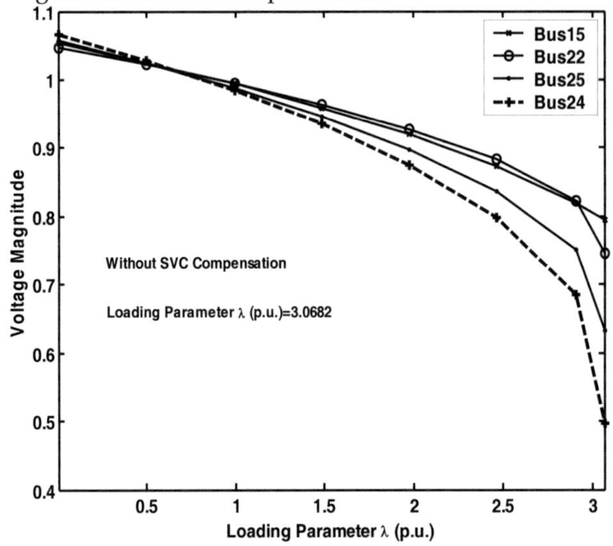

Fig. 13. Critical buses identification using load incrementation without DG and SVC installation (λ=3.0682).

3) Optimal Active Power Planning for DG for Power Loss Reduction

In this step the fuzzy controlled genetic algorithm with the same initial crossover and mutation parameter is used to generate and optimizes the active power of multi distributed generation to minimize power losses. Fig. 15 shows the best solution obtained at different load incrementation.

Loading λ	0%	1%	2%	3%	4%	5%
Dg15	0	0.0130	0.0242	0.0338	0.0482	0.0586
Dg21	0	0.0128	0.0258	0.0396	0.0520	0.0650
Dg22	0	0.0124	0.0172	0.0104	0.0148	0.0450
Dg24	0	0.0018	0.0144	0.0372	0.0442	0.0306
Dg25	0	0.0094	0.0208	0.0348	0.0492	0.0622
Ploss	0.11332	0.11568	0.11076	0.10715	0.10502	0.1036
Cost	1467.9	1469.3	1468.1	1467.2	1466.8	1466.5
PD	5.3000	5.353	5.406	5.459	5.512	5.565
Pg1	1.6495	1.6555	1.6506	1.6466	1.6448	1.6434

Table 5. Results of Active Power Planning with Load Incrementation.

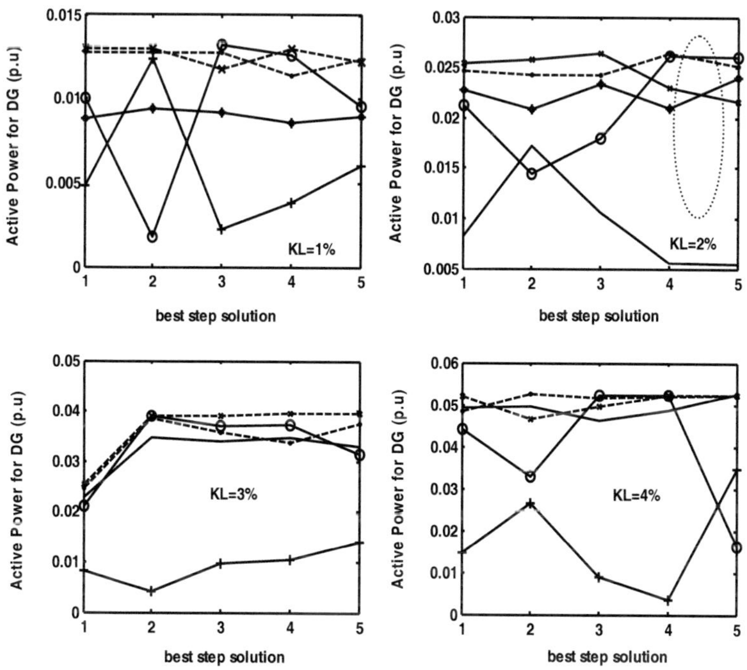

Fig. 15. The best solution (5 run) with load incrementation: (KL=1-4%)

4) System loadability Enhancement with Efficient Reactive Power Planning for Multi SVC Installation

In this second step, the initial SVC data used to control the reactive power are presented in Table 4. To demonstrate the efficiency of the reasoning fuzzy rules designed as a second subproblem to control the reactive power exchanged with the network, the algorithm applied again on the IEEE 25-bus.

From Table 5, it is observed that there is a decrease in power loss due to the integration of DG. Table 7 shows the results of the reactive power planning for the shunt FACTS devices installed at the critical buses. Fig. 16 shows that the system loadability is enhanced (3.4453 p.u) compared to the case (3.0682 p.u) without DG and shunt FACTS devices installation. Fig. 17 shows clearly that the voltage profile is enhanced at the base case (530 MW). Fig. 18 illustrates the exchanged reactive power between the shunt compensators and the network at different loading factor. Fig. 19 shows the active power exchanged between the distributed generation and the network at different loading factor.

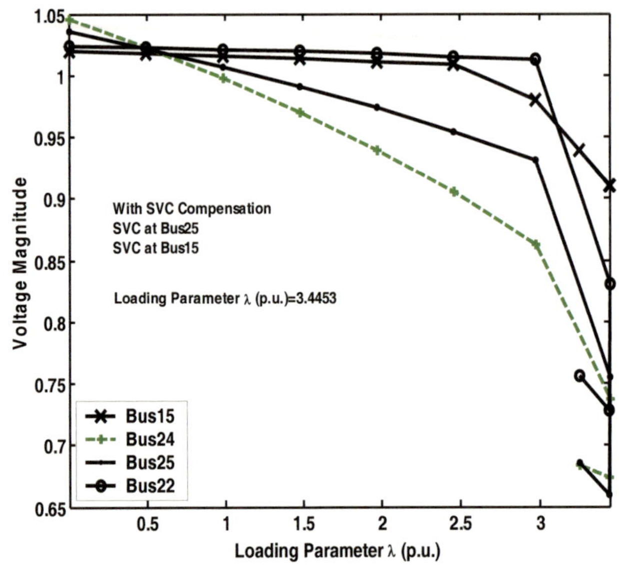

Fig. 16. Voltage profile improvement for the critical buses with SVC installation with load incrementation (λ=3.4453).

Table 6. SVCs data

	Bmin (p.u)	Bmax (p.u)	Binit (p.u)
Susceptance SVC Model	-0.35	0.35	0.035

Reactive Power (p.u)	0%	1%	2%	3%	4%	5%
Q_{svc} 15	0	-0.06545	0.00091	0.06924	0.13723	0.19889
Q_{svc} 21	0	-0.03883	0.03221	0.10459	0.17904	0.24125
Q_{svc} 22	0	-0.05308	-0.04193	-0.02599	-0.01366	-0.01093
Q_{svc} 24	0	0.06026	0.05544	0.04669	0.04390	0.04726
Q_{svc} 25	0	-0.07026	-0.04735	-0.02466	-0.00158	0.02157
Q_{G1}	0.23357	0.32462	0.23159	0.13585	0.03958	-0.04922
Q_{G2}	0.07378	0.07638	0.07898	0.08161	0.08427	0.08682
Q_{G3}	0.34748	0.37618	0.35897	0.34182	0.32468	0.30747

Q_{G4}	0.24699	0.28302	0.25136	0.22124	0.19151	0.15712
Q_{G5}	0.15600	0.19297	0.16237	0.13209	0.10183	0.07083
PD (p.u)	**5.3000**	**5.353**	**5.406**	**5.459**	**5.512**	**5.565**

Table 7. Results of Reactive Power Planning with Load Incrementation.

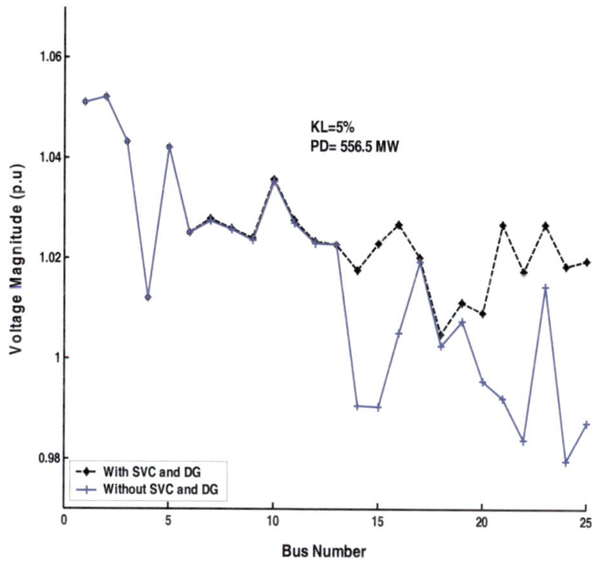

Fig. 17. Voltage profile in two cases; without SVC/DG, and with SVC/DG

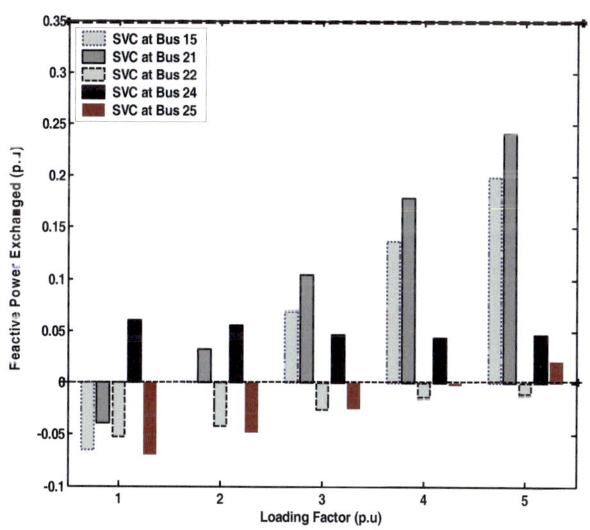

Fig. 18. Reactive power exchanged with the network at different loading factor.

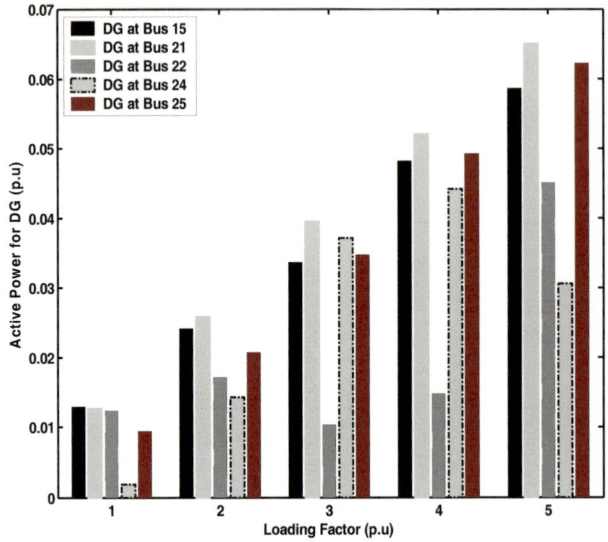

Fig. 19. Active power of DG exchanged with the network at different loading factor.

9. Discussions

It is clear from results depicted in Table 7 that the dynamic voltage control using shunt FACTS Compensators has a significant impact on the potential integration of DG.

The power losses and correspondingly the optimal cost for the standard generation are enhanced at an acceptable technical values considering load incrementation, for example at loading factor KL=5% the power losses reduced to 10.36 MW and the cost maintained to 1466.5 $/hr compared to the base case.

It has found that based on the dynamic reactive index sensitivity introduced the expert engineer can choose economically the size of the shunt Compensators to be installed in a practical network. The proposed new size of shunt dynamic Compensators are depicted in Table 8. The size of the SVC installed at bus 22, 24, 25 reduced from the initial value 0.35 p.u to 0.1 p.u.

Further research is required to include the real cost of DG units into the objective function.

Bus	**Initial SVC Size**	**New SVC Size**	**Marge Security utilization in %**
15	$-0.35 < Q_{svc} < 0.35$	$-0.35 < Q_{svc} < 0.35$	56.83
21	//	$-0.35 < Q_{svc} < 0.35$	68.93
22	//	$-0.10 < Q_{svc} < 0.10$	10.93
24	//	$-0.10 < Q_{svc} < 0.10$	47.26
25	//	$-0.10 < Q_{svc} < 0.10$	21.57

Table 8. SVC Size

10. Conclusion

An approach combining Genetic Algorithm and fuzzy logic expert rules aims to demonstrate the importance of finding the best locations and sizes of a distribution generation to be integrated dynamically in a practical network.
One might think that the larger size of DG or shunt dynamic Compensators, the greater increase in the maximum load, based in experience and results given in this paper that this is not always true. There is a maximum increase on load margin with respect to the compensation level for shunt FACTS devices and active power injected by DG. The objective of the proposed approach is to coordinate and adjust the active power for DG and the reactive power exchanged with dynamic Compensators and the network to minimize fuel cost and to improve the index power quality (voltage deviation, power losses).

11. References

Méndez, V. H.; Rivier, J. & Gomez, T. (2006), Assessment of Energy Distribution Losses for Increasing Penetration of Distribution Generation. *IEEE Trans. Power Systems*, vol. 21, n. 21, 533-5400.

Vovos, P. N.; Kiprakis, A. E.; Wallace, A. R. & Harison, G. P (2007). Centralized and Distribution Voltage Control: Impact on Distribution Generation Penetration. *IEEE Trans. Power Systems*, vol. 22, n. 1, 476-683.

Mahdad, B.; Bouktir, T. & Srairi, K. (2007). Flexible methodology based in fuzzy logic rules for reactive power planning of multiple shunt FACTS devices to enhance system loadability. *Power Engineering Society General Meeting, IEEE,* Page(s):1–6, Digital Object Identifier 10.1109/PES.2007.385750.

Bansal, R. C. (2005). Optimization Methods for Electric Power Systems: An Overview. *International Journal of Emerging Electric Power Systems*, vol. 2, no. 1, 1-23.

Huneault, M.; & Galiana, F. D. (1991). A Survey of the Optimal Power Flow Literature. *IEEE Trans. Power Systems,* vol. 6, no. 2, 762-770.

Keane, A.; & O'Malley. M. (2006). Impact of distribution generation capacity on losses," *in Proc. IEEE,* pp. 1-7.

Wang, C.; & Nehrir, M. H. (2004). Analytical approaches for optimal placement of distribution generation sources in power systems. *IEEE Trans. Power Systems,* vol. 19, no. 4, pp. 2068-2076.

Hanif, A. & Choudhry, M. A. (2006). Investigation smooth power flow control for dispersed generator working parallel to the grid system on the level load. *In Proc. PSCE,* 2241-2248.

Keane, A. & O'Malley, M. (2005). Impact of distribution network constraints on distribution generation capacity. *In Proc. of 40th International Universities power engineering conference, Cork.*

Kuri, B.; Redfern, M. & Li, F. (2004) .Optimization of rating and positioning of dispersed generation with minimum network disruption. *In Proc. IEEE Power Eng. Soc. Gen. Meeting, Denver, CO,* 2074-2078.

El-Khaltam, W.; Bhattacharya, K.; Hegazy, Y. & Salama, M. M. A. (2004). Optimal investment planning of distributed generation in a competitive electricity market. *IEEE Trans. Power Syst.*, vol. 19, no. 3,1674-1684.

Harrison, G. P.; Piccolo, A.; Siano, P. & Wallace, A. R. (2007). Distributed generation capacity evaluation using combined genetic algorithm and OPF. *International Journal of Emerging Electric Power Systems,* vol. 8, Issue. 2.

Mahdad, B.; Bouktir, T. & Srairi. K. (2007). Methodology based in practical fuzzy rules coordinated with asymmetric dynamic compensation applied to the unbalanced distribution network. *International Review of Electrical Engineering (IREE),* vol. 3, no. 2, 145-153, ISSN 1827- 6660, Praise Worthy Prize, Italy.

Bakistzis, A. G.; Biskas, P. N.; Zoumas, C. E.; & Petridis, V. (2002). Optimal power flow by enhanced genetic algorithm. *IEEE Trans. Power Systems,* vol. 17, no. 2, 229-236.

Davis, L. (1989). Adapting operator probabilities in genetic algorithms. *In Proc. 3rd Int. Conf. Genetic Algorithms Applications, J. Schaffer, Ed., SanMateo, CA,* 61-69.

Ng, H. N.; Salama; Chikhani, M. M. A. & Chikhani, A.Y. (2000). Classification of capacitor allocation techniques," IEEE Trans. Power Delivery, vol. 15, no. 1, 387-392.

Yalcinoz, T; Altun, H & Uzam, M. (2001). Economic dispatch solution using genetic algorithm based on arithmetic crossover," *in Proc. IEEE Porto Power Tech. Conf., Porto,* Portugal.

Todorovski, M.; & Rajičič, D. (2006). An initialization procedure in solving optimal power flow by genetic algorithm. *IEEE Trans. Power Systems,* vol. 21, no. 2, 480-487.

Iba, K. (1994). Reactive power optimization by genetic algorithm. *IEEE Trans. Power Systems,* vol. 9, no. 2, pp. 685-692.

Lai, L. L.; Ma, J. T. R.; Yokoyama, & Zhao, M. Improved genetic algorithm for optimal power flow under both normal and contingent operation states. *Elec. Power Energy Syst.,* vol. 19, no. 5, 287-292, 1997.

M. E. H. Golshan, and S. A. Arefifar, "Distribution generation, reactive sources and network-configuration planning for power and energy-loss reduction, " IEE Proc.-Gener. Transm. Distrib., vol. 153, no2, pp. 127- 136, March 2006.

Feurt-Esquivel, C. R.; Acha, E. Tan SG, JJ. Rico. (1998). Efficient object oriented power systems software for the analysis of large-scale networks containing FACTS controlled branches. *IEEE Trans. Power Systems,* vol. 13, no. 2, 464-472.

Bouktir, T.; Slimani, L. & Mahdad, B. (2008). Optimal power dispatch for large scale power system using stochastic search algorithms. *International Journal of Power and Energy Systems,* vol. 28, no. 1, 1-10.

Slimani, L. & Bouktir, T. (2007). Economic power dispatch of power system with pollution control using multiobjective ant colony optimization. International Journal of Power and Computational Intelligence Research, vol. 03, no. 2, 145-153, 2007.

Saini, A.; Chaturvedi, D. K. & Saxena, A. (2006). Optimal power flow solution: a GA-Fuzzy system approach. *International journal of emerging electric power systems,* vol. 5, Issue. 2, 1-21.

Canizares, C. A. (2000). Power flow and transient stability models of FACTS controllers for voltage and angle stability studies. *IEEE Proceeding.*

Dhaoyun, G. & Chung, T. S. (1998). Optimal active power flow incorporating FACTS devices with power flow constraints, *Electrical Power & Energy Systems,* vol. 20, no. 5, 321-326.

16

Flexibility Value of Distributed Generation in Transmission Expansion Planning

Paúl Vásquez[1] and Fernando Olsina[2]
[1]*Consejo Nacional de Electricidad (CONELEC), Ecuador*
[2]*Institute of Electrical Energy (IEE), CONICET, Argentina*

1. Introduction

The efficiency of the classic planning methods for solving realistic problems largely relies on an accurate prediction of the future. Nevertheless, the presence of strategic uncertainties in current electricity markets has made prediction and even forecasting essentially futile. The new paradigm of decision-making involves two major deviations from the conventional planning approach. On one hand, the acceptation the fact the future is almost unpredictable. On the other hand, the application of solid risk management techniques turns to be indispensable.

In this chapter, a decision-making framework that properly handles strategic uncertainties is proposed and numerically illustrated for solving a realistic transmission expansion planning problem.

The key concept proposed in this chapter lies in systematically incorporating flexible options such as large investments postponement and investing in Distributed Generation, in foresight of possible undesired events that strategic uncertainties might unfold. Until now, the consideration of such flexible options has remained largely unexplored. The understanding of the readers is enhanced by means of applying the proposed framework in a numerical mining firm expansion capacity planning problem. The obtained results show that the proposed framework is able to find solutions with noticeably lower involved risks than those resulting from traditional expansion plans.

The remaining of this chapter is organized as follows. Section 2 is devoted to describe the main features of the transmission expansion problem and the opportunities for incorporating flexibility in transmission investments for managing long-term planning risks. The most salient characteristics of the several formulations proposed in the literature for solving the optimization problem are reviewed and discussed along Section 3. The several types of uncertain information that must be handled within the optimization problem are classified and analyzed in Section 4. The proposed framework for solving the stochastic optimization problem considering the value provided to expansion plans by flexible investment projects is presented in Section 5. In Section 6, an illustrative-numerical example based on an actual planning problem illustrates the applicability of the developed flexibility-based planning approach. Concluding remarks of Section 7 close this chapter.

2. The transmission expansion planning

Since the beginning of the power industry, steadily growing demand for electricity and generation commonly located distant from consumption centres have led to the need of planning for adapted transmission networks aiming at transport the electric energy from production sites towards consumption areas in an efficient manner. In the vertically integrated power industry, the responsibility for optimally driving the expansion of transmission networks has typically lied with a centralized planner.

During the last two decades, stimulating competition has been a way to increase the efficiency of utilities as well as to improve the overall performance of the liberalized electricity industry (Rudnick & Zolezzi, 2000; Gómez Expósito). Because of the large economies of scales, a unique transmission company is typically responsible for delivering the power generation to the load points. Under this paradigm, the transmission activity has special significance since it allows competition among market participants. In addition, the transmission infrastructure largely determines the economy and the reliability level that the power system can achieve. For this reasons, planning for efficient transmission expansions is a critical activity. With the aim of solving the transmission expansion planning problem (TEP), a great number of approaches have been devised (Latorre *et al.*, 2003; Lee *et al.*, 2006).

A classic TEP task entails determining *ex-ante* the location, capacity, and timing of transmission expansion projects in order to deliver maximal social welfare over the planning period while maintaining adequate reliability levels (Willis, 1997). Under this traditional perspective, the TEP problem can be mathematically formulated as a large scale, multi-period, non-linear, mixed-integer and constrained optimization problem. In practice, however, such a rigorous formulation is unfeasible to be solved. Planners typically solve the TEP problem under a very simplified framework, e.g. static (one-stage) formulations, where timing of decisions is not a decision variable (Latorre *et al.*, 2003).

2.1 The emerging new TEP problem

The improvement of computing technology with increasingly faster processors along with the option of solving the problem in a distributed computing environment has made possible to handle a bigger number of parameters and variables and even formulate the TEP as a multi-period optimization problem (Youssef, 2001; Braga & Saraiva, 2005). However, jointly with the above mentioned increasing competition brought by the deregulation, relevant aspects such as: the development of new small-scale generation technologies (Distributed Generation, DG), the improvement of power electronic devices (e.g. FACTS), the environmental concerns that makes more difficult to obtain new right-of-way for transmission lines, the lack of regulatory incentives to investing in transmission projects, among others, have increased considerably the dynamic of power markets, the number of variables and parameters to be considered, and the uncertainties involved. Accordingly, the TEP problem is now substantially more complex (Buygi, 2004; Neimane, 2001).

Under this perspective, *ad doc* adjustments of expansion plans or additional contingent investments made in order to mitigate the harmful economic consequences that unexpected events have demonstrated the limited practical efficiency of applying classic TEP models (Añó *et al.*, 2005). In fact, the substantial risks involved in planning decisions emphasize the need of developing practical methodological tools which allow for the assessment and the risk management.

2.2 Nature of transmission investments

Due to some singular characteristics, transmission investments exhibit a distinctive nature with respect to other related investment problems (Kirschen & Strbac, 2004; Dixit & Pindyck, 1994):

Capital intensive: because of the substantial economies of scale, large and infrequent transmission investments are often preferred, involving huge financial commitments.

One-step investments: a substantial fraction of total capital expenditures must be committed before the new transmission equipment can be commissioned.

Long recovering times: transmission lines, transformers, etc. are expected to be paid-off after several years or even decades.

Long-run uncertainties: transmission investments are vulnerable to unanticipated scenarios that can take place in the long-term future. Future demand, fuel costs, and generation investments are uncertain variables at the planning stage.

Low adaptability: transmission projects are typically unable to be adapted to circumstances that considerably differ from the planning conditions. An unadapted transmission system entails considerable loss of social welfare.

Irreversibility: once incurred, transmission investments are considered sunk costs. Indeed, it is very unlikely that transmission equipment can serve other purposes if conditions changes unfavourably. Under these circumstances the transmission equipment could not be sold off without assuming significant losses on its nominal value.

Postponability: In general, opportunities for investing in transmission equipment are not of the type "now or never". Thus, it is valuable to leave the investment option open, i.e. wait for valuable, arriving information until uncertainties are partially resolved. Thus, transmission investment projects can be treated in the same way as a financial call option. The opportunity cost of losing the ability to defer a decision while looking for better information should be properly considered.

Due to the mentioned features, transmission network expansions traditionally respond to the demand growth by infrequently investing in large and efficient projects. Consequently, traditional solutions to the TEP inevitably entail two evident intrinsic weaknesses:

- Because only large projects are economically efficient, planners have a limited number of alternatives and consequently the solutions found provide low levels of adaptability to the demand growth, and
- To drive the expansion, enormous irreversible upfront efforts in capital and time are required.

The huge uncertainties of the problem interact with the irreversible nature of transmission investments for radically increasing the risk present in expansion decisions. Such interaction has been ignored in traditional models at the moment of evaluating expansion strategies. More recently, it has been recognized that conventional decision making approaches usually leads to the wrong investment decisions (Dixit & Pindyck, 1994). Therefore, the interaction between uncertainties and the nature of transmission investments must be properly accounted for.

2.3 New available flexible options

Although the major negative concerns regarding classic TEP models have been analyzed, in this work potential positive aspects are also considered and exploited. In fact, available *technical and managerial embedded options* exhibit some desirable features such as: modularity, scalability, short lead times, high levels of reversibility, and smaller financial commitments. This option can be incorporated as novel decision choices that a planner has available for reducing the planning risks as well as for improving the quality of the found solutions.
In this sense, planners must rely on an expansion model able to capture all major complexities present in the TEP in order to properly manage the involved huge long-term uncertainties and deal with the problem of dimensionality.
The key underlying assumption of conventional probabilistic models is the passive planner's attitude regarding future unexpected circumstances. In fact, available choices for reacting to the several scenarios which could take place overtime are ignored during the planning process. However, in practice planners have the ability to adapt their investment strategies in response to undesired or unanticipated events.
Hence, planning for contingent scenarios by exploiting technical and managerial options embedded in transmission investment projects is a effective mean for satisfactorily dealing with the current TEP problem

2.4 The flexibility value of Distributed Generation

Distributed Generation is defined as a source of electric energy located very close to the demand (Ackerman *et al.*, 2001; Pepermans *et al.*, 2003). Usually, DG investments are neither more efficient nor more economic than conventional generation or transmission expansions, which still enjoy of significant economies of scale such. Nevertheless, important contributions of DG occur when: energy T&D costs are avoided, demand uses it for peak shaving, losses are reduced, network reliability is increased, or when it lead to investment deferral in T&D systems (Jenkins *et al.*, 2000; Willis & Scott, 2000; Brown *et al.*, 2001; Grijalva & Visnesky, 2005).
DG seems a plausible means of improving the traditional way of driving the expansion of the transmission systems. Delaying investments in T&D systems by investing in DG is one of the major motivations and research topics of this work (Brown *et al.*, 2001; Daly & Morrison, 2001; Vignolo & Zeballos, 2001; Dale, 2002; Vásquez & Olsina, 2007).
The fact of considering DG projects as new decision alternatives within the TEP, involves the incorporation of additional parameters such as investment and production costs of DG technologies, firm power, etc.
Based on the typical short lead times of DG projects and their lower irreversibility, the uncertainty present in DG project investment decisions and investment costs can be neglected. Provided that the DG technologies considered in this work are fuel-fired plants, the availability of the DG could be modelled by assessing only availability factors (Samper & Vargas, 2006).

3. State-of-art of the TEP optimization approaches

The successful development of an efficient and practical expansion model primarily depends on considering the following topics: the planner's objectives, the availability and quality of the information to be handled as well as the depth level at which the planner decides to face the problem. In this sense, a set of basic elements that the planner must consider and specify before mathematically formulating the problem are summarized in the Table 1.

Topic	Concern	Recommended Value	Symbol
Scales of time	Planning horizon	10 to 15 years	T
	Decision periods	≥ 1 year	p
	Sub-periods resolution	Weekly, monthly, seasonally	*subp*
	Demand duration curve	Peak, valley, mid-load	$P(t)$, $Q(t)$
Decision alternatives	Alternatives that planner has available for driving the expansion	Expansion strategy	S_k, S_f
		Large transmission projects	$D_k(p)$
		Defer transmission projects	$O_k(p)$
		Invest in DG projects	
		Type of alternative	$[0,1,2,3..n]$
		Investment decision timing	p
		Decision alternative location	$\overline{f}(bus)$
Objective function (C_k) components	Efficiency in investments, operative efficiency, reliability and technical feasibility	Investment costs	C_I, C_{IDG}
		Operative costs	C_G, C_{GDG}
		O&M costs	$C_{O\&M}$
		VOLL or EENS costs	C_{LOL}
		Active power losses costs	-
Constraints	Transmission expansion plans performance assessment subject to:	Power balance	$S_G + S_D = S_L$
		Voltage limits	$V_{j\,min}$, $V_{j\,max}$
		Generators capacity limits	$P_{i\,min}$, $P_{i\,max}$
		DG plants capacity limits	$DG_{i\,min}$, $DG_{i\,max}$
		Transmission lines power flow limits	F_l
		Budgetary constraints	-
Input parameters	Certain	Certain	$S(t)$
	Uncertain	Random	$X(t)$
		Truly uncertain	
		Fuzzy	-

Table 1. Basic elements to be defined before devising a TEP methodology

The current TEP problem can be described as the constant planners' dilemma of deciding on a sequential combination of large transmission projects and new available flexible options, which allows the planners to efficiently adapting their decisions to unexpected circumstances that may take place during the planning period.

Under this novel paradigm, TEP is a multi-period decision-making problem which entails determining *ex-ante* the right type, location, capacity, and timing of a set of available decision options in order to deliver a maximal expected social welfare as well as suitably reducing the existing risks over the planning period.

Probabilistic decision theory, i.e. the probabilistic choice paradigm, is well-known and has been extensively applied in several stochastic optimization problems. However, a probabilistic decision formulation within the TEP is an intractable task and its application has only been feasible when very strong simplifications are adopted by planners (Neimane, 2001). This work proposes a practical framework for treating the TEP. Even though a number of simplifications are still necessary, the main features of the new TEP problem are retained.

The analysis of the state-of-art of the TEP solutions approaches sets as a start point the classic stochastic optimization problem formulation. Under the assumption of inelastic demand behaviour, the optimization problem can be rigorously stated as follows:

$$\underset{S_{opt}\in \bar{S}_f}{opt}\left\{E[OF|_S]\right\} = \underset{S_{opt}\in \bar{S}_f}{opt}\left\{\int_0^T \overbrace{\int \ldots \int_\Omega}^{i} OF(C)dF(C)\right\} \quad (1)$$

where, the performance measure of the optimization is the expected present value of the objective function E[*OF(C)*] evaluated over a planning horizon *T*, for a proposed expansion strategy *S*. $\overline{S_f}$ is the set of all feasible states of the network, *F(C)* is the distribution function of the expansion costs function $C(C_1, C_2, C_3, \ldots, C_i)$. The planning period *T* usually only can take discrete values $t_0, t_1, t_2, t_3, \ldots, t_p$, and Ω is the domain of existence of C(X,S). The expansion costs function depends on several uncertain input parameters $X(x_1(t), x_2(t), x_3(t), \ldots, x_n(t))$ which change over the time, as well as depending on the state of the network, which also varies over the time $S(s_1(t), s_2(t), s_3(t), \ldots, s_d(t))$. It is important to note that the problem is subject to a set of constraints, namely Kirchhoff's laws, upper and lower generation plants capacity limits, transmission lines capacity limits, upper and lower voltage and phase nodes limits, and budgetary constraints, among others, which are represented by means of equality and inequality equations. With these considerations, (1) can be rewritten as follows:

$$\underset{S_{opt}\in \bar{S}_f}{opt}\left\{E[OF]\right\} = \underset{S_{opt}\in \bar{S}_f}{opt}\left\{\overbrace{\int \ldots \int_\Psi}^{n.(p+1)} OF[C(X,S)]d\Phi(X)\right\} \quad (2)$$

subject to:

$$P_A(X,S) + P_B(X,S) = P_L(X,S)$$
$$b_{1\min} \le g_1(X,S) \le b_{2\max}$$
$$b_{1\min} \le g_2(X,S) \le b_{2\max}$$
$$\vdots \qquad \vdots \qquad \vdots$$
$$b_{m\min} \le g_m(X,S) \le b_{m\max}$$

where $\Phi(X)$ is the $n(p+1)$-dimensional function of probability distributions of input parameters and Ψ is the domain of existence of the input parameters X.

Formulating $\Phi(X)$, which incorporates the information about the uncertainties that largely influence the solution, is a complex task as it involves determining probabilities and distribution functions of $n(p+1)$ uncertain parameters. However, the more difficult (and in some cases impossible) task is the formulation of the objective function *OF(C)*. In this sense the most common simplification considered by TEP models is $OF[C(X,S)]=C(X,S)$ and (2) can be rewritten as:

$$\min_{S_{opt}\in S_f}\{E[C|_S]\}=\min_{S_{opt}\in S_f}\left\{\overbrace{\int\ldots\int}^{n.(p+1)}_{\Psi} C(X,S)d\Phi(X)\right\} \tag{3}$$

which implies that the objective function can be entirely described by the expansion costs function. In this case the planning problem is often reduced to the minimization of the expected total expansion costs. Although the complexity of the problem is greatly reduced, such a formulation does not take into account desires of the decision-maker for reducing risks present in the expansion decisions. Eventually, this risk neutral formulation may lead to wrong decisions.

On the other hand, considerable difficulties are related to the computational effort necessary for efficiently assess the multidimensional integral and for proposing the corresponding optimization procedure. The only method for dealing with (3) as strict as possible, given that the $n(p+1)$-dimensional integral must be solved, is applying Monte-Carlo simulation techniques for evaluating the attributes of the objective function.

There are $(n+d)(p+1)$ input parameters in the expansion costs function $C(x_{1,0},\ldots x_{n,0},\ldots,x_{n,p},\ldots s_{1,0},\ldots s_{d,p})$, from which $d(p+1)$ are decision variables. Assuming as I the number of available decision choices in each possible right-of-way d, the number of possible candidate solutions are $I^{d(p+1)}$. Additionally, by denoting as N the number of simulations that requires the Monte Carlo simulation, the number of simulations to be performed depends on the number of periods of time as $N(p+1)$. It is important to mention that N depends on the degree of confidence that the planner demands on the results. Under these considerations, the number of required computations for rigorously evaluating the multidimensional integral and therefore for finding the global optimum is $N(p+1)I^{d(p+1)}$. Unfortunately performing this task in a real multi-period TEP is not possible since the number of simulations dramatically increases with the result of multiplying the possible links and the time periods $d(p+1)$. Due to this fact, researchers have proposed diverse approaches in order to make the TEP feasible and, in some cases, to incorporate the desires of the decision-maker for reducing the planning risks. According to the reviewed literature such simplifications can be categorized as static, deterministic and non-deterministic formulations of the TEP.

3.1 Static formulation

When the planner demands on further simplifying a deterministic formulation, the intertemporal dependences and the dynamic nature of the TEP problem is not considered. Such a formulation is named static. This is a deterministic formulation that entails finding the optimal state of the network for a future fixed year. Consequently, the input parameters X do not change during the whole solving process. In this case, there are $n+d$ input parameters within the expansion costs function $C(x_1,x_2,\ldots,x_n,\ldots s_1,\ldots s_d)$ from which d are decision variables. Assuming as I the number of available decision choices in each possible right-of-way d, the number of possible solutions is I^d. For instance, in a small TEP problem with d = 11 and five decision choices on each right-of-way I = 5, the number of possible combinations is $5^{11} = 4.88 \cdot 10^7$.

3.2 Deterministic formulation

Deterministic models are nowadays widely used in practice for transmission network planning. This type of models assumes that all the input parameters and variables are known with complete certainty and, therefore, there is a unique and known scenario for the evolution of all input parameters. Consequently, there is no need to use probability distribution functions and the complexity of the optimization process is greatly reduced. Thus, deterministic formulation entails finding the optimal state of the network over a planning horizon T, given that the evolution of X along the time is known with certainty.

There are $(n+d)(p+1)$ input parameters inside the expansion costs function $C(x_{1,0},...x_{n,0},...,x_{n,p},...s_{1,0},...s_{d,p})$ from which $d(p+1)$ are decision variables. Assuming as I the number of available decision choices in each possible right-of-way d, the number of possible solutions to be evaluated for finding the global optimum is $I^{d(p+1)}$. For instance, in a small TEP problem with eleven possible new right-of-ways d = 11, five decision choices in each right-of-way I = 5, and only two decision periods p+1 = 2, the number of possible combinations are $5^{11.(1+1)} = 2.38 \cdot 10^{15}$.

In this work, the subject of optimization is the present value of the total expansion costs function C(X,S), evaluated along a planning horizon T, for a proposed expansion strategy S. C(X,S) is a non-linear function subject to a set of constraints, i.e. Kirchhoff's laws, generation plants capacity limits and transfer capacity of transmission lines, among others. Such constraints are represented by means of equality and inequality equations.

$$\min_{S_{opt} \in S_f} \{C|_S\} = \min_{S_{opt} \in S_f} \{C(X,S)\} \tag{4}$$

$$C(X,S) = \sum_{t=0}^{T} \left[\frac{C_I(X,S)}{(1+r)^t} + \frac{C_{Gen}(X,S)}{(1+r)^t} + \frac{C_{O\&M}(X,S)}{(1+r)^t} + \frac{C_{LoL}(X,S)}{(1+r)^t} \right] \tag{5}$$

subject to:

$$\begin{aligned} &P_A(X,S) + P_B(X,S) + ... + P_R(X,S) = P_L(X,S) \\ &b_{1\min} \le g_1(X,S) \le b_{2\max} \\ &b_{1\min} \le g_2(X,S) \le b_{2\max} \\ &\quad \vdots \qquad \vdots \qquad \vdots \\ &b_{m\min} \le g_m(X,S) \le b_{m\max} \end{aligned}$$

where

$C_I(X,S)$: Investment costs of the new expansion decisions.

$C_{Gen}(X,S)$: Production costs of the different generations units.

$C_{O\&M}(X,S)$: Annual O&M costs of the transmission network elements.
$C_{LoL}(X,S)$: Loss of load annual costs.
r : Annual discount rate.

3.3 Non-deterministic formulation

Basically non-deterministic formulations of the TEP problem are able to consider the possible events which could take place in the future by taking into account the uncertainty present in the information. In this category, the TEP problem can be solved either by means of a stochastic optimization-based formulation, where the objective function is typically formulated in term of an expected value or by means of a decision-making framework, which encompasses a deterministic optimization plus a decision tree analysis. Unfolding uncertainties are incorporated as branches and decisions are made on the evaluation of the consequences of deciding on the different expansion alternatives. In this sense, the decision-making framework allows the planners to gain insight into the risks involved in each expansion choice and could even suggest new and improved alternatives.
The dimension of the search space for the different TEP formulations depends on the number of decision choices, the number of decision variables and the number of periods. Additionally, the degree of detail of the model describing the temporal evolution of the PES along the planning horizon, namely demand discretization, time resolution and extent of the planning horizon is another important aspect to take into account since the computational effort for evaluating each combination depends on it.
To reasonably accomplish the challenging task of solving the TEP problem from a non-deterministic perspective, require incorporating and modeling a variety of data of diverse nature. Moreover, due to the large problem size, which is clearly defined by its stochastic, multi-period, multi-criteria and combinatorial nature, substantial efforts are required in order to sustain the viability of the proposed models. In this sense, an adequate treatment of the different types of the information is one of the most important stages before formulating the non-deterministic TEP model.

4. Handling information within the TEP

The process of solving actual planning problems requires handling a large amount of information from which only a small fraction is known with complete certainty. In this section, the major uncertainties affecting the TEP and referred to as variables that affect the outcomes of decisions and which are not known at time of planning, are analyzed and categorized from a descriptive viewpoint. Excluded here are the uncertainties originated in the model's user, i.e. what is not captured by the model but desired by the user, as well as uncertainties originated in the model (i.e. the "right" model structure, modelling techniques and tools).

4.1 Uncertainties present in the TEP

Data about the current state of the network is much more accurate than forecasted data. Furthermore, uncertainties present in forecasted data are very diverse in nature (Neimane,

2001). Therefore, it is recognized the importance of categorizing the uncertain information to be incorporated within TEP models.

In this work, it is assumed that forecasts and characterization of the forecast uncertainty are provided to the planning activity. Instead, the attention of this research work is posed in categorizing all the information to be handled within the TEP and proposing a systematic methodology for properly incorporating uncertain information of various source and nature within the TEP model.

4.2 Certain Information

Certain data are those parameters which can be defined explicitly (Neimane, 2001). This category includes the present network configuration, electrical parameters of the network components, possible expansion choices and their electrical parameters capacity limits of transmission lines, nominal voltages and voltage limits.

4.3 Information subjetc to stochastic uncertainty

Uncertainty in data mostly appears due to the inevitable errors incurred when forecasts are performed. When it is possible to objectively assess the magnitude of such errors with a satisfactory degree of confidence, then the uncertainty is said to be of random nature (Buygi, 2004). The uncertainty of such variables can be adequately represented by means of probability distribution functions. Demand, fuel prices and hydrologic resources evolution are typical examples belonging to this category. In (Vásquez *et al.*, 2008) a well-founded means for modelling random uncertainties is extensively presented.

4.4 Uncertain non-random information

When it is not possible to estimate with a satisfactory degree of confidence the errors incurred when forecasts are performed, information is deemed to be of a non-random nature (Buygi, 2004). Uncertainties in this group are related to human processes (e.g. investors decisions, changes in regulation, planners and managers investment strategies, beliefs or subjective judgments). In fact, the future does not appear to be predictable through extrapolation of historical trends applied to the current environment (Clemons & Barnett, 2003). Thus, non-random uncertainties assessment is derived from decision-makers perception, experience, expertise and reasoning. Inside this group there are two types of uncertainties.

The first type belongs to a large amount of valuable information that only can be expressed in linguistic form, e.g. "satisfactory", "considerable", "large", "small", "efficient", etc. Although this **vague information** has a very subjective nature and usually is based on expert judgment, it can be useful during the decision-making process. Fuzzy sets theory is a well-founded approach for modelling properly these kinds of uncertainties (Buygi, 2004).

The second type of non-random uncertain information is distinguished by holding uncertainties typical of dynamic environments that undergo severe and unexpected changes. This is the case with the TEP environment. According to the literature, these kinds of uncertainties are known as **strategic uncertainties** (Clemons & Barnett, 2003; Brañas et al., 2004; Detre et al., 2006). A specific feature of them is that they are gradually solved as new information arrives over time and, once enough information is known, the uncertainty is solved and disappears definitively (Dillon & Haimes, 1996; Clemons & Barnett, 2003).

Within the TEP problem, this uncertainty affects crucial events that could take place in the future, such as the generation expansion evolution or the delay on the expansion projects completion. Data with strategic uncertainties are considered the most important information to be handled within TEP since they are fundamental drivers of PES evolution and, therefore, of this decision-making problem. For further reading about this topic see (Detre *et al.*, 2006). On the other hand, within the PES planning environment, there are not much bibliographic references about modelling of strategic uncertainties in planning models. In (Neimane, 2001), this type of information has been designated as **truly uncertain information**[1]. Either discrete probability distribution functions or a scenarios technique are proposed for modelling information of this kind.

Taking into account the above mentioned, in this work it is proposed to model truly uncertain information by means of discrete probability distribution functions (PDF) where the probabilities assigned to the occurrence of different scenarios are assumed as known information. In this sense, a reasonable way for dealing with these two types of uncertainties is proposed in (Vásquez *et al.*, 2008).

5. The proposed flexibility-based TEP framework

The described TEP problem can be suitably faced by applying the decision tree technique, which basically consists in decomposing the whole problem into a number w of less complex sub-problems, each one concerned with solving a multi-period deterministic optimization as well as assessing the attributes of the expansion plans.

A sub-problem or complete path is represented by a number P of sequential discrete events. Such events are specified by the assumed discrete nature of strategic uncertainties. Under these conditions, each sub-problem handles only random uncertainties. Therefore, the different feasible expansion plans can be valued by applying a probabilistic analysis of the attributes of the objective function and decisions are made by applying a robustness-based risk management technique.

A master dynamic programming (DP) problem, by means of a backward induction of P sequential decisions, makes it possible to incorporate flexible options and, subsequently, rank the new flexible expansion strategies.

The entire proposed methodology, can be described as follows in five stages and illustrated in Fig. 1:

1. To decompose the TEP problem into w sub-problems
2. To obtain a set of feasible expansion plans for each sub-problem w.
3. To assess the OF attributes of the different expansion plans for each path w.
4. To sequentially incorporate in the expansion plans, starting from the last decision period P, new flexible decisions for each path w.
5. To form flexible expansion strategies, by repeating 3 and 4 with backward induction until $P = 1$.

---

[1] This term refers to relevant non-random uncertain variables, which convey strategic information.

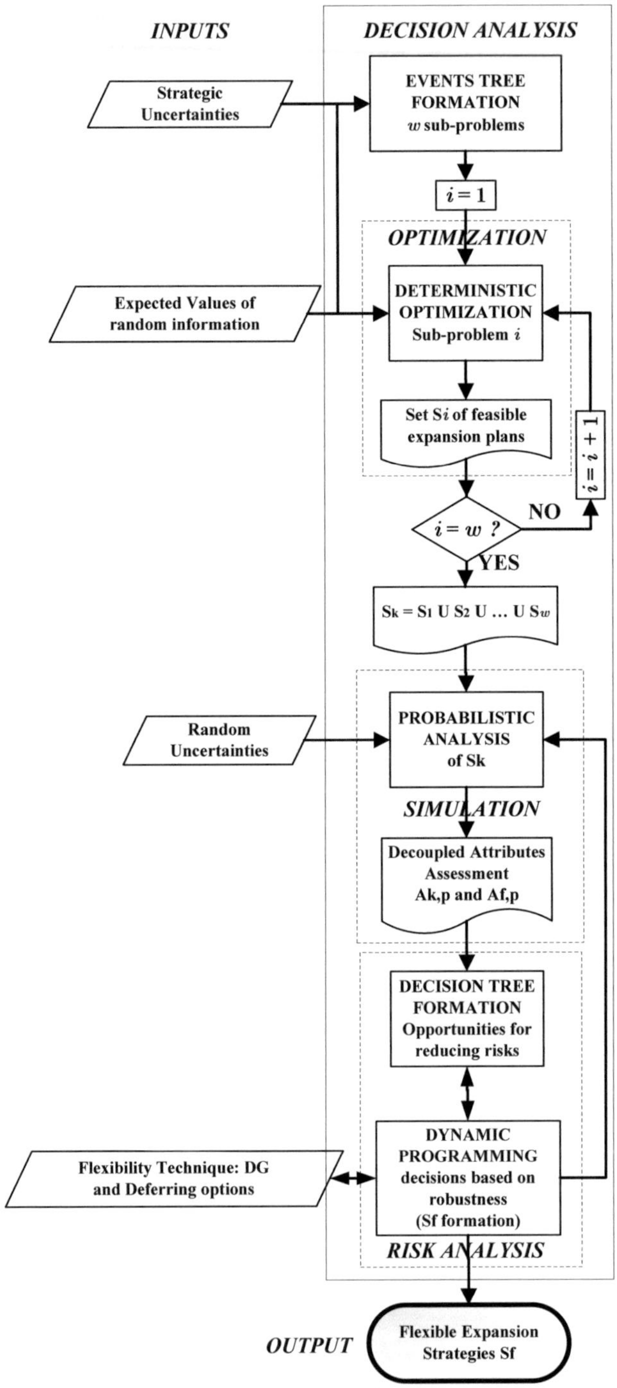

Fig. 1. Complete proposed framework for finding a flexible strategy

5.1 Decomposing the problem

The reason why optimization-based TEP models are inefficient is the presence of uncertainties. In fact, one of the most important concerns within the current TEP problem lies in suitably handling a large amount of uncertain information of diverse natures.
The traditional TEP formulations commonly reduce the future into an assumed probability-weighted certainty equivalent. This fact, in presence of strategic uncertainties implies averaging highly different scenarios. However, in practice equivalent scenarios will never take place since the future can unfold as either favourable or adverse. Therefore, stochastic optimization models formulated in terms of expected values are not suitable approaches for treating the TEP.
Event tree technique is a graphic tool that provides an effective structure for decomposing complex decision-making problems under the presence of uncertainties. The interested reader in decision tree analysis technique is further referred to Dillon & Haimes, 1996 and Majlender, 2003.

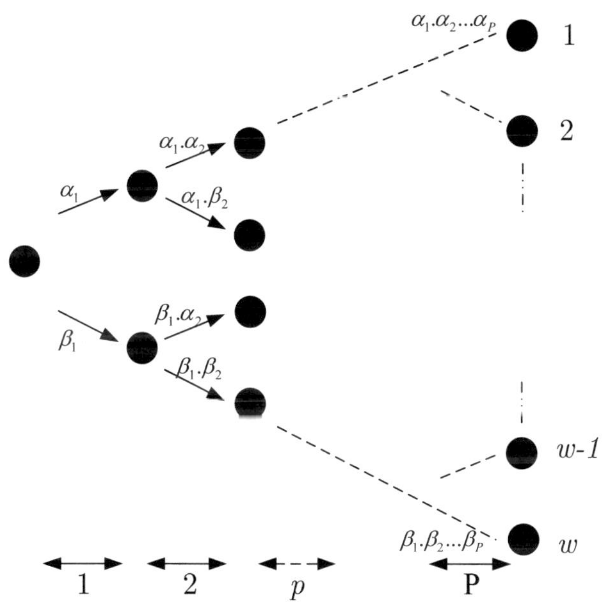

Fig. 2. Example of a binomial event tree

Fig. 2 depicts an example of a resulting events tree formed by assuming that the whole of the problem's strategic uncertainties can unfold into only two discrete scenarios. A complete event tree representing crucial states of the problem along the planning horizon allows getting insight about the diverse future circumstances, which candidate expansion plans should cope with.
Nodes of the event tree represent an explicit feasible scenario obtained as a result of combining all the possible discrete probability distributions of uncertain events along a discrete time p-decision periods. Each event is associated with composed occurrence probability which results from combining the discrete subjective probabilities assigned to the occurrence of a single uncertain event (α_p, β_p, ...) and provided that the occurrence of such probabilities are independent of what happened in previous periods as shown in Fig. 2.

5.2 Obtaining a set of feasible expansion plans

The goal of this stage of the planning process lies in successfully reducing the dimension of the TEP by finding a set of feasible candidate expansion plans which fulfil fundamental constraints of the sub-problem. By reducing the search space, a rigorous economical and risk-based assessment of a reduced set of feasible expansion plans in subsequent stages turns practicable.

Under the scope of this work, it is assumed that the regulatory entity annually executes a centralized TEP task, in which a set of environmental, societal and political long-term energy policies must be achieved. In fact, the previous performance of environmental, societal and political feasibility assessments reduces the large number of decision alternatives to be considered by planners for searching candidate expansion plans for driving the expansion of the transmission grid. It is assumed that a number of possible transmission expansion alternatives have indentified. Despite this, the number of possible combinations of sequential decisions, i.e. the potential solutions, is still enormous. Since only a reduced number of combinations will meet the constraints of the TEP sub-problem, a technical efficiency-based assessment is a plausible means for reducing the search space and finding a set of technically feasible expansion plans.

The TEP sub-problems are formulated as a deterministic multi-period optimization and an evolutionary algorithm has been developed for properly solving such optimization problem (Vásquez, 2009)

Why is deterministic optimization the best choice?

The major foundations of this work for deciding on the deterministic choice lie in the nature of the TEP problem as well as in the problem decomposition proposed in the previous section. In fact, since only a reduced number of combinations will meet the TEP problem's constraints, and given that location, timing and type of the transmission expansion alternatives are discrete and limited in number, feasible candidate solutions are therefore also limited in number and noticeably different from one another. On the other hand, with the proposed decomposition of the TEP into sub-problems, strategic uncertainties have been removed temporarily. In this sense, the only presence of random information, which implies that uncertainties can be forecasted with a satisfactory degree of confidence, allows for a suitable technical assessment where the uncertain input variables are explicitly modelled by means of expected values.

5.3 Assessing the performance of expansion plans

The reduced number of candidate solutions allows a more detailed valuation of the expansion plans. This stage of the planning process entails performing a probabilistic technical-economical performance assessment of all the feasible expansion plans. The performance assessment of an expansion plan is achieved by accounting for a group of decoupled attributes of the objective function. Decoupled attributes A_k denote a measurement of the relative "goodness" of a specific transmission expansion plan S_k in every decision period p expressed by means of its probability distribution $F_{k,p}$. These p probability distribution functions represent the likelihood of the possible future values that the *OF* could acquire over time, characterizing the time-dependent risk profile of selecting the expansion plan S_k.

Stochastic simulation

Stochastic simulation techniques are applied for modelling the randomness of the objective function. In spite of the large computational effort demanded by Monte Carlo methods, the most significant advantage of the simulative approach over analytical probabilistic techniques is the accurate estimation of the tails of probability distribution $F_{k,p}$.

On the other hand, some planning engineers may worry about a possible conflict between the proposed deterministic optimization stage and the subsequent probabilistic and risk analysis stages. In fact, there is no conflict at all provided that all the feasible expansion plans have been found during the deterministic analysis stage. The probabilistic analysis stage is not intended to replace the deterministic TEP models, but to add better information on the merits of the expansion plan and its risk profile. This goal is achieved by assessing the time-decoupled attributes for every feasible expansion plan.

The total attributes of a specific expansion strategy S_k, A_k comprise all the information enclosed in the probability distributions $F_{k,p}$, which describe the possible future performance of S_k provided that all the problem uncertainties (random and strategic) have been taken into account during the simulative process (Neimane, 2001). If such resulting probability distribution function, defined in this work as F_1, can be fit to a Gaussian distribution, A_k can be expressed as follows:

$$A_k = F_{1,k}\left(\bar{C}_k, \sigma_k\right) \tag{6}$$

where,

$\bar{C}_k = \frac{1}{N_k}\sum_{i=1}^{N} C_{k,i}$: OF's Expected Value for S_k

$C_{k,i}$: OF's value of the strategy S_k during the realization i. See Equation (5)

A_k : Total Attributes of the stratey S_k

$F_{1,k}$: OF's Probability Distribution for strategy S_k

N_k : Number of realizations until achieve the required confidence in determining $F_{1,k}$

σ_k : OF's standard deviation for strategy S_k

Although an assessment of A_k provides the information about the performance of an expansion strategy, the planner is unable to visualize the risk evolution over time and the effects on the OF's performance caused by the diverse type of uncertain variables. Nevertheless, having this information is a key issue for properly tackling the TEP. One of the major contributions of this work lies in successfully coping with these two concerns. In first place, Section 4.1 proposed to decompose the problem by applying the event tree technique. In second place, under the assumption that each node of the events tree represents one event unfolded by the combination of strategic uncertainties, a set of decoupled attributes where only random nature uncertainties are present needs to be evaluated. Under this perspective, by performing w Monte-Carlo realizations and then, by means of backward induction and considering the associated cumulative occurrence probabilities, the individual effects of the strategic uncertainties can properly be accounted for, from the last decision period until the first one. At the same time, the diverse time-

decoupled attributes of an expansion strategy F_p are assessed, step by step, until its total attributes F_1 are obtained.
At the end of this valuation process, F_1, which represents the total attributes of the analyzed expansion strategy, is obtained as follows:

$$
\begin{aligned}
F_P\left(\bar{C}_P,\sigma_P\right)\Big|_{w_{P-1}} &= \left[f_{P-1,j}\left(\bar{c}_{P-1,j},\nu_{P-1,j}\right) + \sum_{i=1}^{s_P} \alpha_{P,i}\cdot f_{P,i}\left(\bar{c}_{P,i},\nu_{P,i}\right)\right]\Bigg|_{w_{P-1}} \\
F_{P-1}\left(\bar{C}_{P-1},\sigma_{P-1}\right)\Big|_{w_{P-2}} &= \left[f_{P-2,k}\left(\bar{c}_{P-2,k},\nu_{P-2,k}\right) + \sum_{j=1}^{s_{P-1}} \alpha_{P-1,j}\cdot F_P\left(\bar{C}_P,\sigma_P\right)\right]\Bigg|_{w_{P-2}} \\
\vdots \qquad & \qquad \vdots \\
F_1\left(\bar{C}_1,\sigma_1\right) = A\left(\bar{C},\sigma\right) &= f_{0,v}\left(\bar{c}_{0,v},\nu_{0,v}\right) + \sum_{k=1}^{s_1} \alpha_{1,u}\cdot F_{2,u}\left(\bar{C}_{2,u},\sigma_{2,u}\right)
\end{aligned}
\tag{7}
$$

where,

F_p : *OF*'s decoupled attributes during the period p

$\bar{C}_{p,s}$: *OF*'s Expected Present Value for event s during p

$\sigma_{p,s}$: *OF*'s standard deviation for event s during p

$\alpha_{p,s}$: Ocurrence Probability of event s during the decision period p

s_p : Number of feasible discrete events during p

$f_{p,s}$: Partial *OF*'s Probability Distribution for event s during p

$\bar{c}_{p,s}$: Partial *OF*'s Expected Value for event s during p

$\nu_{p,s}$: Partial *OF*'s standard deviation for event s during p

Fig. 3 graphically shows the increasing uncertainty of the objective over time. As planning horizon extends in time, the risk grows accordingly. Provided that the probabilistic properties of expansion attributes are reasonably described by a Gaussian probability distribution, blue dots correspond to the annual expected values of the expansion costs and the vertical black segments represent the annual standard deviations of the objective function.

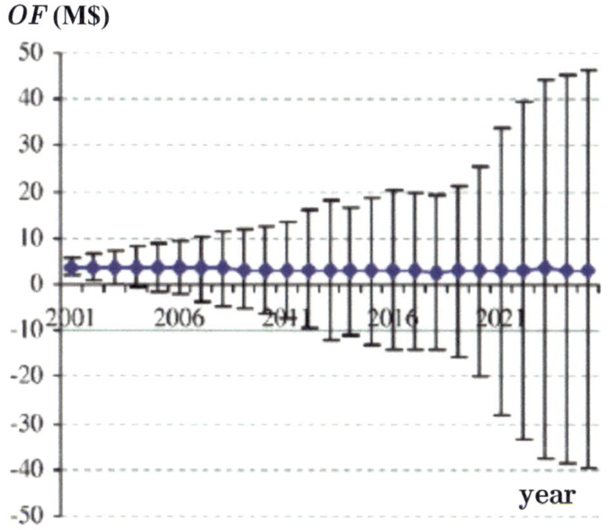

Fig. 3. Graphic representation of the time-decoupled attributes of an underlying asset

The idea of a decoupled assessment of the expansion plans' attributes can be rooted to the Bellman's Principle of Optimality since it allows applying dynamic programming for valuing the flexibility gained when embedded or contingent decision options are incorporated within the planning process (Dixit & Pindyck, 1994). In following sections, this process is explained in detail.

Ranking of expansion strategies and decision-making

Derived from the optimal portfolio selection theory, expansion plan attributes can be ranked based on their efficiency, by means of the Sharpe ratio r_{sharpe} (Nielsen & Vassalou, 2003). This index was proposed by Sharpe in 1966 as the ratio between the expected benefit and the risk, where risk is measured as a standard deviation of the benefit. According to static mean-variance portfolio theory, if investors face an exclusive choice among a number of alternatives, then they can unambiguously rank them on the basis of their robustness (Sharpe ratios). An expansion alternative with a higher Sharpe ratio will enable all investors to achieve a higher expected utility by accepted risk unit.

The inverse of r_{sharpe}, which is known as the coefficient of variation according to Ladoucette & Teugels (2004) and Feldman & Brown (2005) is a useful measure for comparing variability between positive distributions with different expected values. An alternative with a lower coefficient of variation will result in lower risk exposition per unit of expected benefit. In this work, the inverse of the r_{sharpe} will be used to measure the desirability of an expansion strategy.

In order to express in percentage the coefficient of variation, the use of a relative volatility, which is accounted for as the relation between the expected volatility of the underlying asset $\bar{\sigma}_k$ divided by the maximum expected volatility of all the evaluated strategies $\bar{\sigma}_{max}$, is proposed. See (8).

$$r_{sharpe}{}^{-1} = \frac{\bar{\sigma}_k}{\bar{B}_k} = \frac{\dfrac{\bar{\sigma}_k}{\bar{\sigma}_{\max}}}{\dfrac{\bar{C}_{\max}}{\bar{C}_k}} \cdot 100\ (\%) \tag{8}$$

where,

$\bar{B}_k$: Expected Benefit of the underlying asset S_k

$\bar{\sigma}_k$: Underlying asset expected volatility

5.4 Risk management by incorporating flexible Options

An important underlying assumption of the probabilistic optimization approach is the passive planner's attitude regarding the future. In fact, under this modelling paradigm, the diverse available choices that the planner has for reacting upon the occurrence of unexpected events are ignored. However, in practice planners have the ability for adapting their expansion decisions in response to undesired events (Gorenstin *et al.*, 1993; Dixit & Pindyck, 1994; Ku *et al.*, 2003; Vásquez & Olsina, 2007). A well-established way to systematically incorporate this fundamental aspect is the application of a complementary flexibility-based risk analysis stage.

New decision variables, new objective function

The flexibility-based risk analysis stage basically consists in solving an optimization problem of dynamic nature. The decision variables are the type, the timing and the location of the flexible -technical or managerial- options which are embedded in the previously obtained set of feasible expansion plans. Indeed, to solve this problem involves finding expansion strategies that are improved in performance in terms of their total attributes. Such expansion strategies are composed not only of large transmission projects D_p, but also of flexible decision options. Thus, flexible options O_p available in each decision period p, are planned for being advantageously incorporated if strategic uncertainties unfold as unfavourable scenarios.

Like previous stages, a total expansion costs-based objective function, which includes the new components of costs relative to the new flexible choices, is defined. This new *OF* is still subject to the same constraints of the original problem plus the constraints relative to the flexible options, e.g. generation capacity limits of DG plants and feasible locations of DG projects.

Visualizing opportunities for contingent decisions

A graphic illustration (see Fig. 4) of a complete event paths representing crucial states of the problem along the planning horizon together with the time-decoupled attributes information (F_P, F_{P-1}, ..., F_1) suitably represents the dynamic process that this optimization problem involves. In fact, with this information the planner has an insight into the risks associated with the decisions as well as is able to determine the timing when it would be meaningful to incorporate flexible or contingent choices. The problem search space is therefore noticeably reduced.

Given that only one discrete probability function during each period is assumed, the nodes of events tree showed in Fig. 2 represents the planner's opportunity for incorporating flexible or contingent decisions.

5.5 Valuing flexibility and ranking expansion strategies

When an irreversible expenditure D_p is made, i.e. the investment option is exercised, not only the deferment choice disappears but also all the other investment choices (Kirschen & Strbac, 2004). The value of the lost option, analogous to a financial call option, is an opportunity cost, which depends on the project's irreversibility as well as on the existing risk and flexible embedded options at the decision time (Dixit & Pindyck, 1994; Ramanathan & Varadan, 2006). However, classical project appraisal methods overlook this interaction even though, in practice, it evidently affects the planner's decisions.

Since flexibility can only be assessed by comparison (Ku et al., 2003; Gorenstin et al., 1993), the value of a flexible option is assessed by comparing its coefficient of variation with the coefficient of variation of a feasible inflexible reference strategy (*flexibility* = 0) belonging to the set of feasible expansion plans S_k. This basic procedure can be systematically extended into a multi period strategies comparison problem and solved by using the dynamic programming formulation expressed in (9) and (10) (Dixit & Pindyck, 1994; Ramanathan & Varadan, 2006).

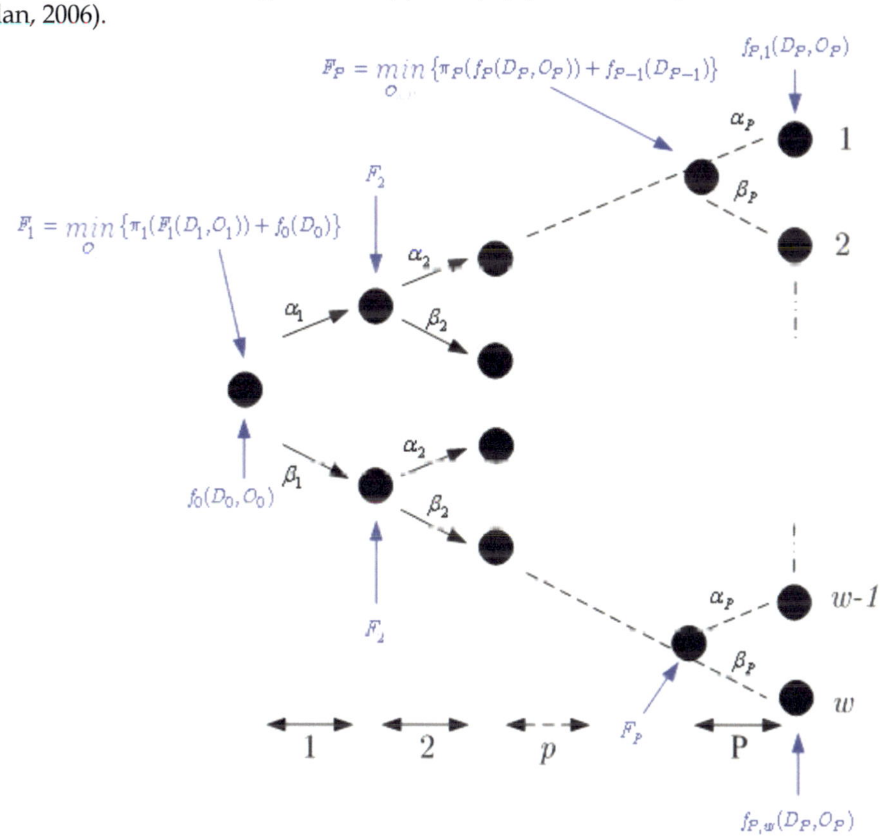

Fig. 4. Decoupled attributes of the objective function and decision tree representation

$$
\begin{aligned}
F_P &= \min_{O_P}\{F_P(D_P,O_P)\} = \min_{O_{n,P}}\{\pi_P(f_P(D_P,O_P)) + f_{P-1}(D_{P-1})\} \\
F_{P-1} &= \min_{O_{P-1}}\{\pi_{P-1}(F_P(D_{P-1},O_{P-1})) + f_{P-2}(D_{P-2})\} \\
&\vdots \qquad\qquad \vdots \\
F_1 &= \min_{O_1}\{\pi_1(F_1(D_1,O_1)) + f_0(D_0)\}
\end{aligned}
$$

where

D_p : Expansion decision to be defferred since period p

O_p : Flexible option incorporated in period p

π_p : Immediate costs valuation function during p

Going backward in dynamic programming allows decomposing a whole sequence of decisions into just two components: the immediate decision and a valuation function which encapsulates the consequences of all the subsequent future decisions.

As shown in Fig. 4, the process of incorporating flexible options starts at the last decision period ($p = P$), which is concerned with deciding for or against incorporating ($\min\{F_P\}$) one of the available flexible choices O_P. This is a classic single-stage optimization problem under the presence of only random uncertainties. As was analyzed in Sections 4.3 and 4.4 of this chapter, this task is proposed to be solved by applying a robustness-based probabilistic decision approach. In fact, by assessing, on one side, the time-decoupled attributes of the static expansion plan F_P (D_P) and, on the other, the time-decoupled attributes of one or more new flexible expansion strategies composed by a flexible option F_P (O_P), the planner can decide about the incorporation or not of such a flexible option in $p = P$, by comparing the two coefficients of variation (r^{-1}_{sharpe} (D_P), r^{-1}_{sharpe} (O_P)).

This solution ($\min\{F_P\}$) provides the information for the penultimate decision in P-1 which, in turn provides the information for deciding in P-2 and so on until $p = 1$ the moment in which a flexible strategy S_f is obtained. This procedure repeated for all the feasible expansion plans can be used for obtaining a set of flexible expansion strategies.

6. Numerical example: power supply capacity expansion planning problem of a the mining firm

In the following, a numerical planning problem built on an actual setting demonstrates the contribution of the proposed flexibility-based framework by enhancing the ability of making contingent expansion decisions along the planning horizon. Investing in DG projects and delaying a large transmission project are flexible options that the planner has available for reducing the planning risks.

Investment, energy procurement, and maintenance costs as well as expected unserved energy costs have been considered for computing the expected total costs of the diverse expansion strategies. Because of the short lead time, DG investment decisions are assumed to be made in the same time interval that the additional capacity is required. On the other hand, due to the large construction time, transmission projects are commenced one year before the additional capacity is required.

Let considers a mining firm which will operate over ten years located in a remote site without public service of electric energy supply. Daily production of the mine is assumed constant. It is known with certainty that the demand for the first to fifth year is 60 MW. Available information in year zero indicates that demand would increase to 120 MW depending on results of a current assessment of mineral reserves. The probability of the higher demand scenario is $p = 0.5$. The probability of power demand remaining at 60 MW is $1\text{-}p = 0.5$. Then, the expected value of demand along the second time period is 90 MW. Fig. 5 depicts the two possible demand paths along the planning horizon, which is set to 10 years. The main question is: How the mining firm should meet, in an optimal way, its current and future requirements of electric energy under consideration of ongoing demand uncertainty? For successfully accomplishing this task, the proposed flexibility-based decision-making framework will be applied, which involves the development of the following stages:

- Identification of a set of feasible expansion strategies.
- Assessments of the corresponding objective function for each feasible expansion strategy.
- Incorporation of flexible decision options in order to conform new expansion strategies.
- Ranking the expansion strategies by properly valuing the flexibility of the options incorporated in 3.

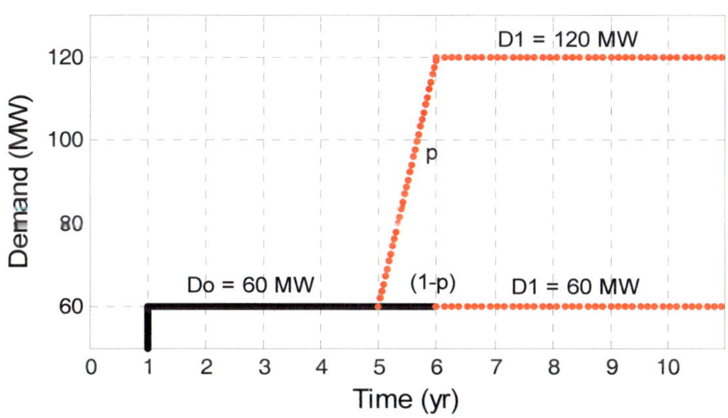

Fig. 5. Future demand scenarios along the operative lifetime of the mining firms

6.1 Obtaining Feasible Expansion Strategies

The large of economies of scale involved indicate that the most efficient expansion strategies have to deal with building 346 km of a new transmission line from the nearest system node instead of installing generation on site. Three technically feasible configuration of transmission lines are obtained as shown in Fig. 6 based admissible voltage limits.

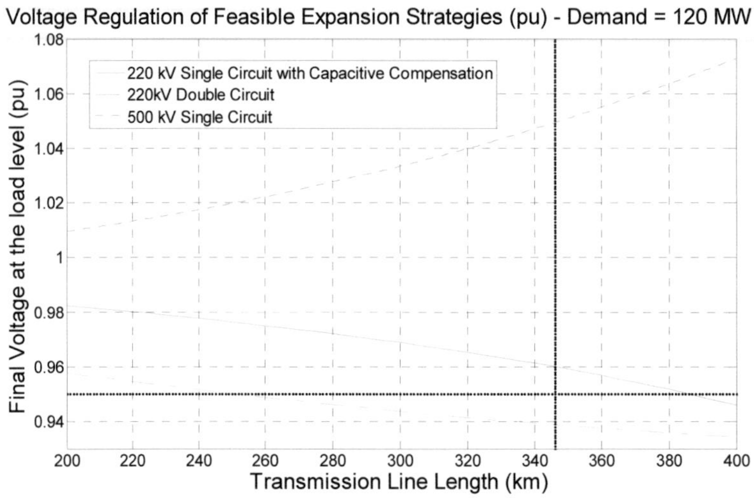

Fig. 6. Convergence and voltage regulation of feasible expansion strategies

The identified strategies are listed as follows:

- To build in year zero a 220 kV single circuit radial transmission network from the nearest system node. In the sixth year a capacitive compensation in node B_2 is installed in order to improve voltage levels (see Fig. 8).
- To build in year zero a 220 kV double circuit radial transmission network from the nearest system node.
- To build in year zero a 500 kV single circuit radial transmission network from the nearest system node.

These three inflexible decision options allow the mining firm to purchase energy from the spot market and therefore meet its expected demand. The next stage in the decision-making process involves valuing and ranking, from a probabilistic viewpoint, the obtained expansion strategies.

6.2 Assessing the attributes of feasible expansion plans

The substantial economy of scales involved in the expansion of the processing plant of the mining firm leads to an increase of electrical demand in large discrete amounts. As remaining relevant variables are assumed to be known with absolute certainty, only the uncertainty affecting the load growth will be resolved over the time. For this reason, the risk profiles of expansions will have the same shape as the forecasted demand evolution (see Fig. 5). Therefore, the assessment of the attributes of the expansion alternatives along the planning horizon can be completely determined without applying the Monte-Carlo technique. The objective function (OF) of the constrained stochastic optimization problem is formulated as follows:

$$\min_{T}\left\{E[C_{T_i}(T)]\right\} = \min_{T}\left\{ C_{I_{TL}}^{S_1} + \left.\sum_{j=1}^{5}\left(\frac{C_{A,j}}{(1.12)^j} + \frac{C_{O\&M,j}}{(1.12)^j} + \frac{C_{E[ENS],j}}{(1.12)^j}\right)\right|_{D_1}^{T} + \right.$$

$$+(1-p)\cdot\left.\sum_{j=6}^{10}\left(\frac{C_{A,j}}{(1.12)^j} + \frac{C_{O\&M,j}}{(1.12)^j} + \frac{C_{E[ENS],j}}{(1.12)^j}\right)\right|_{D_2}^{T} +$$

$$\left. +p\cdot\left.\sum_{j=6}^{10}\left(\frac{C_{A,j}}{(1.12)^j} + \frac{C_{O\&M,j}}{(1.12)^j} + \frac{C_{E[ENS],j}}{(1.12)^j}\right)\right|_{D_2}^{T} \right\} = 302.87\,\text{M\$} \quad (11)$$

subject to:

$S_{G,j} - D_{i,j} - S_{L,j} - S_{NS,j} = 0\ (\text{MVA})$: Balance of power

$0.9 \le V_3 \le 1.1$: Voltage limits

$F_{12} \le T$: Transmission capacity constraint

where

$E[C_T]$: Net Present Value (NPV) of the total expected costs.

$C_{I_{TL}}^{S_1}$: Investment cost in transmission network for strategy S_1.

$C_{A,j}$: Acquisition cost of energy in the spot market in year *j*.

$C_{O\&M,j}$: Operation and maintenance cost of lines and sub-stations incurred in year *j*. These costs are assumed to be 2% and 3% of the respective investment costs.

$C_{E[ENS],j}$: Expected costs of the energy not supplied in year *j*.

$S_{G,j}, D_{i,j}, S_{L,j}, S_{NS,j}$: Spot market power, power demand in *i*-th stage, power losses, and not supplied power in the *j*-th year.

The Value of Lost Load (VOLL) has been estimated at 500 $/MWh and reflects the economic losses incurred when the mining firm stops its production. Discount rate is set to 12%/yr. Because of the length, capacitive compensation is needed for the single circuit 220 kV choice. The fixed investment cost of compensation is 1 M$ and capacity dependent costs are 17 000 $/MW of incremental line capacity. The investment cost functions for 220 kV and 500 kV substations are depicted in Fig. 7.

Costs of transmission lines have been modelled as a linear function of the transmission capacity, as indicated in Table 2.

Table 3 provides the electrical line parameters needed for performing an AC power flow analysis on each alternative in order to verify voltage limits, line flows, losses, etc.

Table 2. Transmission lines investment costs

Voltage	**Fixed costs** $/km	**Capacity costs** $/(MW·km)
220 kV single circuit	90 000	800
220 kV double circuit	135 000	600

500 kV single circuit	200 000	300

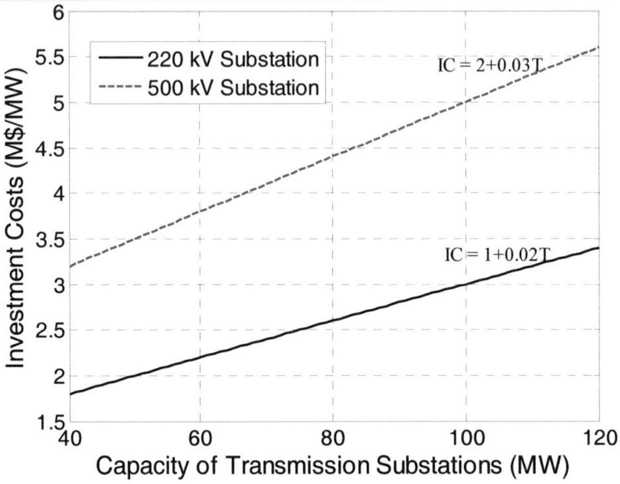

Fig. 7. Investment cost functions of transmission substations.

Voltage	R Ω/km	x Ω/km	B μS/km
220 kV single circuit	0.0481	0.385	2.341
220 kV double circuit	0.0241	0.192	4.682
500 kV single circuit	0.0234	0.279	4.169

Table 3. Electric parameters of transmission lines

In Table 4, reliability parameters of transmission components and DG plants are given, as they are necessary for computing the expected energy not supplied to the mining process. Stochastic behaviour of system components are modelled as two-state Markov reliability model (Billinton & Allan, 1996). Because of the small number of components, exhaustive state enumeration has been applied for the reliability evaluation.

Procurement costs of energy have been computed considering the long-term spot prices that would prevail in node B2 (see Fig. 8) provided that the transmission network was to be built with optimal capacity. The spot price duration curve in node B1 remains constant over the planning period and it is given in Table 5.

Parameter	Market	Line	Transformer	DG Plant
Pr(O)	0.99886	0.99545	0.99825	0.98000
Pr(F)	0.00114	0.00455	0.00174	0.01999

Table 4. Reliability parameters of system components

Duration (%)	6.96	13.87	38.64	32.46	8.33
Price ($/MWh)	82.29	75.7	61.7	57.6	37.03

Table 5. Spot prices during periods

According to (1), the power supply capacity optimization problem is solved when a transmission project, which satisfies technical and economic requirements for all anticipated demand scenarios, minimizes the total discounted expected expansion costs incurred along the planning horizon. In the valuation process, the occurrence probabilities of each demand scenario are considered and the remaining information is assumed to be known with certainty. Fig. 9 shows the performance of the three technically feasible expansion strategies identified before, which meet the uncertain power demand of the mining firm over the planning horizon.

Expansion strategies are ranked considering the minimization of the present value of total expansion costs. Under this perspective, the 220 kV single-circuit transmission line with a capacity of T = 120 MW and capacitive compensation in B_2, which is denominated S_1, would be the strategy that the planner would select under a classic risk-neutral probabilistic choice as it exhibits the lowest expected costs (E[C_T]= 302.87 M$).

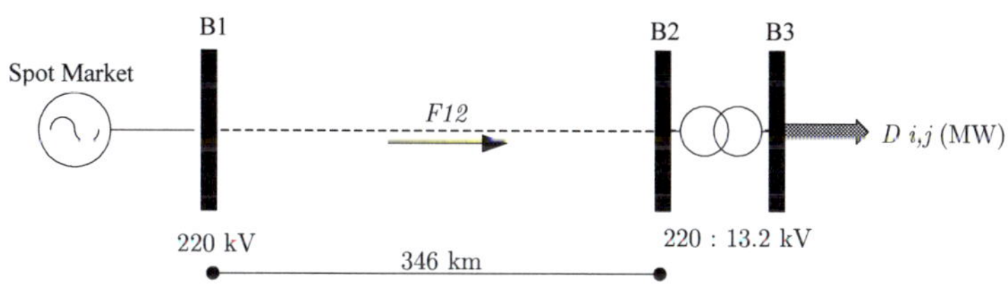

Fig. 8. Expansion strategy S_1, single circuit 220 kV- 120 MW transmission link

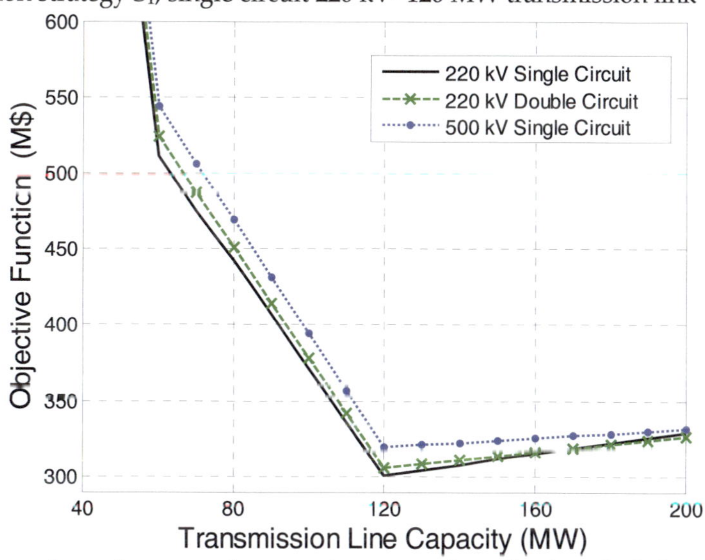

Fig. 9. Present value of expected expansion costs for all identified feasible expansion strategies

6.3 Flexible Expansion Strategies Conformation

For illustrative purposes the only feasible-inflexible expansion strategy considered during the next stages of the planning process is S_1. Unlike the classic probabilistic approach, the proposed valuation method accounts for contingent expansion choices, i.e. DG investments and delay of a large transmission project, that the planner has available in each demand scenario. DG projects are based on installing diesel-fueled gensets. From manufacturer data, nameplate capacity of each DG diesel units is 15 MW. However the maximum power output is derated to 13 MW at the location altitude. Investment costs of diesel units are assumed to be 200 \$/kW. The DG plants fuel specific consumption is 217.98 l/MWh[2] and the fuel price is 0.3 \$/l. In Table 6 below, apart from S_1, are listed five additional expansion strategies with various degrees of flexibility, that planner could consider once better information about demand evolution arrives.

Real Options		
S	**Period 1**	**Period 2**
1	Single Circuit Trans Line 120 MW	
2	1st circuit trans. line 60 MW	2nd. circuit (expand to 120 MW)
3	Single circuit trans. line 60 MW	Single circuit trans. line 60 MW
4	DG 5x15 MVA	Single circuit trans. line 60 MW
5	DG 5x15 MVA	Single circuit trans. line 120 MW
6	DG 5x15 MVA	DG 5x15 MVA

Table 6. Description of possible expansion strategies to be valued

6.4 Decision-Making: Valuation and Ranking of Flexible Expansion Strategies

Valuation of flexible expansion strategies

Strategy S_2 consists of initially building the first circuit of a double circuit 220 kV transmission link with a capacity of 60 MW for satisfying the known demand from year 1 to 5. In the fifth year, if the power demand is increased to 120 MW, i.e. once uncertainty has been unfolded, the planner takes the decision of adding the second circuit, expanding the transfer capacity from 60 to 120 MW. Details of calculation of the discounted expected total expansion costs of S_2 are given by the following expression:

$$\mathrm{E}[C_{T2}] = C_{I_{TL}}^{60MW\,(1st\ circuit)} + \sum_{j=1}^{5}\left(\frac{C_{A,j}}{(1.12)^j} + \frac{C_{O\&M,j}}{(1.12)^j} + \frac{C_{E[ENS],j}}{(1.12)^j}\right)\Bigg|_{D_1=60MW}^{T=60\ MW} +$$

$$+(1-p)\cdot\sum_{j=6}^{10}\left(\frac{C_{A,j}}{(1.12)^j} + \frac{C_{O\&M,j}}{(1.12)^j} + \frac{C_{E[ENS],j}}{(1.12)^j}\right)\Bigg|_{D_2=60MW}^{T=60\ MW} +$$

$$+p\cdot\left[\frac{C_{I_{TL}}^{60\to120(2nd.circuit)}}{(1.12)^5} + \sum_{j=6}^{10}\left(\frac{C_{A,j}}{(1.12)^j} + \frac{C_{O\&M,j}}{(1.12)^j} + \frac{C_{E[ENS],j}}{(1.12)^j}\right)\Bigg|_{D_2=120MW}^{T=120\ MW}\right] \quad (12)$$

$$\mathrm{E}[C_{T2}] = 297.1\mathrm{M\$}$$

[2] A specific fuel consumption of 217.98 l/MWh entails a combined efficiency of the genset (engine and generator) of 43.54%, assuming for diesel fuel an average Net Calorific Value (NCV) of 43 MJ/kg and a density of 0.883 kg/dm³ at 15°C.

Strategy S_3 involves the construction of a 220 kV single circuit transmission line with a capacity 60 MW for satisfying the demand of the first period. In the fifth year, only if the power demand rises to 120 MW, the planner takes the decision of build another 220 kV single circuit transmission line with capacity 60 MW. The present value of the expected total expansion costs for S_3 is computed according to (13).

It is important to notice that the expected total costs of these more flexible strategies are lower than cost of strategy S_1. That is because the total expenditure has been separated into two stages. The second investment is committed in the fifth year only if it is actually needed.

$$E[C_{T3}] = C_{I_{TL}}^{60MW(single)} + \sum_{j=1}^{5}\left(\frac{C_{A,j}}{(1.12)^j} + \frac{C_{O\&M,j}}{(1.12)^j} + \frac{C_{E[ENS],j}}{(1.12)^j}\right)\Bigg|_{D_1=60MW}^{T=60\ MW} +$$

$$+(1-p)\cdot\sum_{j=6}^{10}\left(\frac{C_{A,j}}{(1.12)^j} + \frac{C_{O\&M,j}}{(1.12)^j} + \frac{C_{E[ENS],j}}{(1.12)^j}\right)\Bigg|_{D_2=60MW}^{T=60\ MW} +$$

$$+p\cdot\left[\frac{C_{I_{TL}}^{60MW(single)}}{(1.12)^5} + \sum_{j=6}^{10}\left(\frac{C_{A,j}}{(1.12)^j} + \frac{C_{O\&M,j}}{(1.12)^j} + \frac{C_{E[ENS],j}}{(1.12)^j}\right)\Bigg|_{D_2=120MW}^{T=120\ MW}\right]$$

$$E[C_{T3}] = 295.1\text{M\$} \tag{13}$$

In the following, some investment policies including DG projects are analyzed. Strategy S_4 comprise the installation of five 15 MW diesel generators for satisfying the power demand of the mining firm during the first period. Then, only if the demand effectively grows to 120 MW, a single circuit 220 kV transmission line with a capacity of 60 MW is built in year five, to meet the mining demand along the second period.

The investment cost of a diesel DG plant is 3 M$. Additionally, costs fuel storage facilities are assumed to be 1.5 M$. Maintenance costs are computed as a percentage of investment costs. For DG plants they are set as 5% and, for fuel storage facilities they are assumed to be 3% of its investment costs. Lifetime of DG generators is assumed to be 20 years. Linear depreciation has been used for assessing recovering value of DG equipment. In this case the mining firm sells off the five DG plants at the closing of the mining project. Assuming that 20% of the investments cost are required to uninstall the generation plant, the recovery net costs can be assessed and included. The present value of the expected total expansion costs for S_4 is computed according to (14).

$$E[C_{T4}]=\frac{C_{I_{DG}}^{60MW}}{(1.12)^1}+\sum_{j=1}^{5}\left(\frac{C_{G,j}}{(1.12)^j}+\frac{C_{O\&M,j}}{(1.12)^j}+\frac{C_{E[ENS],j}}{(1.12)^j}\right)\Bigg|_{D_1=60MW}^{DG=60\ MW}+$$
$$+(1-p)\cdot\left[-\frac{C_{recDG}^{60\ MW}}{(1.12)^{11}}+\sum_{j=6}^{10}\left(\frac{C_{G,j}}{(1.12)^j}+\frac{C_{O\&M,j}}{(1.12)^j}+\frac{C_{E[ENS],j}}{(1.12)^j}\right)\Bigg|_{D_2=60MW}^{DG=60\ MW}\right]+$$
$$+p\cdot\left[\frac{C_{I_{TL}}^{60\ MW(single)}}{(1.12)^5}-\frac{C_{recDG}^{60\ MW}}{(1.12)^{11}}+\sum_{j=6}^{10}\left(\frac{C_{G+A,j}}{(1.12)^j}+\frac{C_{O\&M,j}}{(1.12)^j}+\frac{C_{E[ENS],j}}{(1.12)^j}\right)\Bigg|_{D_2=120MW}^{DG+T=120}\right] \quad (14)$$

$$E[C_{T4}]=296.4\,M\$$$

where

$C_{I_{DG}}^{60MW}$: Investment cost of 60 MW DG power plants.

C_G : Annual generation costs of 60 MW DG power plants.

C_{G+A} : Generation and acquisition costs incurred when operate at the same time the DG plants and the trans. line.

C_{rec}^{60MW} : Recovery net value of DG equipments.

Similarly, expansion strategy S_5 consists in installing five 15 MVA diesel DG plants in the first year. If power demand escalates to 120 MW, a single 220 kV transmission line with capacity 120 MW is built to cover the energy needs during the second period. Under S_5, the five DG plants are sold off in the sixth year. The present value of the expected total expansion costs for S_5 is computed according to (15).

$$E[C_{T5}]=\frac{C_{I_{DG}}^{60MW}}{(1.12)^1}+\sum_{j=1}^{5}\left(\frac{C_{G,j}}{(1.12)^j}+\frac{C_{O\&M,j}}{(1.12)^j}+\frac{C_{E[ENS],j}}{(1.12)^j}\right)\Bigg|_{D_1=60MW}^{DG=60\ MW}+$$
$$+(1-p)\cdot\left[-\frac{C_{recDG}^{60\ MW}}{(1.12)^{11}}+\sum_{j=6}^{10}\left(\frac{C_{G,j}}{(1.12)^j}+\frac{C_{O\&M,j}}{(1.12)^j}+\frac{C_{E[ENS],j}}{(1.12)^j}\right)\Bigg|_{D_2=60MW}^{DG=60\ MW}\right]+$$
$$+p\cdot\left[\frac{C_{I_{TL}}^{120\ MW(single)}}{(1.12)^5}-\frac{C_{recDG}^{60\ MW}}{(1.12)^{6}}+\sum_{j=6}^{10}\left(\frac{C_{A,j}}{(1.12)^j}+\frac{C_{O\&M,j}}{(1.12)^j}+\frac{C_{E[ENS],j}}{(1.12)^j}\right)\Bigg|_{D_2=120MW}^{T=120MW}\right] \quad (15)$$

$$E[C_{T5}]=294.04\,M\$$$

Similarly as with S_4 and S_5, strategy S_6 entails installing five 15 MVA diesel DG plants in the first period, and then, adding five new 15 MVA diesel DG for covering the mining peak load in the second period only if it unfolds as the high demand scenario. The expected total costs of strategy S_6 can be evaluated according to (16).

$$
\begin{aligned}
E[C_{T6}] = & \frac{C_{I_{DG}}^{60MW}}{(1.12)^1} + \sum_{j=1}^{5}\left(\frac{C_{G,j}}{(1.12)^j}+\frac{C_{O\&M,j}}{(1.12)^j}+\frac{C_{E[ENS],j}}{(1.12)^j}\right)\Bigg|_{D_1=60MW}^{DG=60\,MW} + \\
& +(1-p)\cdot\left[-\frac{C_{recDG}^{60MW}}{(1.12)^{11}} + \sum_{j=6}^{10}\left(\frac{C_{G,j}}{(1.12)^j}+\frac{C_{O\&M,j}}{(1.12)^j}+\frac{C_{E[ENS],j}}{(1.12)^j}\right)\Bigg|_{D_2=60\,MW}^{DG=60\,MW}\right] + \\
& +p\cdot\left[\frac{C_{I_{DG}}^{60MW}}{(1.12)^6} - \frac{C_{recDG}^{120MW}}{(1.12)^{11}} + \sum_{j=6}^{10}\left(\frac{C_{G,j}}{(1.12)^j}+\frac{C_{O\&M,j}}{(1.12)^j}+\frac{C_{E[ENS],j}}{(1.12)^j}\right)\Bigg|_{D_2=120\,MW}^{DG=120\,MW}\right]
\end{aligned}
\tag{16}
$$

$$E[C_{T6}] = 291.5\text{M\$}$$

Components of OF	$E[C_{T1}]$	$E[C_{T2}]$	$E[C_{T3}]$	$E[C_{T4}]$	$E[C_{T5}]$	$E[C_{T6}]$
Net Investment (M$)	72.36	67.52	66.12	28.61	32.52	16.37
O&M (M$)	8.46	7.59	7.03	5.57	5.16	5.25
Generation & Procurement (M$)	221.9	221.9	221.9	262.2	256.3	269.83
E[ENS] (M$)	0.163	0.067	0.087	0.020	0.034	0.006
TOTAL (M$)	**302.88**	**297.07**	**295.13**	**296.40**	**294.01**	**291.45**
Standard Deviation (M$)	50.2	56.5	71.6	68.9	65.5	61.8
Risk, r_{sharpe}^{-1} (%)	**70.11**	**77.40**	**97.44**	**94.17**	**88.80**	**83.06**

Table 7. Summary of decision variables for the proposed expansion strategies

Ranking of Flexible Expansion Strategies

Depending on the planner´s attitude regarding risk, the decision could be made either based on the minimum total expansion costs criterion (S_6) or by choosing the option with lowest risk (S_1). For instance, by comparing the present values of the expected present costs of the six alternatives it is concluded that flexible strategies S_4, S_5, and S_6, though much more expensive in terms of operating costs, are better than the economies of scale provided by S_1. A breakdown of the costs incurred by each alternative is provided in Table 7.

Fig. 10 illustrates the effect of uncertainty on the expected costs of each strategy. The economic efficiency of S_1 increases when the probability for the occurrence of the higher demand scenario is high. On the other hand, investing for retaining flexibility is more convenient if there is a low probability for the occurrence of the higher demand scenario.

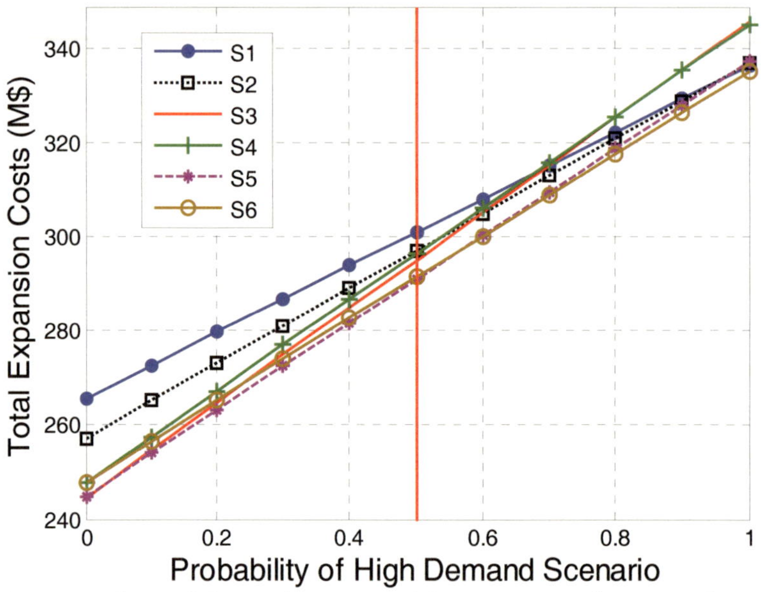

Fig. 10. Sensitivity analysis of the performance of the expansion choices to the probability of a high demand scenario

Valuing flexibility of the embedded options

Decomposing the problem based on the nature of the uncertainties allows the proposed framework to account for the value of the gained flexibility by comparison using discrete Sharpe ratios r_{sharpe}. Nevertheless, in this specific example, it has been assumed that the planner has made his decision based on the expected NPV of the total expansion costs. In fact, a suitable way for valuing the flexibility of strategies S_2, S_3, S_4, S_5, and S_6 with regard to S_1 is to ask how much lower should be the investment costs of S_1 (CI_{TL0}) to make this investment policy the preferred alternative. For instance, the flexibility of S_6 could be assessed by equating (11) and (16) and solving for transmission investment costs CI_{TL1}.

Fig. 11 shows that investment costs of the single-circuit 220 kV transmission line should be about 13% lower to prefer investment strategies S_1 to strategy S_6. Therefore, the flexibility value provided by S_6 can be computed as follows: $CI_{TL_1} = CI_{TL_0} + RO_6; \quad RO_6 = Flexibility_{S_6} = 60.95 - 72.36 = 11.4 M\$$

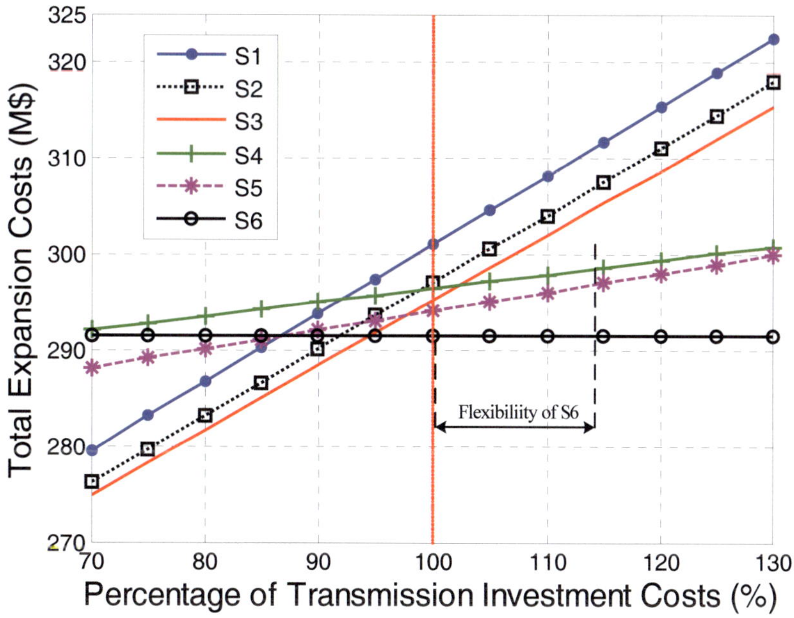

Fig. 11. Flexibility value for diverse transmission investment costs

7. Conclusion

Risk management techniques based on strategically incorporating flexible investment projects represent a tool for consistently dealing with risks present in transmission planning problems. The larger the irreversibility of the conventional expansion investments and the uncertainties affecting future conditions more efficiency can be achieved by the proposed planning approach.

In a numerical example, it has been verified that ignoring the value of flexible choices can lead to wrong investment decisions. Despite the large economies of scale of the traditional transmission expansion projects, it has been shown that the optimal investment strategy would be to preserve the wait option by investing in more expensive DG projects, while deferring the decision of building the transmission line until uncertainties are resolved.

In order to accomplish the goal of properly integrating DG investments and exploiting their potential benefits requires an efficient cost-based evaluation. A very important emerging issue is therefore the pricing and reward of system benefits provided by DG projects. In fact, electricity prices as seen by final consumers are substantially higher than prices at centralized generation levels. This difference is due to the added costs of T&D services to transport electricity from the point of production to consumption. Distribute generation, however, located close to demand, delivers electricity directly with limited requirement for use of T&D network (Djapic et al., 2007). Ignoring this particular feature in the valuation process could result a false DG non competitiveness perception regarding traditional expansion decisions.

Other important topic that calls for further investigation is the proper valuation of the planning flexibility provided by option of relocating DG facilities across large networks, as uncertainty on demand growth unfolds divergently among the different system zones.

8. References

Ackerman, T.; Andersson, G.; Soder, L. (2001). Distributed Generation: A definition, *Electric Power Systems Research*, Vol. 57, No. 3, pp. 195-204, 2001.

Añó, G.; Frezzi, P.; Vargas, A. (2005). Planificación de la expansión de sistemas de transporte de energía eléctrica en mercados competitivos: Estado del arte, *Proceedings of XI Regional Iberoamerican Meeting of CIGRE* (XI ERIAC), May 2005, Paraguay.

Billinton, R; Allan, R. (1996). *Reliability Evaluation of Power Systems*, Ed. Plenum Press, ISBN-10: 0306452596, New York.

Braga, A. S. D; Saraiva, J. T. (2005). A multiyear dynamic approach for transmission expansion planning and long-term marginal costs computation, *IEEE Transactions on Power Systems*, Vol. 20, pp. 1631-1639, 2005.

Brañas, P.; Jiménez, F.; Morales, A. (2004). Strategic uncertainty and risk attitudes: The experimental connection, Cuadernos de Economía, Vol. 27, Universidad de Granada (La Cartuja), Granada, Spain.

Brown, R. E.; Jiuping, P.; Xiaorning, F.; Koutlev K. (2001). Siting distributed generation to defer T&D expansion, *IEEE PES Transmission and Distribution Conference and Exposition*, Vol. 2, pp. 622-627, 2001.

Buygi, M. O. (2004). *Transmission Expansion Planning in Deregulated Power Systems*, Ph.D. Thesis, Faculty of Electric and Information Systems, Technical University of Darmstadt, Germany, pp. 5-14.

Clemons, E. K. & Barnett, S. (2003). Strategic uncertainty and alternative futures: evaluating our options in the post-september 11 world, Operations and Information Management Department, Wharton School of the University of Pennsylvania, online: http://opim.wharton.upenn.edu/~clemons /files/conflict_scenarios_v3_6_doc.pdf

Dale, L. (2002). Distributed Generation and Transmission, *Proceedings of IEEE PES Winter Meeting*, Vol. 1, pp. 132 - 134.

Daly, P. A. & Morrison, J. (2001). Understanding the potential benefits of distributed generation on power delivery systems, *Proceedings of IEEE Rural Electric Power Conference*, pp.A2/1-A213, Little Rock, AR, USA.

Detre, J. D.; Briggeman, B. C.; Boehlje, M.; Gray, A. W. (2006). Score-carding and heat mapping: tools and concepts for assessing strategic uncertainties, *International Food and Agribusiness Management Review*, Vol. 09, online: http://www.ifama.org/tamu/iama/conferences/2005conference/Papers &Discussions/1044_Paper_Final.pdf.

Dixit, A. K.; Pindyck, R. S. (1994). *Investment under Uncertainty*, Princeton University Press, pp. 3 -54, New Jersey.

Dillon, R.; Haimes, Y. (1996). Risk of extreme events via multi-objective decision trees: Application to tele-communications, *IEEE Transactions on Systems, Man and Cybernetics - Part A: Systems and Humans*, Vol. 26, No. 2., pp. 262-271, 1996.

Djapic, P.; Ramsay, C.; Pudjianto, D.; Strbac, G.; Mutale, J.; Jenkins, N.; Allan, R. (2007). Taking an active approach, *IEEE Power and Energy Magazine*, Vol. 5, pp. 74-76, 2007.

Feldman, J.; Brown, R. L. (2005). Risk and Insurance, Education and Examination Committee of the Society of Actuaries, USA.

Gómez Expósito, A. (2002). *Análisis y Operación de Sistemas de Energía Eléctrica*, McGraw-Hill, pp. 33-66, Spain.

Gorenstin, B.G.; Campodonico, N.M.; Costa, J. P. & Pereira, M.V.F. (1993). Power system expansion planning under uncertainty, *IEEE Transactions on Powers Systems*, Vol. 8, pp. 129-135, 1993.

Grijalva, S. & Visnesky, A.M. (2005). Assessment of DG Programs Based on Transmission Security Benefits, *Proceedings of IEEE PES General Meeting*, pp. 1441–1446, 2005.

Jenkins, N.; Allan, R.; Crossley, P.; Kirschen, D.; Strbac, G. (2000). *Embedded Generation*, Institution of Engineering and Technology, London, ISBN: 0852967748, pp. 1-20.

Kirschen, D. & Strbac, G. (2004). *Fundamenntals of Power Systems Economics*, John Wiley and Sons, ISBN-10: 0470845724, pp. 228 - 264.

Ku, A.; Wagle, P.; Miller, A.; Christian, J.; Sioshansi, F. P.; Hinrichs, L.; Eydeland, A.; Wolyniec, K.; Dyner, I.; Larsen, E.; Lomi, A. (2003). *Risk and Flexibility in Electricity*, Risk Books, Incisive RWG Ltd., pp. 1-32.

Ladoucette, S. A. & Teugels, J. L. (2004). Risk measures for a combination of quota-share and drop down excess-of-loss reinsurance treaties, *EURANDOM Reports Netherlands*, online: http://www.eurandom.nl/

Latorre, G.; Cruz, R. D.; Areiza, J. M.; Villegas, A. (2003). Classification of publications and models on transmission expansion planning, *IEEE Transactions on Power Systems*, Vol. 18, No. 2, pp. 938 - 946, 2003.

Lee, C. W.; Ng, S. K. K.; Zhong, J.; Wu, F. F. (2006). Transmission expansion planning from past to future, *Proceedings of IEEE PES Power Electric Systems Conference and Exposition, PSCE*, pp. 257-265, 2006.

Majlender, P. (2003). Strategic investment planning by using dynamic decision trees, *IEEE Proceedings of the 36th Hawaii International Conference on Systems Sciences*.

Neimane, V. (2001). *On Development Planning of the Electricity Distribution Networks*, Ph.D. Thesis, Royal Institute of Technology, pp. 12-14, Stockholm, Sweden.

Nielsen, L. T. & Vassalou, M. (2003). Sharpe Ratios and Alphas in Continuous Time, *Journal of Financial and Quantitative Analysis*, Vol. 39, No. 1, pp. 7-8, 2003.

Pepermans, G.; Driesenb, J.; Haeseldonckxc, D.; Belmansc, R.; D'haeseleer, W. (2003). Distributed Generation: definition, benefits and issues, *Energy Policy*, Vol. 33, pp. 787-798, 2003.

Philipson, L. & Willis, H. L. (1999). *Understanding Electric Utilities and De-Regulation*, Marcel Dekker Inc., pp.1-70, New York.

Ramanathan, B. & Varadan, S. (2006). Analysis of transmission investments using real options, *Proceedings of IEEE PES Power Systems Conference and Exposition PSCE '06*, pp. 266-273, Atlanta, USA, 2006.

Rudnick, H. & Zolezzi, J. (2000). Planificación y expansión de la transmisión en mercados eléctricos competitivos, *Proceedings of VII Symposium of Specialists in Electric Operational and Expansion Planning* (VII SEPOPE), May 23-28, Curitiba, Brazil, online: http://www2.ing.puc.cl/~power/ paperspdf/sepope%20planif.pdf.

Samper, M. & Vargas, A. (2006). Caracterización de las Principales Tecnologías de generación distribuida, CIDEL, Congreso Internacional de Distribución Eléctrica 239, Buenos Aires, Argentina.

Vásquez, P. & Olsina, F. (2007). Valuing flexibility of DG investments in transmission expansion planning, *Proceedings of IEEE PES Power Tech 2007*, July 2007, Lausanne, Switzerland.

Vásquez P., Styczynski Z.A., Vargas A. (2008). Flexible decision making-based framework for coping with risks existing in transmission expansion plans, *Proceedings of IEEE PES Transmission and Distribution Conference and Exposition*, Bogotá, Colombia.

Vásquez, P. (2009). Flexibility-Based Decision-Making Framework for Facing The Current Transmission Expansion Planning Problem, Ph.D Thesis, Institute of Electrical Energy, National University of San Juan, ISBN: 978-987-05-6401-0, San Juan, Argentina.

Vignolo, M. & Zeballos, R. (2001). Transmission networks or distributed generation?, *IASTED International Conference on Power and Energy Systems*, Proceedings 6., Rhodes, Greece.

Willis, H. L. (1997). *Power Distribution Planning - Reference Book*, ABB Power T&D Company Inc. Cary, North Carolina, ISBN: 0-8247-4875-1, pp.807-862, 901-960.

Willis, H. L. & Scott, W. G. (2000). *Distributed Power Generation*, Marcel Dekker, Inc., ISBN: 0-8247-0336-7, pp. 1-34, 97-150, New York.

Youssef, H. K. M. (2001). Dynamic transmission planning using a constrained genetic algorithm, *International Journal of Electrical Power & Energy Systems*, Vol. 23, pp. 857-862, 2001.

Acronyms

DG	Distributed Generation
DP	Dynamic Programming
EENS	Expected Value of Energy Not Supplied
FACTS	Flexible AC Transmission Systems
NPV	Net Present Value
O&M	Operation and Maintenance
PDF	Probability Distribution Function
PES	Power Electric System
TEP	Transmission Expansion Planning
T&D	Transmission and Distribution (Systems)
TS	Transmission System
VOLL	Value of Lost Load

17

State identification of underdetermined grids

Martin Wolter
Leibniz Universität Hannover
Germany

1. Introduction

Methods of grid state identification are mathematical algorithms to obtain information on state variables of electric power system grids. These procedures are based fundamentally on the acquisition of measured, forecasted or estimated values, which are universally referred to as influencing variables, or, more simply, as measurands. In combination with knowledge of grid topology and electrical equipment parameters, identification methods aim at the best possible determination of the grid state.

Depending on the ratio of the number of available independent measurands and the quantity of state variables, the equation system to be solved is either overdetermined, determined or underdetermined. Thus, different procedures of grid state identification are applied.

Independent of the identification approach the basic proceeding of grid state determination remains the same. Starting at an initial guess the grid state identificator returns a nodal voltage vector which is used to calculate secondary grid values like nodal currents, nodal powers or terminal currents and powers. Calculation results are compared to the available information on influencing variables. The grid state identificator now iteratively adjusts the voltages to minimize the difference between measured and calculated values.

If the input vector is redundant and contains more (real-valued) measurands than twice the number of nodes, the equation system is overdetermined. This sort of problem is characteristic of high voltage and ultra high voltage transmission grids. Recording of more measurands than needed for uniquely solving the equation system is used to protect grid state identification against measurement uncertainties and outages of measurement facilities.

Typically, the number of measurands is more than twice as high as the amount which would have been necessary to uniquely calculate a grid state. To solve an overdetermined equation system state estimation algorithms based on Gaussian Least-Squares-Optimization or artificial neural networks are applied which find a solution vector of nodal voltages that minimizes the mean square error.

If the number of independent (real-valued) measurands is exactly twice the number of nodes, the equation system is uniquely solvable. In case of measuring only voltages and currents, a direct solution without any iteration is possible. If powers are used as influencing variables, an iterative solution approach is applied using a Newton-Raphson or fixed-point

iteration method. The special case of using solely nodal powers as influencing variables is called power flow calculation.
If the input vector contains less real-valued measurands than twice the number of nodes, the equation system is underdetermined. This sort of problem is characteristic of medium and low voltage distribution grids due to in the past exact knowledge of the grid state was unnecessary. So, installation of measurement facilities was not enforced to save costs. Nowadays, changed basic conditions necessitate adequate knowledge of equipment utilization. Unfortunately, underdetermined equation systems are not uniquely solvable but offer an infinite number of solutions instead. Thus, estimation approaches are needed to either limit the solution space or to estimate the most probable grid state.

2. Motivation

Considering particularly distribution grids, in a liberalized market the behavior and utilization of the electric power system is strongly influenced by the interests of market participants, namely on the one hand producers and consumers of energy and on the other hand the grid operators. The interests of all parties are mostly driven by economical considerations.
It is obvious that the goals of different market participants are divergent. As a consequence, energy management systems (EMS) are developed to economically combine all objectives. They are supposed to find strategies for grid operation and utilization which are convenient for all involved parties.
Decentralized energy management systems offer concepts to local utilities to face future challenges under changing conditions of the energy sector. Therefore, the energy management covers all the processes prospectively involved in the future of distributed energy supply to have the continuing ability to safely operate distribution grids and to securely, effectively and economically supply customers. This includes methods to optimize power acquisition as well as strategies to maintain and improve power quality.
On the one hand the above mentioned aims are achieved by influencing load and generation i.e. by optimizing assignment of feeders and adaptable loads taking into account technical and economical constraints.
On the other hand grid operators are supposed to invest in their grid infrastructure, implying expansion of power grid, information technology and measurement facilities. This is mainly due to the increasing amount of decentralized generation, resulting in dramatically changed demands on distribution grids which were originally planned for predictable and unidirectional power flow from the hand-over points to the end customers and to inferior grids. The high variability of load and generation - especially of feeders based on renewable sources - results in fast changing, hardly determinable and possibly illegal equipment utilization. In the future, even more DG sources will feed into distribution grids. These new challenges strictly necessitate intelligent energy management systems.
To enable energy management systems to find equitable operation strategies and to detect overstressing of lines and transformers, knowledge of grid state and equipment utilization is a matter of vital importance. Unfortunately, measurement technique is sparsely spread in distribution grids. Hence, grid state is not completely observable and has to be estimated. That is why research on state identification methods of underdetermined power systems is done.

3. Boundary Load Flow

In general, underdetermined equation systems offer an infinite number of solutions, whereof in power systems only a part of them are actually possible. The boundary load flow method limits the solution space to these thinkable solutions by introducing constraints for all influencing variables. Thereby, the amount of solutions remains infinite, indeed. Unlike probabilistic power flow computation, power density functions and correlations are unknown.

So, in contrast to the other grid state identification methods the boundary load flow approach does not aim at finding one determined grid state. Instead, this method finds maxima and minima of state variables which actually are of less informative value than consequential resultant equipment utilization bounds. Thus, the latter is aspired.

Terminal currents are regarded as objectives due to continuous cable utilization is limited by currents. Nevertheless, dispersion of nodal powers is likewise valuable information and can be calculated in almost the same manner.

If drag indicators are installed at substation transformers to record maximum and minimum utilization, these values can be used as interval bounds. Otherwise current bounds can be obtained from transformer rating plates. In doing so, for each node upper and lower bounds of currents and powers are available. At switching stations all bounds are zero.

3.1 Boundary Current Flow

As mentioned above, for each node a current interval can be determined.

$$I_{\mathrm{K},\nu,\mathrm{act}} \in \left[I_{\mathrm{K},\nu,\min}, I_{\mathrm{K},\nu,\max} \right] \tag{1}$$

To calculate voltage alteration against nodal currents, Eq. (1) is reformulated

$$\underline{\boldsymbol{Y}}_{\mathrm{KK}}\, \Delta\underline{\boldsymbol{u}}_{\mathrm{K}} = \Delta\underline{\boldsymbol{i}}_{\mathrm{K}} \tag{2}$$

The slack voltage is independent of load situation, so $\Delta\underline{\boldsymbol{u}}_{\mathrm{K,sl}}$ is always 0. This constraint is inserted into Eq. (2) and replaces the former slack line of $\underline{\boldsymbol{Y}}_{\mathrm{KK}}$.

$$\begin{bmatrix} \underline{y}_{1,1} & \cdots & \underline{y}_{1,\mathrm{sl-1}} & \underline{y}_{1,\mathrm{sl}} & \underline{y}_{1,\mathrm{sl+1}} & \cdots & \underline{y}_{1,n} \\ \vdots & \ddots & \vdots & \vdots & \vdots & \ddots & \vdots \\ \underline{y}_{\mathrm{sl-1},1} & \cdots & \underline{y}_{\mathrm{sl-1},\mathrm{sl-1}} & \underline{y}_{\mathrm{sl-1},\mathrm{sl}} & \underline{y}_{\mathrm{sl-1},\mathrm{sl+1}} & \cdots & \underline{y}_{\mathrm{sl-1},n} \\ 0 & \cdots & 0 & 1 & 0 & \cdots & 0 \\ \underline{y}_{\mathrm{sl+1},1} & \cdots & \underline{y}_{\mathrm{sl+1},\mathrm{sl-1}} & \underline{y}_{\mathrm{sl+1},\mathrm{sl}} & \underline{y}_{\mathrm{sl+1},\mathrm{sl+1}} & \cdots & \underline{y}_{\mathrm{sl+1},n} \\ \vdots & \ddots & \vdots & \vdots & \vdots & \ddots & \vdots \\ \underline{y}_{n,1} & \cdots & \underline{y}_{n,\mathrm{sl-1}} & \underline{y}_{n,\mathrm{sl}} & \underline{y}_{n,\mathrm{sl+1}} & \cdots & \underline{y}_{n,n} \end{bmatrix} \begin{bmatrix} \Delta\underline{U}_{\mathrm{K},1} \\ \vdots \\ \Delta\underline{U}_{\mathrm{K,sl-1}} \\ \Delta\underline{U}_{\mathrm{K,sl}} \\ \Delta\underline{U}_{\mathrm{K,sl+1}} \\ \vdots \\ \Delta\underline{U}_{\mathrm{K},n} \end{bmatrix} = \begin{bmatrix} \Delta\underline{I}_{\mathrm{K},1} \\ \vdots \\ \Delta\underline{I}_{\mathrm{K,sl-1}} \\ 0 \\ \Delta\underline{I}_{\mathrm{K,sl+1}} \\ \vdots \\ \Delta\underline{I}_{\mathrm{K},n} \end{bmatrix} \tag{3}$$

or shorter

$$\underline{\boldsymbol{Y}}_{\mathrm{KK,red}}\, \Delta\underline{\boldsymbol{u}}_{\mathrm{K,red}} = \Delta\underline{\boldsymbol{i}}_{\mathrm{K,red}} \tag{4}$$

Inversion of Eq. (3) and insertion into the terminal voltage equation results in

$$\Delta\underline{\boldsymbol{u}}_{\mathrm{T}} = \boldsymbol{K}_{\mathrm{KT}}^{\mathrm{T}}\, \underline{\boldsymbol{Y}}_{\mathrm{KK,red}}^{-1}\, \Delta\underline{\boldsymbol{i}}_{\mathrm{K,red}} \tag{5}$$

Finally, Eq. (5) is inserted into the terminal current equation.

$$\Delta\underline{\boldsymbol{i}}_{\mathrm{T}} = \underline{\boldsymbol{Y}}_{\mathrm{T}}\, \boldsymbol{K}_{\mathrm{KT}}^{\mathrm{T}}\, \underline{\boldsymbol{Y}}_{\mathrm{KK,red}}^{-1}\, \Delta\underline{\boldsymbol{i}}_{\mathrm{K,red}} \tag{6}$$

or shorter

$$\Delta \underline{\boldsymbol{i}}_{\mathrm{T}} = \underline{\boldsymbol{D}}_{\mathrm{I,I}}\, \Delta \underline{\boldsymbol{i}}_{\mathrm{K,red}} \tag{7}$$

with the current dispersion matrix $\underline{\boldsymbol{D}}_{\mathrm{I,I}}$. Considering grids without capacitive shunts, $\underline{\boldsymbol{D}}_{\mathrm{I,I}}$ is real-valued.

To find maximum and minimum of each coefficient of $\Delta \underline{\boldsymbol{i}}_{\mathrm{T}}$, Eq. (6) combined with the nodal current bounds formulates a continuous constrained linear optimization problem which can be solved by well known methods i.e. the Simplex-Algorithm.

3.2 Boundary Power Flow

Accordant to current bounds, maxima and minima of substation powers can be obtained from drag indicators or substation transformer rating plates. In the latter case

$$-\boldsymbol{s}_{\mathrm{r}} \leq \boldsymbol{s}_{\mathrm{K,act}} \leq \boldsymbol{s}_{\mathrm{r}} \tag{8}$$

Again, a linear dependency of nodal voltages on nodal powers is needed, but contrarily to boundary current flow calculation, power is a quadratic function against voltages.

$$\underline{\boldsymbol{s}}_{\mathrm{K}} = 3\, \underline{\boldsymbol{U}}_{\mathrm{K}} \left(\underline{\boldsymbol{Y}}_{\mathrm{KK}}\, \underline{\boldsymbol{u}}_{\mathrm{K}} \right)^{*} \tag{9}$$

Thus, accordant to power flow calculation, Eq. (9) is linearized by first-order Taylor-series expansion, such that

$$\boldsymbol{J} \begin{bmatrix} \Delta \boldsymbol{u}_{\mathrm{K,r}} \\ \Delta \boldsymbol{u}_{\mathrm{K,i}} \end{bmatrix} = \begin{bmatrix} \boldsymbol{J}_{\mathrm{A}} & \boldsymbol{J}_{\mathrm{B}} \\ \boldsymbol{J}_{\mathrm{C}} & \boldsymbol{J}_{\mathrm{D}} \end{bmatrix} \begin{bmatrix} \Delta \boldsymbol{u}_{\mathrm{K,r}} \\ \Delta \boldsymbol{u}_{\mathrm{K,i}} \end{bmatrix} = \begin{bmatrix} \Delta \boldsymbol{p} \\ -\Delta \boldsymbol{q} \end{bmatrix} \tag{10}$$

Split-up into real and imaginary part doubles the equation system size, so two constraints for the slack node voltages have to be inserted into Eq. (10). $\boldsymbol{J}_{\mathrm{A}}$ and $\boldsymbol{J}_{\mathrm{D}}$ are modified in the same way as demonstrated in Eq. (3), whereas slack lines of $\boldsymbol{J}_{\mathrm{B}}$ and $\boldsymbol{J}_{\mathrm{C}}$ are replaced by zeros. Inversion now results in

$$\begin{bmatrix} \Delta \boldsymbol{u}_{\mathrm{K,r}} \\ \Delta \boldsymbol{u}_{\mathrm{K,i}} \end{bmatrix} = \boldsymbol{J}_{\mathrm{red}}^{-1} \begin{bmatrix} \Delta \boldsymbol{p} \\ -\Delta \boldsymbol{q} \end{bmatrix} = \begin{bmatrix} \boldsymbol{L}_{\mathrm{A}} & \boldsymbol{L}_{\mathrm{B}} \\ \boldsymbol{L}_{\mathrm{C}} & \boldsymbol{L}_{\mathrm{D}} \end{bmatrix} \begin{bmatrix} \Delta \boldsymbol{p}_{\mathrm{red}} \\ -\Delta \boldsymbol{q}_{\mathrm{red}} \end{bmatrix} \tag{11}$$

At the actual operation point, it is imperative that

$$\begin{aligned} \boldsymbol{L}_{\mathrm{A}} &= \boldsymbol{L}_{\mathrm{D}} \\ \boldsymbol{L}_{\mathrm{B}} &= -\boldsymbol{L}_{\mathrm{C}} \end{aligned} \tag{12}$$

The Jacobian of the last power flow iteration step offers a good approximation for $\boldsymbol{J}$ which virtually fulfills Eq. (12).

Due to Eq. (12) it is assumed that

$$\underline{\boldsymbol{L}} = \boldsymbol{L}_{\mathrm{A}} + \mathrm{j}\, \boldsymbol{L}_{\mathrm{B}} \tag{13}$$

such that

$$\Delta \underline{\boldsymbol{u}}_{\mathrm{K}} = \underline{\boldsymbol{L}}\, \Delta \underline{\boldsymbol{s}}_{\mathrm{K,red}} \tag{14}$$

Following Eq. (5) and Eq. (6) a linear dependency of terminal currents on nodal powers is given

$$\Delta \underline{\boldsymbol{i}}_{\mathrm{T}} = \underline{\boldsymbol{Y}}_{\mathrm{T}}\, \boldsymbol{K}_{\mathrm{KT}}^{\mathrm{T}}\, \underline{\boldsymbol{L}}\, \Delta \underline{\boldsymbol{s}}_{\mathrm{K,red}} \tag{15}$$

or shorter

$$\Delta \underline{\boldsymbol{i}}_{\mathrm{T}} = \underline{\boldsymbol{D}}_{\mathrm{I,S}}\, \Delta \underline{\boldsymbol{s}}_{\mathrm{K,red}} \tag{16}$$

with the power dispersion matrix $\underline{\boldsymbol{D}}_{\mathrm{I,S}}$.

In combination with the nodal power constraints, Eq. (16) formulates a continuous constrained linear optimization problem which can be solved by well known methods, i.e. the Simplex-Algorithm, just like the boundary current flow.

4. Sensitivity analysis

Sensitivity analysis is used to systematically determine the effect of any arbitrary influencing variable on all state variables. Therefore, nonlinear system equations are linearized at a known operating point to achieve a linear dependency of state variables on influencing variables.
In the case of electric power systems, to simplify matters, typically complex nodal voltages are used as state variables whereas all measurable values can be considered as influencing variables, for example nodal powers, current magnitudes or even voltage magnitudes. Due to the required linear dependency a power flow calculation is not necessary.
'Known' operating points are system states in which all state variables and all influencing variables are known and consistent. They can be achieved by power flow calculation using a nodal power vector for example with all zero elements, rated powers of transformers or a presumption of load situation according to chapter 5. The easiest and clearest way is the first one which is therefore used in the following sections.

4.1 Sensitivity Matrices

As mentioned above normally only information on magnitudes of currents, voltages and apparent powers is available. Solely at hand-over-points, plant farms or special customers active and reactive powers are measured separately and appropriately signed. Taking this into account sensitivity matrices for these three cases have to be developed.

With

$$\boldsymbol{Y}_{\mathrm{KK,r}} = \mathrm{Re}\{\underline{\boldsymbol{Y}}_{\mathrm{KK}}\} \tag{17}$$

and

$$\boldsymbol{Y}_{\mathrm{KK,i}} = \mathrm{Im}\{\underline{\boldsymbol{Y}}_{\mathrm{KK}}\} \tag{18}$$

The nodal current equation can be reformulated

$$\begin{bmatrix} \boldsymbol{Y}_{\mathrm{KK,r}} & -\boldsymbol{Y}_{\mathrm{KK,i}} \\ \boldsymbol{Y}_{\mathrm{KK,i}} & \boldsymbol{Y}_{\mathrm{KK,r}} \end{bmatrix} \begin{bmatrix} \boldsymbol{u}_{\mathrm{K,r}} \\ \boldsymbol{u}_{\mathrm{K,i}} \end{bmatrix} = \begin{bmatrix} \boldsymbol{i}_{\mathrm{K,r}} \\ \boldsymbol{i}_{\mathrm{K,i}} \end{bmatrix} = \begin{bmatrix} \boldsymbol{I}_{\mathrm{K}} \cos(\boldsymbol{\varphi}_{\mathrm{K}}) \\ \boldsymbol{I}_{\mathrm{K}} \sin(\boldsymbol{\varphi}_{\mathrm{K}}) \end{bmatrix} \tag{19}$$

or shorter

$$\boldsymbol{Y}_{\mathrm{KK,ext}}\, \boldsymbol{u}_{\mathrm{K,ext}} = \boldsymbol{i}_{\mathrm{K,ext}} \tag{20}$$

Deletion of both slack lines results in

$$\boldsymbol{Y}_{\mathrm{KK,red}}\, \boldsymbol{u}_{\mathrm{K,red}} = \boldsymbol{i}_{\mathrm{K,red}} \tag{21}$$

Inversion of Eq. (21) now leads to

$$\begin{bmatrix} \boldsymbol{u}_{\mathrm{K,red,r}} \\ \boldsymbol{u}_{\mathrm{K,red,i}} \end{bmatrix} = \boldsymbol{Y}_{\mathrm{KK,red}}^{-1}\, \boldsymbol{i}_{\mathrm{K,red}} = \boldsymbol{Z}_{\mathrm{KK,red}} \begin{bmatrix} \boldsymbol{I}_{\mathrm{K,red}} \cos(\boldsymbol{\varphi}_{\mathrm{K,red}}) \\ \boldsymbol{I}_{\mathrm{K,red}} \sin(\boldsymbol{\varphi}_{\mathrm{K,red}}) \end{bmatrix} \tag{22}$$

A linear dependency of $\boldsymbol{u}_{\mathrm{K,red}}$ on $\boldsymbol{i}_{\mathrm{K,red}}$ and $\boldsymbol{\varphi}_{\mathrm{K,red}}$ is achieved by first-order Taylor series expansion.

$$\boldsymbol{u}_{\mathrm{K}} \approx \boldsymbol{u}_{\mathrm{K},0} + \Delta\boldsymbol{u}_{\mathrm{K}} \tag{23}$$

$\boldsymbol{u}_{\mathrm{K},0}$ is the known operation point voltage and $\Delta\boldsymbol{u}_{\mathrm{K}}$ the linearized deviation

$$\begin{bmatrix} \Delta\boldsymbol{u}_{\mathrm{K,r}} \\ \Delta\boldsymbol{u}_{\mathrm{K,i}} \end{bmatrix} = \boldsymbol{A}_{\mathrm{UK,IK}} \begin{bmatrix} \Delta\boldsymbol{\varphi}_{\mathrm{K}} \\ \Delta\boldsymbol{i}_{\mathrm{K}} \end{bmatrix} \tag{24}$$

with the Jacobian $\boldsymbol{A}_{\mathrm{UK,IK}}$ which is already the requested sensitivity matrix

$$\boldsymbol{A}_{\mathrm{UK,IK}} = \frac{\partial\left(\boldsymbol{Z}_{\mathrm{KK,red}} \begin{bmatrix} \boldsymbol{I}_{\mathrm{K,red}} \cos\left(\boldsymbol{\varphi}_{\mathrm{K,red}}\right) \\ \boldsymbol{I}_{\mathrm{K,red}} \sin\left(\boldsymbol{\varphi}_{\mathrm{K,red}}\right) \end{bmatrix} \right)}{\partial\begin{bmatrix} \boldsymbol{\varphi}_{\mathrm{K}}^{\mathrm{T}} & \boldsymbol{i}_{\mathrm{K}}^{\mathrm{T}} \end{bmatrix}} \tag{25}$$

$\boldsymbol{A}_{\mathrm{UK,IK}}$ can be separated into four sub matrices

$$\boldsymbol{A}_{\mathrm{UK,IK}} = \begin{bmatrix} \boldsymbol{A}_{\mathrm{Ur,\varphi I}} & \boldsymbol{A}_{\mathrm{Ur,I}} \\ \boldsymbol{A}_{\mathrm{Ui,\varphi I}} & \boldsymbol{A}_{\mathrm{Ui,I}} \end{bmatrix} \tag{26}$$

Coefficients of all sub matrices can easily be calculated

$$\boldsymbol{A}_{\mathrm{Ur,\varphi i}} = -\boldsymbol{Z}_{\mathrm{KK,r}}\, \boldsymbol{I}_{\mathrm{K,0}} \sin\left(\boldsymbol{\varphi}_{\mathrm{K,0}}\right) - \boldsymbol{Z}_{\mathrm{KK,i}}\, \boldsymbol{I}_{\mathrm{K,0}} \cos\left(\boldsymbol{\varphi}_{\mathrm{K,0}}\right) \tag{27}$$

$$\boldsymbol{A}_{\mathrm{Ur,I}} = \boldsymbol{Z}_{\mathrm{KK,r}} \cos\left(\boldsymbol{\varphi}_{\mathrm{K,0}}\right) - \boldsymbol{Z}_{\mathrm{KK,i}} \sin\left(\boldsymbol{\varphi}_{\mathrm{K,0}}\right) \tag{28}$$

$$\boldsymbol{A}_{\mathrm{Ui,\varphi i}} = -\boldsymbol{Z}_{\mathrm{KK,i}}\, \boldsymbol{I}_{\mathrm{K,0}} \sin\left(\boldsymbol{\varphi}_{\mathrm{K,0}}\right) + \boldsymbol{Z}_{\mathrm{KK,r}}\, \boldsymbol{I}_{\mathrm{K,0}} \cos\left(\boldsymbol{\varphi}_{\mathrm{K,0}}\right) \tag{29}$$

and

$$\boldsymbol{A}_{\mathrm{Ui,I}} = \boldsymbol{Z}_{\mathrm{KK,i}} \cos\left(\boldsymbol{\varphi}_{\mathrm{K,0}}\right) + \boldsymbol{Z}_{\mathrm{KK,r}} \sin\left(\boldsymbol{\varphi}_{\mathrm{K,0}}\right) \tag{30}$$

This leads to the following linear dependencies

$$\Delta\boldsymbol{u}_{\mathrm{K,r}} = \boldsymbol{A}_{\mathrm{Ur,\varphi i}}\, \Delta\boldsymbol{\varphi}_{\mathrm{K}} + \boldsymbol{A}_{\mathrm{Ur,I}}\, \Delta\boldsymbol{i}_{\mathrm{K}} \tag{31}$$

and

$$\Delta\boldsymbol{u}_{\mathrm{K,i}} = \boldsymbol{A}_{\mathrm{Ui,\varphi i}}\, \Delta\boldsymbol{\varphi}_{\mathrm{K}} + \boldsymbol{A}_{\mathrm{Ui,I}}\, \Delta\boldsymbol{i}_{\mathrm{K}} \tag{32}$$

If information on phase angles is not available, the first summands of Eq. (31) and Eq. (32) have to be omitted.

$$\begin{bmatrix} \Delta\boldsymbol{u}_{\mathrm{K,r}} \\ \Delta\boldsymbol{u}_{\mathrm{K,i}} \end{bmatrix} = \begin{bmatrix} \boldsymbol{A}_{\mathrm{Ur,I}} \\ \boldsymbol{A}_{\mathrm{Ui,I}} \end{bmatrix} \Delta\boldsymbol{i}_{\mathrm{K}} \tag{33}$$

Alternatively sensitivity matrices can be set up using polar coordinates. Therefore, the nodal current equation has to be reformulated into nonlinear equations of phase angles

$$\begin{bmatrix} \varphi_{\mathrm{K}1} \\ \vdots \\ \varphi_{\mathrm{K}n} \end{bmatrix} = \boldsymbol{f}_{\varphi\mathrm{K}}\left(\begin{bmatrix} \delta_{\mathrm{K}1} \\ \vdots \\ \delta_{\mathrm{K}n} \end{bmatrix}, \begin{bmatrix} U_{\mathrm{K}1} \\ \vdots \\ U_{\mathrm{K}n} \end{bmatrix} \right) \tag{34}$$

and magnitudes

$$\begin{bmatrix} I_{\mathrm{K}1} \\ \vdots \\ I_{\mathrm{K}n} \end{bmatrix} = \boldsymbol{f}_{\mathrm{IK}}\left(\begin{bmatrix} \delta_{\mathrm{K}1} \\ \vdots \\ \delta_{\mathrm{K}n} \end{bmatrix}, \begin{bmatrix} U_{\mathrm{K}1} \\ \vdots \\ U_{\mathrm{K}n} \end{bmatrix} \right) \tag{35}$$

First-order Taylor series expansion around the operation point results in

$$\begin{bmatrix} \varphi_{K1} \\ \vdots \\ \varphi_{Kn} \\ \hline I_{K1} \\ \vdots \\ I_{Kn} \end{bmatrix} \approx \begin{bmatrix} \varphi_{K1,0} \\ \vdots \\ \varphi_{Kn,0} \\ \hline I_{K1,0} \\ \vdots \\ I_{Kn,0} \end{bmatrix} + \begin{bmatrix} \Delta\varphi_{K1} \\ \vdots \\ \Delta\varphi_{Kn} \\ \hline \Delta I_{K1} \\ \vdots \\ \Delta I_{Kn} \end{bmatrix} \tag{36}$$

with

$$\begin{bmatrix} \Delta\varphi_{K1} \\ \vdots \\ \Delta\varphi_{Kn} \\ \hline \Delta I_{K1} \\ \vdots \\ \Delta I_{Kn} \end{bmatrix} = \begin{bmatrix} \dfrac{\partial \boldsymbol{f}_{\varphi}}{\partial [\boldsymbol{\delta} \mid \boldsymbol{u}]^{T}} \\ \hline \dfrac{\partial \boldsymbol{f}_{I}}{\partial [\boldsymbol{\delta} \mid \boldsymbol{u}]^{T}} \end{bmatrix} \begin{bmatrix} \Delta\delta_{K1} \\ \vdots \\ \Delta\delta_{Kn} \\ \hline \Delta U_{K1} \\ \vdots \\ \Delta U_{Kn} \end{bmatrix} \tag{37}$$

and the Jacobian

$$\boldsymbol{J}_{I} = \begin{bmatrix} \dfrac{\partial \boldsymbol{f}_{\varphi}}{\partial [\boldsymbol{\delta} \mid \boldsymbol{u}]^{T}} \\ \hline \dfrac{\partial \boldsymbol{f}_{I}}{\partial [\boldsymbol{\delta} \mid \boldsymbol{u}]^{T}} \end{bmatrix} \tag{38}$$

which is set up as shown in Eq. (39)

$$\boldsymbol{J}_{I} = \left[\begin{array}{ccc|ccc} \dfrac{\partial \varphi_{K1}}{\partial \delta_{K1}} & \cdots & \dfrac{\partial \varphi_{K1}}{\partial \delta_{Kn}} & \dfrac{\partial \varphi_{K1}}{\partial U_{K1}} & \cdots & \dfrac{\partial \varphi_{K1}}{\partial U_{Kn}} \\ \vdots & \ddots & \vdots & \vdots & \ddots & \vdots \\ \dfrac{\partial \varphi_{Kn}}{\partial \delta_{K1}} & \cdots & \dfrac{\partial \varphi_{Kn}}{\partial \delta_{Kn}} & \dfrac{\partial \varphi_{Kn}}{\partial U_{K1}} & \cdots & \dfrac{\partial \varphi_{Kn}}{\partial U_{Kn}} \\ \hline \dfrac{\partial I_{K1}}{\partial \delta_{K1}} & \cdots & \dfrac{\partial I_{K1}}{\partial \delta_{Kn}} & \dfrac{\partial I_{K1}}{\partial U_{K1}} & \cdots & \dfrac{\partial I_{K1}}{\partial U_{Kn}} \\ \vdots & \ddots & \vdots & \vdots & \ddots & \vdots \\ \dfrac{\partial I_{Kn}}{\partial \delta_{K1}} & \cdots & \dfrac{\partial I_{Kn}}{\partial \delta_{Kn}} & \dfrac{\partial I_{Kn}}{\partial U_{K1}} & \cdots & \dfrac{\partial I_{Kn}}{\partial U_{Kn}} \end{array}\right] \tag{39}$$

After removing both slack lines in Eq. (39), inversion leads to

$$\boldsymbol{B}_{UK,IK} = \boldsymbol{J}_{I,red}^{-1} = \left[\begin{array}{c|c} \boldsymbol{B}_{\delta,\varphi} & \boldsymbol{B}_{\delta,I} \\ \hline \boldsymbol{B}_{U,\varphi} & \boldsymbol{B}_{U,I} \end{array}\right] \tag{40}$$

Phase angles and magnitudes of nodal voltages can now be approximated.

$$\begin{bmatrix} \Delta\delta_{K1} \\ \vdots \\ \Delta\delta_{Kn} \end{bmatrix} = \boldsymbol{B}_{\delta,\varphi} \begin{bmatrix} \Delta\varphi_{K1} \\ \vdots \\ \Delta\varphi_{Kn} \end{bmatrix} + \boldsymbol{B}_{\delta,I} \begin{bmatrix} \Delta I_{K1} \\ \vdots \\ \Delta I_{Kn} \end{bmatrix} \tag{41}$$

and

$$\begin{bmatrix} \Delta U_{\mathrm{K1}} \\ \vdots \\ \Delta U_{\mathrm{K}n} \end{bmatrix} = \boldsymbol{B}_{\mathrm{U},\varphi} \begin{bmatrix} \Delta \varphi_{\mathrm{K1}} \\ \vdots \\ \Delta \varphi_{\mathrm{K}n} \end{bmatrix} + \boldsymbol{B}_{\mathrm{U,I}} \begin{bmatrix} \Delta I_{\mathrm{K1}} \\ \vdots \\ \Delta I_{\mathrm{K}n} \end{bmatrix} \tag{42}$$

Due to the extensive nonlinearity of square-root and arc tangent functions, the approximation error is expected to be greater compared to Cartesian approximation. Again, if information on phase angles is not available, the first summand has to be omitted in Eq. (41) and Eq. (42) which leads to additional approximation errors.
Calculation of apparent power sensitivity matrices is similar and sensitivity matrices based on active and reactive power can calculated in terms of the power flow Jacobian.

4.2 Terminal values

In addition to nodal values, terminal measurands may be used for sensitivity analysis as well. Calculation of terminal-based sensitivity matrices proves to be more difficult due to the obtained Jacobians, which are set up analogously to nodal measurand matrices, are not square and thus not invertible. Sole exception is a single branch line with two nodes. So as a rule, grids consist of more terminals than nodes and thus the initial equation system is underdetermined. One reasonable solution approach out of an infinite number is the calculation of pseudo-inverses according to the method of Moore-Penrose.
For each influencing variable $\boldsymbol{g}$ starting with the initial equation

$$\Delta \boldsymbol{g} = \boldsymbol{J}\, \Delta \boldsymbol{u} \tag{43}$$

the method of Moore-Penrose generates a matrix $\boldsymbol{M}$ with the dimension of $\boldsymbol{J}^{\mathrm{T}}$ such that

$$\boldsymbol{M}\, \Delta \boldsymbol{g} = \Delta \boldsymbol{u} \tag{44}$$

$\boldsymbol{M}$ has to satisfy the constraints

$$\begin{aligned} \boldsymbol{J}\,\boldsymbol{M}\,\boldsymbol{J} &= \boldsymbol{J} \\ \boldsymbol{M}\,\boldsymbol{J}\,\boldsymbol{M} &= \boldsymbol{M} \\ \boldsymbol{J}\,\boldsymbol{M} &= \left(\boldsymbol{J}\,\boldsymbol{M}\right)^{*\mathrm{T}} \\ \boldsymbol{M}\,\boldsymbol{J} &= \left(\boldsymbol{M}\,\boldsymbol{J}\right)^{*\mathrm{T}} \end{aligned} \tag{45}$$

and particularly stands out from all other possible solutions due to

$$\left\| \Delta \boldsymbol{u} \right\| = \left\| \boldsymbol{M}\, \Delta \boldsymbol{g} \right\| \tag{46}$$

is minimized.
In general, due to equivocality terminal measurands are rather ineligible for grid state identification based on sensitivity analysis. Furthermore, simulations proved that approximation error is vastly higher compared to the approach solely based on nodal values. Thus, in the following chapters nodal measurands are focused. Nevertheless, the algorithm works analogously using terminal values. If selected nodal and terminal values are not interdependent, it is even possible to combine them and use both for state identification, although in the majority of cases interdependency is given.

4.3 State identification

Provided that at each possible measuring location at the most one device is installed and only magnitudes are recorded, a measuring vector $\boldsymbol{m}$ of dimension $4k \times 1$ can be introduced

$$\boldsymbol{m} = \begin{bmatrix} \boldsymbol{m}_{\mathrm{iK}} \\ \boldsymbol{m}_{\mathrm{sK}} \\ \boldsymbol{m}_{\mathrm{pK}} \\ \boldsymbol{m}_{\mathrm{qK}} \end{bmatrix} \tag{47}$$

Each measuring device can either record

- magnitude and phase angle or solely magnitude of current
- magnitude and phase angle or solely magnitude of apparent power
- active or reactive power or both active and reactive power

Sub vectors of $\boldsymbol{m}$ contain measured values at the corresponding positions. Residual elements are either filled with predicted values or are set to zero. Sequence of sub vectors is irrelevant and if possibly phase angles are measured, $\boldsymbol{m}$ has to be accordingly extended.
Depending on the sequence of sub vectors, all possible sensitivity matrices can be combined to

$$\begin{bmatrix} \Delta\boldsymbol{u}_{\mathrm{K,r}} \\ \Delta\boldsymbol{u}_{\mathrm{K,i}} \end{bmatrix} = \boldsymbol{A}\,\boldsymbol{m} \tag{48}$$

with the sensitivity matrix

$$\boldsymbol{A} = \begin{bmatrix} \boldsymbol{A}_{\mathrm{Ur,I}} & \boldsymbol{A}_{\mathrm{Ur,S}} & \boldsymbol{A}_{\mathrm{Ur,P}} & \boldsymbol{A}_{\mathrm{Ur,Q}} \\ \boldsymbol{A}_{\mathrm{Ui,I}} & \boldsymbol{A}_{\mathrm{Ui,S}} & \boldsymbol{A}_{\mathrm{Ui,P}} & \boldsymbol{A}_{\mathrm{Ui,Q}} \end{bmatrix} \tag{49}$$

Alternatively in polar coordinates

$$\begin{bmatrix} \Delta\boldsymbol{\delta}_{\mathrm{K}} \\ \Delta\boldsymbol{u}_{\mathrm{K}} \end{bmatrix} = \boldsymbol{B}\,\boldsymbol{m} \tag{50}$$

with

$$\boldsymbol{B} = \begin{bmatrix} \boldsymbol{B}_{\delta,\mathrm{I}} & \boldsymbol{B}_{\delta,\mathrm{S}} & \boldsymbol{B}_{\delta,\mathrm{P}} & \boldsymbol{B}_{\delta,\mathrm{Q}} \\ \boldsymbol{B}_{\mathrm{U,I}} & \boldsymbol{B}_{\mathrm{U,S}} & \boldsymbol{B}_{\mathrm{U,P}} & \boldsymbol{B}_{\mathrm{U,Q}} \end{bmatrix} \tag{51}$$

The grid state can now be estimated using the developed linearized models by determining the deviation from the known operation point. As mentioned above, in this work the idling grid is used as operation point due to the entire consistent and complex-valued grid state can be easily obtained and no modifications of the measured values are needed.
In Cartesian coordinates the grid state is estimated as follows

$$\begin{bmatrix} \boldsymbol{u}_{\mathrm{K,r,est}} \\ \boldsymbol{u}_{\mathrm{K,i,est}} \end{bmatrix} = \begin{bmatrix} \boldsymbol{u}_{\mathrm{K,r,0}} \\ \boldsymbol{u}_{\mathrm{K,i,0}} \end{bmatrix} + \boldsymbol{A}\,\boldsymbol{m} \tag{52}$$

and in polar coordinates respectively

$$\begin{bmatrix} \boldsymbol{\delta}_{\mathrm{K,est}} \\ \boldsymbol{u}_{\mathrm{K,est}} \end{bmatrix} = \begin{bmatrix} \boldsymbol{\delta}_{\mathrm{K,0}} \\ \boldsymbol{u}_{\mathrm{K,0}} \end{bmatrix} + \boldsymbol{B}\,\boldsymbol{m} \tag{53}$$

At the operation point, the difference between actual and estimated grid state is zero. The estimation error increases with growing deviation from the operation point. Thus, error estimation should be applied.

5. Nodal Load Decomposition

Grid state identification algorithms depend on some information on grid condition. In low and medium voltage grids this information is not available in sufficient amount to calculate a unique grid state. One possibility to bypass lack of information is to supplement available measurands by assumptions.
Further literature offer miscellaneous methods to forecast power time-series over a various amount of time. For the purpose of grid state identification a forecasting method is needed which solely gets by on historic and present measured time-series and information on grid topology. So it is totally satisfactory to roughly assume nodal voltage values. Therefore, measured time series are used to back-reference on load type mixture. With this additional information forecasting of nodal active power is possible using analytic load profiles. To predict reactive power as well, corresponding time-series are introduced.

5.1 Active power profiles

To simplify estimation of energy consumption of costumers within one year, the VDEW announced eleven different analytic load profiles merged into groups of domestic, industrial and agricultural type:

- common industry
- industry on workdays (8 am to 6 pm)
- industry on workdays (evening)
- continuous industry
- stores and barbers
- bakeries
- industry on weekends
- common domestic
- common agriculture
- milk production
- miscellaneous agriculture

These profiles are obtained by averaging over a huge amount of customers of the same type within several years and do not represent the stochastic behavior of one single load. The eleven linear independent load profiles are supplemented by profiles describing off-peak storage heating, street lighting, heat pumps and other detachable loads.

5.2 Nodal load decomposition

The year is split up into three seasons: 'winter', 'transition' and 'summer'. Each season consists of three reference days: 'Workday', 'Saturday' and 'Sunday'. For each load type and for the reference days of all seasons, the VDEW announced a set of 96 values to be assigned to any quarter of an hour of that specific day, forming 99 unique daily time-series. Depending on the year, the federal state and the holidays they can be merged to 11 yearly time-series, one for each load type, and are normalized to $1000\,\frac{\text{kWh}}{\text{a}}$.

Nodal load extraction offers the possibility of determining load composition from measured active power time-series of nodes or hand-over points. Therefore, a measured active power time series $\boldsymbol{p}_{\text{meas}}$ of the point of interest has to be available and should provide information on consumption of at least one year. Furthermore, the measuring time period and the set of μ load profiles have to be known. This set can contain all analytic load profiles as well as off-peak heating, street lighting and other specific or detachable load profiles. Additionally, it is possible to add some series of measured loads or feeders to the set to optimize extraction quality.

For each 35040 quarters of an hour within one year it is assumed that the resulting active power can be described as a sum of the corresponding values of all load types

$$P(d,t) = \sum_{s=1}^{3} w_s(t) \sum_{p=1}^{\mu} P_{d,s,p}(t \bmod 96)\, w_p \tag{54}$$

d indicates the day type workday, Saturday or Sunday. t is the time which has elapsed since 1st of January, 0:00 o'clock in quarters of hours. Thus, within a year t ranges from 0 to 35039 or 0 to 35135 in leap years, respectively. w_s is a seasonal interpolation factor indicating the affiliation of t to the seasons. This might be a block function clearly assigning each time step to one season. Better results are achieved using fading crossovers. All interpolation methods, i.e. splines, are suitable as long as

$$\sum_{s=1}^{3} w_s(t) = 1 \tag{55}$$

for all time steps. Furthermore, in Eq. (54) w_p is the profile weighting vector indicating how many of each normalized profile types are connected to the node.

Due to the linear dependency of $P(t)$ on w_p Eq. (54) can be written in matrix notation

$$\boldsymbol{p}_{\mathrm{cal}} = \boldsymbol{L}\, \boldsymbol{w}_{\mathrm{p}} \tag{56}$$

To find the best fitting profile weighting vector $\boldsymbol{w}_{\mathrm{p}}$ for a measured time-series of active power, the difference between $\boldsymbol{p}_{\mathrm{cal}}$ and $\boldsymbol{p}_{\mathrm{meas}}$ has to be minimized.

$$\min_{\boldsymbol{w}_{\mathrm{p}}} (\Delta \boldsymbol{p}) = \min_{\boldsymbol{w}_{\mathrm{p}}} \left(\boldsymbol{L}\, \boldsymbol{w}_{\mathrm{p}} - \boldsymbol{p}_{\mathrm{meas}} \right) \tag{57}$$

Due to Eq. (57) is overdetermined results are obtained by linear least-squares optimization. Thus, $\boldsymbol{w}_{\mathrm{p}}$ can be directly calculated.

$$\boldsymbol{w}_{\mathrm{p}} = \left(\boldsymbol{L}^{\mathrm{T}} \boldsymbol{L} \right)^{-1} \boldsymbol{L}^{\mathrm{T}} \boldsymbol{p}_{\mathrm{meas}} \tag{58}$$

Extraction quality decreases with increasing amount of fluctuating feeders which are hardly to forecast. It can be improved by out-counting weather-dependent feeders like wind turbines before load extraction which mostly is possible due to data availability. Alternatively, measuring period should be extended or average time series of several years should be used.

Besides, weighting shifts among agricultural load profiles might be possible due to similarity to themselves. For the same reason influence of this shifting on the results is almost negligible.

5.3 Determination of Reactive Power Profiles

Chronological sequence admits the presumption that customers exhibit not only a distinct time-series of active power but also of reactive power depending on time and weekday. Thus, it is reasonable to assume that reactive power as well can be expressed as a sum of load profiles at any discrete time interval.

$$Q(d,t) = \sum_{s=1}^{3} w_s(t) \sum_{p=1}^{\mu} Q_{d,s,p}(t \bmod 96)\, w_p \tag{59}$$

In Eq. (59) d, t, w_p and w_s are the same as before. For each node w_p is obtained by load type mixture extraction. So, in this case reactive power profiles $\boldsymbol{Q}_{\mathrm{d,s,p}}$ are unknown. $Q(d,t)$ is the measured reactive power time-series.
Due to linear dependency of $Q(d,t)$ on $\boldsymbol{Q}_{\mathrm{s,p}}$ it is possible to reorder Eq. (59) in the style of Eq. (56). This leads to a linear equation system of 9504 unknowns.

$$\boldsymbol{q}_{\mathrm{meas}} = \boldsymbol{W}_{\mathrm{s,p}}\, \boldsymbol{q}_{\mathrm{p}} \tag{60}$$

Coefficients of $\boldsymbol{Q}_{\mathrm{s,p}}$ are now found in vector $\boldsymbol{q}_{\mathrm{p}}$. $\boldsymbol{W}_{\mathrm{s,p}}$ is a sparse matrix containing the elements of $w_s(t)$. Seasonal weighting factors are linear dependent, so for each node

$$\mathrm{rank}\left(\boldsymbol{W}_{\mathrm{s,p}}\right) \le 96 \cdot 3 \cdot 3 = 864 < 9504 \tag{61}$$

Thus, the resulting equation system is underdetermined. Furthermore, in addition to some unavoidable outage of measurement and to compensate statistical spread at least μ measured yearly times-series of different nodes with linear independent load type mixture are needed. To obtain reliable results utilization of more that 2μ time series is reasonable. In case exactly 2μ yearly time-series are considered, the equation system to be solved is

$$\begin{bmatrix} \boldsymbol{q}_{\mathrm{meas},1} \\ \vdots \\ \boldsymbol{q}_{\mathrm{meas},2\mu} \end{bmatrix} = \begin{bmatrix} \boldsymbol{W}_{\mathrm{s,p},1} \\ \vdots \\ \boldsymbol{W}_{\mathrm{s,p},2\mu} \end{bmatrix} \boldsymbol{q}_{\mathrm{p}} \tag{62}$$

Contrarily to active power, the contribution of capacitive shunts to the reactive power at the hand-over point is not negligible and has to be considered. Using an equipment model with concentrated parameters, for each node the contribution of connected capacitive shunts can be calculated

$$Q_{\mathrm{K,C}} = 3\,\omega C_{\mathrm{K}}\, U_{\mathrm{K}}^2 \tag{63}$$

where C_{K} is the sum of all capacitors attached to the node. Voltages are virtually never known in distribution grids. To anyway obtain binding approximations of time-series it is assumed that the capacitive contribution is practically constant due to the voltage deviation from nominal rating is less than 10 %. Thus, the overall contribution of capacitive shunts Q_{C} can be estimated.

$$Q_{\mathrm{C}} \approx 3\,\omega \sum_{i=1}^{\mathrm{k}} C_{\mathrm{K},i}\, U_{\mathrm{nN}}^2 \tag{64}$$

Due to Q_{C} is treated as base load, this results in an offset of the aspired load profiles on the ordinate. In different grids this offset is unknown and has to be calculated prior to the state identification process by least-squares analysis. Furthermore, current-dependent contribution of inductive series-elements is on the one hand small and on the other hand hardly determinable in underdetermined grids. Hence it is ignored.
In the following figures reactive power time-series for the previously mentioned reference days for the domestic, an industrial and an agricultural VDEW load type are shown. They are obtained using 52 measured yearly time-series of hand-over point active and reactive powers out of 2006 and 2007. This results in an equation system of dimension

$1.82 \cdot 10^6 \times 9504$. After solving the least-squares problem a polynomial smoothing is applied using a 12th order polynomial.
Relating reactive power profiles i.e. to 1000 kVArh is not reasonable due to this would result in loss of load amount coherence.
Although no starting vector is given and no information on active power profiles is used during calculation, particularly the domestic and the industrial reactive power profiles strongly resembles the accordant VDEW profiles which can be seen as validation of the obtained results.

5.4 State identification

To obtain a complete power vector $\underline{s}_{\mathrm{in}}$, the measured nodal powers $\underline{s}_{\mathrm{meas}}$ are supplemented by estimated nodal powers $\underline{s}_{\mathrm{est}}$. For each unobserved node active and reactive powers are forecasted.

$$\underline{S}_{\mathrm{est}}(t) = \sum_{s=1}^{3} w_s(t) \left(\sum_{p=1}^{\mu} P_{d,s,p}(t \bmod 96) + \mathrm{j} \sum_{p=1}^{\mu} Q_{d,s,p}(t \bmod 96) \right) w_p + \mathrm{j}\, Q_{\mathrm{offset}} \tag{65}$$

Q_{offset} is calculated accordant to Eq. (58)

$$Q_{\mathrm{offset}} = \left(\boldsymbol{o}^{\mathrm{T}} \boldsymbol{o}\right)^{-1} \boldsymbol{o}^{\mathrm{T}} \left(\boldsymbol{Q}_{\mathrm{meas}} - \sum \boldsymbol{Q}_{\mathrm{profiles}}\right) \tag{66}$$

with

$$\boldsymbol{o} = \frac{1}{\sum C_{\mathrm{K}}} \begin{bmatrix} C_{\mathrm{K},1} \\ \vdots \\ C_{\mathrm{K},n} \end{bmatrix} \tag{67}$$

and proportionally distributed to the nodes

$$\boldsymbol{q}_{\mathrm{offset}} - \boldsymbol{o}\, Q_{\mathrm{offset}} \tag{68}$$

$\underline{s}_{\mathrm{meas}}$ and $\underline{s}_{\mathrm{est}}$ are merged to $\underline{s}_{\mathrm{in}}$ which is used for a power flow calculation. The resulting nodal voltages determine the aspired grid state.

6. Grid reduction approach

The goal of grid state identification is to find the best estimation of grid condition using as few additional measurands as possible. Therefore sensitivity analysis targets on searching the most valuable measurands to keep estimation error within a predefined tolerance and investment cost on an acceptable ratio. On the other hand nodal load assumptions can be used to complete the underdetermined input vector, so the grid state can easily be estimated by power flow calculation. In contrast, grid reduction tries to more intensely exploit existent measurands particularly with regard to terminal values due to this information is almost discounted by sensitivity analysis. Therefore grid reduction offers a way to reconstruct lost information on voltage and current angles at bus bars. It is even possible that there is no need of installing new devices.
Typically only magnitudes of voltages at central busbars and current magnitudes of connected lines are measured. By reducing the grid to these central busbars and some auxiliary nodes the entire complex-valued state of the reduced grid can be estimated. These results offer inference on condition of the whole original grid. On the one hand they can be

used for other grid state identification methods like sensitivity analysis and on the other hand they set the fundament for a method to recursively back-reference to the original grid state.
State identification by grid reduction is done by five consecutive steps which are shown in the following subchapters.

6.1 Deletion of unobserved branch lines

The first step is to eliminate branch lines with unobserved nodes at both ends such that the whole grid now only contains tie-lines starting and ending at observed bus bars. Nominal powers or currents of removed nodes are added to the remaining node on the tie-line the branch previously was connected to. This results in a new grid topology $\boldsymbol{K}_{\mathrm{KT,simp}}$, $\underline{\boldsymbol{Y}}_{\mathrm{T,simp}}$ and $\underline{\boldsymbol{Y}}_{\mathrm{KK,simp}}$ and new nodal and terminal influencing vectors

$$\begin{aligned}\underline{\boldsymbol{g}}_{\mathrm{K,simp}} &= \boldsymbol{K}_{\mathrm{K,simp,org}}\,\underline{\boldsymbol{g}}_{\mathrm{K,org}}\\ \underline{\boldsymbol{g}}_{\mathrm{T,simp}} &= \boldsymbol{K}_{\mathrm{T,simp,org}}\,\underline{\boldsymbol{g}}_{\mathrm{T,org}}\end{aligned} \tag{69}$$

where $\underline{g}$ universally represents power or current measurands.

Indeed, doing so results in a little error due to grid losses on the removed branch line are ignored, but this error is negligible. Nevertheless, maximum estimation errors will be encountered at unobserved branch lines during back-referencing due to missing information.

6.2 Grid reduction

In the new grid all tie-lines start and end at observed central switching stations but contain a varying number of unobserved nodes. In the next step all these nodes on each tie-line are removed and replaced by a single auxiliary node at a certain position on each line which carries the whole load of all removed nodes.
This method differs from conventional grid simplification methods at which the first and the last unobserved node of each line are used as auxiliary nodes and the rest of the tie-line is removed. The cumulated load is distributed to these nodes such that the inner lines are idle. Thus, grid losses can not be considered. Furthermore, due to missing information, a uniquely determinable power distribution is impossible using the mentioned previous approach.

Only tie-lines with more than one unobserved node have to be processed in this step. To obtain the position of the auxiliary node, first of all the parameters $r_\nu, x_\nu, c_\nu, g_\nu$ and the length l_ν of each complete tie-line ν between the observed nodes have to be calculated from the parameters of its μ subsections.

$$r_\nu = \frac{\sum_\mu \left(r_{\nu,\mu}\, l_{\nu,\mu}\right)}{l_\nu},\quad x_\nu = \frac{\sum_\mu \left(x_{\nu,\mu}\, l_{\nu,\mu}\right)}{l_\nu},\quad c_\nu = \frac{\sum_\mu \left(c_{\nu,\mu}\, l_{\nu,\mu}\right)}{l_\nu},\quad g_\nu = \frac{\sum_\mu \left(g_{\nu,\mu}\, l_{\nu,\mu}\right)}{l_\nu} \tag{70}$$

and

$$l_\nu = \sum_\mu l_{\nu,\mu} \tag{71}$$

Each processed tie-line can be described as follows

$$\boldsymbol{K}_{\mathrm{K,T},\nu} = \begin{bmatrix} 1 & 0 & 0 & 0 \\ 0 & 1 & 1 & 0 \\ 0 & 0 & 0 & 1 \end{bmatrix} \tag{72}$$

and

$$\underline{\boldsymbol{Y}}_{\mathrm{T},\nu} = \left[\begin{array}{cc|cc} \underline{y}_{\nu,\mathrm{left,aa}} & \underline{y}_{\nu,\mathrm{left,ab}} & 0 & 0 \\ \underline{y}_{\nu,\mathrm{left,ba}} & \underline{y}_{\nu,\mathrm{left,bb}} & 0 & 0 \\ \hline 0 & 0 & \underline{y}_{\nu,\mathrm{right,aa}} & \underline{y}_{\nu,\mathrm{right,ab}} \\ 0 & 0 & \underline{y}_{\nu,\mathrm{right,ba}} & \underline{y}_{\nu,\mathrm{right,bb}} \end{array}\right] \tag{73}$$

For ease of better understanding it is assumed that the left line is connected to the node with the smaller number and the right line to the node with the higher number, respectively. The auxiliary node is placed between left and right line.

This method needs assumptions of active and reactive power of each unobserved node and thus depends on estimation quality. A rough estimation can be obtained by substation transformer rating plates. On the other hand, forecasted values offer a better approximation. Due to the fact that nodal voltages and terminal powers at the ends of each tie-line are known, the system is decoupled and all tie-lines can be calculated separately. Thus, in addition to the forecasted values at each unobserved node, it is assumed, that nodal powers of the observed nodes equal the negative value of the connected observed terminal. Now, the rule of torque can be applied.

$$l_{\nu,\mathrm{left}} = \frac{\sum_{\mu}\left(S_{\nu,\mu}\, d_{\nu,\mu}\right)}{\sum_{\mu} S_{\nu,\mu}} \tag{74}$$

Analogously to the definition of left and right line, it is assumed that $d_{\nu,\mu}$ is the distance to the node with the smaller number. The parameters of the auxiliary lines are calculated as described above.

The decoupled tie-lines are now reassembled to the reduced grid which is described by $\boldsymbol{K}_{\mathrm{KT,red}}$, $\underline{\boldsymbol{Y}}_{\mathrm{T,red}}$ and $\underline{\boldsymbol{Y}}_{\mathrm{KK,red}}$. For the purpose of better overview, nodes are renumbered such that

$$\boldsymbol{K}_{\mathrm{KT,red}} = \begin{bmatrix} \boldsymbol{K}_{\mathrm{KT,red,obs}} \\ \boldsymbol{K}_{\mathrm{KT,red,unobs}} \end{bmatrix} \tag{75}$$

The reduced grid exhibits the following properties:

- no unobserved branches,
- a maximum of one unobserved node at each tie-line,
- all tie-lines start and end at observed nodes.

6.3 Calcuation of the reduced grid state

In the reduced grid unknown values which might be used as state variables are:

- voltage magnitudes at auxiliary nodes,
- all voltage angles except at the slack node,
- terminal current magnitudes connected to auxiliary nodes,
- all terminal current phase angles without exception.

Nodal currents at auxiliary nodes are not used as state variable since they depend on terminal currents

$$\underline{\boldsymbol{i}}_{\mathrm{K,unobs}} = \boldsymbol{K}_{\mathrm{KT,unobs}}\, \underline{\boldsymbol{i}}_{\mathrm{T,unobs}} \tag{76}$$

It is obvious that the above mentioned variables can be divided into two groups: nodal and terminal values. Both groups offer the ability to uniquely describe the grid state. Therefore, two state vectors containing the unknown values are introduced.

$$\boldsymbol{x}_{\mathrm{red}} = \begin{bmatrix} \boldsymbol{u}_{\mathrm{K,unobs}} \\ \boldsymbol{\delta}_{\mathrm{K,unobs}} \end{bmatrix} \tag{77}$$

and

$$\boldsymbol{y}_{\mathrm{red}} = \begin{bmatrix} \boldsymbol{i}_{\mathrm{T,unobs}} \\ \boldsymbol{\varphi}_{\mathrm{T,unobs}} \end{bmatrix} \tag{78}$$

They are completed by the measured values such that

$$\boldsymbol{x}_{\mathrm{u}} = \begin{bmatrix} \boldsymbol{u}_{\mathrm{obs}} \\ \boldsymbol{u}_{\mathrm{K,unobs}} \end{bmatrix}, \quad \boldsymbol{x}_{\delta} = \begin{bmatrix} 0 \\ \boldsymbol{\delta}_{\mathrm{K,unobs}} \end{bmatrix} \tag{79}$$

and

$$\boldsymbol{y}_{\mathrm{i}} = \begin{bmatrix} \boldsymbol{i}_{\mathrm{obs}} \\ \boldsymbol{i}_{\mathrm{T,unobs}} \end{bmatrix}, \quad \boldsymbol{y}_{\varphi} = \boldsymbol{\varphi}_{\mathrm{T,unobs}} \tag{80}$$

Thus, complex-valued state variables are obtained by

$$\underline{\boldsymbol{u}}_{\mathrm{K,x}} = \boldsymbol{X}_{\mathrm{u}}\, \mathrm{e}^{\mathrm{j}\,\boldsymbol{x}_{\delta}} \tag{81}$$

and

$$\underline{\boldsymbol{i}}_{\mathrm{T,y}} = \boldsymbol{Y}_{\mathrm{i}}\, \mathrm{e}^{\mathrm{j}\,\boldsymbol{y}_{\varphi}} \tag{82}$$

respectively. According to Kirchhoff's first law, the two constraints

$$\underline{\boldsymbol{i}}_{\mathrm{K,obs}} + \boldsymbol{K}_{\mathrm{KT,obs}}\, \underline{\boldsymbol{i}}_{\mathrm{T,y}} = 0 \tag{83}$$

and

$$\underline{\boldsymbol{i}}_{\mathrm{K,obs}} + \boldsymbol{K}_{\mathrm{KT,obs}}\, \underline{\boldsymbol{Y}}_{\mathrm{T,red}}\, \boldsymbol{K}_{\mathrm{KT,red}}^{\mathrm{T}}\, \underline{\boldsymbol{u}}_{\mathrm{K,x}} = 0 \tag{84}$$

have to be fulfilled. Unfortunately the equation systems of both groups are still underdetermined. But as both groups should describe the same grid state, it is mandatory to match both states, resulting in two additional equations

$$\underline{\boldsymbol{i}}_{\mathrm{T,y}} - \underline{\boldsymbol{Y}}_{\mathrm{T}}\, \boldsymbol{K}_{\mathrm{KT,red}}^{\mathrm{T}}\, \underline{\boldsymbol{u}}_{\mathrm{K,x}} = 0 \tag{85}$$

and

$$\underline{\boldsymbol{u}}_{\mathrm{K,x}} - \underline{\boldsymbol{u}}_{\mathrm{K}}\left(\underline{\boldsymbol{i}}_{\mathrm{T,y}}\right) = 0 \tag{86}$$

with

$$\underline{\boldsymbol{u}}_{\mathrm{K}}\left(\underline{\boldsymbol{i}}_{\mathrm{T,y}}\right) = \begin{bmatrix} \underline{U}_{\mathrm{slack}} \\ \underline{\boldsymbol{Y}}_{\mathrm{KK,red,red}}^{-1}\left(\boldsymbol{K}_{\mathrm{KT,red}}\, \underline{\boldsymbol{i}}_{\mathrm{T,y}} - \underline{\boldsymbol{y}}_{\mathrm{KK,red,slack}}\, \underline{U}_{\mathrm{slack}}\right) \end{bmatrix} \tag{87}$$

$\underline{\boldsymbol{Y}}_{\mathrm{KK,red,red}}^{-1}$ is the slack reduced inverse and $\underline{\boldsymbol{y}}_{\mathrm{KK,red,slack}}$ the slack column of $\underline{\boldsymbol{Y}}_{\mathrm{KK,red}}$.

Eq. (83) to Eq. (87) now formulate a nonlinear, overdetermined equation system which can be solved with any least-squares method adequate to conventional power system state estimation.

The result of the reduced grid state calculation is a consistent, complex-valued state vector with all inferior values. For the subsequent steps the following information is needed anyway:

- nodal voltages at central busbars,
- currents of terminals connected to central busbas,
- currents of auxiliary nodes.

Additionally, in some cases voltages of auxiliary nodes might be used.

6.4 Recursive inference on the original grid state

In this step the complex-valued grid state of the previous grid without unobserved branches is calculated. Again, the resulting equation system is underdetermined and thus back-referencing is done recursively. The topology and results obtained in chapter 6.3 are referred to as 0th recursion step.

Based on the grid of the previous recursion step, additional nodes are added and a new grid state is calculated until all nodes of all tie-lines are added. Due to nodal voltages at central busbars are already known, tie-lines are again decoupled and can be processed separately. In the following, a recursion step is described.

Adding a node inherits adding at least one line. This line is situated between a previously added node and the new one. Due to the original grid is reconstructed, the node position and line parameters can be obtained directly from $\underline{\boldsymbol{Y}}_{\mathrm{T,simp}}$.

Depending on the number of nodes still not added to the tie-line, one of the following algorithms is applied.

If only one node has to be added, the last two original line segments defining the position of the node have to be added. The tie-line is now completely reconstructed. The voltage of the new node is the voltage of the auxiliary node of the previous recursion step.

If two nodes have to be added, the last three original line segments have to be added. Thereby, the positions of the new nodes are defined. Due to the fact that voltages and currents of the terminals connected to the outer nodes are entirely known, the transfer functions $\underline{\boldsymbol{H}}$ of the outer lines are used to calculate the voltages of the new nodes.

$$\begin{bmatrix} \underline{U}_{\mathrm{T2}} \\ \underline{I}_{\mathrm{T2}} \end{bmatrix} = \underline{\boldsymbol{H}}_{2,1} \begin{bmatrix} \underline{U}_{\mathrm{T1}} \\ \underline{I}_{\mathrm{T1}} \end{bmatrix} \tag{88}$$

and

$$\begin{bmatrix} \underline{U}_{\mathrm{T5}} \\ \underline{I}_{\mathrm{T5}} \end{bmatrix} = \underline{\boldsymbol{H}}_{5,6} \begin{bmatrix} \underline{U}_{\mathrm{T6}} \\ \underline{I}_{\mathrm{T6}} \end{bmatrix} \tag{89}$$

respectively. The tie-line is now completely reconstructed.

In all other cases to keep the equation system determinable, a maximum amount of three nodes can be processed. So, if there are exactly three nodes to add, they and the respective lines are taken from the original grid. Otherwise, only the two outer lines are taken and all other nodes and lines are reduced to two auxiliary lines and one auxiliary node. Considering the outer nodes, which have been determined in the previous recursion step, as nodes 1 and 5 and the new nodes as 2 to 4, respectively, this results in the following equation system

$$\begin{bmatrix} \underline{y}_a & \underline{y}_b & & & \\ \underline{y}_b & \underline{y}_c & \underline{y}_d & & \\ & \underline{y}_d & \underline{y}_e & \underline{y}_f & \\ & & \underline{y}_f & \underline{y}_g & \underline{y}_h \\ & & & \underline{y}_h & \underline{y}_i \end{bmatrix} \begin{bmatrix} \underline{U}_1 \\ \underline{U}_2 \\ \underline{U}_3 \\ \underline{U}_4 \\ \underline{U}_5 \end{bmatrix} = \begin{bmatrix} \underline{I}_1 \\ \underline{I}_2 \\ \underline{I}_3 \\ \underline{I}_4 \\ \underline{I}_5 \end{bmatrix} \tag{90}$$

$\underline{U}_1$, $\underline{U}_5$, $\underline{I}_1$ and $\underline{I}_5$ are already known. $\underline{U}_2$ and $\underline{U}_4$ can be obtained either from the first and last line of Eq. (90) or as well from Eq. (88) and Eq. (89). Thus, Eq. (90) is reformulated

$$\begin{bmatrix} \underline{y}_d & -1 & 0 & 0 \\ \underline{y}_e & 0 & -1 & 0 \\ \underline{y}_f & 0 & 0 & -1 \end{bmatrix} \begin{bmatrix} \underline{U}_3 \\ \underline{I}_2 \\ \underline{I}_3 \\ \underline{I}_4 \end{bmatrix} = - \begin{bmatrix} \underline{y}_b \underline{U}_1 + \underline{y}_c \underline{U}_2 \\ \underline{y}_d \underline{U}_2 + \underline{y}_f \underline{U}_4 \\ \underline{y}_g \underline{U}_4 + \underline{y}_h \underline{U}_5 \end{bmatrix} \tag{91}$$

Eq. (91) still is underdetermined. But in the previous recursion step, nodes 2 to 4 were summarized as auxiliary node carrying a current $\underline{I}_{aux}$. Thus, as a constraint

$$\sum_{r=2}^{4} \underline{I}_r = \underline{I}_{aux} \tag{92}$$

This results in a square equation system which is determinable if capacitive shunts are not ignored.

$$\begin{bmatrix} \underline{y}_d & -1 & 0 & 0 \\ \underline{y}_e & 0 & -1 & 0 \\ \underline{y}_f & 0 & 0 & -1 \\ 0 & 1 & 1 & 1 \end{bmatrix} \begin{bmatrix} \underline{U}_3 \\ \underline{I}_2 \\ \underline{I}_3 \\ \underline{I}_4 \end{bmatrix} = - \begin{bmatrix} \underline{y}_b \underline{U}_1 + \underline{y}_c \underline{U}_2 \\ \underline{y}_d \underline{U}_2 + \underline{y}_f \underline{U}_4 \\ \underline{y}_g \underline{U}_4 + \underline{y}_h \underline{U}_5 \\ -\underline{I}_{aux} \end{bmatrix} \tag{93}$$

If there are more than three nodes to be added, $\underline{I}_3$ will be $\underline{I}_{aux}$ in the next recursion step, otherwise the reconstruction of the tie-line is complete.

6.5 Recovery of unobserved branch lines

Due to missing information, the reconstruction of unobserved branch lines can not be done without assumptions. A reasonable way to allocate the load of the remaining node, obtained during recursive back-referencing, to all nodes of the branch is to distribute it proportional to the estimated nodal power which was used to define positions of auxiliary nodes. If this information is not available, other assumptions can be applied i.e. uniformly distributed nodal powers.

Assumptions are solely used for recovery of unobserved branch lines and the determination of auxiliary node positions. Besides the grid reduction approach entirely gets by without any assumed information.

It is obvious, that at unobserved branches the estimation error strongly depends on assumption accuracy. Due to the actual branch state is virtually never correctly assumed, in normal case this faulty distribution pattern results in higher errors compared to the state identification process at tie-lines.

7. Summary and Conclusion

The behavior and utilization of the electric power system is strongly influenced by the divergent interests of different market participants, particularly in the distribution level. To find a convenient grid operation strategy, decentralized energy management systems are developed.

Knowledge of the grid state is essential for these management systems to work. For historical reasons measurement facilities are sparsely spread in distribution grids. Thus, grid states are not uniquely determinable solely by measured values and have to be estimated using on the one hand supplementary predicted values and on the other hand intelligent methods to infer on probable grid states.

Methods of grid state identification are mathematical algorithms to obtain information on state variables of electric power systems grids. These procedures are based fundamentally on the acquisition of measured, forecasted or estimated values influencing the grid state.

In underdetermined grids the system state is not entirely observed. Therefore, in combination with knowledge of grid topology, equipment parameters and technical constraints, state identification algorithms offer methods to best possibly infer on the actual grid state. They result in a deterministic and consistent state vector.

In this work four different methods of state identification of underdetermined distribution grids are developed and demonstrated using various examples. They vary in the use of available information, the objective and the way, the grid state is determined.

- The **boundary load flow algorithm** is applied to calculate limits of equipment utilization. Contrarily to the other developed procedures, this method does not return a determined grid state but a set of maximum and minimum terminal current bounds indicating whether equipment is always operated safely or is potentially overstressed. In the latter case grid operation might carry risk and thus, further determining state identification algorithms and an energy management strategy should be applied. So, the boundary load flow may be used as preprocessing method. Furthermore, this algorithm is applicable for power flow decomposition problems and offers information on optimal grid expansion or retreat strategies.

- The second type of grid state identification uses **sensitivity analysis** to estimate nodal voltages. This method is based on a linearized system model to calculate the influence of all measured values on state variables. Therefore, depending on the coordinate system, different sensitivity matrices are developed determining the linear dependency of nodal voltages on currents and powers. In addition these matrices are used to calculate sensitivity indexes, a valuation of the usefulness of each nodal measurand. Thus, if grid operators aim at improving approximation of grid state, which is only practicable by installing new measuring devices, sensitivity indexes offer information on optimal measurement expansion strategies under technical and economical constraints much faster than conventional optimization problem solvers. Due to the fact that the grid reduction approach offers more accurate state approximations and is not based on a linearized system model, the main field of application of sensitivity analysis is the search for optimal measurands.

- The availability of measured values is mostly insufficient to uniquely calculate a grid state. Thus, measurands have to be supplemented by a **prediction of nodal values**. Therefore, active nodal power can be easily forecasted using existing analytic load profiles. Due to reactive power profiles have not been available so far, they are developed in this work based on measured time-series of hand-over points to superior grids within several years. Measured nodal powers and power estimations at each unobserved node are merged to a complete and complex-valued nodal power vector which can be used to obtain the grid state by power flow calculation. This method should be applied in case of utilization of the grid reduction approach is not possible due requirements are not fulfilled, only a few measurands are missing in the nodal power vector and the time-dependent behavior of loads is roughly predictable.
- The final grid state identification method which is called **grid reduction approach** uses both measured nodal and terminal values as input variables. Contrarily to the method based on sensitivity analysis this approach extensively exploits typical properties of distribution grid topology and thus is able to more accurately estimate the grid state even without installation of new measurement facilities which is a further advantage. To estimate nodal voltages, the grid is reduced to the observed central busbars and a minimum of auxiliary nodes on remaining tie-lines. Thereby the equation system size is reduced until the system is solvable. After the reduced grid state is calculated a recursive algorithm is applied to back-reference to the original grid by re-adding a small number of nodes in each recursion step. The obtained results are more accurate than the ones obtained by sensitivity analysis. Furthermore, the grid reduction approach does not need a linearized model and thus approximation error is independent of the deviation from an assumed operation point.

It is clearly demonstrated that state identification methods are of practical importance for grid operators but as well for grid planners. In the latter case especially the boundary load flow method is valuable to indicate necessary grid modifications and to find cost-effective strategies of expansion, maintenance and even retreat. Additionally, this method helps making decisions where to connect new loads or DG sources. Furthermore, grid planners may apply sensitivity analysis to identify significant influencing variables and to expand the measurement topology accordingly.

On the other side, during operating conditions knowledge of the grid state is essential for grid control systems. On the one hand, the generated reactive power profiles offer the ability of entire long-term forecasting of load situations to local utility operators and on the other hand particularly the grid reduction approach returns close approximations of the current actual system state which is used for resource scheduling and allocation.

Thus, it is obvious that state identification of distribution grids offers an important contribution to face the challenges of future distributed energy supply and establish a sound basis for highly automated decentralized energy management systems.

8. References

Abur, A.; Expósito, A. G.: Power system state estimation: theory and implementation, Dekker, New York, 2004.

Albert, A.: Regression and the Moore-Penrose Pseudoinverse, Academic Press Inc., New York, London, 1972.

Dimitrovski, A.; Tomsovic, K.: Boundary Load Flow Solutions, IEEE Transactions of Power Systems, Vol. 19, No. 1, February 2004.

EWE AG (Hrsg.): 10 Bullensee-Thesen und abgeleitete Handlungsempfehlungen zur zukünftigen Energieversorgung (3. Auflage), Oldenburg, 2007.

Meier, H.; Fünfgeld, C.; Adam, C.; Schieferdecker, B.: Repräsentative VDEW-Lastprofile, VDEW, Frankfurt (Main), 1999.

Merkt, B.: Beitrag zur Zustandsidentifikation von elektrischen Energieversorgungsnetzen, Dissertation, Leibniz Universität Hannover - Institut für Energieversorgung und Hochspannungstechnik - Fachgebiet Elektrische Energieversorgung, Hannover, August 2007.

Oeding, D.; Oswald, B. R.: Elektrische Kraftwerke und Netze (6. Auflage), Springer, Berlin, 2004.

Reese, C.: Beitrag zur Entwicklung eines dezentralen Energiemanagementsystems, Diplomarbeit DA-473, Leibniz Universität Hannover - Institut für Energieversorgung und Hochspannungstechnik - Fachgebiet Elektrische Energieversorgung, Hannover, October 2007.

VDEW (Hrsg.): Repräsentative VDEW-Lastprofile, VDEW-Verlag, Frankfurt (Main), 1999.

Walters, H.; Parker, L. R.; Morgan, S. L.; Deming, S. N.: Sequential Simplex Optimization, CRC Press, Boca Raton, 2000.

Wang, Z; Alvarado, F. L.: Interval Arithmetic in Power Flow Analysis, Proceedings of the IEEE International Conference on Power Industry Computer Application, May 1991

Wolter, M.: A new approach on state identification of underdetermined MV grids based on grid reduction, Proceedings of the 4th IASTED Asian Conference on Power & Energy Systems (AsiaPES2008), Langkawi, Malaysia, April 2008.

Wolter, M.: Cost-effective grid state identification of mean voltage grids by using Sensitivity Analysis, Proceedings of the 7th IASTED International Conference on Power and Energy Systems (EuroPES 2007), Spain, August 2007.

Wolter, M.: Entwicklung eines MicroGrid Simulator, Diplomarbeit DA-469, Leibniz Universität Hannover - Institut für Energieversorgung und Hochspannungstechnik - Fachgebiet Elektrische Energieversorgung, Hannover, August 2006.

Wolter, M.: Load type mixture extraction from measured time series of hand-over point active power, Proceedings of the 7th IASTED International Conference on Power and Energy Systems (EuroPES 2007), Spain, August 2007.

Wolter, M.: Grid State identification of distribution grids, Shaker Verlag, Aachen, 2008.

Wolter, M.: Grid reduction approach for state identification of distribution grids, Proceedings of the IEEE PES Annual General Meeting 2009, Calgary, Canada, September 2009.